Biotechnology and Ecology of Pollen

# Biotechnology and Ecology of Pollen

Proceedings of the International Conference on the Biotechnology and Ecology of Pollen, 9–11 July 1985 University of Massachusetts, Amherst, MA, U.S.A.

Edited by
David L. Mulcahy, Gabriella Bergamini Mulcahy
and Ercole Ottaviano

With 141 Figures

Springer-Verlag
New York Berlin Heidelberg Tokyo

David L. Mulcahy
Gabriella Bergamini Mulcahy
Department of Botany
University of Massachusetts
Amherst, MA 01003
U.S.A.

Ercole Ottaviano
Department of Genetics
and Microbial Biology
University of Milano
Milano 20133
Italy

Library of Congress Cataloging in Publication Data
Main entry under title:
Biotechnology and ecology of pollen.
Proceedings of the International Conference on Biotechnology and Ecology of Pollen, held at University of Massachusetts, July 9–11, 1985.
Bibliography: p.
Includes index.
1. Pollen—Congresses. 2. Biotechnology—Congresses.
I. Mulcahy, David L. II. Mulcahy, Gabriella Bergamini.
III. Ottaviano, Ercole. IV. International Conference
on Biotechnology and Ecology of Pollen (1985 : University
of Massachusetts at Amherst) V. Series.
QK658.B57 1986 582′.0463 85-30357



Printed and bound by Halliday Lithograph, West Hanover, Massachusetts.
Printed in the United States of America.

9 8 7 6 5 4 3 2 1

ISBN 0-387-96267-0 Springer-Verlag New York Berlin Heidelberg Tokyo
ISBN 3-540-96267-0 Springer-Verlag Berlin Heidelberg New York Tokyo

# DEDICATION

This volume is affectionately and respectfully

dedicated to the memory of

G. Sarfatti

and

R. A. Brink

for their outstanding contributions to the study of pollen

and to

Helen Vogel Mulcahy

for her many contributions to two of the editors.

# Preface

## In Recognition of the Forgotten Generation

D. L. MULCAHY[1]

Pollen was long believed to serve primarily a single function, that of delivering male gametes to the egg. A secondary and generally overlooked value of pollen is that it serves to block the transmission of many defective alleles and gene combinations into the next generation. This latter function comes about simply because pollen tubes carrying defective haploid genotypes frequently fail to complete growth through the entire length of the style. However, the beneficial consequences of this pollen selection are diluted by the fact that the same deleterious genotypes are often transmitted through the egg at strictly mendelian frequencies (Khush, 1973). Gene expression in the pollen might thus at least appear to be a phenomenon of trivial consequence. Indeed, Heslop-Harrison (1979) rightly termed the gametophytic portion of the angiosperm life cycle, the "forgotten generation." This neglect, however, came about despite subtle but constant indications that pollen is the site of intense gene activity and selection. For example, Mok and Peloquin (1975) demonstrated that relatively heterozygous diploid pollen shows heterotic characteristics whereas relatively homozygous diploid pollen does not. This was proof positive that genes are expressed (that is, transcribed and translated) in the pollen.

---

[1] Department of Botany, University of Massachusetts
Amherst, MA 01003, USA

However, the implications for pollen biology of even this recent and well known study were not widely recognized. In order to study the consequences of gene expression in the pollen, we have to go back to work of Ter-Avanesian, first published in 1949. He determined that progeny resulting from rapidly growing pollen tubes differed from those of unselected pollen tubes, and, as shown in later investigations, these differences have a genetic basis. This single conclusion holds two important implications. First of all, since the pollen used in each of Ter-Avanesian's experiments came from single plants, the genetic differences among which he was selecting must reflect post-meiotic gene expression. Secondly, because selecting among different pollen genotypes modifies the resultant sporophytic generation, we must conclude that some of the genes which are expressed, and thus subject to selection, in the pollen, must be expressed also in the sporophtye. In other words, the set of genes which are expressed in the gametophyte must overlap the set expressed in the sporophyte. Subsequent studies confirmed both of these conclusions.

When the overlapping relationship between gametophytic and sporophytic genotypes was first quantified (Tanksley, Zamir, and Rick, 1981), it was determined that 18 out of 30, or 60%, of the structural genes which are expressed in the sporophyte of Lycopersicon esculentum are expressed also in the pollen. In addition to these 18 genes expressed in both parts of the life cycle, one other was expressed only in the pollen. Thus 18/19, or 94.7% of the genes which are expressed in the pollen are expressed also in the sporophyte. More recently Mascarenhas and colleagues have measured genetic overlap in other species. Extending their survey to a very large fraction of the structural genes expressed in either phase of the life cycle, they detected 20,000 types of mRNA in pollen and 30,000 in the stem of Tradescantia paludosa. In a conclusion which was strikingly close to that reached by Tanksley, Zamir, and Rick, they found that approximately 60% of the structural genes which are expressed in the sporophyte are expressed also in the pollen, and fewer than 15% of the genes expressed in the pollen are not expressed in the sporophyte (Willing and Mascarenhas, 1984; Mascarenhas, et al., this volume). Comparable figures on genetic overlap were obtained by Sari Gorla, et al., (this volume) with Zea mays and by G. F. Moran (CSIRO Div. of Forest Research, Canberra, ACT, Australia) with Pinus radiata (G. Moran, pers. comm.).

Before evaluating the implications of these measurements of overlap, recall that the extraordinary adaptability of microorganisms is based on two characteristics: haploidy and large population sizes. Both of these characteristics are found, not only in microorganisms, but also in pollen. Thus gametophytically expressed genotypes should be as responsive to selective pressures as are those of microorganisms. More significantly, the extensive overlap between gametophytically and sporophytically expressed structural genes means that also 60% of the sporophytic genome is subject to a microbial type of selection. Certainly this holds important implications for both basic and applied studies. In fact, it may represent a rapid, inexpensive, and effective method of screening genotypes for stress resistance. Obviously, this would hold substantial implications for both basic and applied studies.

When many of the participants of the present symposium met at Lake Garda in Italy, in 1982, most of the above facts on gene expression in pollen were known, although precise data were few. Other topics discussed then were self-incompatibility, the movement of pollen in populations, pollen transformation, physiology, and competition. Today, we possess clear evidence that a majority of the structural genes which are expressed in the sporophytic portion of the life cycle are expressed, and thus exposed to selection also in the pollen. Pollen should thus provide a major access point in natural and artificial manipulation of the sporophytic genome. Examples of this type of selection, dealing with heavy metals, herbicides, and pathotoxins are presented in this volume. Other subjects in the realm of pollen biology and biotechnology are still undecided. Self-incompatibility (at least in the eyes of some!) is an unsettled issue. Pollen mediated transformation is a field of intense activity because of its potential applications, and pollen physiology continues to be a rich lode for basic and applied studies. It now holds, among other things, the promise of effective gametocides, and thus, new hybrid crops. In the field of ecology, as much as in any other, recent studies have provided precise measurements of phenomena only hypothesized three years ago. Here too, the significance of pollen competition and selection seems to be at least as great as early speculations suggested. Furthermore, phenomena such as negative interactions between pollen tubes within styles, first hinted at in 1980 (see Sari-Gorla and Rovida, 1980) are now clearly indicated in several studies (see Snow; d'Eeckenbrugge; and Hessing; all in this volume.

If our next meeting, perhaps four years from now, is marked by progress comparable to that made in the past three years, pollen biology and biotechnology will be ready to make substantial contributions in some both basic and applied areas of study.

## References


Heslop-Harrison, J. 1979. The forgotten generation: some thoughts on the genetics and physiology of angiosperm gametophytes. In: D.R. Davies and D.A. Hopwood (editors). Proceedings of the fourth John Innes Symposium. John Innes Inst., Norwich.

Khush, G.S., 1973. Cytogenetics of aneuploids. N.Y., Academic Press.

Mok, D.W.S. Peloquin, S.J. 1975. Three mechanisms of 2n pollen formation in diploid potatoes. Canad. Jour. Gen. Cytol. 17, 217-225.i

Sari-Gorla, M., Rovida, E., 1980. Competitive ability of maize pollen. Intergametophytic effects. Theor. Appl. Genet. 57, 37-42.

Tanksley, S., Zamir, D., Rick, C.M. 1981. Evidence for extensive overlap of sporophytic and gametophytic gene expression in *Lycopersicon esculentum*. Science 213, 453-455.

Ter-Avanesian, D.V. 1949. The influence of the number of pollen grains used in pollination. Bull. Appl. Genet. Plant Breeding Leningrad 28, 119-133.

Willing, R. P., Mascarenhas, J.P. 1984. Analysis of the complexity and diversity of mRNSa from pollen and shoots of *Tradescantia palludosa*. Plant Physiol. 75, 865-868.

# Acknowledgments

Since the strength of a scientific meeting depends primarily on the presentations of the participants, the first acknowledgements must be to the speakers at this meeting. They are to be congratulated, not only on the quality of the talks themselves, but, to an even greater extent, on the very substantial scientific progress made in the last three years.

Beyond the scientific presentations, several industries, Ciba-Geigy Incorporated, Del Monte Corporation, Funk Seeds International, Monsanto Agricultural Products Co., Pfizer Central Research, and Pioneer Hi-bred International Inc., provided essential financial support. Other support came from the National Science Foundation sections of population biology and developmental biology, and also from the University of Massachusetts, at Amherst. Ultimately, the actual progress of the meeting was, to a very great extent, dependent on the energy, ability, and cooperation of the staff and students in the Department of Botany, University of Massachusetts. These include Sally Klingener, Douglas MacMillan (an outstanding candidate for graduate school), Carol Mardeusz, and graduate students Patty Shesgreen, Jennifer Ramstetter, Douglas Robertson, Jeanne Curry, Lisa Webber-Barnes, and Jean Starobin. To each of these, institutions and colleagues alike, we extend, not only thanks, but, once more, congratulations.

The Editors

# Contents

GAMETOPHYTIC ECOLOGY

International Conference on Biotechnology and Ecology of Pollen
9 - 11 July 1985, University of Massachusetts, Amherst, MA 01003 USA

Row 1: Annicke Daframmond, John Sanford

Row 2: I. Negrutiu, Frank A. Loewus, Tito Schiva, Enrico Pe, Tony Perdue, M. V. Parthasarthy, Andrew Burns, Kenichi Hida

Row 3: Thomas Lee, Carla Frova, Paul Willing, Norm Weeden, Roger Cox, Tom Mitchell-Olds, Carl Schlicting

Row 4: Andy Stephenson, Peter Ascher, Victoria J. Apsit, Graciela Jaschek, Pirangelo Landi, Willie H. T. Loh, Varien R. Tilton, Sandra Russell, Karen Pittman, Juanita Ladyman, James Flynn, Pat Heslop-Harrison

Row 5: George Coppens d´Eeckenbrugge, Colette Nitsch, Maureen Stanton, Lorne Wolfe, Kamal Bawa, James Beach, Jeffery Stinson, T. M. Klein, Patricia Gradziel, Tom Gradziel, R. W. Robinson, Dvora Lapushner, Anton den Nijs

Row 6: Gunter Wricke, Stan Peloquin, Seppo Sorvari, Mark Hessing, Thierry Gaude, Christian Dumas, Jonathan Irvine, Sherret S. Chase, Chris Flick, John W. Cross, Folkert Hoekstra

Row 7: Tom Ferrari, William Feder, Georgio Binelli, Charles Simon, Robert N. Bowman, A. Kheypour, M. E. Nasrallah, W. John Cress, Richard G. Olmstead, David W. Meinke, Robert W. Briggs, Patricia M. O´Neill, Philip J. Harris, Mohan B. Singh

Row 8: Bruce Baldi, Willian Jensen, Lloyd Mogensen, Dana Moxley, Nello Bagni, Robin Scribailo, Amy Iezzoni, Garry A. Smith, Allison Snow, Steven N. Handel, Sophie C. Ducker, R. Wiermann, John F. Jackson

Row 9: Patrick J. Shiel, Joseph P. Mascarenhas, Ercole Ottaviano, Tijs Visser, Giovanni Iapichino, Mark Hutton, Brent Loy, Dennis Rowe, Regina K. Higgins, Lori Marshall, Dennis Hourcade, Gabriella Mulcahy

Row 10: Walton C. Galinat, Mirella Sari Gorla, Mauro Cresti, His Wilms, Joke Janse, Scott Russell, H. C. Sharma, Jon Lovett Doust, D. B. Walden, Michael N. Todhunter, C. A. McConchie, D. Mulcahy

Row 11: Hedera L. Porter, Raoul J. Bino, Paul Pfahler, M. A. van Herpen, H. F. Linskens, C. P. Malik, Minoru Yamada, Mario Cappadocia, Joyce Miller, Douglas MacMillan, Sherri Brown

Row 12: David W. Brown, Andrea Sonnia, Noel E. Pallais, Paul Comtois, B. Pfahler, K. Hinata, K. R. Shivanna, J. M. J. de Wet, Wayne E. Vian, Ray J. Mathia, Thomas H. Boyle

# Gene Expression In Pollen

# Expression of β-Galactosidase Gene in Pollen of Brassica Campestris

M.B. SINGH AND R.B. KNOX[1]

In 1982, we reported the occurrence of new molecular marker of gametophytic transcription i.e. gal gene in pollen of Brassica campestris. It was found that in certain plants, the anthers shed pollen half of which is β-galactosidase deficient. The cytochemical test for enzyme activity involved incubation of the pollen in an artificial cytochemical substrate, 5-bromo-4-chloro-3-indoxyl β-galactoside (Singh and Knox, 1984). Genetic studies of the inheritance of the deficiency indicated it is due to the presence of the single recessive gene gal in pollen. This locus is heritable through both female and male gamete transmission (Singh and Knox, 1985). Genotypes are available whose anthers showed the following pollen phenotypes: normal (Gal/gal), heterozygous i.e. 50% deficient (Gal/gal) and deficient (gal/gal).

In the present report we examine the molecular nature of the genetic defect responsible for β-galactosidase deficiency. Using immunoblotting and immunofluorescence techniques we demonstrate that deficiency of β-galactosidase activity in gal pollen possibly occurrs due to defective processing of the enzymatic protein.

## 1. β-Galactosidase as an antigen

Preliminary studies on the isolation and purification of β-galactosidase from Brassica pollen show it as a minor component of the total cellular proteins, and is quite unstable in a purified form, making monitoring of purification very difficult (Singh and Knox, unpublished results).

Expecting that β-galactosidase from one dicot plant source may have strong structural homologies to that from another source, we decided to prepare pabs to a commercially available enzyme preparation from cotyledons of jack bean, Canavalia ensiformis. Li et al. (1975) showed that this β-galactosidase is homogenous, giving a single band with PAGE after Coomassie blue staining. The presence of antibodies to β-galactosidase was demonstrated by a dot immunobinding assay (see Hawkes et al., 1982).

---

[1] Plant Cell Biology Research Centre, School of Botany, University of Melbourne, Parkville, Victoria, 3052, Australia.

When the polyclonal antiserum were mixed with a diluted series of purified enzyme, and the antigen- antibody complex precipitated by addition of Protein A-Sepharose followed by assay of β-galactosidase in the supernatant, the absence of enzyme activity confirmed that the antiserum does, in fact, contain antibodies to the enzyme. Similar observations were made when jack bean cotyledon, *Brassica campestris* leaf and pollen extracts were used in place of diluted purified jack bean β-galactosidase.

## 2. Monospecific polyclonal antibodies (pabs) to β-galactosidase

For immunofluorescence analysis it is pre requisite to have antibodies which are monospecific to particular antigen. These monospecific antibodies can be prepared by antigen affinity chromatography. The chromatographic procedure involves immobilization of highly purified antigen on a solid support usually dextran beads. A highly purified pollen β-galactosidase was not available, so we obtained monospecific pabs by using the Western blotting method.

Recently, in several animal systems, Blank et al. (1983) have shown that following SDS-PAGE of cell extracts, the detergent SDS can be removed from the gel by repeated washing of the gel with 20% isopropanol in 10 mM Tris buffer pH 7.4. This procedure results in renaturation of enzyme molecules in the gel with restored catalytic activity, detected in DSD-PAGE gels using artificial substrates.

Using this method, successful renaturation of β-galactosidase in SDS-PAGE gels was obtained. We stained the SDS-PAGE gels for β-galactosidase activity using the artifical indoxyl substrate after photography restained the same gels with coomassie blue protein dye. The turquoise blue enzym stained band could be easily identified in such double stained gels. (Fig. 1). In this way it was possible to ascertain which protein in SDS-PAGE gels is β-galactosidase. By this procedure it was found that enzyme present in sporophytic tissues i.e. leaves have a higher molecular weight (74 Kd) as compared to that of pollen enzyme (63 Kd),

A number of SDS-PAGE gels of pollen and leaf proteins were run and, in case, small strips were removed to carry out double enzyme/protein staining and the proteins from the rest of the gel were electro-blotted on to nitrocellulose paper. One strip each was cut from two sides of the blot and stained for proteins with amido black, or the india ink method (Hancock and Tsang, 1983). The position of the β-galactosidase protein band was identified by comparison with the earlier stained acrylamide gel. After blocking the remaining protein binding sites, the nitrocellulsoe blot was incubated for 16 h in 1:50 dilute anti-β-galactosidase serum. Subsequently the β-galactosidase band was cut out, and the paper shredded into small pieces. The bound antibodies on the paper were eluted, giving a monospecific antibody preparation. An aliquot was tested by probing a strip of pollen proteins using Western blot analysis. The antibodies bound only to a single band corresponding to the β-galactosidase (Fig. 2).

The monospecific antibodies were used for immunofluorescent analysis of β-galactosidase in cryostat sections (4 μm) in thickness) of pollen (Fig. 2). Both *Gal* and *gal* pollen grains showed positive fluorescence for β-galactosidase. This indicated that *gal* pollen possesses antigenically active but enzymically inactive molecules.

The presence of antigenic activity was also confirmed by probing the Western blot of proteins from gal/gal pollen. Thus, immunofluorescence analysis has provided some evidence suggesting that post-transcriptional processes are implicated in the genetic defect in pollen β-galactosidase expression.

## References


Blank, A., Silber, J.R., Thelan, M.P. and Dekker, C.A. (1983) Detection of enzymatic activities in SDS-polyacrylamide gels: DNA polymerases as model enzymes. Anal Biochem 135: 423-430.

Hancock, K. and Tsang, V.C.W. (1983) India ink staining of proteins on nitrocellulose paper. Anal Biochem 133: 157-162.

Hawkes, R., Niday, E., and Gordon, J. (1982) A dot-immunobinding assay for monoclonal and other antibodies. Anal Biochem 119 142-147.

Li, S.C., Mazzotta, M.Y. and Chien, S.F. (1976) Isolation and characterization of Jack bean β-galactosidase. J Biol Chem 250: 6786-6791.

Singh, M.B. and Knox, R.B. (1984) Quantitative cytochemistry of β-galactosidase in normal (Gal) and enzyme deficient (gal) pollen of Brassica campestris: An application of the indigogenic method. Histochem J 16: 1273-1296.

Singh, M.G., O'Neill, P.M. and Knox, R.B. (1985) Initiation of post-meiotic β-galactosidase synthesis during microsporogenesis in oil seed rape. Plant Physiol 77: 225-228.

Singh, M.B. and Knox, R.B. (1985) Gene controlling β-galactosidase deficiency in pollen of oil seed rape. The Journal of Heredity 76 (in press).

Singh, M.G. and Knox, R.B. (1985) Immunofluorescence applications in plant cells. In: Robards, A.W. (ed.) Botanical Microscopy 1985. Oxford, University Press, Oxford (in press).

Smith, E.D. and Fischer, P.A. (1984) Identification, developmental regulation, and response to heat shock of two antigenically related forms of a major nuclear envelope proteins in Drosophila embryos: application of an improved method for affinity purification of antibodies using polypeptides immobilized on nitrocellulose blots. J Cell Biol 99: 20-28.

Towbin, H., Staehlin, T. and Gordon, J. (1979) Electrophoretic transfer of proteins from polyacrylamide gels to nitrocellulose sheets: procedure and some applications. Proc Natl Acad Sci USA 76: 4350.4354.

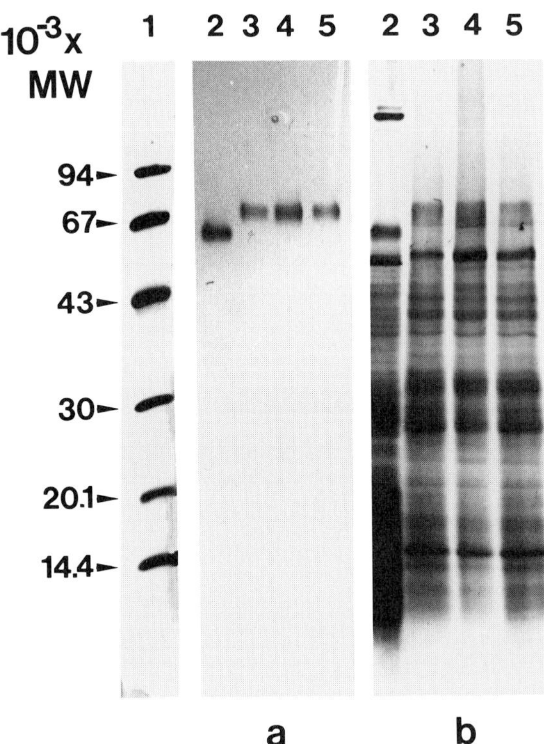


Fig. 1. Identification of β-galactosidase bands in SDS-PAGE gels following renaturation treatments. Lanes 2 to 5: staining for β-galactosidase activity; 2, Pollen from Gal/gal plant; 3-5, Leaf proteins from Gal/gal, Gal/gal and gal/gal plants. Fig. 1b shows coomassie blue staining of same gel as in 1a.

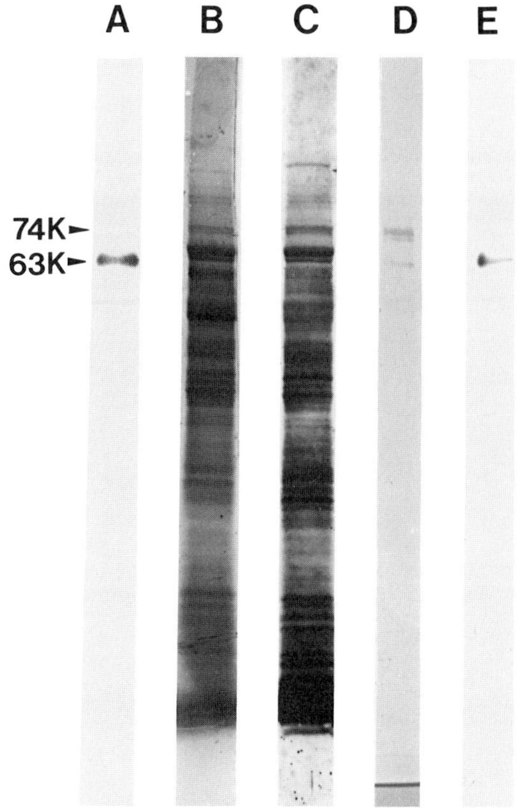


Fig. 2. Immunoblotting of pollen and leaf β-galactosidases. A: SDS-PAGE gel of pollen proteins stained with enzyme reaction mixture (indoxyl substrate, Singh and Knox, 1984); B: as A, but stained with coomassie blue; C: Western blot of gel shown in B stained with india ink method (Hancock and Tsang, 1983); D: Western blot of <u>Brassica</u> leaf proteins probed with pabs to jack bean β-galactosidase; E: Western blot of Brassica pollen proteins probed with monospecific pab isolated by western blotting method.

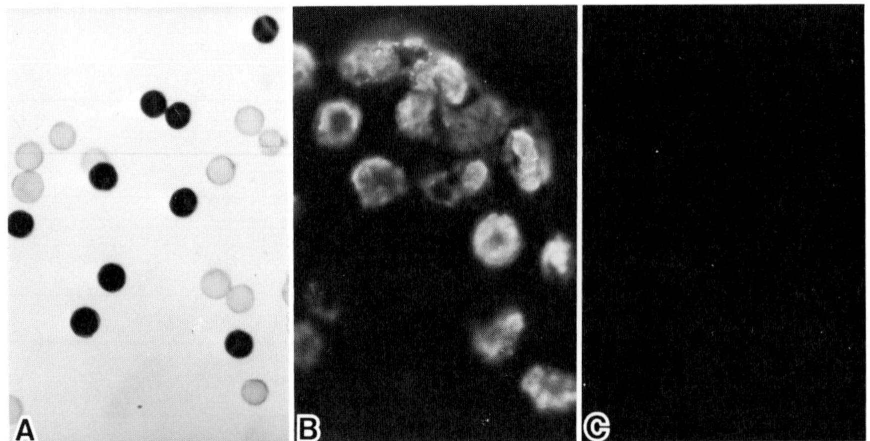


Fig. 3. Immunofluorescence analysis of β-galactosidase in pollen of oil seed rape, Brassica campestris. A: enzyme cytochemistry in Gal/gal pollen using the indoxyl method; B: detection of the enzyme as an antigen in frozen semi-thin sections of Gal/gal pollen using monospecific pab; C: control for B omitting primary antibody.

# Identification of Duplicate Loci and Evidence for Post-meiotic Gene Expression in Pollen

N.F. WEEDEN[1]

## 1 Introduction

Male and female gametophytes represent a crucial portion of the life cycle in plants. In gymnosperms and angiosperms such stages are highly specialized and relatively brief; however, other tracheophytes possess a more developed gametophyte which often exists as a free-living entity. One would expect such gametophytes to express all enzymes required for an autotropic existence. Indeed, isozyme studies on ferns (Gastony and Gottlieb, 1982) have demonstrated that many of the enzymes of intermediate metabolism are present in gametophytes and that these were synthesized by gametophytic tissue. Parallel investigations on conifers and herbaceous angiosperms have produced similiar results (O'Malley et al., 1978; Weeden and Gottlieb, 1979; Mulcahy et al., 1979, 1981; Adams and Joly, 1980; Tanksley et al., 1981). It appears that, for many vascular plants, approximately 60% of the genes expressed in the sporophyte are also expressed during the gametophytic stage.

---

[1]Department of Horticultural Sciences, NYSAES, Cornell University, Geneva, New York.

The haploid nature of gametophytic tissue implies that each allele at expressed loci will be exposed to the intense selection pressure operating during this stage. Those alleles producing slightly defective proteins would be expected to be rapidly eliminated as a result of poor performance by the gametophytes possessing them. Duplication of a locus, theoretically, can free one of the pair to diverge and either be silenced or develop new functions (Ohno, 1970). Therefore, plants exhibiting duplications of genetic material might be expected to display significant differences between the isozyme phenotypes of pollen and somatic tissue. I report here a comparison of isozyme expression in apple pollen and leaf tissue. Relatively few woody angiosperms have been subjected to such an analysis, but more importantly, apple (2n = 34) is probably an ancient polyploid and may be a suitable organism to examine for loss or divergence of gene expression in pollen.

## 2 Procedures

Pollen was collected from ten apple cultivars immediately before or just after anthesis. The pollen was placed in extraction buffer (ca 1 mg/0.2 ml). After 14 hours the pollen in the extraction buffer was collected by centrifugation and crushed in a drop of buffer between two glass microscope slides. Isozyme phenotypes of the two extracts (leachate from pollen and crushed pollen after soaking) were compared to those from fresh leaf material using procedures described in Chyi and Weeden (1984) and Weeden and Lamb (1985). Isozyme assays were identical to, or minor modifications of, standard recipes.

## 3 Results

The isozymes investigated in both leaf and pollen tissue are listed in the first column of Table I. In most cases, the isozyme was known to be specified by a single Mendelian gene distinct from genes coding other forms of the enzyme. In all, 32 isozymes were studied of which 26 exhibited expression in both pollen and leaf extracts. The highest activities were obtained in extracts from soaked pollen. Pollen leachate contained certain isozymes such as GPI-2 and -3 but lacked many others and, thus, was not very satisfactory for the purposes of the present study.

Seven pairs of what appeared to be duplicate loci were observed (Table I). In all cases these pairs were identified because intergenic hybrid dimers were observed to form between the products of the loci. Six of the seven pairs were present in both leaf and pollen extracts (Table I), with only the IDH isozymes being absent or unobservable in pollen. These IDH isozymes were cytosolic so that their disappearance can not be related to the absence of an organelle.

Among the other isozyme systems examined, two esterases, a peroxidase and the plastid TPI were the only isozymes present in leaf extracts which were not also expressed in pollen. There was no example of a pollen-specific isozyme. Differences in the relative intensity of isozyme bands were observed between the two tissues, but these were mainly the plastid-specific isozymes of GPI and 6PGD, and could be attributed to the relative scarcity of plastids in pollen cells.

Evidence for post-meiotic expression of loci required dimeric enzymes which displayed heterozygous phenotypes in

Table I. Expression of isozymes in apple pollen and somatic tissues.

| Isozyme | Structure | Duplicate system | Tissue expressed | Post-meiotic expression |
|---|---|---|---|---|
| AAT-1 | dimeric | no | both | yes |
| AAT-2 | dimeric | no | both | yes |
| AAT-3 | dimeric | yes | both | no data |
| AAT-4 | dimeric | yes | both | no data |
| DIAP-1 | unknown | unknown | both | no data |
| DIAP-2 | dimeric | yes | both | yes |
| DIAP-3 | dimeric | yes | both | yes |
| DIAP-4 | unknown | unknown | both | no data |
| EST-1 | unknown | unknown | leaf | no data |
| EST-2 | unknown | unknown | leaf | no data |
| GPI-1 | unknown | unknown | both | no data |
| GPI-2 | dimeric | yes | both | yes |
| GPI-3 | dimeric | yes | both | yes |
| IDH-1 | dimeric | yes | leaf | no data |
| IDH-2 | dimeric | yes | leaf | no data |
| LAP-1 | unknown | unknown | both | no data |
| MDH-1 | dimeric | yes | both | no data |
| MDH-2 | dimeric | yes | both | no data |
| MDH-3 | unknown | unknown | both | no data |
| ME-1 | tetramer | no | both | yes |
| PRX-1 | unknown | unknown | both | no data |
| PRX-2 | unknown | unknown | both | no data |
| PRX-3 | unknown | unknown | leaf | no data |
| PRX-4 | unknown | unknown | both | no data |
| 6GPD-1 | dimeric | no | both | yes |
| 6GPD-2 | dimeric | yes | both | no data |
| 6GPD-3 | dimeric | yes | both | no data |
| PGM-1 | monomer | unknown | both | no data |
| PGM-2 | monomer | unknown | both | no data |
| TPI-1 | unknown | unknown | leaf | no data |
| TPI-2 | dimeric | yes | both | yes |
| TPI-3 | dimeric | yes | both | yes |

leaf tissue. Fortunately, apple cultivars show a high degree of isozyme polymorphism, and five appropriate instances of heterozygosity were found (Table I). In all five cases the absence of intragenic hybrid dimers in the pollen extracts (e.g. Fig. 1) demonstrated that the polypeptides present in the pollen had been synthesized after the first meiotic division.

## 4 Discussion

The results of this investigation indicate that a

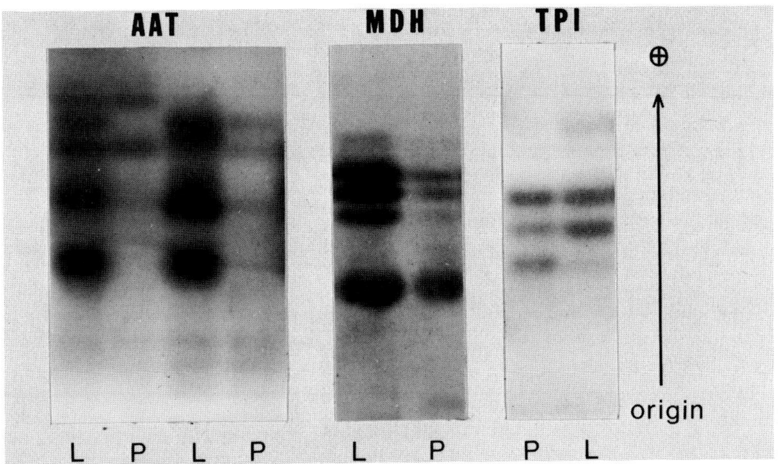


Fig. 1. Leaf (L) and pollen (P) isozyme phenotypes for aspartate aminotransferase (AAT), malate dehydrogenase (MDH), and triose phosphate isomerase (TPI). The AAT phenotypes show two examples of the loss of the intragenic hybrid dimer in the non-duplicated system, AAT-1. The MDH comparison is an example of duplicate loci (MDH-1 and MDH-2) displaying a hybrid dimer in both leaf and pollen tissue. The TPI phenotypes are from an individual heterozygous at TPI-3. The middle band of the lower triplet is relatively fainter in pollen due to the absence of the intragenic heterodimer. Note also the absence of the most anodal isozyme, TPI-1, in the pollen extract.

majority of the protein products involved in cellular metabolism were present in apple pollen tissue and were products of the same genes that are active in somatic tissues. That this overlap of gene products was not a result of a carry-over of proteins synthesized before meiosis was demonstrated in all 5 systems from which such information could be obtained. These results corroborate the current dogma that the male gametophyte in angiosperms has retained a considerable degree of genetic and metabolic independence.

The apparent duplication of approximately half of the loci specifying multimeric enzymes provided further evidence for a polyploid origin of the apple and related taxa.

However, the duplicate loci did not exhibit significant divergence either in pollen or leaf tissue. In no case did the relative expression of duplicate loci differ in pollen compared to leaf tissue. The differences in isozyme phenotype which were observed could, in general, be explained either by the respective isozyme system (esterase and peroxidase isozymes are known to often display tissue specificity) or by organellar location (TPI).

## References

Adams WT, Joly RJ (1980) Genetics of allozyme variants in loblolly pine. J. Hered. 71:33-40

Chyi YS, Weeden NF (1984) Relative isozyme band intensities permit the identification ofthe 2N gamete parent for triploid apple cultivars. HortScience 19:818-819

Gastony GJ, Gottlieb LD (1982) Evidence for genetic heterozygosity in a homosporous fern. Amer. J. Bot. 69:634-637

Mulcahy DL, Mulcahy GB, Robinson RW (1979) Evidence for postmeiotic genetic activity in pollen of *Cucurbita* species. J. Hered. 70:365-368

Mulcahy DL, Robinson RW, Ihara M, Kesseli R (1981) Gametophytic transcription for acid phosphatases in pollen of *Cucurbita* species hybrids. J. Hered. 72:353-354

Ohno S (1970) Evolution by Gene Duplication, New York, Springer

O'Malley DM, Allendorf FW, Blake GM (1979) Inheritance of isozyme variation and heterozygosity in *Pinus ponderosa*. Biochem. Genet. 4:297-320

Tanksley SD, Zamir D, Rick CM (1981) Evidence for extensive overlap of sporophytic and gametophytic gene expression in *Lycopersicon esculentum*. Science 213:453-455

Weeden NF, Gottlieb LD (1979) Distinguishing allozymes and isozymes of phosphoglucoisomerases by electrophoretic comparisons of pollen and somatic tissues. Biochem. Genet. 17:287-296

Weeden NF, Lamb RC (1985) Identification of apple cultivars by isozyme phenotypes. J. Amer. Soc. Hort. Sci. 110:509-515

# Gametophytic Gene Expression in Embryo-lethal Mutants of *Arabidopsis thaliana*

DAVID W. MEINKE AND ANN D. BAUS*

The genetic control of embryo development in higher plants has been approached in part through the isolation and characterization of embryo-lethal mutants (Meinke 1986). The most extensive studies have dealt with defective kernel mutants of corn (Sheridan and Neuffer 1982), embryo-lethal mutants of *Arabidopsis* (Müller 1963; Meinke 1985), and variant cell lines of carrot unable to complete somatic embryogenesis *in vitro* (Breton and Sung 1982). *Arabidopsis thaliana* (Cruciferae) has been used as a model plant system for various studies in developmental and molecular genetics because it produces many seeds per plant and has a short generation time, low chromosome number, well-characterized mutants (Rédei 1975), an established linkage map (Koornneef et al. 1983), and an unusually small genome with little repetitive DNA (Leutwiler et al. 1984). Recessive embryo-lethal mutants of *Arabidopsis* isolated following EMS seed mutagenesis have been shown previously to differ with respect to the stage of developmental arrest, the color of arrested embryos and aborted seeds, the percentage and distribution of aborted seeds in heterozygous siliques, the extent of abnormal embryo development, the response of mutant embryos in culture, the development of homozygous mutant plants, the formation of protein and lipid bodies, and the accumulation of seed storage proteins (Meinke and Sussex 1979a,b; Meinke 1982, 1985; Meinke et al. 1985; Marsden and Meinke 1985).

* Department of Botany and Microbiology, Oklahoma State University, Stillwater, OK 74078 U.S.A.

One question that needs to be addressed in the analysis of developmental mutants is whether expression of the mutant gene is limited to a single stage of the life cycle. One approach that has been used to address this question in a variety of animal systems has been to isolate temperature-sensitive mutants and determine whether mutant individuals develop normally at the restrictive temperature once they complete embryo development at a permissive temperature. This same approach should be applicable to plant systems, but unfortunately no temperature-sensitive embryo-lethal mutants of Arabidopsis have yet been found (Meinke 1985). Expression of mutant alleles at other stages of development has instead been examined by studying the morphology of homozygous mutant plants grown in culture (Meinke et al. 1985) and by analyzing the distribution of aborted seeds along the length of heterozygous siliques (Meinke 1982). This second approach has proven to be a simple but powerful test for gametophytic expression of mutant alleles.

## Distribution of Aborted Seeds in Heterozygous Siliques

Siliques on plants heterozygous for an embryo-lethal mutation contain approximately 25% aborted seeds following self-pollination (Figure 1). Any significant deviation from a random distribution of aborted seeds along the length of heterozygous siliques must be caused either by gametophytic expression of the mutant allele or by the presence of a second mutant allele that disrupts pollen germination, pollen-tube growth, or the distribution of mutant ovules along the length of the silique prior to fertilization (Meinke 1982). Ten embryo-lethal mutants of Arabidopsis have been identified (Table 1) in which aborted seeds are distributed non-randomly along the length of heterozygous siliques (Meinke 1982, 1985). Although the presence of a linked gametophytic factor cannot be eliminated, the available evidence suggests that in most if not all of these mutants, the non-random distribution of aborted seeds is caused by gametophytic expression of the same mutant allele that prevents the completion of normal embryo development. This conclusion is consistent with the results of other studies that have shown extensive overlap

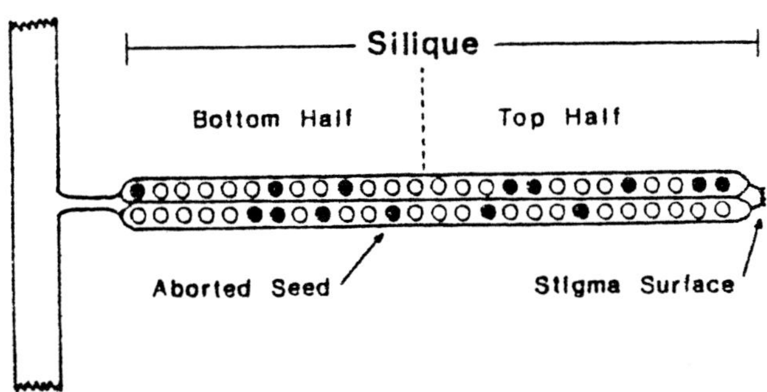


Figure 1. Drawing of a silique from a plant heterozygous for a recessive embryo-lethal mutation. Phenotypically normal seeds are composed of two genotypic classes (+/m and +/+) that can be distinguished by screening plants in the next generation. Aborted seeds are distributed randomly along the length of this particular silique. (Adapted from Meinke 1982)

Table 1. Non-random distribution of aborted seeds in heterozygous siliques from 10 embryo-lethal mutants of *Arabidopsis thaliana* (Meinke 1982, 1985).

| Mutant[a] | Siliques screened | Percent Aborted seeds | Percent top half[b] | Chi-square | Lethal phase |
|---|---|---|---|---|---|
| 124D | 100 | 23.9 | 61.5 | 71.3*** | Preglobular |
| 127AX-A | 41 | 23.6 | 58.9 | 14.7*** | Preglobular |
| (18°C) | 40 | 22.2 | 62.3 | 26.1*** | " |
| 113J-4A | 43 | 21.3 | 64.5 | 33.6*** | Preglobular |
| (18°C) | 44 | 20.8 | 70.9 | 67.8*** | " |
| 79A | 98 | 23.9 | 59.9 | 48.5*** | E. globular |
| 95A-2B | 57 | 24.5 | 40.6 | 19.6*** | E. globular |
| (18°C) | 78 | 22.0 | 51.6 | 0.6 | " |
| 109F-5D | 60 | 21.0 | 62.8 | 35.9*** | Globular |
| (18°C) | 42 | 23.6 | 56.6 | 7.3** | " |
| 112G-1A | 60 | 20.7 | 57.7 | 13.9*** | Globular |
| (18°C) | 40 | 24.0 | 50.3 | 0.01 | " |
| 115D-4A | 40 | 18.8 | 57.9 | 8.5** | Green blimp |
| (18°C) | 40 | 19.1 | 61.5 | 18.2*** | " |
| 126E-B | 40 | 23.8 | 65.9 | 49.0*** | Heart-linear |
| (18°C) | 40 | 12.3 | 84.0 | 114.2*** | " |
| 130BA-2 | 41 | 24.1 | 58.1 | 12.9*** | Cotyledon |
| (18°C) | 64 | 24.1 | 57.7 | 16.1*** | " |

** Significantly different from 50% at P=0.01; *** at P=0.001
[a] Most mutants were grown at both 24°C and 18°C. In several cases the distribution of aborted seeds was clearly affected by temperature.
[b] Percentage of total aborted seeds positioned in the top half of the silique. This should equal 50% if aborted seeds are distributed randomly throughout the silique.

between sporophytic and gametophytic gene expression in higher plants (Ottaviano et al. 1980; Tanksley et al. 1981; Willing and Mascarenhas 1984). The significance of studies with embryo-lethal mutants is that the mutant allele in question is already known to perform an essential function during at least one stage of the life cycle. The high frequency of embryo-lethal mutants exhibiting a non-random distribution of aborted seeds in heterozygous siliques suggests that many of the genes that are required for the completion of embryo development are also expressed prior to fertilization.

## Reciprocal Crosses Between Heterozygous and Wild-Type Plants

Several different lines of evidence suggest that the effect of the mutant allele is either on pollen germination or pollen-tube growth and not on the distribution of mutant ovules (Meinke 1982). This assumption has now been tested for one mutant (126E-B) with a particularly non-random distribution of aborted seeds by making reciprocal crosses between heterozygous (+/m) and homozygous wild-type (+/+) plants and analyzing the distriubtion of +/m and +/+ seeds in the resulting siliques (Baus and Meinke, manuscript in preparation). The results of these studies (Table 2) clearly demonstrate that the non-random distribution of aborted seeds in mutant 126E-B is caused by gametophytic gene expression that disrupts the growth of mutant pollen tubes and not the distribution of mutant ovules.

Reciprocal cross-pollinations were performed by either (1) saturating the stigma surface of a mature bud prior to anthesis with pollen such that subsequent self-pollination did not result in self-fertilization; or (2) emasculating a mature bud prior to anthesis and applying pollen to the stigma surface. Similar results were produced by these two methods. The two genotypic classes of seeds (+/+ and +/m) produced by these crosses were distinguished by screening the resulting plants for the presence of aborted seeds following self-pollination. Maps showing the distribution of +/+ and +/m seeds in siliques produced by these crosses were made by planting seeds in order of their location within the silique.

Table 2. Results of reciprocal crosses between heterozygous 126E-B (+/m) and homozygous wild-type (+/+) plants.[a]

| | Plant Used as Pollen Source | |
|---|---|---|
| | 126E-B (+/m) | Wild-type (+/+) |
| Expected distribution of +/m seeds in silique[b] | Non-random | Random |
| Siliques produced by cross | 18 | 9 |
| F-1 plants screened | 827 | 357 |
| Percent +/m seeds in top half of original silique | 85.5% | 52.0% |
| Chi-Square | 82.5*** | 0.16 |
| Expected frequency of +/m and +/+ seeds in silique[c] | More +/+ than +/m | Equal numbers |
| +/+ seeds in silique | 68.9% | 41.7% |
| +/m seeds in silique | 20.1% | 42.0% |
| Seeds not screened | 11.0% | 16.3% |

*** Significantly different from 50% at P=0.001.
[a] Results of pollination method #1 as described in text.
[b,c] Distribution and frequency expected if the mutant allele affects only the growth of mutant pollen tubes.

Over 1500 plants were screened to produce maps of 41 siliques. The results showed a non-random distribution of +/+ and +/m seeds in these siliques when 126E-B was used as a male parent, and a random distribution when a wild-type plant was used as a male parent (Table 2). Although this study was undertaken primarily to examine gametophytic expression of the 126E-B mutant allele, a similar approach could be used in future experiments to study gametophytic expression of mutant alleles that disrupt known biochemical pathways (Somerville and Ogren 1983; Browse et al. 1985).

Acknowledgements

We thank Linda Franzmann, Chris Monnot, and David Patton for their assistance. This research was supported by NSF Grant No. PCM82-15667 to D.W. Meinke and by funds from the College of Arts and Sciences, Oklahoma State University.

References


Breton A, Sung ZR (1982) Temperature-sensitive carrot variants impaired in somatic embryogenesis. Dev Biol 90: 58-66.

Browse J, McCourt P, Somerville CR (1985) A mutant of *Arabidopsis* lacking a chloroplast-specific lipid. Science 227: 763-765.

Koornneef M, van Eden J, Hanhart CJ, Stam P, Braaksma FJ, Feenstra WJ (1983) Linkage map of *Arabidopsis thaliana*. J Hered 74: 265-272.

Leutwiler LS, Hough-Evans BR, Meyerowitz EM (1984) The DNA of *Arabidopsis thaliana*. Mol Gen Genet 194: 15-23.

Marsden MPF, Meinke DW (1985) Abnormal development of the suspensor in an embryo-lethal mutant of *Arabidopsis thaliana*. Amer J Bot (in press).

Meinke DW (1982) Embryo-lethal mutants of *Arabidopsis thaliana*: Evidence for gametophytic expression of the mutant genes. Theor Appl Genet 63: 381-386.

Meinke DW (1985) Embryo-lethal mutants of *Arabidopsis thaliana*: Analysis of mutants with a wide range of lethal phases. Theor Appl Genet 69: 543-552.

Meinke DW (1986) Use of embryo-lethal mutants to study plant embryo development. In: Miflin BJ (ed) Oxford Surveys of Plant Molecular and Cell Biology, Vol 3. Oxford University Press (manuscript in preparation).

Meinke DW, Franzmann L, Baus A. Patton D, Weldon R, Heath JD, Monnot C (1985) Embryo-lethal mutants of *Arabidopsis thaliana*. In: Freeling M (ed) Plant Genetics, UCLA Symposia on Molecular and Cellular Biology, Vol 35. Alan Liss, New York (in press)

Meinke DW, Sussex IM (1979a) Embryo-lethal mutants of *Arabidopsis thaliana*: A model system for genetic analysis of plant embryo development. Dev Biol 72: 50-61.

Meinke DW, Sussex IM (1979b) Isolation and characterization of six embryo-lethal mutants of *Arabidopsis thaliana*. Dev Biol 72: 62-72.

Müller AJ (1963) Embryonentest zum Nachweis rezessiver Letalfaktoren be *Arabidopsis thaliana*. Biol Zentralbl 82: 133-163.

Ottaviano E, Sari-Gorla M, Mulcahy DL (1980) Pollen tube growth rates in *Zea mays*: implications for genetic improvement of crops. Science 210: 437-438.

Rédei GP (1975) *Arabidopsis* as a genetic tool. Ann Rev Genet 9: 111-127.

Sheridan WF, Neuffer MG (1982) Maize developmental mutants: Embryos unable to form leaf primordia. J Hered 73: 318-329.

Somerville SC, Ogren WL (1983) An *Arabidopsis thaliana* mutant defective in chloroplast dicarboxylate transport. Proc Natl Acad Sci (USA) 80: 1290-1294.

Willing RP, Mascarenhas JP (1984) Analysis of the complexity and diversity of mRNAs from pollen and shoots of *Tradescantia*. Plant Physiol 75: 865-868.

# Pollen Competitive Ability in Maize Selection and Single Gene Analysis

E. OTTAVIANO, P. SIDOTI AND M. VILLA

## 1. INTRODUCTION

Pollen competitive ability is one of the major components of gametophytic fitness; it depends mainly on germination time, tube growth rate and fertilization ability. The resulting gametophytic selection can produce significant evolutionary changes (1,2) and can be used to develop efficient methods of plant breeding (3,4,5).

The phenomenon is assumed to rely on two main factors: i) genetic variability of gametophytic origin and ii) gametophytic-sporophytic genetic overlap. Data supporting this assumption have been obtained by studying maize pollen tube elongation in vitro (6) tomato and maize isoenzyme patterns in pollen and in sporophytic tissues (7,2) and mRNAs from pollen and shoots of *Tradescantia* (8).

However, very little is known with regard to the genetical control of pollen competitive ability components. Most of the mutations affecting gametophytic development and resulting in male sterility, concern genes expressed in the sporophytic phase (9). The exception is represented by $Rf_3$, the restorer of cytoplasmic S-male sterility. On the other hand *ga* alleles (gametophytic factors) seem to represent only special cases controlling the mating system in some maize populations.

The effect on the sporophytic generation of pollen competitive ability selection is indicated by correlations found between this character and sporophytic traits (10,3). In maize

---

Dipartimento di Biologia, Sezione di Genetica e Microbiologia, Via Celoria 26 - 20133 Milano, Italy, tel. (02) 23.08.23.

selection for competitive ability produced positive and correlated responses for sporophytic traits (11). However, in view of the selection procedure adopted, the results furnished only an indication that selection response is due to variability of genes expressed in the gametophytic phase.

This chapter refers to selection experiments showing the genetical basis and the significance of pollen competitive ability. It also describes a procedure to select mutants affecting pollen development and pollen function.

## 2. SELECTION FOR POLLEN COMPETITIVE ABILITY

Selection for pollen competitive ability can be studied by means of two different methods: i) pollination of flower at different pollen densities (12),ii) varying the distance to be covered by the pollen tube in the style (11). The second approach can be easily used in maize where the silk length varies according to the position of the flower on the ear, increasing from the top to the base. Within plant selection is applied when the pollen from the same heterozygote plant is used to pollinate a single plant, either for selfing or for crossing. Differences in the progeny due to the position of the kernels on the ear (apex or base) reveal response to gametophytic selection due to genes expressed in the postmeiotic male gametophytic phase.

Recently detailed information has been obtained by means of this procedure in a recurent selection scheme, where the sporophytes (plants) were chosen strictly at random (13). After two cycles of selection two populations were produced: Base population (high gametophytic selection intensity) and Apex population (low gametophytic selection intensity). Response to selection was evaluated in 60 S2 and 160 FS (full-sib) families derived from the two populations. The S2's (30 base and 30 apex) were used to study pollen competitive ability and the FS's (80 base and 80 apex) to evaluate the correlated response of sporophytic traits: 50 kernel weight (50-KW) kernel number per row (KNR) and number of kernel rows per ear (RN).

Gametophytic competitive ability of each S2 family was evaluated in comparison with a standard inbred line by means of a mixed pollination technique. The value is expressed as the coefficient of regression (b) describing the variation of uncoloured kernels (the standard produces coloured aleurone) from the apex to the base of the ear. The results are summarized in table 1.

Table 1. EFFECTS OF GAMETOPHYTIC RECURRENT SELECTION

| Families from | b mean | b range | 50 KW mean | 50 KW range | KNR mean | KNR range | RN mean | RN range |
|---|---|---|---|---|---|---|---|---|
| 1) BASE POPUL. | 0.31 | -5.34<br>5.58 | 15.67 | 11.97<br>19.02 | 46.0 | 39.9<br>51.7 | 15.6 | 13.2<br>18.1 |
| 2) APEX POPUL. | -1.24 | -6.14<br>2.67 | 14.53 | 11.22<br>17.88 | 46.8 | 38.7<br>52.0 | 16.3 | 13.7<br>19.3 |
| DIFFERENCES (%) | | | 7.85* | | $-1.68^{n.s.}$ | | $-4.04^{n.s.}$ | |

* : significant difference (P<0.05).

The mean value of gametophytic competitive ability in the progeny produced at higher selection intensity is higher than that of the progeny produced at low selection intensity, showing that the variability of the character is largely based on genes expressed in the gametophytic phase. Sporophytic effects could also play an important role; however, considering the selection procedure used, they should not contribute to the selection response obtained.

A positive correlated response is observed for mean kernel weight (50 KW) a character which reflects growth in the endosperm. KNR and RN which are mainly related to developmental processes do not reveal significant differences. Considering that the base population was intercrossed for several generations and therefore linkage should not play an important role in the correlated response, the results obtained indicate that there are genes controlling basic physiological processes of growth which show sporophytic-gametophytic genetic overlap.

## 3. ENDOSPERM GAMETOPHYTIC MUTANTS

Genetic dissection of male gametophytic development and functions requires single gene gametophytic mutants. However, not much has been done to produce this material, mainly because within species morphological variability of the pollen is very poor and efficient selection methods for viable gametophytic mutants are difficult to set up.

The correlation between pollen competitive ability and kernel weight, endosperm-gametophytic genetic overlap of some genes controlling starch synthesis (14) and the results obtained in Arabidopsis (15) suggest an indirect method of approaching this problem. It is based on the screening of viable defective endosperm (de) mutants, which are then evaluated for gametophytic expression. The first step is accomplished quite easily; in fact several de mutants have been described (16,17). The second consists in the evaluation of the gametophytic competitive ability of the mutants, on the basis of distortions from mendelian segregation ratios or by using pollen mixture techniques.

Following this strategy the F2 segregation of 34 different single gene de mutants, all in homozygote genetic backgrounds, were analyzed. To study the effect on pollen tube growth rate, F2 ears were divided into three sectors of equal length. The results are summarized in table 2 where P1, P2 and P3 are the proportions of de kernels in each sector from the top to the base of the ear and P the proportion on the entire ear.

Table 2. PROPORTION (P) OF de KERNELS ON F2 EARS OF de/+ PLANTS

| | $P \cong .25$ (i) | $P > .25$ (ii) | $P < .25$ (iii) | $P1 > P2 > P3$ (iv) | Tot. |
|---|---|---|---|---|---|
| Number of Mutants | 14 | 4 | 11 | 5 | 34 |

The type of segregation on which the classification of mutants is based was assessed by the chi square test. Mutants of class (i) do not reveal gametophytic effects, while those of class (ii) call for further investigation of the genetical basis and endosperm dosage effects. For the mutants of class (iii) the proportion of de kernels was significantly lower than expected, but the proportion did not change from sector to sector. This result would be found when the gene effects pollen development; in this case the viability of pollen carrying de allele would be reduced. Class (iv) includes mutants showing a very clear reduction (table 3) of proportion of <u>de</u> kernels from the apex to the base of the ear, a pattern which can be explained as a reduction of tube growth rate of <u>de</u> pollen grains.

Table 3. PROPORTION (Pi) IN THREE EAR SECTORS

| MUTANTS | N. OF KERNELS | P1 (APEX) | P2 | P3 (BASE) | CHI SQUARE 3:1 | CHI SQUARE P1=P2=P3 |
|---|---|---|---|---|---|---|
| de-B1 | 2039 | .2598 | .1903 | .1726 | ** | ** |
| de-B3 | 1447 | .2477 | .1641 | .1677 | ** | ** |
| de-B18 | 1990 | .2970 | .2209 | .1897 | n.s. | ** |
| de-B37 | 1779 | .3185 | .1997 | .1652 | * | ** |
| de-B127 | 1691 | .2680 | .1450 | .0969 | ** | ** |

* : $P < 0.05$; ** : $P < 0.01$.

Therefore two classes of endosperm-gametophytic mutants were selected, one affecting the first stage of the gametophytic phase, i.e., pollen development, and the second affecting post-pollination functions. It is important to note that for the mutants of this second group a constant reduction of dry matter accumulation during endosperm development was found (17); moreover some of these mutants showed a considerable reduction of indole-3-acetic acid in the endosperm (18). The results obtained confirm the efficiency of the method used and provide further confirmation of the extent of the gametophytic-sporophytic genetic overlap.

ACKNOWLEDGEMENTS

Research work supported by C.N.R., Italy. Special grant I.P.R.A., Subproject 1, Paper N. 537. The Authors are grateful to F.Salamini for furnishing most of the <u>de</u> material.

REFERENCES

1) Mulcahy D.L. (1979) The rise of the Angiosperms: a genecological factor. Science, 206: 20-23.

2) Sari Gorla M., Frova C., and Ottaviano E. (1985). The extent of gametophytic-sporophytic gene expression in maize (in press).

3) Ottaviano E., Sari Gorla M. and Mulcahy D.L. (1980) Pollen tube growth rate in Zea mays: implications for genetic improvement of crops. Science, 210: 437-438.

4) Zamir D. (1983) Pollen gene expression and selection: applications in plant breeding. S.D. Tanksley and T.J. Orton (Eds), Isozymes in Plant Genetics and Breeding, part A: 313-329.

5) Mulcahy D.L. (1983) Manipulation of gametophytic populations. Efficiency in Plant Breeding. Lange W., Zeven A.C., Hogenboom N.G. (Eds). Pudoc Wageningen.

6) Sari Gorla M., Ottaviano E. and Faini D. (1975) Genetic variability of gametophytic growth rate in maize. Theor. Appl. Genetics, 46: 289-294.

7) Tanksley D.S., Zamir D. and Rick C.M. (1981) Evidence for extensive overlap of sporophytic and gametophytic gene expression in Licopersicon esculentum. Science, 213: 453-455.

8) Willing R.P. and Mascarenhas J.P. (1984) Analysis of complexity and diversity of mRNAs from pollen and shoots of Tradescantia. Plant Physiol., 75: 865-868.

9) Majodelo S.D.P., Groyan C.O., Servella P.A. (1966) Morphological expression of genetic male sterility in maize. Crop. Sci., 6: 379-380.

10) Mulcahy D.L. (1971) A correlation between gametophytic and sporophytic characteristics in Zea mays L. Science, 171: 1155-1156.

11) Ottaviano E., Sari Gorla M., and Pe E. (1982) Male gametophytic selection in maize. Theor.Appl.Genetics,63: 249-254.

12) Ter-Avanesian P.V. (1978) The effect of varying the number of pollen grain used in fertilization. Theor. Appl. Genetics, 52: 77-79.

13) Ottaviano E. (1985) Male gametophytic selection in maize II (in press).

14) Bryce W.H., Nelson O.F.(1979) Starch-synthesizing enzymes in the endosperm and pollen of maize. Plant Physiol. 63:312.

15) Meinke D.W. (1982) Embryo-lethal mutants of Arabidopsis taliana: Evidence for gametophytic expression of mutant genes. Theor. Appl. Genetics 63: 381-386.

16) Coe E.H., Neuffer M.G. (1977) The genetic of corn. In Corn and Corn Improv. G.F. Sprague Ed. pp. 111-223. Am. Soc. Agr. Madison.

17) Manzocchi L.A., Daminati M.G., Gentinetta E., Salamini F. (1980) Viable defective endosperm mutants in maize I. Maydica XXY: 105-116.

18) Torti G., Lombardi L., Manzocchi L.A. (1984) Indole-3-acetic acid content in viable defective endosperm mutants of maize. Maydica XXIX: 335-343.

# Extent of Gene Expression at the Gametophytic Phase in Maize

M. Sari Gorla[1], C. Frova[1], E. Redaelli[1]

## INTRODUCTION

The male gametophyte of higher plants is metabolically active; during microspore maturation, grain germination and tube growth, the most important physiological processes are those related to respiration, reserve mobilization, nutrient uptake, biosynthesis first of intine and then of pollen tube wall (3). Pollen is also able to respond to environment; this ability is expressed both in physiological interaction with stylar tissues during tube development, and in relation to environmental stresses, toxic products, pathogens (8). The responses are generally correlated to sporophytic behaviour (7,9).

With regard to genetic control of gametophytic functions, the main sporophytic role consists probably in regulating the haploid genome during ontogenesis and providing a nutritional support to microspores. As in every tissue, microspore differentiation has to be promoted by the repression of many sporophytic genes and by the parallel expression of pollen specific characteristics; for instance, the gametophytic growth pattern is different from that of other tissues (2).

However there is evidence that a good part of the genome is expressed in both phases of life cycle; selection applied at the gametophytic level induces responses for adaptive traits in the resulting sporophyte (14); in mature pollen pre-existing mRNA is present,but new transcripts are synthesized, and genes

[1]Dipartimento di Biologia, Sezione di Genetica e Microbiologia, via Celoria 26, 20122 Milano, Italy, tel. (02) 23.08.23.

active during the last phase of microspore maturation and during pollen germination and tube growth are the same;the analysis of gametophytic transcrips demonstrates that sequences specific to pollen, not expressed in the sporophyte, and sequences expressed both in pollen and sporophyte exist (4,13).

The evaluation of the extent and the type of genes shared between gametophytic and sporophytic generations is important in order to utilize pollen as a model system for physiological process studies or as a biological indicator, and especially to assess the weight of gametophytic competition on the sporophytic generation (6).

One approach to this problem consists in analyzing the type of expression of single genes by means of isozymes, used as genetic markers. The first account based on this procedure was given by Tanksley et al. (11) in tomato. Owing to the random distribution of the coding genes in the entire genome, their great polymorphism and their involvement in different metabolic pathway steps, they can furnish both a spatial and functional sampling of the genome. They also allow the type of expression of the gene to be directly visualized. In fact, the presence of enzymatic activity in pollen may be the result of haploid transcription, mRNA of sporophytic origin translation, or may be due to a product transcribed and translated by the sporophyte. The expected electrophoretic pattern in pollen extracts from heterozygous F/S plants, is different according to the gametophytic or sporophytic origin of the transcripts.

## EXPERIMENTAL PROCEDURE

Thirty-four loci were studied, the majority of which code for a multimeric protein. In these cases the type of gene expression can be deduced by comparing gametophytic and sporophytic zymograms from plants heterozygous for electrophoretic mobility. Assuming that the active molecule originates by random association of the monomers and that the different multimeric forms are equally active, in the sporophyte a multibanded pattern is expected, where the bands have different intensity, according to $(F+S)^n$ development, $\underline{n}$ being the number of subunits of the active molecule. Thus, for a dimer the expectation is 1:2:1, for a tetramer 1:4:6:4:1, and so on. However, in pollen, if transcription is postmeiotic, the information for but one variant being present in each grain, only two bands are expected in any case (Fig. 1).

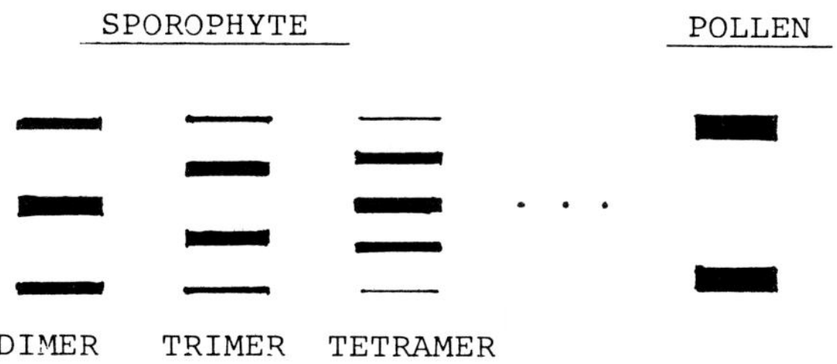


Fig.1. Expected electrophoretic patterns for sporophytic tissues and pollen from heterozygous F/S plants.

The reliability of this method has been controlled (1); using translocation between A and B chromosomes, hyperploid plants, heterozygous for electrophoretic mobility, have been obtained; these produced partially diploid, F/S pollen,which showed a zymogram identical to that of the sporophyte, thus demonstrating that the lack of hybrid bands in normal gametophyte zymograms can be attributed only to post-meiotic gene expression. Since this procedure gave positive results in the three cases tested (ADH-1, GOT-1, PHI), the general method was applied to all other enzymes.

The extracts from sporophytic tissues (scutellum, coleoptile) and pollen, were submitted to starch or polyacrylamide gel electrophoresis (10), and the enzymatic activity was revealed by dipping the gel slab, after running, into a proper reaction mixture (12). This contains the enzyme substrate and, since the primary product of the reaction is generally not directly visible, another reagent is added, to give a secondary product. This leads, spontaneously or in the presence of a cofactor,to a coloured, non diffusable precipitate formation, which provides an activity band.

## RESULTS AND CONCLUSIONS

The quantitative estimate concerned structural genes, involved in the genetic control of basic metabolic functions; the studied enzymes were oxidoreductases (Alcohol dehydrogenase, Catalase, Glutamate dehydrogenase, Isocitric dehydrogenase, Malate dehydrogenase, 6-P-Gluconate Dehydrogenase, Superoxide dismutase), transferases (Glutamate oxaloacetate transaminase, Phosphoglucomutase, UDP-glucose pyrophosphorylase), hydrolases (Acid phosphatase, β-glucosidase, Endopeptidase, Invertase), isomerases (Phosphohexose isomerase), which do not have a highly specialized role.

Table 1. ENZYME CATEGORIES.

| Haplo-diploid expression | Tissue specific | Sporophytic only | Present in pollen but gene expression not detected |
|---|---|---|---|
| ACP-1 | β-GLU | CAT-2 | GOT-3 |
| ADH-1 | INV | CAT-3 | SOD-1 |
| CAT-1 | | 6PGD-1 | SOD-2 |
| GDH | | | SOD-3 |
| GOT-1 | | | SOD-4 |
| GOT-2 | | | UDPGpp |
| IDH-1 | | | MDH-1 |
| IDH-2 | | | MDH-2 |
| MDH-4 | | | MDH-3 |
| 6PGD-2 | | | MDH-5 |
| PHI | | | ADH-2 |
| | | | ADH-3 |
| | | | ACP-2 |
| | | | ENP |
| | | | PGM-1 |
| | | | PGM-2 |

The results allow the enzymes to be grouped in different categories (Table 1):

1) Enzymes present in pollen and sporophyte with haplo-diploid expression; they represent the majority of the analyzed loci (ACP-1, ADH-1, CAT-1, GDH, GOT-1, GOT-2, IDH-1, IDH-2, MDH-4, 6PGD-2).

2) Enzymatic activity present in both pollen and sporophyte but performed by different tissue-specific enzymes; this was observed in only two cases (β-GLU, INV).

3) Enzymes present in the sporophyte only (CAT-2, CAT-3, 6PGD-1); they are enzymes determined by multiple loci, of which

only one form was expressed in the gametophyte.

4) Enzymes whose presence was detected in pollen,but whose coding gene expression it was not possible to test. GOT-3, SOD, UDPGpp were invariant; MDH revealed a too complex multiple pattern (MDH-4 was investigated by isoelectricfocusing); ADH-2, ACP-3 did not give a clear response; ACP-2, ENP, PGM are monomeric forms. The analysis of the latter is still in course, since the procedure we have developed, based on B-A translocations, is quite complex. It consists in producing hypoploid, F/S heterozygous plants, that will give only one type of live pollen; thus, if the origin of the enzyme is haploid, a one-banded electrophoretic pattern is expected.

A different pattern for pollen and sporophyte was observed for GDH and ACP-2; since probably the same gene is involved, these forms appear to be due to post-genetic modifications of the gene product; for ACP-2 these results are very similar to what was observed in the human monomeric Acid Phosphatase (5).

Out of the thirty-four enzymes examined, twenty-nine were present in pollen (0.853); of these, two proved to be gametophyte-specific (0.069), while eleven of the thirteen tested for gene expression showed haplo-diploid expression (0.846). Thus it is possible to estimate the portion of genome shared between the two phases at 72% (0.846 x 0.853), while 6% and 22% represented the gametophytic and sporophytic domains. This estimate agrees with data obtained by Willing and Mascarenhas by mRNA analysis in Tradescantia (13) and corn (4), and by Tanksley in tomato (11).

Information about mechanisms and genes involved in regulation of gametophytic gene expression is not yet available. However, it can be concluded that a large part of the genome is expressed in the haploid phase and of this a good portion is common to the sporophyte. The extensive overlap between gametophytic and sporophytic phases may constitue the biological basis of the observed sporophytic responses to gametophytic selection and thus may have been a significant factor in the evolution of the Angiosperms (6).

ACKNOWLEDGMENTS

Research work supported by C.N.R., Italy, Special Grant I.P.R.A., Subproject 1, Paper N. 538.

REFERENCES


1. Frova C, Sari Gorla M, Ottaviano E, Pella C (1983) Haplo-diploid gene expression in maize and its detection. Biochem Genet 21: 923-931.

2. Heslop-Harrison H (1980) The forgotten generation: some thoughts on the genetic and physiology of Angiosperm gametophytes. In: Davies DR, Hopwood DA (eds) The Plant Genome. The John Innes Institute, Norwich, pp 1-14.

3. Hoekstra FA, Bruinsma J (1978) Reduced independence of the male gametophyte in Angiosperm evolution. Ann Bot 42: 759-762.

4. Mascarenhas NT, Bashe D, Eisenberg A, Willing RP, Xiao CM, Mascarenhas JP (1984) Messenger RNAs in corn pollen and protein synthesis during germination and pollen tube growth. Theor Appl Genet 68: 323-326.

5. Moss WD (1982) Isoenzymes. Chapman and Hall,London pp 100-101.

6. Mulcahy DL (1979) The rise of Angiosperms: a genecological factor. Science 206: 20-23.

7. Ottaviano E, Sari Gorla M, Mulcahy DL (1980) Pollen tube growth rate in _Zea mays_: implication for genetic improvement of crops. Science 210: 437-438.

8. Pfahler PL, Linskens KF, School HW, Wilcox M (1981) Surfactant effects on petunia pollen germination _in vitro_. Bull Environ Contam Toxicol 26: 567-570.

9. Searcy KB, Mulcahy DL (1983) The parallel expression of metal tolerance in pollen and sporophyte of _Silene dioica_ (L.) Clairv., _S. alba_ (Mill.) Krause and _Mimulus guttatus_ DC. Theor Appl Genet 69: 597-602.

10. Shields CR, Orton TJ, Stuber CW (1983) An outline of general resource needs and procedures for the electrophoretic separation of active enzymes from plant tissue. In: Tanksley SD, Orton TJ (eds) Isozymes in Plant Genetics and Breeding, Part A, Elsevier, Amsterdam, pp. 443-468.

11. Tanksley SD, Zamir D, Rick CM (1981) Evidence for extensive overlap of sporophytic and gametophytic gene expression in _Licopersicon esculentum_. Science 213: 453-455.

12. Vallejos EC (1983) Enzyme activity staining. In: Tanksley SD, Orton TJ (eds) Isozymes in Plant Genetics and Breeding, Part A. Elsevier, Amsterdam, pp. 469-516.

13. Willing RP, Mascarenhas JP (1984) Analysis of the complexity and diversity of RNAs from pollen and shoots of _Tradescantia_. Plant Physiol 75: 865-868.

14. Zamir D, Tanksley SD, Jones RA (1982) Haploid selection for low temperature tolerance of tomato pollen. Genetics 101: 129-137.

# Male Gametophyte Response to High Temperature in Maize

C. FROVA, G. BINELLI AND E. OTTAVIANO

## 1. INTRODUCTION

Gametophytic selection for important sporophytic traits such as resistence to environmental stresses and to pathogenic toxins has been proposed as a very efficient tool for the improvement of plant breeding methods[1,2,3], on the grounds of the existence of an extensive genetic overlap between the gametophytic and sporophytic phases[4,5,6].

High temperature stresses are very common during the corn growing season, particularly at the time of pollen maturation and germination, and a wide genetic variability at male gametophytic level with regard to the response to such stresses has been reported[7] and can be directly observed in the field in terms of pollen viability and seed yield.

Aside from these observations little is known about the mechanisms involved in gametophyte response to heat stress. At the sporophytic level the induction of specific proteins, hsp's,has been observed in all plants analyzed[8,9] but such an event was not found in mature pollen[10]. However no data are available about hsp's during pollen maturation, from meiosis to anthesis. During this phase, expecially in trinucleate pollen, many important metabolic processes relating to germination and pollen tube growth are concluded. Thus one could suppose that the response to environmental stresses would be critical in this stage of pollen life, and influence competitive ability during the fertilization process.

Dipartimento di Biologia, Sezione di Genetica e Microbiologia, Via Celoria 26 - 20133 Milano, Italy, tel. (02) 23.08.23.

In this chapter we report data showing that hsp's are induced, at different stages of pollen grain development. Response to heat-shock for pollen germination and pollen tube elongation was used to select the material for this study.

## 2. MATERIAL AND METHODS

Percent of germination and pollen tube growth "in vitro" were studied following the procedures described by Ottaviano et al.[11]; 8 inbred lines were analyzed. The experiment was conducted over 2 days; a total of 800 pollen grains per treatment were counted for germination and 30 tubes per treatment were measured for tube elongation. Pollen tubes were filmed by a video camera and their length automatically reckoned by a computer.

The labeling of sporophytic proteins was conducted in whole 5-days-old seedlings treated as described by Cooper and Ho[9] with the following modifications: distilled water was used as incubation medium and labeling was carried on for 3 hours. Four seedlings per genotype and 2 temperature treatments (25° and 40°C) were used.

For the analysis of gametophytic proteins, 70 immature anthers from a single plant per treatment were put into 2 ml. of distilled water under agitation at 25°C in the dark for 90'; 100 μCi ($^{35}$S) methionine were added after the first 45'. After 90' the anthers in the medium were brought to the desired temperature for 3 hours. Two temperature treatments (25° and 37°C) were used. The anthers were then rinsed, transversely cut and transferred in a 1.5 ml. Eppendorf tube containing 1 ml. Tris-Cl buffer 0.05M pH 7.5, and pollen shed by agitation. Anthers were then carefully removed and pollen centrifuged at 8000 g for 15', collected and extracted by the same procedure used for root tips. Proteins were analyzed by one-dimensional SDS polyacrylamide gel electrophoresis: a 12% polyacrylamide separating gel was used and the same number of cpm was loaded into each lane. After running, the gel was fixed in a 7% $CH_3COOH$ solution, dried and exposed at -80°C.

## 3. RESULTS AND DISCUSSION

High temperature clearly reduced both germination percentage and tube growth in all the genotypes analyzed. The effect was greater on tube growth, a reduction in length of over 60% compared with the 25°C control, than on germination, which on the whole was reduced by no more than 40%. However, within this general trend, different inbred lines showed a considerable variability in their response to heat-shock: interaction between lines and temperature resulted highly significant (ANOVA) on original data. As shown in Fig. 1 reduction of percent germination and tube length ranged from 15% to 38% and from 65% to 77% respectively for different lines. One line was extremely sensitive: germination dropped to 11% and tube length to 16% compared with the control.

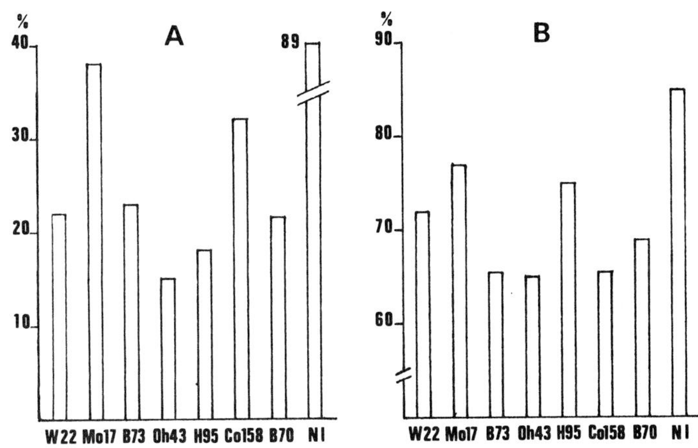


Fig.1: Proportional reduction in 40°C treated pollen related to the 27°C control. A: germination; B:tube elongation.

To investigate hsp synthesis, inbreds NI72, B70, Mo17 were studied. Immature pollen was taken at 8, 12 days after meiosis and at anthesis and analyzed. At anthesis no incorporation of ($^{35}S$) methionine into proteins could be detected either at 25°or 37°C. It seems that no protein synthesis takes place just prior to dehiscence. Incorporation observed at 8-12 days after meiosis, although low if compared with incorporation in root tips under the same conditions, indicates that protein synthesis is present at this stage. Although hsp's were detected at 8 days after meiosis, the pattern resulted clearer at 12 days after meiosis. The results of electrophoretic analysis are shown in Fig.2. Both heat-shock and heat-stroke proteins were found. Inbreds B70 and NI72 showed two new bands, approximately 72 and 81 kD in the electrophoretic pattern of heat-shocked immature pollen.

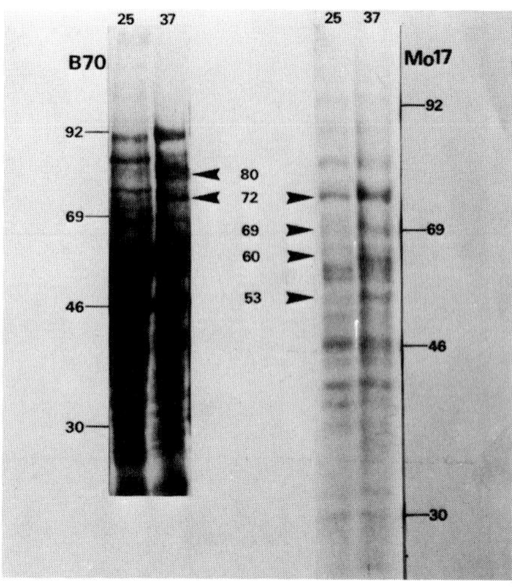


Fig.2. Protein synthesis of immature pollen (12 days after meiosis) at 25 and 37°C. Arrows indicate hsp's.

Inbred Mo17 showed four hsp's, 72,69,60,53 kD. Gametophytic hsp's 80 and 72 kD coincided with two sporophytic hsp's, while hsp's 69,60,53 kD seem to be peculiar to the gametophyte (Fig.3).

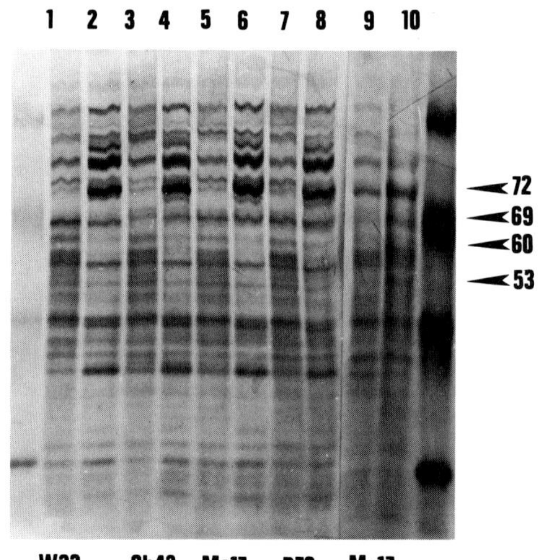


Fig.3. Protein synthesis of root tips (lanes 1-8) and immature pollen (lanes 9-10) at 25° and 37°C. ➤: hsp's.

Thus the response to heat-shock appears different in developing pollen and in the sporophyte, both in the number and in the size of hsp's induced. The data show that gametophytic genetic variability in the response to high temperature stress exists also at the molecular level, although they do not establish correlations between specific hsp's and the degree of tolerance to heat-shock expressed by different genotypes in terms of germination and tube growth.

Finally it seemed interesting to explore the existence of hsp genetic variability at sporophytic level. In contrast with the findings of Baszcynski et al.[12], a preliminary screening of 16 lines revealed several differences, as shown in Fig. 4. Moreover the resolving power of one-dimensional SDS polyacrylamide gel electrophoresis, which was used in this study, is limited, and therefore some additional variability might have been missed.

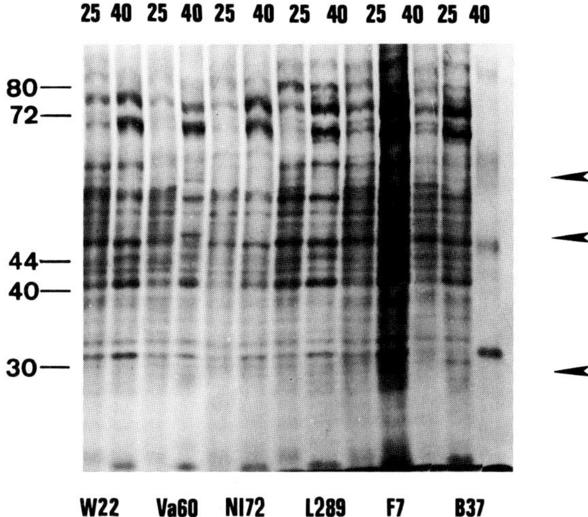


Fig.4. Protein synthesis of root tips at 25° and 40°C. Common hsp's are indicated by bars. ➤: variant hsp's.

A previous report by Mascarenhas and Altschuler[10] showed that in Tradescantia germinating pollen no hsp's were detectable. Our data show that at least some hsp's are synthesized during maize pollen development whthin the anther, while their detections was not possible at full maturity.

This accords well with what has been proposed by Hoekstra[13] to explain the high evolutionary rate of Angiosperms: many synthetic processes could have shifted from early germination or tube growth to the maturation stage within the anther. Evidence of this is the fact that pollens germinating very quickly (e.g., many trinucleate pollens) utilize preformed proteins and mRNAs previously synthesized[13,14] in the course of pollen development during emergence and early tube growth: hsp's could thus be no exception.

ACKNOWLEDGMENTS

Research work supported by CNR, Italy. Special Grant I.P.R.A., Subproject 1, Paper N. 539.

REFERENCES

1. Sacher RF, Mulcahy DL, Staples RC (1983) In: Mulcahy DL, Ottaviano E (eds) Pollen: Biology and Implications for Plant Breeding. Elsevier Biomedical, pp. 329-334.

2. Zamir D, Tanksley SD, Jones RA (1982) Genetics 101: 129-137.

3. Laughnan JR, Gabay SG (1973) Crop Science 13: 681-684.

4. Tanksley SD, Zamir D, Rich CM (1981) Science 213: 453-455.

5. Willing RP, Eisenberg A, Mascarenhas JP (1984) Incompatibility Newsletter 16: 11-12.

6. Sari-Gorla M, Frova C, Redaelli R (1985) This volume.

7. Herrero MP, Johnson RR (1980) Crop Science 20: 795-800.

8. Barnett T, Altschuler M, McDaniel CM, Mascarenhas JP (1980) Dev Genet 1: 331-340.

9. Cooper P, Ho THD (1983) Plant Physiol 71: 215-222.

10. Mascarenhas JP, Altschuler M (1983) In: Mulcahy DL, Ottaviano E (eds) Pollen: Biology and Implications for Plant Breeding. Elsevier Biomedical, pp. 3-8.

11. Ottaviano E, Sari-Gorla M, Pè E (1982) Theor Appl Genet 63: 249-254.

12. Baszcynski CL, Boothe JG, Walden DB, Atkinson BG (1982) Maize Genet Coop News Letter 56: 112-113.

13. Hoekstra FA (1983) In: Mulcahy DL, Ottaviano E (eds) Pollen: and Implications for Plant Breeding. Elsevier Biomedical, pp. 35-42.

14. Mascarenhas NT, Bashe D, Eisenberg A, Willing RP, Xiao CM, Mascarenhas JP (1984) Theor Appl Genet 68: 323-326.

# Genes and Their Expression in the Male Gametophyte of Flowering Plants

JOSEPH P. MASCARENHAS[1], JEFFREY S. STINSON, R. PAUL WILLING AND M. ENRICO PE'

The pollen grain at the time of release from the anther contains a store of presynthesized poly(A)RNAs, i.e., messenger RNAs (mRNAs) (Mascarenhas and Mermelstein 1981, Tupy 1982; Mascarenhas N.T. et al 1984). These poly(A)RNAs have been shown to code in cell free translation systems for proteins that are similar to the proteins synthesized during pollen germination and tube growth (Frankis and Mascarenhas 1980; Mascarenhas N.T. et al 1984).

RNA Cot analyses using radioactively labeled complementary DNAs (cDNAs) made to poly(A)RNA from mature *Tradescantia paludosa* pollen and hybridized to unlabeled poly(A)RNA in excess from pollen have shown that the mRNAs present in the mature pollen grain are the products of approximately 20,000 different genes. In addition heterologous hybridizations have shown that a large fraction (greater than 60%) of the genes expressed in pollen are also expressed in vegetative tissues (Willing and Mascarenhas 1984). Similar results have also been obtained with pollen of corn (Willing and Mascarenhas, manuscript in preparation).

[1]Department of Biological Sciences
State University of New York at Albany
Albany, NY 12222, USA

To make further progress in our knowledge of the regulation of pollen development it was important to isolate the genes that are expressed in pollen, especially those that are unique to pollen, to identify and characterize them in detail in order to obtain an understanding of their developmentally specific regulation. This information would moreover, be of value for applied applications such as the intelligent use of pollen selection as a tool in plant breeding and the generation of male sterile plants.

Two libraries of cDNA sequences complementary to poly(A)RNA from mature pollen of _T. paludosa_ and _Zea mays_ have been constructed by the procedures of recombinant DNA (Mascarenhas et al, manuscript in preparation). Each library consists of several thousand clones which were initially selected on the basis of colony hybridizations to $^{32}$P-labeled cDNAs made to poly(A)RNA from pollen and the same colony filters were subsequently hybridized to $^{32}$P-cDNAs made to poly(A)RNA from vegetative tissues. A few of the clones are pollen specific, that is, are expressed only in pollen. The majority of the clones are however, expressed in both pollen and vegetative tissues. The pollen specificity of certain clones and the sharing of others with vegetative tissues has been confirmed for several clones by northern hybridizations (Thomas 1980) and dot blot hybridizations (White and Bancroft 1982). Several of the clones from both _Tradescantia_ and corn have been characterized with respect to their insert size and size of the mRNA they are complementary to.

The availability of the cloned libraries has enabled us to ask questions with a reasonable expectation of obtaining answers, something that could not be done before. Two examples of the types of questions being asked and a summary of the results obtained are presented.

1. _Are the genes expressed in pollen members of large families of genes or are they represented as one or a few copies in the genome?_

To answer this question Southern hybridizations (Southern 1975) have been carried out. Restriction endonuclease digested corn nuclear DNA was analyzed by agarose gel electrophoresis, the DNA transferred to nylon membranes and hybridized to each of several pollen specific clones labeled with $^{32}P$-dCTP followed by autoradiography. The results of these experiments indicate that the pollen specific clones are represented in the genome by single genes or by families of a very few members.

2. <u>The stage of pollen development when the synthesis is initiated of the mRNAs that are found in mature pollen and the pattern of accumulation of the mRNAs in the pollen grain</u>.

The synthesis of the enzyme alcohol dehydrogenase (ADH) which is known to be transcribed and translated after meiosis Schwartz 1971; Freeling 1976) is first detectable in the developing maize microspore soon after the tetrads begin to break apart but before the microspores have completely separated (Stinson and Mascarenhas 1985). There is a constant rate of increase in ADH activity thereafter until the microspores become vacuolate prior to microspore mitosis. Soon after microspore mitosis the rate of ADH accumulation increases until generative cell division when it levels off (Stinson and Mascarenhas 1985). These results would suggest that the transcription of the alcohol dehydrogenase gene from the haploid genome must occur very soon after meiosis is completed. A similar situation has been found for the synthesis of the enzyme β-galactosidase in <u>Brassica</u> <u>campestris</u> (Singh et al 1985).

At the present time there is however, no direct information concerning the pattern of synthesis of the mRNAs that are present in the mature pollen grain.

To determine the pattern of accumulation of mRNAs in developing pollen, RNA was isolated from microspores and pollen of <u>Tradescantia</u> at various stages of development beginning with tetrads formed soon after meiosis. The RNAs were analyzed by electrophoresis in agarose gels, followed by their transfer to nitrocellulose membranes and hybridizations to different cloned DNAs labeled with $^{32}P$ by nick translation. The autoradiograms of such Northern hybridizations were scanned in an LKB Ultroscan Laser Densitometer to quantitate the amounts of specific mRNAs at the different stages of pollen development.

The clones used were Tpc44, Tpc70 and pSAc3. Tpc44 and

Tpc70 are pollen specific clones that hybridize to mRNAs of 2000 and 620 nucleotides (NT) respectively from *Tradescantia* pollen. The clone pSAc3 is a genomic clone for soybean actin. It hybridizes to a mRNA of 1750 NT and was kindly provided by Dr. Richard B. Meagher.

The results of the Northern analyses are presented in Figure 1. The mRNAs complementary to both the pollen specific probes (Tpc44 and Tpc70) are first detectable after microspore mitosis during early pollen interphase. They continue to accumulate thereafter reaching a maximum concentration in the mature pollen grain. These results provide very direct evidence for the haploid transcription of these two mRNAs. Similar results have been obtained for several other pollen specific clones for both *Tradescantia* and corn (Mascarenhas *et al* manuscript in preparation).

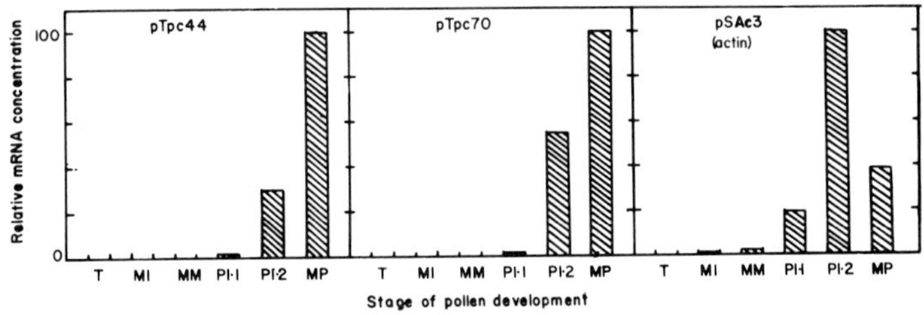


Figure 1. Relative specific mRNA levels at different stages of *Tradescantia* pollen development determined by densitometric scans of Northern blots using two pollen specific probes (Tpc44 and Tpc70) and an actin probe (pSAc3).
T: tetrads, immediately following meiosis
MI: microspore interphase
MM: microspore mitosis
PI-1: early pollen interphase; the vegetative nucleus becomes diffuse while the generative nucleus begins to elongate.
PI-2. late pollen interphase, the generative nucleus has become condensed and has a distinct crescent shape
MP: mature pollen, collected from recently dehisced anthers

The pattern of synthesis and accumulation of actin mRNA (Fig. 1) is quite different. Actin mRNA is first seen during microspore interphase, accumulates thereafter reaching a maximum at late pollen interphase and shows a substantial decrease in concentration in the mature pollen grain.

The results in figure 1 show that there are at least two different groups of mRNAs in pollen with respect to their synthesis during microsporogenesis. Those like actin which are synthesized soon after meiosis, reach their maximum during late pollen interphase and decrease substantially in the mature pollen grain. The second group represented by Tpc44 and Tpc70 are only synthesized after microspore mitosis but reach their maximum accumulation in the mature pollen grain. This pattern of accumulation might suggest primary functions for these mRNAs during the terminal stages of pollen development and during germination and early pollen tube growth.

The cDNA clones are currently being used to isolate genomic clones with the aim of characterizing the fine structure of pollen expressed genes.

Acknowledgement: Supported by NSF Grants PCM82-03169 and DCB-8501461.

References

Frankis, R, Mascarenhas JP (1980) Messenger RNA in the ungerminated pollen grain. Ann Bot 45: 595-599

Freeling M (1976) Intragenic recombination in maize: pollen analysis methods and the effects of the parental $\underline{Adh1}^{+}$ isoalleles. Genetics 83: 707-717

Mascarenhas JP, Mermelstein J (1981) Messenger RNAs: their utilization and degradation during pollen germination and tube growth. Acta Soc Bot Polon 50: 13-20

Mascarenhas NT, Bashe D, Eisenberg A, Willing RP, Xiao CM, Mascarenhas JP (1984) Messenger RNAs in corn pollen and protein synthesis during germination and pollen tube growth. Theoret Appl Genet 68, 323-326

Schwartz D (1971) Genetic control of alcohol dehydrogenase - a competition model for regulation of gene action. Genetics 67: 411-425

Singh MB, O'Neill P, Knox RB (1985) Initiation of postmeiotic β-galactosidase synthesis during microsporogenesis in oilseed rape. Plant Physiol 77: 225-228

Stinson J, Mascarenhas JP (1985) Onset of alcohol dehydrogenase synthesis during microsporogenesis in maize. Plant Physiol 77: 222-224

Southern E (1975) Detection of specific sequences among DNA fragments separated by gel electrophoresis. J Mol Biol 98: 503-517

Thomas PS (1980) Hybridization of denatured RNA and small DNA fragments transferred to nitrocellulose. Proc Nat Acad Sci USA 77: 5201-5205

Tupy J (1982) Alterations in polyadenylated RNA during pollen maturation and germination. Biol Plant 24: 331-340

White BA, Bancroft FC (1982) Cytoplasmic dot hybridization. Simple analysis of mRNA levels in multiple small cell or tissue samples. J Biol Chem 257: 8569-8572

Willing RP, Mascarenhas JP (1984) Analysis of the complexity and diversity of mRNAs from pollen and shoots of *Tradescantia*. Plant Physiol 75: 865-868

# Pollen In Biotechnology

# Chromosome Engineering with Meiotic Mutants

S. J. PELOQUIN[1]

## Introduction

Meiotic mutants which result in the formation of 2n pollen are being utilized in potato research involving breeding, evolution, germplasm transfer, and half-tetrad analysis (Peloquin, 1982). Hybrid 2x clones that form 2n pollen (pollen with the unreduced chromosome number) by either the genetic equivalent of first division restitution with crossing-over, FDR-CO, (parallel spindles mutant, ps) or FDR without crossing-over (parallel spindles in combination with the synaptic 3 mutant, sy 3) provide the genetic stocks for these investigations (Mok and Peloquin, 1975a; Okwuagwu, 1981).

The value of these stocks resides in the genetic consequences of 2n pollen formation. Approximately 80 percent of the heterozygosity of the 2x parent is transmitted to the 4x (tetraploid) progeny following 4x x 2x and 2x x 2x crosses when 2n pollen is formed by FDR-CO. Even more exciting is the fact that with FDR-NCO the intact genotype of the 2x parent is present in all 2n pollen, providing the unique opportunity of transmitting 100 percent of the heterozygosity and epistasis of the 2x parent to the 4x offspring (Peloquin, 1983). Thus, the meiotic mutants provide pwoerful tools for chromosome engineering in plant breeding (Mendiburu and Peloquin, 1977). Meiotic mutants similar to those found in potato have been detected in alfalfa (McCoy, 1982) and red clover (Parrott and

[1] Departments of genetics and Horticulture, University of Wisconsin-Madison.

Smith, 1984). Indirect evidence, obtaining 4x progeny from 4x x 2x crosses, indicates meiotic mutants which result in the production of 2n pollen occur in many dicot species, for example azaleas and blueberries. This report is concerned with progress made in potato breeding with meiotic mutants.

Tuber Yield of 4x Progeny From 4x x 2x Crosses

Several investigators have reported that tuber yields of some 4x progenies, obtained from crossing USA or Canadian 4x potato cultivars with 2x FDR-CO hybrids, significantly exceeded the yield of the cultivar parents (DeJong et al., 1981; DeJong and Tai, 1977; McHale and Lauer, 1981; Mendiburu et al., 1974; Mok and Peloquin, 1975b). Recently, high parent heterosis for total tuber yield and marketable tuber yield was obtained in 4x families from crosses between European cultivars and 2x hybrids (Masson, 1985). The mean total yield of 80, 4x families exceed the mean yield of the 4x parents by 19 percent, and that of the 10 best families exceeded the 4x parents by about 35 percent. Heterosis for marketable yield was 10 percent for all families and approximately 30 percent for the ten best families.

The total tuber yield and marketable tuber yield of 31, 4x x 2x hybrids, 9 USA cultivars and 10 European cultivars were compared at two locations in Italy, one irrigated and one not irrigated (Concilio, 1985). The mean yield of the 4x hybrids with irrigation was 678 q/ha with a range of 212-1,207 compared to the European cultivars with a mean of 667 and a range of 480-870. The percent marketable yield was 86 for the 4x hybrids and 79 for the European cultivars. The 4x hybrids were very superior to USA and European cultivars in both total yield and marketable yield under nonirrigated conditions. The total mean yield and range in yield was 370 q/ha (157-606) for the 4x hybrids, 269 (118-441) for USA cultivars and 267 (169-353) for European cultivars. Surprisingly, marketable yield was 87 percent for the 4x hybrids and 60 percent for European cultivars.

A somewhat surprising and new result was the heterosis for total tuber solids (specific gravity) of the 4x hybrids as compared to their 4x and 2x parents. In two separate experiments one involving USA cultivars and the other mainly European cultivars the mean total solids of many 4x families exceeded that of parental clones. Buso (unpublished) found, in using USA cultivars, in a 5 (4x) x 4 (2x) partial diallel that heterosis in total solids for the 20 families was about 112 percent. More important, for the 5 best families heterosis was 124 percent. Similar results were obtained with European cultivars (Masson, 1985). He found that heterosis for total solids in 44, 4x families obtained from 4x x 2x cross was 111 percent, and for the 5 best families 122 percent.

The 4x x 2x breeding scheme is also an important approach in the production of potatoes from botanical seed. The 4x hybrid seed is superior to open pollinated seed (mainly from self pollination by bumblebees and hybrid seed obtained from intercrossing cultivars (Macaso and Peloquin, 1983; Kidane-Mariam et al., 1985). Most important, the tuber yield of field transplants from hybrid seed exceeds that of those from open pollinated seed by 30-70 percent. The production of potatoes from true potato seed offers an alternative to countries that do not have the conditions to produce virus free tubers, and cannot afford to buy clean tubers from other countries.

### Relationships Between 4x Progeny and 2x Parents in 4x x 2x Crosses.

Selection among 2x clones for use as parents in 4x x 2x crosses is very important for this breeding scheme to be effective and to be improved. Thus, it is important to determine correlations between 2x parents and 4x progeny for particular traits. The relationships between 4x progeny and their 4x and 2x parents has been evaluated for maturity, tuber eye depth, and yield (Schroeder, 1983). Parent-family correlations were done for 11, 4x cultivars and 26, 2x clones. For maturity, family means were highly correlated with the 2x parents, but not with the 4x parents. Correlations for depth of eye gave similar results, means of families were significantly correlated only to the 2x parents. Unfortunately, no significant correlations existed between the total yield of either parent and the total yield of the families. This is particularly disappointing in regard to the 2x parents, since it was hoped that one could simply evaluate the yields of 2x clones and use the highest yielders as parents in 4x x 2x crosses. Since 2x parents cannot be selected based on their own phenotypes for yield, progeny testing must be done to determine their value. Preliminary evidence indicates that crossing a 2x clone to one early maturing 4x cultivar and one late maturing cultivar, and evaluating the progenies for yield, is the minimum testing one should do to have reasonable estimations of the yield potential of a 2x clone in obtaining 4x progeny from 4x x 2x crosses.

It was of considerable interest to compare the 4x progeny from 4x x 2x FDR-CO with those from 4x x 2x FDR-NCO. Twenty families resulting from crossing two 4x cultivars with five FDR-CO 2x clones and with five FDR-NCO 2x clones were evaluated (Masson, 1985). The within family variances for total tuber yield, marketable tuber yield, weight of undersized tubers, and number of marketable tubers were significantly higher with FDR-CO. The variance ratio FDR-NCO/FDR-CO was from 0.65 to 0.69 for the four traits. This is not an unexpected result, since the gametes from the FDR-NCO clones have identical genotypes, so variation in the 4x progeny is entirely due to the 4x parent. In contrast there would be considerable variation in the gametes from the FDR-CO 2x clones, since at least one chiasma is formed in each of the 12 bivalents, so there would be variation both from intrachromosomal and interchromosomal recombination. However, variation between gametes would be much less than in a 2x clone without parallel spindles, since with this meiotic mutant all the heterozygosity from the centromere to the first crossover present in the 2n clone is maintained intact in the gamete, and one-half of the parental heterozygosity from the first to second crossover is present in the gamete.

## The Occurrence and Frequency of 2n Pollen

The production of 2n pollen by haploid-species hybrids is dependent on the gene frequency for the *ps* allele in both haploid and species parents. In turn the gene frequency in the haploids is related to that of the cultivar parents of the haploids. An analysis of 56 USA cultivars revealed that the *ps* gene frequency was 0.69 in these clones (Iwanaga and Peloquin, 1982). Further, more than 65 percent of the cultivars were simplex, Pspspsps. Assuming chromosome segregation for ps in the tetraploid parent, 50 percent of the haploids from these cultivars would be psps and 50 percent Psps. Even if the cultivar were duplex, PsPspsps, 67 percent of the haploids would be Psps and 17 percent psps.

Plants that produced 2n pollen were found in almost all 2x species evaluated (Quinn et al., 1974). More extensive studies of a few species have determined that the gene frequences for ps in these species is very high. The frequency of 2n pollen producing plants among more than 500 plants representing 56 Plant Introductions of three, wild tuber-bearing, *Solanum* species *S. gourlayi*, *S. infundibuliforme*, and *S. spegazzini* was determined. The gene frequency for the ps allele was 0.46, 0.37, and 0.29, respectively (Camadro and Peloquin, 1980). Recently, similar results have been obtained in three other wild, 24-chromosome species; *S. berthautii* (0.27), *S. chacoense* (0.28), and *S. sparsipilum* (0.42) (Hermundstad, 1984).

With the high gene frequencies for ps in both haploids and wild species, it should not be difficult to obtain haploid-wild species hybrids with 2n pollen. In fact we have found it relatively easy to identify haploid-wild species hybrids with adequate levels of 2n pollen (Hermundstad and Peloquin, 1985a).

Chromosome engineering through the 4x x 2x breeding scheme continues to be valuable as a method of increasing yield, and of transferring the allelic diversity and valuable traits of the wild tuber-bearing 24-chromosome species into cultivated 48-chromosome potatoes.

Acknowledgment. Paper No. 2839 from the Laboratory of Genetics. Research supported by College of Agricultural and Life Sciences; International Potato Center; SEA USDA CGRO 84-CRCR-1-1389; and Frito-Lay, Inc.

References

Camadro EL, Peloquin SJ (1980) The occurrence and frequency of 2n pollen in three diploid Solanums from northwest Argentina. Theor Appl Genet 56:11-15.

Concilio L (1985) Evaluation of yield and other agronomic characteristics of TPS families and advanced clones from different breeding schemes. MS Thesis University of Wisconsin-Madison 63p.

DeJong H, Tai GCC, Russell WA, Johnston GR, Proudfoot KG (1981) Yield potential and genotype-environment interactions of tetraploid-diploid (4x-2x) potato hybrids. Amer Potato J 58:191-199.

DeJong H, Tai GCC (1977) Analysis of tetraploid-diploid hybrids in cultivated potatoes. Potato Res 20:111-121.

Hermundstad SA (1984) Production and evaluation of haploid Tuberosum-wild species hybrids. MS Thesis University of Wisconsin-Madison 65p.

Hermundstad SA, Peloquin SJ (1985) Male fertility and 2n pollen production in haploid-wild species hybrids. Amer Potato J 62:479-487.

Iwanaga M, Peloquin SJ (1982) Origin and evolution of cultivated tetraploid potatoes via 2n gametes. Theor Appl Genet 61:161-169.

Kidane-Mariam HM, Arndt GC, Macaso AC, Peloquin SJ (1985) Comparisons between 4x x 2x hybrid and open-pollinated true-potato-seed families. Potato Res 28:35-42.

Macaso AC, Peloquin SJ (1983) Tuber yields of families from open pollinated and hybrid true potato seed. Am Potato J 9:645-651.

Masson MF (1985) Mapping, combining abilities, heritabilities and heterosis with 4x x 2x crosses in potatoes. Ph.D. Thesis University of Wisconsin-Madison 119p.

McCoy TJ (1982) The inheritance of 2n pollen formation in diploid alfalfa Medicago sativa. Can J Genet Cytol 24:315-323.

McHale NA, Lauer FI (1981) Breeding value of 2n pollen diploid from hybrids and Phureja in 4x-2x crosses in potatoes. Amer Potato J 58:365-374.

Mendiburu AO, Peloquin SJ (1977) The significance of 2n gametes in potato breeding. Theor Appl Genet 49:53-61.

Mendiburu AO, Peloquin SJ, Mok DWS (1974) Potato breeding with haploids and 2n gametes. <u>In</u> Kasha KJ (ed) Haploids in higher plants. Guelph Ontario p249-258.

Mok DWS, Peloquin SJ (1975a) Three mechanism of 2n pollen formation in diploid ptoatoes. Can J Genet Cytol 17:217-225.

Mok, DWS, Peloquin SJ (1975b) Breeding value of 2n pollen (diplandroids) in 4x-2x crosses in potatoes. Theor Appl Genet 46:307-314.

Okwuagwi CO (1981) Phenotypic evaluation and cytological analysis of 24-chromosome hybrids for analytical breeding in potatoes. Ph.D. Thesis University of Wisconsin-Madison 170p.

Parrott WA, Smith RR (1984) Production of 2n pollen in red clover. Crop Sci 24:469-472.

Peloquin SJ (1982) Meiotic mutants in potato breeding. <u>In</u> Redei G (ed) Stadler Genetic Symposium 14:99-109.

Peloquin SJ (1983) Genetic engineering with meiotic mutants. <u>In</u> Mulcahy DL, Ottaviano E (ed) Elsivier Inc p311-316.

Quinn AA, Mok DWS, Peloquin SJ (1974) Distribution and significance of diplandroids among the diploid Solanums. Amer Potato J 51:16-21.

Schroeder, SH (1983) Parental value of 2x, 2n pollen clones and 4x cultivars in 4x x 2x crosses in potato. Ph.D. Thesis University of Wisconsin-Madison 192p.

# Intergeneric Crosses Between Zea and Pennisetum Reciprocally by In Vitro Methods

C. NITSCH, K. MORNAN AND M. GODARD[1]

1 INTRODUCTION. New technology is opening up different methods of introducing desirable characters into cells. In the plant kingdom, it can be done either by using a vector that brings selected genes into plant protoplasts which can then be grown into plants. This method is successful with some dicotyledons. Another way is by fusing protoplasts from different plants. In this case, fusion of the two genomes is obtained and the plant has to sort out to become viable. The "pomato" is an example (Melchers et al. 1978).

With cereals, however, no way as yet has been found to modify their genomes since it has not yet been possible to regenerate plants from single cells. The idea which suggested this work was: If any two cells are able to fuse, why shouldn't gametic cells also fuse. Having worked on the androgenesis of maize and that of pearl millet, we observed the fusion of the gametes and the development of the zygote with the hope that maize traits will be transmitted into the pearl millet and vice versa.

This paper is to show how we have been able to modify a pure line, dwarf and male sterile, by using corn pollen on pearl millet and reciprocally, pearl millet pollen on corn.

2 MATERIAL AND METHODS. Pearl Millet: Pennisetum americanum, variety $23D_2A_1$, was first selected at Tifton (Georgia) for cattle grass. It is an inbred line with two genetic markers: dwarfness and male sterility. When it is grown in the greenhouse of the Phytotron at Gif (France), with the culture conditions indicated below, the plants are about 1 m high with 4 tillers, the spike is about 13 cm long and 2 cm wide.

Corn: Variety 6512411, also a pure line, was selected at the French National Agronomy Station (INRA). It has several markers; besides others, it has waxy light brown kernels. Grown in our greenhouse conditions, the plants are 1.2 m high, the cob 10 cm long and 4 cm wide.

Culture conditions: Plants are grown in pots on a mixture of peat and perlite 1:1 volume. All plants are watered daily with the same amount of nutrient solution (J.P. Nitsch). Day temperature is 27°C, night 17°C, photoperiod is 16 hr day natural light with the addition of incandescent light on dull days. All spikes and cobs were bagged as soon as they appeared in order to prevent any pollen floating in the air.

[1]Centre d'Etude de Biologie Appliquée CNRS F-91190 Gif/Yvette France

Fertilization and embryo rescue by in vitro culture: Pearl millet plants, growing in the greenhouse, have been pollinated with corn pollen when the stigmas were about 5 mm long, that is 3 to 4 da before anthesis. Corn pollen was placed onto the stigmas for 3 da. One week after the initial pollination, the spike was cut off and taken to the laboratory. When it remained on the plant, it died. The spikes were sterilized by submerging them briefly into 70% alcohol into which a few drops of a wetting agent was added. They were then given a 3 min wash in a 7% calcium hypochlorite solution and rinsed twice with sterile distilled water. The immature seeds were taken out under sterile conditions. Generally, each spike would have only 30 or fewer seeds. The embryos of these seeds were very small (0.3 to 0.4 mm) and were placed into small petri dishes (5 cm in. diam.) containing 1.5 ml of liquid medium described by C. Nitsch et al. (1982) for Pennisetum embryo culture to which l-Arginin and l-Asparagin at $3x10^{-4}$ Mol. had been added. The medium was filter-sterilized by 0.22 micron milipore filters. The cultures were placed under red fluorescent light at $25^oC$ 12 hours day $T^o$ and $20^oC$ night $T^o$. One week to 10 da later the young plantlets were put into tubes containing a solid medium made of Knop's nutrient with 0.5% sucrose, 0.2% activated charcoal and 0.7% Difco agar. These plantlets were grown under the same condition except that they were exposed to white light from both incandescent and fluorescent lamps. When they reached about 15 cm, in approximately a wk, they were transplanted into pots in the greenhouse. For controls, the pearl millet were pollinated with their maintainer $23D_2A_2$, the embryos extracted the same way as above, were grown under the same conditions. When the reciprocal cross was performed, the fertilization and the embryo rescue procedure was the same except that the embryos were extracted and put in culture 12 da after the first application of pearl millet pollen onto the corn silks.

3 RESULTS AND DISCUSSION. Pearl millet ($23D_2A_1$ x 6512411): Penetration of maize pollen into the pearl millet egg cell was followed using the anilin blue fluorescence method described by R.C. Child (1966). Microscopic observation of the embryo sac was done by K. Mornan (1985). If the pearl millet is pollinated with the pollen of the maintainer ($23D_2A_2$) the nuclear fusion between the pollen nucleus and the nucleus of the egg cell is observed 5 hr after the pollen has been placed on the stigmas. When maize pollen is put on Pennisetum stigmas, K. Mornan observes that nearly 100% of the maize pollen has germinated, 6 hr later, 55% of the styles contain corn pollen. After 24 hr, 75% of that pollen reached the embryo sac. However, it is only after 72 hr that the nuclear fusions have been observed. Chromosomes have been observed in the root tips of the young plantlets when they were taken

out of the test tube and potted. Though the majority of the cell showed 14 chromosomes, great variation from cell to cell was observed (Table 1).

Table 1.

| Per 100 cells | Chromosome number |
|---|---|
| 70 | 14 |
| 2 | 7 |
| 8 | 17 (plate 1 E) |

Twenty percent of the cells showed a chromosome number between 8 and 16; we have also observed fragments of chromosomes in cells. After one mo growing in the greenhouse, another root tips chromosome count was done. At that time most cells had the same (14) chromosome number. Observation of pairing of the chromosomes at meiosis is now being done in the laboratory.

Morphological changes are visible on the plant quite early after it is placed in the greenhouse. Six wk after the day of pollination the "cross" shows a very striking vigor when compared with the inbred line (plate 1 B). As the plants grow, some new traits can be observed, e.g., aerial roots appear on the nodes. This character is very often observed in corn but we have never seen it on pearl millet (plate 1 C). These roots appear on nodes very high above the ground (1.50 m). In general, the plants coming from the cross are very tall compared with the mother plant. The dwarf trait of the mother plant disappears as does the male sterility. The plant which had not lost these two characters was haploid. Late flowering plants are very different from the normal *Pennisetum* although it looks quite like corn, the flower that comes out has in all cases been *Pennisetum*-like. We have made a rather extensive analysis of the morphological traits of one of the crosses. Figure 1 indicates the height and the fertility of the 28 plants of one cross. Two of them (2s and 2t) are haploid; chromosome count of the root tips showed 7 at the time it was flowering. All other plants are taller than either parent. The average height of 10 of each parent is: 0.88 m for *Pennisetum* and 1.20 m for the maize; the average height of the cross, not including the two haploids, is 2.06 m. All the plants are fertile though not all have the same fertility level. Figure 2 indicates the same measurements taken from one S2 obtained by self fertilizing the 2r plant of the first generation. A clear segregation is seen. One plant (2rAF 14) is haploid. In that experiment, the average height is 0.90 m for *Pennisetum*, 1.16 for corn, 1.44 for their descendants, several plants (8 or 10) have a height comparable to the parents -- 8 plants are fertile, 9 (if we exclude the haploid) are sterile. However, there is no correlation between the dwarfness and the sterility.

*Zea mays (6512411) X* $23D_2A_1$: When *Pennisetum* pollen was placed on the corn silk, contrary to what happens with the corn pollen on the *Pennisetum*

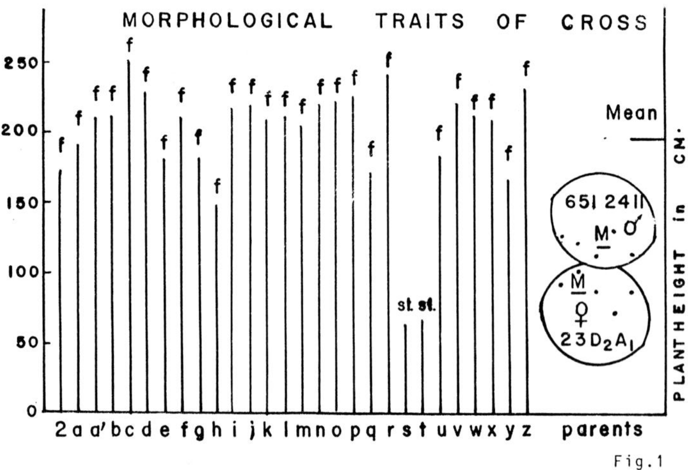


Fig.1

MORPHOLOGICAL TRAITS OF ONE PLANT (2r) SELF FERTILIZED

. 2r

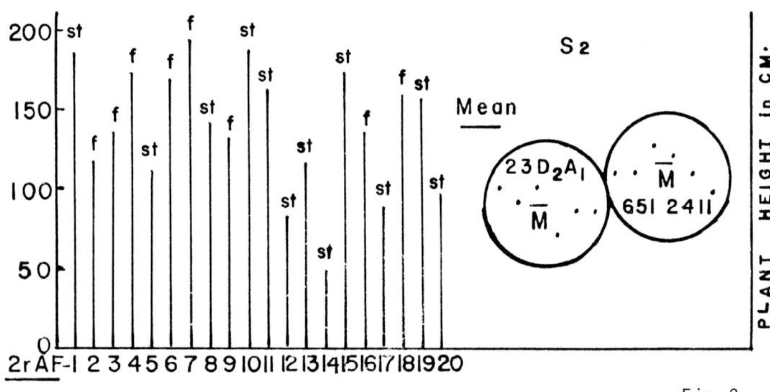


Fig.2

stigmas, no fertilization occurred. However we have been able to fertilize corn by using pollen originating from a Pennisetum that had previously been pollinated by maize pollen. The maximum number of embryos which have been extracted from each cob varies between 0 to 12. The penetration of the pollen has not yet been followed. In order to see if some traits could be introduced in the inbred line of maize we were using, we used as the pollinator the inbred line of Pennisetum which had been pollinated with black mexican sweet for the first time and with golden maize. The genealogy of the complex cross is represented in Figure 3. On plate 1 E we see, as for the previous case, chromosomes from root tips showing the 7 chromosomes of Pennisetum and the 10 chromosomes from the maize. In all cases, we have

PLATE n°1

23D2A1 X 6512411

AERIAL ROOTS

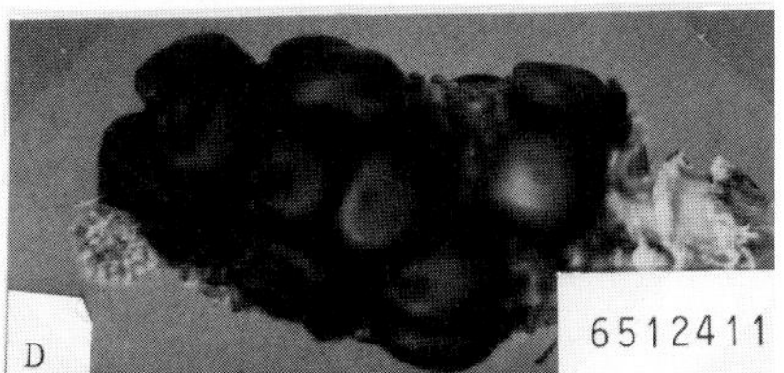


Root tips chromosomes 15 days after pollination.

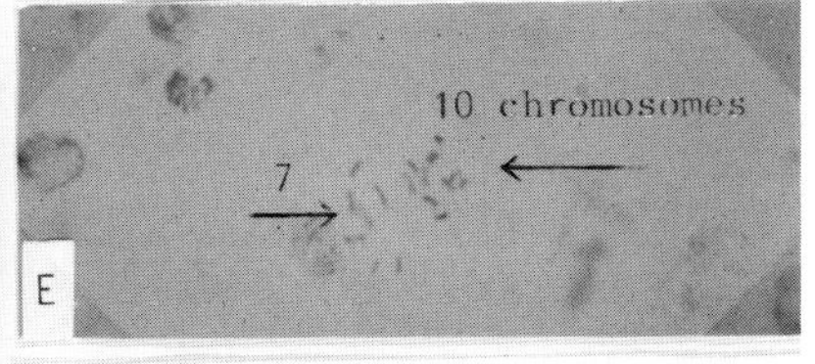


6512411 X (23D2A1 X 222)

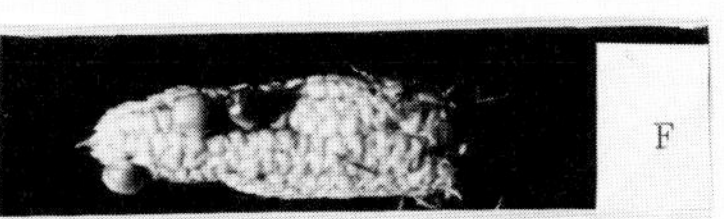

obtained maize plants with 20 chromosomes at the adult state as it was found for the reciprocal cross.

Fig. 3

| MAIZE | PENNISETUM |
|---|---|
| 6512411 | $23D_2A_1$ |
| Inbred line | Inbred line |
| dwarf, waxy | dwarf, male sterile |
| brown seeds (plate 1D) | |

6512411 X $23D_2A_1$ = no cross

BMS = Black Mexican Sweet = black seeds
GOLDEN = Golden 13 = sweet corn = golden seeds

BMS X $23D_2A_1$ = no cross
BMS X ($23D_2A_1$ X Golden) = MM222 = Golden + black seeds

6512411 X ($23D_2A_1$ X MM222) = 2 golden = 1 black seeds (plate 1F)

The maize 6512411 which is an inbred line with brown seeds gave only 3 seeds: 1 black and 2 golden. These characters, black and golden, have been introduced in the corn by the *Pennisetum* pollen.

4 CONCLUSION. This work is the first evidence that a cross between Maize and Pearl millet and reciprocal can be achieved. Other interspecific and intergeneric crosses have already been described in the literature (Y. Cauderon, 1981). This is a new case where new traits can be transferred from one genera to another. The area where corn can be cultivated could be enlarged if the drought resistance of the *Pennisetum* is introduced into it. On the other hand, introducing some of the maize proteins into the *Pennisetum* would be extremely valuable to improve the nutrient quality of pearl millet, the main cereal of India and Africa. The use of *Pennisetum* as a bridge to introduce traits into maize is also to be kept in mind.

REFERENCES

Cauderon Y (1981) Compte rendu des scéances de l'Académie d'Agriculture de France 12: 1001-1012.

Child RC (1966) Res. Sta. Ann. Rep.: 115-120.

Mornan K (1985) Thèse Université de Montpellier France.

Melchers G, Sacristan MD, Holder AA (1978) Carlsbergs Research Communications 43: 203.

Nitsch C, Anderson S, Godard M, Neuffer MG, Sheridan WF (1982) Production of haploid plants of *Zea mays* and *Pennisetum* through androgenesis in variability in plants regenerated from tissue culture. Earle E, Demarly Y (eds.), Praeger.

# Gametophyte Transformation in Maize (Zea mays, Gramineae)

J.M.J. de Wet, A.E. de Wet, D.E. Brink, A.G. Hepburn and J.A. Woods[1]

## 1 Introduction

Genetic transformation, with or without living biological systems has been achieved with several species of higher plants. The most efficient living delivery system is the Ti plasmid of *Agrobacterium tumefaciens*. It functions as a gene transfer vector in a range of dicotyledoneous taxa (Horsch et al. 1984; 1985), and in some monocotyledoneous species belonging to the Amaryllidaceae and Liliaceae (Hooykaas-Van Slogteren, Hooykaas and Schilperoort, 1984). Direct transformation of plant protoplasts incubated with plasmids consisting of the protein coding region of the bacterial APH (3') II gene under control of CaMV gene VI expression signals has been achieved by Paszkowski et al. (1984).

A majority of transformable species, however, can not yet be regenerated into functional plants from protoplasts, cells or callus in culture. This is particularly true of domesticated cereals. Problems associated with regenerating functional plants from transformed cells in culture can be overcome by transforming the male or female gametophyte and allowing normal fertilization to take place. Seeds are produced, and these germinate to develop into transformed plants.

The female gametophyte of plants can be transformed in situ by microinjection (Soyfer, 1980; Zhou et al. 1983). Exogenous DNA can also be introduced into the developing male gametophyte and transported to the embryo sac as a part of regular fertilization. Hess (1985) demonstrated that phages transducing the *E. coli* beta-galactosidase or galactose-1-phosphate uridyltransferase genes are transferred to the offspring of *Petunia* selfed with phage treated pollen. Pollen incubated with naked

[1]Department of Agronomy, University of Illinois, 1102 South Goodwin Avenue, Urbana, Illinois 61801

DNA of higher plants also induce zygote transformation. Offspring showing various degrees of anthocyanin accumulation were produced when Hess (1980) self pollinated a white flowered *Petunia* with pollen treated with total cellular DNA of a red flowered genotype. Although no genetically stable transformants were obtained, the data suggest functional gene transfer as exogenous DNA by treated pollen.

2 Studies with maize

Maize is an ideal model system to use in studying gametophyte transformation (de Wet et al. 1985). The plant is monoecious, produces an abundance of pollen, the female inflorescence consists of several hundred florets with their stigmas held together at the tip of the ear by modified leaves, phenotypically uniform inbreds are available and the genetics of maize is fairly well understood. We use selections of inbreds B73 and Oh43 as experimental material because their pollen grains germinate readily under experimental conditions. Their pollen tubes seem to lack a true cell wall at pollen germination, and DNA fragments are absorbed into the developing male gametophyte from a pollen germination medium in the presence of calcium. Germinating pollen releases nucleases, and DNA in the germinating medium is partially degraded (Matousek and Tupy, 1984). Large amounts of DNA in the presence of sodium chloride, however, allow for survival of intact DNA fragments upto 20 kbp in length for as long as 15 minutes after the initial incubation.

The initial attempt to transform gametophytes of maize involved inbred B73 (2*n*=20) as recipient and *Tripsacum dactyloides* (L.) L. (2*n*=72) as DNA donor. Maize inbred B73 crosses with *T. dactyloides* as pollen parent only when the styles are shortened to about 1 cm in length and hybrid embryos are cultured on a standard growth medium 20 days after fertilization. Hybrids (10 *Zea* + 18 or 36 *Tripsacum* chromosomes) are male sterile but cytologically non-reduced female gametes are produced and these can function sexually. When they are backcrossed with maize for five to eight generations until all *Tripsacum* chromosomes are eliminated, highly tripsacoid maize is recovered (de Wet et al. 1978). Tripsacoid maize differs from *Tripsacum* in phenotype and zein proteins.

Tripsacoid traits allow for selection of transformed plants following self pollination of B73 with pollen germinated in the presence of *Tripsacum* DNA. The 568 ears of B73 (227,000 florets) which were selfed with DNA treated pollen produced 5172 kernels. An equal number of ears selfed with untreated pollen produced 172,000 kernels. Low seedset was a result of pollen germination, not the treatment with Tripsacum DNA. A large percentage of pollen tubes break during fertilization and numerous surviving pollen tubes fail to penetrate stigmas. Pollinating 568 ears with germinated pollen which was not treated with DNA yielded 5879 kernels, and an equal number of ears selfed with germinated pollen incubated with DNA of B73 yielded 4985 kernels.

Seeds of the experiment and three controls (normally selfed, selfed with germinated pollen not incubated with DNA and germinated pollen incubated with B73 DNA) were sown directly into nurseries. Twenty thousand (20,000) kernels of normally selfed B73 germinated (99.7% germination) to produce normal plants, indicating that accidental outcrossing did not occur in our experimental field. The 5879 kernels obtained from selfing B73 with germinated but untreated pollen produced 5869 seedlings which matured into phenotypically normal plants. Male gametophyte selection, as a result of pollen tube breakage or failure to penetrate stigmas, did not affect the phenotype of the offspring. The 5172 kernels from the Tripsacum DNA treatment produced 4362 seedlings (84.4% germination), and the 4985 kernels from the B73 DNA treatment yielded 4421 seedlings (88.7% germination). The reason for this reduction in germination percentage is not clear.

The DNA treatments had a mutagenic effects on the male and/or female gemetophyte of maize inbred B73. Among the 4362 offspring obtained when B73 was selfed with Tripsacum DNA treated pollen, 211 (4.8%) were noticeably weaker than their siblings, and these weak seedlings either died because of chlorophyl deficiencies or other developmental abnormalities (187), or developed into small and barren mature plants (24). All but three of the remaining 4210 seedlings developed into phenotypically normal and fertile mature B73 plants. The three variants tillered to produce a main stem and two well developed secondary culms. The three stems on each of these plants produced terminal and lateral inflorescences. The lateral inflorescence on each culm consisted of a polystichous female section tipped by 5-10 male spikelet pairs. One of these plants had male terminal inflorescences. The remaining two plants had solitary female spikelets at one to three nodes below the male spikelets on branches of terminal inflorescences. These are typically tripsacoid characteristics, and could not have resulted from accidental outcrossing. The three tripsacoid plants and three of their normal siblings were selfed to study combinations of zein proteins.

Selfed kernels of normal B73 plants and of two tripsacoid plants were characterized by standard B73 zein band patterns. The eighty-five selfed kernels of the remaining tripsacoid plant tested were variable in respect to zein proteins. Twenty-five had an easily identifiable tripsacoid band, 42 others had zeins differing from standard B73 in one or more bands, and 18 resembled B73 in zeins. Individual maize inbreds are highly consistent in zeins (Wilson, 1984). The presence of novel zeins in the offspring of tripsacoid B73 plants derived from *Tripsacum* DNA treated pollen suggests transfer of a functional gene. Combination of zeins in plants assumed to be transformed by *Tripsacum* DNA are similar to zeins present in tripsacoid maize recovered from hybrids between B73 and *T. dactyloides*.

Inbred B73 DNA also had an effect on the genotype of offspring when B73 was selfed with treated pollen. Among the 4421 seedlings obtained, 238 were less vigorous than their siblings. Fifty one (51) weak seedlings developed chlorophyll deficiencies and died, and another 150 green seedlings developed other developmental abnormalities and died. The remaining 37 weak seedlings developed into small and barren plants. The 4183 vigorous seedlings developed into phenotypically normal B73 mature plants. Selfed offspring from these plants resembled B73 in phenotype, or rarely segregated for seedling weakness and endosperm mutations. Weak as well as robust seedlings tested were diploid maize ($2n=20$) as determined from root tip squashes.

In a second experiment resistance to common rust was transferred from maize inbred B14A to inbred DP194 using the pollen germination technique. Maize inbred DP194 (modified Oh43) is susceptible to common rust caused by *Puccinia sorghi*. Maize inbred B14A is highly resistant to infection by common rust. Resistance to rust is inherited as a simple dominant over susceptibility. Controls were normally selfed DP194 (20,000 offspring), DP194 selfed with germinated pollen (6,000 offspring) or selfed with pollen germinated in the presence of DP194 DNA (4,000 offspring). Controls yielded not a single rust resistant plant among the 30,000 offspring tested in a greenhouse as seedlings.

Inbred DP194 selfed with B14A treated pollen yielded 7979 kernels of which 6303 germinated in a greenhouse. Among these, 284 were less vigorous than their siblings and 232 died before reaching the four-leaf stage of development. The surviving 6061 seedlings were innoculated with rust and allowed to grow for four weeks. Seventy-two seedlings were more vigorous than their normal siblings, and two of these were resistant to rust. These two plants were transplanted to a nursery and selfed. The 61 selfed offspring segregated into 51 resistant and 10 susceptible seedlings. Fifty-five of these plants were again selfed. Three susceptible plants produced only susceptible offspring (116 seedlings tested), 13 resistant plants produced only resistant offspring (485 seedlings tested), and 39 resistant plants produced offspring which segregated into 1094 resistant and 393 susceptible plants, a close fit to the expected 3:1 ratio of resistant to susceptible for heterozygous ($rp_1^d/Rp_1^d$) parents. The experiment was repeated, but no resistant seedlings were obtained among 5390 offspring of DP194 selfed with pollen which was germinated in the presence of B14A DNA.

The possibility of pollen contamination can not be ruled out in the rust experiment. Several locally planted hybrids and inbreds are rust resistant. The two rust resistant plants, however, did not phenotypically resemble outcrosses, neither did they segregate extensively in later generations for traits other than rust resistance and height. We are convinced that our genetic data show genetic transformation following fertilization by sperm of germinated pollen which was incubated with naked DNA.

Transformation can only be unequivocally demonstrated by the presence in the genomes of genetically altered plants of alien DNA sequences supplied to germinating pollen. Such experiments are under way. Plasmids we use contain a selection marker based on the APH (3') II gene (aminoglycoside phosphotransferase) of the bacterial transposon Tn5. When the bacterial expression sequences at the 5' end of the gene are replaced by eukaryotic expression controls derived from Agrobacterium tumefaciens opine synthase genes, or from the 19S or 35S promotor region of CaMV by either transcriptional or translational fusion, and transcription termination and poly(A) addition signals from either one of these sources are added to the 3' end of the gene, it is capable following transformation of conferring antibiotic resistance on a wide range of higher and lower eukaryotic cells (Horsch et al. 1984;Paszkowski et al. 1984). These biochemical markers allow for selection of transformed plants, and provide unique DNA sequences which are detectable in the genome through Southern blot techniques.

We are also investigating the efficiency of DNA uptake by germinating pollen. In earlier experiments we have used elevated concentrations of $Ca^{++}$ in the pollen germination medium to facilitate DNA uptake. We are now testing polyethylene glycol (Krens et al. 1982) and electric field pulses (Hashimoto et al. 1985) as promotors of DNA uptake, and liposome encapsulation of test plasmids to protect the transforming DNA from nucleases released by germinating pollen (Caboche and Deshayes, 1985). These studies will provide conclusive evidence to either reject or accept the concept of gametophyte transformation using pollen as a transfer agent of naked DNA.

References

Caboche M, Deshayes A (1985) Utilization de liposomes pour la transformation de protoplastes de mesophylle de tabac par un plasmide recombinant de E. coli leur conferant la resistance a la kanamycine. Compt Rend Acad Sci Paris 299: 663-666.

de Wet JMJ, Bergquist RR, Harlan JR, Brink DE, Cohen CE, Newell CA, de Wet AE (1985) Exogenous gene transfer in maize (Zea mays) using DNA treated pollen. In: Chapman GP (ed.). Experimental manipulation of ovule tissue. Plenum Press, New York. In press.

de Wet JMJ, Harlan JR, Randrianasolo AV (1978) Morphology of tripsacoid maize. Amer J Bot 65: 741-747.

Hashimoto H, Morikawa H, Yamada Y, Kimura A (1985) A novel method for transformation of intact yeast cells by electroinjection of plasmid DNA. Appl Microbiol Biotechnol 21: 336-339.

Hess D (1980) Investigations on the intra- and interspecific transfer of anthocyanin genes using pollen as a vector. Zeitschr Pflanzenphysiol 98: 321-337.

Hess D (1985) Gene transfer in plants using pollen as vectors: Bacterial transferase activity expressed in Petunia progenies. In: Chapman GP (ed.). Experimental manipulation of ovule tissue. Plenum Press, New York. In press.

Hooykaas-Van Slogteren GMS, Hooykaas PJJ, Schilperoort RA (1984) Expression of Ti plasmid genes in monocotyledonous plants infected with Agrobacterium tumefaciens. Nature 311: 763-764.

Horsch RB, Fraley RT, Rogers SG, Sanders PR, Lloyd A, Hoffmann N (1984) Inheritance of functional foreign genes in plants. Science 223: 496-498.

Horsch RB, Fry JE, Hoffman NL, Eichholtz D, Rogers SG, Fraley RT (1985) A simple and general method for transferring genes into plants. Science 227: 1229-1231.

Krens F, Molendijk L, Wullems G, Schilperoort R (1982) In vitro transformation of plant protoplasts with Ti-plasmid DNA. Nature 296: 72-74.

Matousek J, Tupy J (1984) Purification and properties of extracellular nuclease from tobacco pollen. Biol Plantarum (Praha) 26: 62-73.

Paszkowski J, Shillito RD, Saul M, Mandak V, Hohn T, Hohn B, Potrykus I (1984) Direct gene transfer to plants. EMBO J 3: 2717-2722.

Soyfer JN (1980) Hereditary variability of plants under the action of exogenous DNA. Theor Appl Genet 58: 225-235.

Wilson CM (1984) Isoelectric focusing of zein in agarose. Cereal Chem 61: 198-200.

Zhou G, Wang J, Zeng Y, Huang J, Qian S, Liu G (1983) Introduction of exogenous DNA into cotton embryos. Meth Enzymol 101: 433-481.

# Attempts to Transform for Kanamycin-Resistance in Mature Pollen of Tobacco

I. NEGRUTIU[1], E. HEBERLE-BORS[2], I. POTRYKUS[1]

The pollen grain, a rather unique structure within the plant, is a natural cellular "vector" and has obvious potential for genetic transformation as it may provide a generally applicable approach to gene transfer in plants.

One area of interest concerning the use of irradiated pollen in gene/chromosome transfer has been discussed elsewhere (Negrutiu et al, 1984). Gene- or total DNA-mediated transformation of pollen has been attempted during the last two decades (reviewed by Hess, 1977): plasmids (with R factor against kanamycin (Km) isolated from E. coli), phages ($\lambda$ pgal$^+$ and $\lambda$ plac$^+$), Ti plasmids (Jackson et al, 1980) and high molecular weight genomic plant DNA with marker genes (de Wet et al, 1984 and ref. therein, Sanford et al, 1985 and this volume) were used. No clearcut transformation events were observed, suggesting both that the available methodologies had too low a resolution in tracing the fate and/or expression of foreign genes in the pollen tube and/or putative transformants, and that if pollen transformation occurs, it must be a very rare event.

Two new experimental factors were recently introduced: (1) use of selectable foreign genes (such as the Tn5 aminoglycoside phosphotransferase II, conferring Km-resistance under strong plant transcriptional signals (cf. Paszkowski et al, 1984) produced consistent evidence that genes can be introduced, maintained, and expressed in plants, and transmitted to progeny in a mendelian fashion; (2) direct gene transfer was achieved in intact yeast and E. coli cells by simple chemical procedures (Ito et al, 1983; Klebe et al, 1983).

We report here on the use of the direct gene transfer approach in (a) a study of plasmid stability in the presence of germinating tobacco pollen; and (b) attempts to introduce the plasmid into mature germinating pollen and screening for Km-resistance in the seed population obtained after pollination with such pollen.

## Materials and Methods

Pollen harvest and storage. N. tabacum cv. Petit Havana clone $SR_1$ was used as source of flowers. The flowers were collected at anthesis and kept for 24-48 h at 8°C in sealed boxes. Pollen was harvested directly in GM or stored for 7 to 21 days at -70° or in liquid $N_2$.

[1] Friedrich Miescher-Institute, Basel, Switzerland

[2] Inst Applied Microbiology, Univ. of Agriculture & Forestry, Vienna, Austria

Germination media and other solutions. Germination medium (GM) of Brewbaker and Kwack (1963), with or without mineral salts (with boric acid only), $W_5$ (154 mM NaCl, 125 mM $CaCl_2 \cdot 2H_2O$, 5 mM KCl, 5 mM glucose, pH 5.7), ML Cal5 (0.4 M mannitol and 15 mM $CaCl_2 \cdot 2H_2O$, pH 5.7) and 0.5 x PBS (4g NaCL, 0.1g KCl, 0.1g $KH_2PO_4$, 0.57g $Na_2HPO_4$ per liter) in 10% sucrose were used. The PEG solution was made by dissolving 30 g of PEG 1500 (Merck) in 100 ml GM, or mixtures of GM with $W_5$ or PBS.

Procedure of pollen incubation with DNA. 20 mg pollen were inoculated in 1 ml GM. DNA was added or the pollen washed and then treated with DNA either immediately or 30 min later. After washing, the pellet was resuspended either in GM or in $W_5$, PBS, or mixtures of GM with $W_5$ or PBS. When a heat shock was applied (5 min at 45°C, 1 min on ice) it preceeded the incubation with DNA. DB (2,6-dichlorobenzonitril, Fluka) was added either after the last wash and before the heat shock or simultaneously with the plasmid. pABDI concentration was 100 µg/ml (circular), or 25-50 µg/ml (linear), in which latter case calf thymus DNA (75 µg/ml) was added as carrier. PEG solution had a final concentration of 10%. The incubation time in DNA + PEG was 30 to 120 min. Electroporation equipment and treatment was as described by Shillito et al (1985): high voltage pulses (1.5-5 kV/cm, 3-9 pulses; 30 nF, 3.5 to 10 sec between pulses) were given 10 to 20 min before the end of the treatment. The pollen was sedimented in an Eppendorf tube; the supernatant was analysed by agarose gel electrophoresis, and the pellet washed once in GM and used for pollination.

Seed harvest and screening for Km-resistance. Seeds were collected under sterile conditions, weighed, and spread onto agar plates (2-fold diluted Murashige and Skoog basal medium containing 300 mg/l Km-sulphate). Germination occurred within 3 to 5 days (25°C, 2000 lux, 16h light). Sensitive seedlings stopped their growth immediately after germination, their cotyledons turning white within 8 to 12 days.

Plasmid and gel electrophoresis. pABDI was linearized with SmaI. Supernatants from DNA-treated pollen were collected and the reaction stopped by addition of EDTA (50 mM final conc). The samples were stored at -20°C and analysed in 0.7% agarose gels.

## Results

1. Response of germinating pollen to conditions that allow transformation of plant protoplasts and yeast cells.

Several such parameters were examined: (1) the salt composition and concentration before and during DNA treatment; (2) DNA form and concentration; (3) the treatment with PEG; (4) the physiological state of the pollen; (5) heat shock and electroporation; DB treatment.

$Ca^{2+}$, $Mg^{2+}$, $Li^{+}$ salts are known to affect transformation efficiency (Ito et al, 1983). In the case of pollen, such salts and several others when supplied separately and at concentrations above 15 mM in the GM strongly affected pollen tube elongation. LiCl produced bursting of pollen tubes. Mixtures of salts were either beneficial (GM), or reversibly blocked pollen tube growth ($W_5$), or damaged the pollen (PBS).

Pollen tube growth was reversibly blocked at concentrations above 25 µg/ml DNA. PEG at a final concentration of 10 to 25% had a similar effect. DB is known (Meyer and Herth, 1978) to specifically and reversibly inhibit cell wall formation and cytokinesis in plant protoplasts. It was tested in a concentration range of 0.5 to 12 mg/l. Bursting of pollen tubes occurred above 4 mg/l.

Electric pulses above 3 kV/cm significantly affected the viability of pollen. Incubation at 45° for 5 min accellerated pollen tube growth. Pollen stored at -70° or in liquid $N_2$ germinated as controls whenever directly incubated in GM.

2. Plasmid stability in the presence of geminating pollen.

Circular pABDI was used to asses for stability under a variety of incubation conditions in agreement with observations made in the previous section. Protoplast cultures of the same species served as controls. The transformation protocol was as described in Materials and Methods.

Under standard conditions of germination and pollen growth (GM at pH 6), the plasmid was completely degraded within 5 to 10 min of incubation (Fig. 1). Protection occurred when: (a) the pH of the GM and PEG solution was above 7; (b) high salt media (e.g. W5) were used after germination was initiated; (c) EDTA was added to GM (≥10 mM final conc.); (d) pollen was washed (at least twice) during the first 10-15 min of cultures; (e) a combination of the above treatments.

Protoplasts responded in a similar way; both cell types converted most of the supercoiled forms into linear and circular molecules.

Incubation media, heat shocks, DB, order of addition of DNA and PEG, time of incubation with DNA, or storage of pollen at very low temperatures had no effects on plasmid stability.

Further experiments were performed with linearized pABDI and carrier DNA.

3. Screening for Km-resistance in seed population obtained after pollination with pABDI-treated pollen.

The following sequence of treatments was developed in attempting pollen transformation: (1) germination of pollen (stored under different conditions); (2) two washes in GM during the first 10 to 20 min in culture; (3) incubation with DB (in some experiments this was done simultaneously with DNA and PEG treatment); (4) heat shock at 45°C/5 min; (5) DNA and PEG treatment; (6) Electropulsing (not in all experiments); (7) washing out of DNA and PEG; (8) pollination.

Variations were made in (a) the duration of the germination phase (from a few to 45 min); (b) the length of incubation with DB and/or DNA + PEG; (c) the salt content in the DNA and PEG incubation solutions; (d) DNA concentration; (e) voltage and number of pulses. The data are given in Table 1; DNA-treated pollen, even under conditions of high and very high pollen abortion or reduced pollen tube elongation (Fig. 3) (mainly due to the strength of DB action, prolonged incubation with DNA and PEG, and/or high voltage pulses), can fertilize and produce large enough numbers of seeds. During each treatment, the pollen samples were examined microscopically in order to control and decide on the timing and strength of various parameters. Furthermore, tests on pollen tube elongation subsequent to the treatment were performed by overnight incubation in GM in parallel with pollination (Fig. 3c). Plasmid samples from each individual treatment were analysed by agarose gel electrophoresis and in all samples intact linear plasmid molecules were recovered (Fig. 2).

In total, $3.99 \times 10^5$ seeds were screened on Km-plates; no resistant seedlings were recovered.

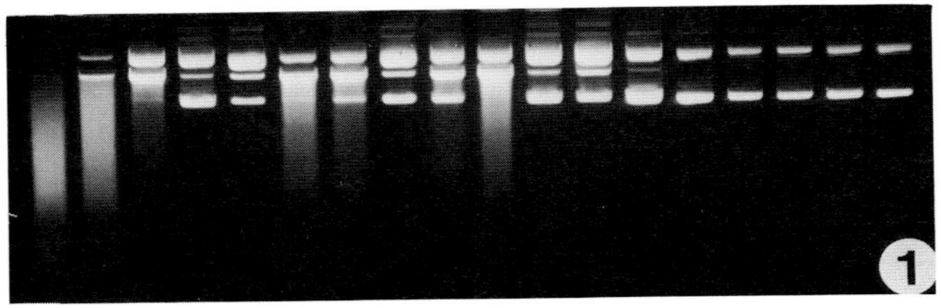

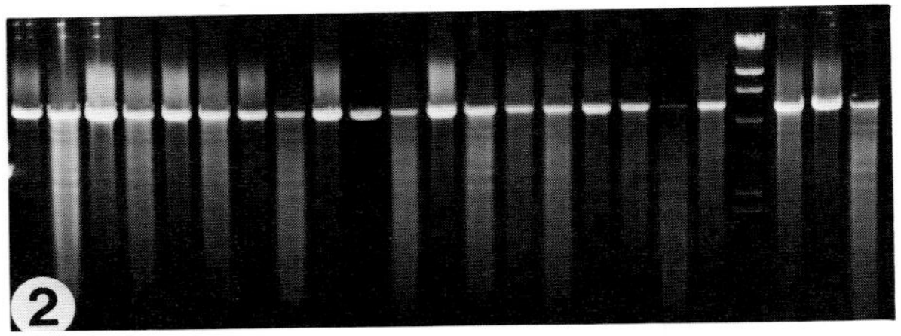

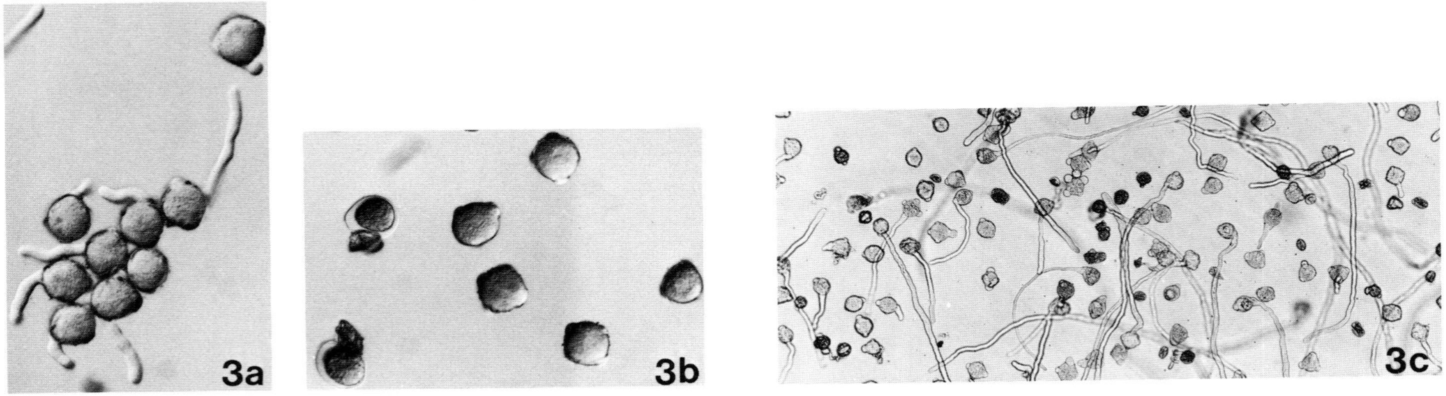

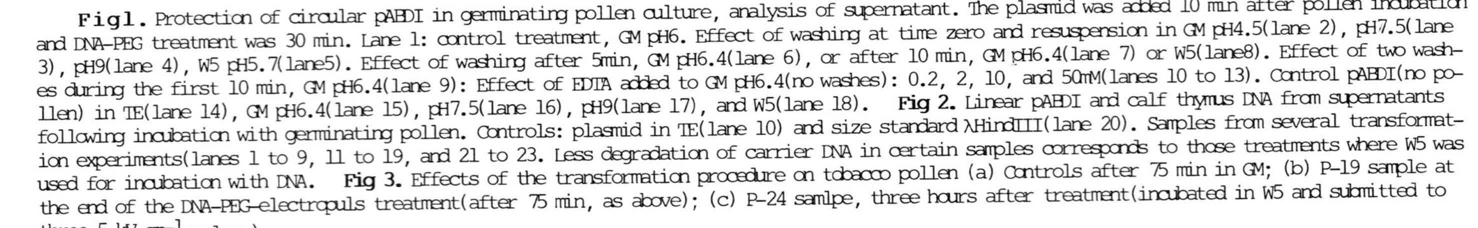

**Fig1.** Protection of circular pABDI in germinating pollen culture, analysis of supernatant. The plasmid was added 10 min after pollen incubation and DNA-PEG treatment was 30 min. Lane 1: control treatment, GM pH6. Effect of washing at time zero and resuspension in GM pH4.5(lane 2), pH7.5(lane 3), pH9(lane 4), W5 pH5.7(lane5). Effect of washing after 5min, GM pH6.4(lane 6), or after 10 min, GM pH6.4(lane 7) or W5(lane8). Effect of two washes during the first 10 min, GM pH6.4(lane 9): Effect of EDTA added to GM pH6.4(no washes): 0.2, 2, 10, and 50mM(lanes 10 to 13). Control pABDI(no pollen) in TE(lane 14), GM pH6.4(lane 15), pH7.5(lane 16), pH9(lane 17), and W5(lane 18). **Fig 2.** Linear pABDI and calf thymus DNA from supernatants following incubation with germinating pollen. Controls: plasmid in TE(lane 10) and size standard λHindIII(lane 20). Samples from several transformation experiments(lanes 1 to 9, 11 to 19, and 21 to 23. Less degradation of carrier DNA in certain samples corresponds to those treatments where W5 was used for incubation with DNA. **Fig 3.** Effects of the transformation procedure on tobacco pollen (a) Controls after 75 min in GM; (b) P-19 sample at the end of the DNA-PEG-electropuls treatment(after 75 min, as above); (c) P-24 samlpe, three hours after treatment(incubated in W5 and submitted to three 5 $kV \cdot cm^{-1}$ pulses).

**Table 1.** Summary of transformation conditions used in the treatment of germinating tobacco pollen. Heat shocks was applied to all samples. A total of 850 mg pollen was treated with DNA and 399,000 seeds were germinated on Km-plates.

| Pollen amounts and conservation | Preincubation conditions and DB treatment | DNA conc., pulsing and total incubation time | No. of seeds tested |
|---|---|---|---|
| 1st series: 26 samples, 350 mg pollen (77% stored at 6°C, 23% at -70°C) | -GM, GM + $Ca(NO_3)_2$, GM + $MgCl_2$; GM + W5; W5<br>-DB at 0.5-1 mg/l (in 34% of the samples simultaneously with the DNA, in 66% before the DNA treatment) | - 25 µg/ml plasmid + 75 µg/ml carrier (70% of the samples) and 35 to 50 µg/ml plasmid + carrier (30% of the samples)<br>- pulsing in 60% of the samples;<br>- 50 to 90 min | 216,000<br>No Km-resistant seedlings |
| 2nd series: 22 samples; 500 mg pollen (45% from liquid $N_2$, 45% from -70°C, 10% from 6°C) | -GM + W5; GM + PBS; W5<br>-DB at 0.75-1.5 mg/ml, 8 to 12 min before DNA treatment | - 35 µg/ml plasmid + 70 µg/ml carrier<br>- pulsing in all samples<br>- 45 to 120 min | 183,000<br>No Km-resistant seedlings |

**Discussion**

This paper describes the extrapolation of conditions known to enable transformation of plant protoplasts and/or intact yeast cells to mature, germinating pollen of tobacco. We were able to show that (1) such pollen can be submitted during early stages of germination to the relatively drastic effects of DNA-mediated transformation protocols without an irreversible loss of fertilization ability for at least part of the pollen population; (2) under such conditions and by combining treatments which, in other cellular systems, are known to block cell wall synthesis and to enhance DNA uptake, we attempted direct gene transfer for Km-resistance into and/or via pollen. No Km-resistant plants were recovered under the transformation conditions tested.

Our failure to obtain transformants via mature pollen can have several explanations:

(1) Degradation of the plasmid before uptake. Matousek and Tupy (1983) have shown that sugar-unspecific nucleases diffuse out of pollen during the first minutes of culture. Previously published reports on pollen transformation never explicitely demonstrated protection of the used DNA against extensive degradation. This can be ruled out in our case, as conditions protecting the plasmids have been worked out and systematically applied during this work.

(2) The exogenous DNA was not, or very poorly taken up by the pollen tube. Although short exposure to DB was included to weaken or break down newly synthesized pollen cell walls, and that DNA uptake through plasmalemmas seems to be a relatively rapid process, salt media and/or PEG treatment and high DNA concentration appeared to block the elongation of pollen tubes. This may limit or impair DNA uptake. Studies on plasmid uptake into pollen have already been discussed (Hess, 1977); we also noticed

in a comparative study of plasmid uptake into pollen and protoplasts that (a) protoplasts take up more plasmid than the pollen, that this uptake is always greater in salt media, and that (b) more supercoiled and oligomeric forms of the plasmid are associated with the pellet fraction of the sample than in the supernatant (data not shown). We are aware that the value of these observations remain questionable as they may describe only or mainly cell wall attached plasmid molecules.
(3) Fate of the plasmid in the pollen tube. One should keep in mind that 12 to 24 h or more may be needed before fertilization by DNA-treated pollen occurs. Little is known about the eventual degradation of foreign DNA within a pollen tube, and on the capacity of highly condensed sperm nuclei to accept foreign DNA. In binucleate species such as tobacco the second mitotic division of the generative nucleus takes place during the growth of the pollen tube and thus integration may occur. Only one of these nuclei-after fusion with the egg-cell is expected to ensure transmission of the foreign DNA into the embryo. Obviously, incorporation into the vegetative- and second sperm nuclei is of little practival value as they are lost at fertilization or during seed maturation. Furthermore, the callose plugging of the pollen tube during growth along the style, the "sieving" of the sperm cells through the synergies, and the tight fusion of sperm cell with the egg cell (this conference) strongly reduce the chances of plasmid transfer to the egg cell via cytoplasm transport.

Having these things in mind, and the fact that only protoplasts have been transformed by the direct gene transfer approach thus far, the transformation frequency in this system is expected to be very low, if any at all. Our figures indicate frequencies below $2.5 \times 10^{-6}$.

Work is in progress to analyse eventual non-specific effects of the foreign DNA as an indirect evidence of DNA uptake: plants produced from seeds grown under nonselective conditions are being grown to maturity and analysed from morphological abnormalities and/or hybridization signals with the donor plasmid.

Circumventing at least part of the mentioned difficulties may require more sophisticated transformation protocols, direct *in situ* pollination of ovules etc. Embryogenic microspore cultures, available in an increasing number of species, may constitute an alternative approach to mature pollen transformation.

## Literature


Brewbaker JL, Kwack BH (1963) Am. J. Bot. 50: 859-864

Hess D (1977) In: Plant, Cell, Tissue and Organ Culture (Reinert J, Bajaj YPS eds) Springer V, Berlin, pp. 506-533

Ito H, Fukuda Y, Murata K, Kimura A (1983) J. Bacteriol. 153: 163-168

Jackson JF, Verburg BML, Linskens HF (1980) Acta Bot. Neerl. 29: 277-283

Klebe RJ, Harris JV, Sharp DZ, Douglas MG (1983) Gene 25: 333-341

Matousek J, Tupy J (1983) Pl. Sci. Letters 30: 83-89

Meyer Y, Herth W (1978) Planta 142: 253-262

Negrutiu I, Jacobs M, Cattoir-Reynaerts A (1984) Plant Mol. Biol. 3: 289-302

Paszkowski Y, Shillito R, Saul M, Mandak V, Hohn T, Hohn B, Potrykus I (1984) EMBO J. 3: 2717-2722

Sanford JC, Skubik KA, Reisch BI (1985) Theor. Appl. Genet. 69: 571-574

Shillito RD, Saul M, Paszkowski Y, Müller M, Potrykus I (1985) Submitted to EMBO J.

de Wet JMJ, Bergquist RR, Harlan JR, Brink DE, Cohen CE, Newell CA, de Wet AE (1985) In: Experimental manipulation of ovule tissue (Chapman GP ed.) Plenum Press, NY, in press

# Attempted Pollen-mediated Transformation using Ti-plasmids

J. C. SANFORD AND K. A. SKUBIK[1]

## 1. Introduction

There has been an increasing awareness that pollen is a desirable target for genetic transformation. This is because: 1) pollen is readily accessible as free cells in large numbers, and 2) pollen naturally gives rise to whole plants through the normal plant reproductive process.

These two points raise the hope that a simple, broadly applicable transformation methodology might be developed which could circumvent the need to regenerate plants from somatic tissues - which is still not possible in many important crops.

The purpose of this research was to investigate the feasibility of transforming genetic markers, borne on native and modified Ti-plasmids of *Agrobacterium tumefaciens*, into plants via plant pollen. Co-cultivation and DNA-incubation methods were both employed.

## 2. Materials and Methods

*Nicotiana langsdorffii* was selected as a model species due to its continuous and prolific flowering, its relatively homozygous and consequently stable plant phenotype, and its amenable pollen biology. Various experimental treatments were applied to pollen, as it was germinated *in vitro* in artificial media (10% sucrose, 100 ppm boric acid, and 300 ppm calcium nitrate). After various periods of *in vitro* germination on an aeration wheel, pollen was collected on a millipore filter (5 um), and was scraped off the filter and transferred to stigmata of immasculated flowers. After such pollination, flowers were covered with gelatin capsules to prevent pollen desiccation or pollen contamination. Resulting seed were either germinated in non-sterile

[1] Department of Horticultural Sciences, Hedrick Hall, Cornell University, Geneva, NY 14456

soil, or were surface sterilized and germinated on artificial hormone-free media (3% sucrose, MS salts, pH 5.7, 1.3% agar).

The resulting seedlings were screened for various characters, including abnormal morphogenesis, hormone-independent growth of explants, and nopaline synthase activity. Viable plants of interest were maintained either in culture, or where possible, were transplanted to the greenhouse where they were further evaluated and, on occasion, self-pollinated. Resulting $F_2$ seed were germinated _in vitro_ in the same manner as with the $F_1$ seed. DNA was extracted from plants which were considered likely transformants. Southern blot DNA hybridizations are presently being performed, using a T-DNA fragment as a probe.

## 3. Results

We have observed that during co-cultivation, _Agrobacterium tumefaciens_ will colonize the surface of pollen tubes of tobacco, but fails to bind to pollen of monocots such as corn or lily. Pollen of _N. langsdorffii_ was co-cultivated with _A. tumefaciens_ for varying lengths of time on an aeration wheel. After 30-60 minutes, bacterial colonization of the pollen tubes was usually quite advanced, and pollen tubes had extended 10-60 microns in length, and the pollen maintained good seed-setting ability.

Pollen of _N. langsdorffii_ was also germinated in the presence of Ti-plasmid DNA. DNA was not found to be inhibitory to pollen germination or growth, and incubations of up to 90 minutes resulted in adequate seed set.

Normal seed germination procedures in non-sterile soil produced plants which were all phenotypically normal, regardless of experimental pollen treatment. 800 such plants were tested, and were found to be negative for nopaline synthase activity. However, a class of semi-lethal seedlings from certain experimental treatments were recovered by germination on sterile media.

_In vitro_ seed germination on hormone-free media produced normal control seedlings, while a high rate of abnormal seedlings arose from a variety of experimental pollen treatments. The nature of the experimental treatments and the frequency of abnormal seedlings are summarized in Table 1. As can be seen, both co-cultivation treatments and DNA incubation treatments produced high rates of abortive seed and high rates of abnormal seedlings. The nature of the abnormal seedlings was suggestive of hormone over production. Abnormalities included subtle effects such as distorted leaves, loss of apical dominance, enlarged hypocotyls, reduced root growth (fig. 1), and more gross defects such as tumorous growth in the epicotyl region and abortion of radicle development (fig. 2), tumorous growth in the hypocotyl region and "hairy" root formation (fig. 3), shoot proliferation from the hypocotyl region, shoot growth from root tips, and simple globular undifferentiated growth. Many abnormal seedlings were in a semi-lethal condition and eventually died. Other abnormal seedlings eventually reverted to relatively normal growth patterns and were transplanted into the greenhouse. Other seedlings degenerated into classic hormone-independent teratomas (fig. 4).

Further phenotypic abnormalities were observed in the plants which were transplanted to the greenhouse. Abnormalities included a male sterile dwarf, an apparent leaf-variegation chimera, and numerous plants having varying degrees of heterostyly, deformed leaves, reduced fertility, and occasional deformed flowers. Some of the relatively normal plants were self pollinated, and the progeny were germinated on sterile media as before. Progeny from one plant (see fig. 1), segregated for the original semi-lethal seedling phenotype (380 normal: 70 abnormal: plus many inviable or aborted seed). This strongly suggests that some of the observed phenotypic abnormalities have a genetic, heritable basis.

All abnormal plants have been tested for nopaline synthase activity, and all plants have been found to be nopaline-negative. Preliminary Southern DNA hybridization clearly indicate that most of the abnormal seedlings which have reverted to normal phenotype do not contain T-DNA. One phenotypically normal plant has shown evidence of homology to a T-DNA probe in both a dot-blot and in a Southern, but the preliminary Southern was poor and did not resolve into clearly defined bands.

## 4. Discussion

These results demonstrate that high rates of abnormal seedlings can be obtained by either incubating pollen in Ti-plasmid DNA, or by co-cultivation of pollen with *Agrobacterium tumefaciens*, prior to pollination. Most of the observed abnormalities were consistent with the effects of hormone over-production, which may indicate variable or temporal expression of 'onc' genes from Agrobacterial T-DNA. Obviously, tumorous growth arising from the seedlings as well as teratomic explants from abnormal seedlings would be consistent with 'onc' gene expression. In addition, failure of radicle development, reduced apical dominance, and leaf distortions could also be regarded as "mildly-teratomic" phenotypes. Whole-plant mutations involving dwarfism, male sterility, and varigation might be attributed to either chemical mutagenic effects or DNA insertional mutations. In addition, the specific phenomenon of heterostyly has been previously reported in progeny of transformed plants regenerated from teratomas. While many of the abnormal plants could not be brought to flower, an example of $F_2$ transmission of the semi-lethal seedling phenotype indicates that at least some of these abnormalities are genetically based. Taken together, our results are highly suggestive of pollen-mediated transformation, but with variable and unstable expression of 'onc' genes.

The absence of nopaline synthase activity in the abnormal plants could indicate that these plants are not transformants but are mutations induced at high frequency by the experimental treatments. However, the nopaline promoter is known to be repressed under certain circumstances due to methylation. In mammalian egg transformation experiments, methylation of foreign DNA is known to occur extensively and generally irreversibly - making the use of many methylation-sensitive promoters useless for egg transformation research. A similar phenomenon may be occurring in pollen, relative to the nopaline promoter, precluding the expression of this gene when introduced directly into a germline cell. Such methylation might also help explain variable expression of the 'onc' genes.

The absence of T-DNA in many plants which initially appeared transformed must indicate that the abnormal plants arose by teratomic or mutagenic effects of the treatment, or that there has been selective loss of T-DNA bearing cells during development. Teratomic effects might arise by temporal expression of 'onc' genes in the early embryo, without T-DNA insertion into a chromosome. Mutagenic effects might be caused by chemical effects or by random insertional mutations caused by short oligonucleotides arising from various fragments of the Ti-plasmid outside the T-DNA region. Selective loss of T-DNA bearing cells is also a possible explanation of these results. If T-DNA insertion occurs after the first zygotic division, chimeric embryos will result. Subsequently, somatic cell selection is expected to favor cells with correct hormone balance. Such chimerism would be consistent with the somatic instability and sectorial sports we have observed in several of our mutant phenotypes, and could explain why many of our abnormal seedlings eventually produce plants which are phenotypically normal and T-DNA negative.

Acknowledgements:

We thank Dr. M. D. Chilton and Dr. T. Burr for the bacterial strains used in these studies. We also thank Nancy Adams, Laura Child, Karen Knauerhase, Kim Reisch and Cindy Smith for various forms of technical assistance. This research was supported, in part, by a grant from Procter and Gamble Co.

Table 1. Various DNA incubation and co-cultivation pollen treatments, and their effects on seed viability and frequency of abnormal progeny

| Treatment | Minutes of incubation | Number Viable seed (and as % of total seed) | Number Abnormal seed (and as % of total seed) |
|---|---|---|---|
| control (normal hybridization) | 0 | >1000 (95) | 0 (0) |
| 10 ug/ml DNA[1] | 60 | 24 (28) | 10 (42) |
| 10 ug/ml DNA[1] + 25 ug/ml PLO | 60 | 3 (10) | 3 (100) |
| 20 ug/ml DNA[1] + 50 ug/ml PLO | 60 | 22 (15) | 11 (50) |
| 10 ug/ml DNA[1] +300 ug/ml EDTA | 30 | 90 (60) | 14 (16) |
| co-cultivation with A208[2] | overnight[5] | 90 (40) | 31 (34) |
| co-cultivation with 14aA208[3] | 90 | 260 (84) | 22 (8) |
| co-cultivation with 14aA208[3] | overnight[5] | 70 (39) | 6 (9) |
| co-cultivation with K-47[4] | 60 | 115 (48) | 6 (5) |
| co-cultivation with K-47[4] | overnight[5] | 88 (36) | 21 (24) |

[1] Ti-plasmid DNA was extracted from *Agrobacterium tumefaciens* strain A208 (a wild-type nopaline strain bearing the pTiT37 plasmid) using a PEG precipitation protocol. Quantification was based on gel electrophoresis. tRNA was generally also present at high levels.

[2] Wild-type *Agrobacterium tumefaciens* strain A208, compliments of M. D. Chilton.

[3] Partially disarmed strain of *Agrobacterium tumefaciens* A208, compliments of M. D. Chilton. (see Barton et al., 1983, Cell 32:1033).

[4] Wild-type strain of *Agrobacterium rhizogenes*, compliments of T. Burr

[5] Overnight incubation involved a preliminary co-cultivation at room temperature until binding was observed, overnight incubation at 4°C, then addition of fresh pollen, and further co-cultivation at room temperature.

Fig. 1 - loss of apical dom.

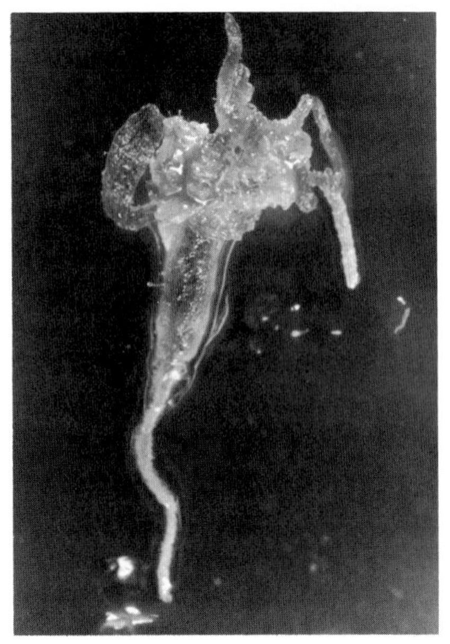

Fig. 2 - tumorous epicotyl

Fig. 3 - tumorous hypocotyl

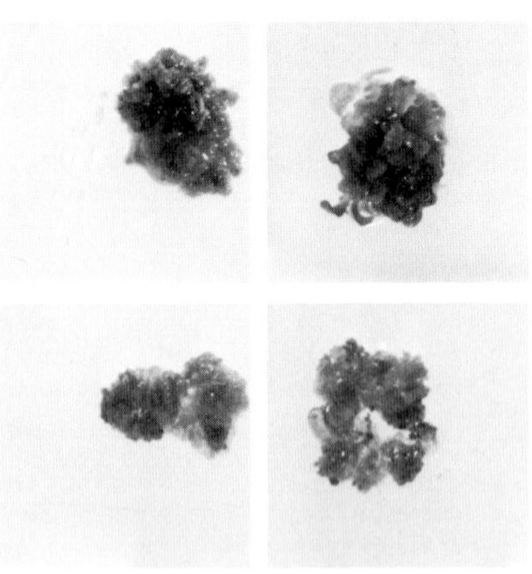

Fig. 4 - classic teratomas

# Dissolution of Pollen Intine and Release of Sporoplasts

BRUCE G. BALDI, VINCENT R. FRANCESCHI AND FRANK A. LOEWUS[1]

## 1 Introduction

Methods for preparation of plant protoplasts are well established but the extension of such methods to the release of exine-free gametophytes from pollen appears to be limited (Bajaj 1974; Bajaj and Davey 1974; Bhojwani and Cocking 1972; Power 1973; Takegami and Ito 1975; Zhu et al. 1984). We have discovered that 4-methylmorpholine N-oxide monohydrate ($MMNO \cdot H_2O$) is an effective solvent of the intine layer when pollen grains of Lilium longiflorum Thunb. (trumpet lily) are dispersed in $MMNO \cdot H_2O$ at its melting point, 75°C (Loewus et al. 1985). Exine-free gametophytes, which we term sporoplasts, are quickly released from their exine enclosures. With time, MMNO also disperses the empty exine 'shells' into immiscible droplets. Prolonged heating ruptures the sporoplasts to produce empty sporoplast envelopes or 'ghosts' which remain intact in the $MMNO \cdot H_2O$ melt.

In order to avoid the high temperature required to melt $MMNO \cdot H_2O$ and consequent denaturing effects on sporoplasts, we have modified our procedure to include cyclohexylamine and a sucrose-based medium. With this new procedure, sporoplasts, still enclosed by intine, are released in less than one hour

---

1 The Graduate Program in Plant Physiology, Department of Botany and Institute of Biological Chemistry, Washington State University, Pullman, Washington 99164-6340, USA

at 20°C. Further treatment by sucrose density gradient separates the sporoplasts from exine 'shells.' When these intine-enclosed sporoplasts are subjected to the standard cellulase/pectinase treatment as used for protoplast release, the intine is rapidly dissolved leaving intact membrane-enclosed sporoplasts. Some properties of these particles are described.

## 2 Preparation of Sporoplasts

Floral buds of *Lilium longiflorum* Thunb. cv. Nellie White were harvested (July 1984) in commercial bulb-growing fields approximately one day before anthesis. Anthers were removed and allowed to dehisce (7 days). Loose pollen was separated from thecae in a 10 mesh/inch brass sieve and stored in loosely-capped plastic vials (20 g/vial) at -18°C until used.

Pollen (25 mg/ml) was suspended in a solution consisting of MMNO (60% aqueous solution, No. 25,882-2, Aldrich Chem. Co.), cyclohexylamine (97-99%, No. C10,465-5, Aldrich Chem. Co.) and 0.29 M sucrose-based medium (Dickinson 1968), proportions 2:1:1, v/v, a treatment which released intine-enclosed sporoplasts from their exine at 20°C. At 90 min, about 90% of the pollen grains had released sporoplasts.

## 3 Separation of Sporoplasts and Exine 'Shells'

Treated pollen (0.5 g in 20 ml of MMNO-cychexylamine-sucrose medium) was diluted with sucrose-based medium (20 ml) and sedimented by centrifugation (2,600 g, 10 min, 4°C). The pellet was washed with sucrose-based medium (20 ml) and finally resuspended in 8 ml of the same medium (Dickinson 1968). Two ml aliquots were placed on a discontinuous sucrose gradient (2 ml, 0.70 M; 6 ml, 0.76 M; 2 ml, 1.78 M) and spun at 2,600 g, 15 min, 4°C in a duPont/Sorvall HB-4 rotor. Based on 90% sporoplast emergence, about 50% of the pollen grains gave complete release of sporoplasts which were recovered in the region of the 0.3 to 0.70 M sucrose interface. Sporoplasts still partially attached to exine concentrated in 0.79 M sucrose and empty exine 'shells' accumulated in the 0.79-1.78 M sucrose interface.

4 Dissolution of Intine

Sporoplasts released by the 20°C treatment retained their intine layer (Fig. 1A, 1B). Intine was readily removed by

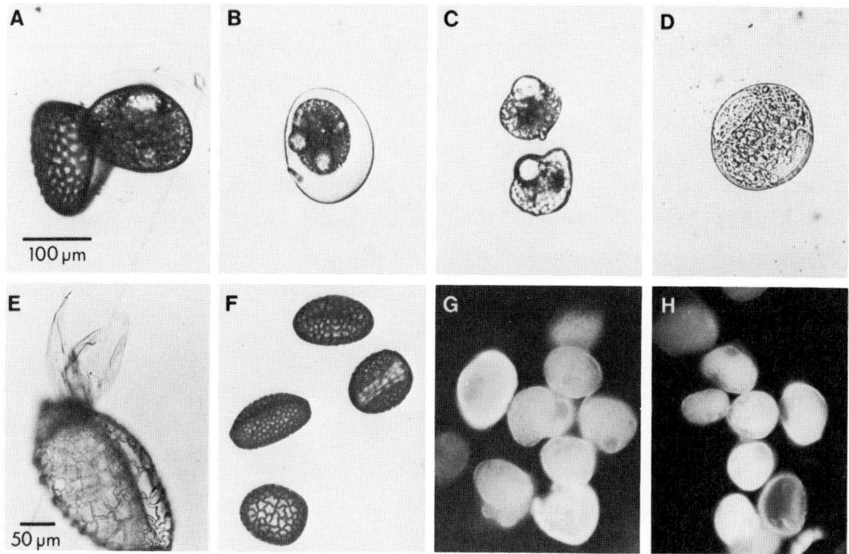


Fig. 1. Sporoplasts of Lilium longiflorum Thunb. cv. Nellie White pollen. A. Intine-enclosed emerging sporoplast. B. Intine-enclosed sporoplast after release from exine. C. Sporoplast after enzymatic removal of intine. D. Sporoplast after dissolution of intine in MMNO·$H_2O$. E. Sporoplastic outer membrane or 'ghost' still partially attached to exine 'shell.' F. Empty exine 'shells.' G. Fluorochromatic response of enzymatically de-intined sporoplasts before adding fluorescein diacetate. H. Fluorochromatic response of enzymatically de-intined sporoplasts after 24 hours in fluoresein deacetate. Scale as indicated in (A), except (E).

treatment with a mixture of 2% (w/v) Cellulysin (Calbiochem/ Hoechst) and 0.5% (w/v) Mascerase (Calbiochem/Hoechst) in 0.1 M ammonium acetate, pH 5.0. After 1 hour at 29°C in this enzyme mixture, 90% of the sporoplasts were completely free of intine (Fig. 1C). Alternatively resuspension of intine-enclosed sporoplasts in MMNO·$H_2O$ at 80°C for 1 hour also dissolved the intine layer (Fig. 1D). Continued heating caused sporoplasts to rupture and empty their contents leaving an outer membrane 'ghost' (Fig. 1E). Empty exine 'shells' are shown in Fig. 1F.

## 5 Some Histochemical Color Reactions of Sporoplasts

Color reactions of intine-enclosed sporoplasts and empty exine 'shells' after staining in aqueous media with various reagents are summarized in Table 1. Sporoplasts and accompanying intine stain strongly for polysaccharides as indicated by the color reactions of congo red (ß-glucans), ruthenium red (acidic polysaccharides), hydroxylamine/$FeCl_3$ (pectin methyl ester), and methylene blue (acidic pectin). The distinctive large oil drops found in nearly all sporoplasts stain darkly with osmium tetroxide. In the absence of exine, both nuclei stain deeply with Schiff's reagent after hydrolysis with HCl.

In addition to these stain procedures, enzymatically de-intined sporoplasts were treated with fluorescein diacetate (Heslop-Harrison and Heslop-Harrison 1970) to determine the integrity of the plasma membrane and the presence of intracellular esterase. After 24 hours in this reagent sporoplasts produced a strong fluorochromatic response (xenon lamp, blue excitation light, 515 nm barrier filter) that was absent in untreated controls (Fig. 1G,H). Although this test does not necessarily establish the biological competence of the sporoplast, it does demonstrate diffusion of the substrate into the cell, hydrolysis of the diacetate, and retention of the fluorescein within the confines of the plasma membrane.

In other experiments (data not given) we obtained weak but reproducible evidence of the uptake and oxidative conversion of myo-[2-$^3$H]inositol to insoluble, and presumably polysaccharidic, products (Loewus and Loewus, 1983) after intine-enclosed or intine-free sporoplasts were shaken with the labeled substrate in pentaerythritol medium (Dickinson 1968) for 20 hours at 29°C. We also obtained a very low but measurable conversion of L-[6-$^{14}$C]galactono-lactone to

Table 1. Color reactions of sporoplasts to various stains

| Stain | Sporoplast | Intine | Exine 'shell' |
| --- | --- | --- | --- |
| None | pale yellow | colorless | pale yellow |
| Congo red[a] | red | pink | red-brown |
| Ruthenium red[b] | deep red, oil drops colorless | pink | faint pink |
| Hydroxylamine/$FeCl_3$[c] | faint yellow-brown, oil drops yellow | flocculant brown precipitate | light brown |
| Methylene blue[d] | oil drops yellow, nuclei pale blue, membranes blue | blue | blue-green |
| Osmium tetroxide vapor[e] | oil drops brown to black | colorless | light brown |
| Schiff's reagent[f] | selective staining of vegetative and generative nuclei | colorless | unstained |

[a]Lillie 1977 p. 147. [b]Jensen 1962 p. 201. [c]Jensen 1962 p. 202. [d]Lillie 1977 p. 423. [e]Thompson 1966 p. 342. [f]Franceschi and Horner Jr. 1979.

L-ascorbic acid (Leung and Loewus, 1985) with enzymatically treated intine-free sporoplasts (data not given).

In summary, we have demonstrated the chemical release of sporoplasts from L. longiflorum pollen at a low temperature and its separation from the empty exine 'shell.' The chemical treatment that is used damages membranes and metabolic processes. Hopefully, additional studies directed toward the composition of the medium used in the releasing sporoplasts will uncover innocuous reagents which will allow biological studies to proceed. Apart from this problem, the present methodology offers a rapid, convenient route to the production of sporoplasts in quantity and to new investigations on the subcellular composition of pollen.

Acknowledgements. Wendy Weeks assisted in the preparation of sporoplasts and studies of their properties. We thank the Electron Microscopy Center at Washington State University for use of their equipment. Pollen was obtained through cooperation of the Oregon Propagating Company. Supported by NSF grant PCM84-04157. Scientific Paper No. 7190, Project 0266, College of Agriculture Research Center, Washington State University, Pullman, WA 99164.

## References

Bajaj YPS (1974) Isolation and culture studies on pollen tetrad and pollen mother-cell protoplasts. Plant Sci Lett 3:93-99.

Bajaj YPS, Davey MR (1974) The isolation and ultrastructure of pollen protoplasts. In: Linskens, HF (ed) Fertilization in Higher Plants. North Holland, pp 73-80.

Bhojwani SS, Cocking EC (1972) Isolation of protoplasts from pollen tetrads. Nature New Biol 239:29-30.

Dickinson DB (1968) Rapid starch synthesis associated with increased respiration in germinating lily pollen. Plant Physiol. 43:1-8

Franceschi VR, Horner HT, Jr (1979) Nuclear condition of the anther tapetum of *Ornithogalum caudatum* during microsporogenesis. Cytobiologie 18:413-421.

Heslop-Harrison J, Heslop-Harrison Y (1970) Evaluation of pollen viability by enzymatically induced fluorescence: intracellular hydrolysis of fluorescein diacetate. Stain Tech 45:115-120.

Jensen WA (1962) Botanical Histochemistry. WH Freeman, San Francisco, CA.

Leung CT, Loewus FA (1985) Ascorbic acid in pollen: conversion of L-galactono-1,4-lactone to L-ascorbic acid by *Lilium longiflorum*. Plant Sci (in press).

Lillie RD (1977) HJ Conn's Biological Stains, Ninth Ed, Williams and Wilkins, Baltimore, MD.

Loewus FA, Baldi BG, Franceschi VR, Meinert LD, McCollum JJ (1985) Pollen sporoplasts: dissolution of pollen walls. Plant Physiol 78:in press.

Loewus FA, Loewus MW (1983) *myo*-Inositol: its biosynthesis and metabolism. Annu Rev Plant Physiol 34:137-161.

Power JB (1973) Isolation of mature tobacco pollen protoplasts. In: Street HE (ed) Plant Tissue and Cell Culture. Blackwell, pp 118-120.

Takegami MH, Ito M (1975) Studies on the behavior of meiotic protoplasts. III Features of nuclear and cell division for lily protoplasts in the cell-wall-free state. Bot Mag Tokyo 88:187-196.

Thompson SW (1966) Selected Histochemical and Histopathological Methods. CC Thomas, Springfield, IL.

Zhu C, Xie Y, Hu S (1984) Isolation and cultural behavior of pollen tube subprotoplasts in *Antirrhinum majus* L. Acta Bot Sinica 26:459-465.

# Sporophytic Screening and Gametophytic Verification of Phytotoxin Tolerance in Sugarbeet (Beta vulgaris L.)

G. A. Smith[1]

## 1 INTRODUCTION

The rationale for developing crop cultivars that can tolerate phytotoxic agents such as herbicides has been discussed at length. Perhaps, the most convincing argument for such development is that, with traditional methods, the cost of developing a new crop cultivar is only 1 to 5% of the developmental cost of a new herbicide (Faulkner, 1982). However, selection for herbicide resistance or tolerance in the field by traditional breeding techniques is very time (years) and labor-intensive and, at best, very inexact. Low frequency of resistant individuals, low heritability of tolerance, escapes -- all associated with high environmental variability -- account for severely limited progress.

It has been suggested that a substantial number of genes expressed in the gametophyte (pollen) also are expressed in the sporophyte (Mulcahy 1971, Ottaviano et al. 1980). Recent evidence in maize (*Zea mays* L.) (Mulcahy 1971, 1979; Mulcahy and Mulcahy 1975; Ottaviano et al. 1980) and in tomato (*Lycopersicon esculentum*) (Tanksley et al. 1981) have supported the hypothesis that a portion of the genes (60% in tomato) expressed in the sporophyte also are expressed in the gametophyte.

The primary objective of our study was to develop and evaluate in vitro techniques for identifying genotypes within heterogeneous seedling populations that are tolerant to specific herbicides, and to use meristematic cloning procedures to synthesize

[1]USDA-ARS, Crops Research Laboratory, Colorado State University, Fort Collins, CO 80523

clones genetically tolerant to the herbicide. Our in vitro selection procedures were developed initially with the sporophytic generation, and then tested in the gametophytic generation in accordance with the proposal that a correlated effect between generations might result from selection of genes expressed in both stages (Mulcahy 1979).

## 2 PROCEDURES

In vitro techniques were developed with which herbicide tolerant genotypes could be identified from heterogeneous seedling populations. The specific details of the techniques have been described (see Smith and Moser 1985). In brief, the techniques involve subjection of young seedling populations grown on MS media to known concentrations of a challenging agent. We chose a common sugarbeet herbicide ethofumesate (Nortron) as our initial challenging agent. In vitro selection criteria for identification of tolerant genotypes included stunting, irregular morphology, root growth and central bud development. Concentrations of the herbicide used were equal to at least four times the normal field recommended dosage rates. Individual seedlings recognized as tolerant were cloned (see Moser and Smith 1985) to obtain enough plant copies for sexual seed production. Selected clones were then intercrossed and seed obtained for testing. Such seed was planted in the greenhouse in soil containing concentrations of ethofumesate ranging from 0 to 4 times the recommended dosage rates. From such tests, clonal lines were identified as having tolerance greater than that of the original population. Other 'tolerant' clonal lines along with a clonal line developed by the same in vitro techniques but never exposed to ethofumesate were brought to flower and pollen studies conducted. Pollen from the clonal lines was collected and cultured in modified Brewbaker and Kwack (BK) solution to which ethofumesate had been added. Following 24 hrs of incubation, pollen was examined for germination and tube morphology.

In further research, pollen was collected from various clones and incubated in BK solution containing high concentrations of ethofumesate. Plants with the highest pollen germination were intercrossed to provide a 'tolerant' seed source. Plants with the lowest pollen germination were intercrossed to provide a 'non-tolerant' seed source. These seed sources were then grown in the greenhouse in soil containing various concentrations of ethofumesate.

## 3 RESULTS IN THE GAMETOPHYTE FROM PLANTS SELECTED IN THE SPOROPHYTE

Results from the pollen germination of seven clones cultured in media containing 10 and 20 mg/L ethofumesate are presented in Table 1. Five of six clones developed from seedlings screened and selected as sporophytes, produced pollen that germinated significantly better at 10 mg/L than did pollen from the UNC clone which was developed from an unscreened selection. These five clones (N3-3, N3-6, N15-8, N15-10 and N15-18) averaged

Table 1. Germination of pollen from each of seven clones cultured in media containing 10 and 20 mg/L ethofumesate. Values for each clone are given as percent of the clones' control. The $LSD_{.05}$ = 10.2 for comparison of entries within the same level of herbicide. UNC = unselected.

| | Pollen germination, % of control | |
|---|---|---|
| Clone | 10 mg/L | 20 mg/L |
| N3-3 | 50±6 | 12±6 |
| N3-6 | 78±3 | 41±4 |
| N15-2 | 36±9 | 9±2 |
| N15-8 | 68±4 | 15±2 |
| N15-10 | 63±7 | 25±5 |
| N15-18 | 68±3 | 26±4 |
| UNC | 34±0.1 | 9±5 |

65.3% of their respective controls compared to 33.8% for the UNC clone. Pollen from the best clonal selection, N3-6, germinated 44% better than pollen from the unscreened UNC clone (78% of control vs 34% of control). Pollen from clones N3-6, N15-10 and N15-18 also germinated significantly better than the UNC clone at 20 mg/L (average = 31% of control vs 9% of control). As in a previous experiment, one clone (N15-2) produced pollen with germination about equal to the UNC clone at either herbicide concentration. This suggests that a selection made in the sporophyte was not tolerant.

## 4 RESULTS IN THE SPOROPHYTE FROM PLANTS SELECTED IN THE GAMETOPHYTE

Table 2 presents emergence, leaf development and root and top weight data of seedling progeny from plants selected as 'tolerant' and 'non-tolerant' in the gametophyte. Seedling tests were conducted in soil containing 6 or 12 mg ai/kg. These rates compare with a commercial use recommendation of 3 mg ai/kg. Largest differences were between clones selected as 'tolerant' and those selected as 'non-tolerant' in the gametophyte when evaluated at the higher herbicide concentration in

Table 2. Comparison of 'tolerant' and 'non-tolerant' lines as determined by pollen germination in medium containing high concentrations of ethofumesate.

| | | Entries | | |
|---|---|---|---|---|
| Character | Conc. | Tol-1[†] | Non tol-1 | UNS |
| Seedling emergence[‡] | L[§] | 96.6 | 94.2 | 100.0 |
| | H | 98.3 | 73.1 | 98.0 |
| Leaf development | L | 86.2 | 38.8 | 35.1 |
| | H | 53.4 | 4.1 | 24.3 |
| Root dry weight - 28 days | L | 79.5 | 57.2 | 57.3 |
| | H | 93.4 | 56.7 | 83.4 |
| Top dry weight - 28 days | L | 76.6 | 67.6 | 75.2 |
| | H | 67.7 | 47.2 | 72.5 |

[†]Tol-1 = 'tolerant' seed based on pollen test;
Non tol-1 = 'non-tolerant' seed based on pollen test;
UNS = unscreened seed from original population.

[‡]% of check = % of untreated check within each entry.

[§]L = 6 mg ethofumesate per kg of soil; H = 12 mg ethofumesate per kg of soil.

the sporophyte. The most significant effects were found for normal leaf development at 14 days. Central bud development and subsequent leaf development are known to be the 'target' indicators for ethofumesate damage. It was also noted that differences for 8-day seedling emergence of the unscreened seedlings was not significantly different than the screened clone progeny. This might have been expected since exposure time was short and maximum uptake of ethofumesate is through the hypocotyl and roots.

## 5 DISCUSSION

Once the seed from cloned selections was obtained, selected clonal progeny were compared with plants from seed of the original population via root dry weight, and central bud development. In addition to greenhouse progeny evaluation, verification of in vitro selection accuracy was accomplished by gametophyte (pollen germination) studies. Our preliminary gametophytic-sporophytic association studies determined the effect of ethofumesate on pollen germination and development. Once satisfactory procedures for pollen collection, incubation and ethofumesate concentration were established, it was possible to compare pollen from selected and unselected clonal families. The results indicate that in vitro selection of germinated seedlings in the presence of the proper concentration of challenging agent can be effective in identifying genotypes tolerant to ethofumesate. Such identification was accomplished in fully differentiated tissue, but without the necessity of mature plants.

Perhaps of most significance was the finding that gametophytic studies, via pollen germination, indicated an association between genes operating in the sporophyte and those in the gametophyte. The commonality of gene expression in the sporophyte and gametophyte indicates that further research is needed to determine if selection might be made in the gametophyte via pollen germination. Although results from this research indicate that such an approach is plausible, further refinement of techniques should promote more consistency. Some of our evidence indicated that not all sporophytes selected for ethofumesate tolerance demonstrate this tolerance in the gametophyte. For example, clone N15-2, which had been screened at 12 mg/L, did not produce pollen germination results better or worse than pollen from unscreened clones from the same original population when tested at 10 and 20 mg/L of ethofumesate. We suspect that an errant selection was made in the sporophyte which underlines the importance of accurate selection of sporophytes. On the other hand, because perfect selection in the sporophyte cannot be guaranteed (most likely with any selection scheme), verification in the gametophyte should certainly increase selection accuracy.

## REFERENCES


Faulkner JS (1982) Breeding herbicide-tolerant crop cultivars conventional methods. In: LeBaron HM, Gressel J (eds) Herbicide resistance in plants. Wiley and sons, New York, pp 235-256

Moser HS, Smith GA (1985) Factors affecting in vitro multiplication and rooting of sugarbeet. (in press)

Mulcahy DL (1971) Correlation between gametophytic and sporophytic characteristics in *Zea mays* L. Science 171:1155-1156

Mulcahy DL (1979) The rise of the angiosperms: a genecological factor. Science 206:20-23

Mulcahy DL, Mulcahy GB (1975) The influence of gametophytic competition on sporophytic quality in *Dianthus chinensis*. Theor Appl Genet 46:277-280

Ottaviano E, Sarri-Gorla M, Mulcahy DL (1980) Pollen tube growth rates in *Zea mays*: Implications for genetic improvement of crops. Science 42:437-438

Smith GA, and Moser HS (1985) Sporophytic-gametophytic herbicide tolerance in sugarbeet. Theor Appl Gen. (in press)

Tanksley SD, Zamir D, Rick CM (1981) Evidence for extensive overlap of sporophytic and gametophytic gene expression in *Lycopersicon esculentum*. Science 213:452-455

# Predicting Species Response to Ozone Using a Pollen Screen

WILLIAM A. FEDER[1]

## 1 Introduction

Ozone is a major cause of injury to vegetation in Europe and the USA. Injury to a broad spectrum of Gymnosperms and Angiosperms after exposure to low $0_3$ dosages was documented. (Feder *et al* 1969,1970;Gentile *et al* 1971;Omrod *et al* 1976;Davis & Wood 1972;Miller *et al* 1969;and Berry 1973.) Sensitive plant species are injured by 0.04-0.06ppm $0_3$ for 4hr. Petunia, tobacco, potato, and morning glory and tree species like white pine and ponderosa pine were in this $0_3$-sensitive group. (Feder 1970; Miller *et al* 1969;Berry 1973.) Sensitivity to $0_3$ varies among CVs and within species. Variability of $0_3$ response was noted for corn (Cameron *et al* 1970);tomato (Gentile *et al* 1971);bean (Davis & Kress 1974);Petunia (Feder *et al* 1969);alfalfa (Howell *et al* 1971); white pine (Berry 1973);Ponderosa pine (Miller *et al* 1969);and forest trees (Jensen 1973.)

Feder *et al* (Feder 1968;Feder & Sullivan 1969;Feder 1981; Riley & Feder 1974;and Krause *et al* 1975) showed that pollen germination and tube elongation were depressed by exposing pollen *in vitro* to $0_3$ dosages of 0.04-0.10ppm for 1-5hr. Sensitivity to $0_3$ was found to vary within pollen populations from a single species or CV, i.e. each population described a sensitivity curve in terms of % germination and/or tube elongation. The shape of these sensitivity distribution curves could be altered by changing the $0_3$ dosage (Feder & Sullivan 1969.)

The shape of the curves varied with the pollen source, i.e.

---

[1] Suburban Experiment Station, University of Massachusetts, Waltham, 240 Beaver St., Waltham, Massachusetts

pollen taken from an $O_3$-sensitive parent was affected more by exposure to $O_3$ than pollen from an $O_3$-tolerant parent. The 2 pollen populations produced 2 different distribution curves, each corresponding to the $O_3$ response of the respective parent. Two Petunia and 2 tobacco CVs whose $O_3$ responses were well documented were used.(Feder & Sullivan 1969) Data showed that the haploid $O_3$ responses mimicked the diploid responses and vice versa. These results were supported by comparison of mRNAs from pollen and sporophytic tissues of _Tradescantia paludosa_ and _Zea mays_ (Willing _et al_ 1984) who concluded that in _Tradescantia_ and _Zea_ about 85% of genes expressed in pollen are expressed in the sporophyte. Only 15% of the genes are unique to pollen. They postulated that selecting for genes in the haploid phase could positively affect sporophytic success. Could the pollen-$O_3$ response predict the sporophytic-$O_3$ response? If so, this information would be useful in establishing risk to pollutant stress and in predicting the success of species or communities chronically exposed to that stress, i.e.-pollen response might provide a simple, quick screen for establishing long-term response to a pollutant. The present study was done to provide additional data to substantiate the pollen/sporophyte/$O_3$ response.

## 2 Material and Methods

Pollen from the $O_3$-sensitive _Nicotiana tabacum_ CV Bel-W3 and the $O_3$-sensitive _Petunia hybrida_ CVs White Bountiful,White Cascade,Pinwheel,and Tiffany, and from the $O_3$-tolerant _N tabacum_ CV Bel-B and _P hybrida_ CV Blue Lagoon was used. These CVs were used because their $O_3$ response was understood. Plants were grown in charcoal-filtered air. Pollen was collected by removing the anthers and crushing them in a sterile Petri plate. Pollen was dusted onto squares of split dialysis membrane which had been placed on the surface of 1% Difco agar discs.(Feder & Sullivan 1969). 10% sucrose was added to the tobacco pollen medium; 15% sucrose and 300 ppm boron were added to the Petunia pollen medium. $O_3$ was generated and pollen exposed as previously described.(Feder 1968;Feder & Sullivan 1969) After 5.5hr fumigation, the discs were removed and placed in sterile humidity chambers and incubated for 24hr @ 26.7C. Germination % and tube length were recorded. Controls were exposed to charcoal-filtered air under the identical conditions. % germination was measured by counting 3 random groups of 100 pollen grains on each of 3 discs for each dosage. Tube length was measured in mm for the same number of grains.

## 3 Results

Table 1 shows the difference in % germination among pollen populations from several CVs exposed to $O_3$. Pollen populations from $O_3$-sensitive CVs had a lower % germination than those from $O_3$-tolerant CVs. Unexposed pollen had about the same % germination for all CVs tested. Bel W-3 was the most $O_3$-sensitive CV tested. Pollen germination response to $O_3$ followed the parent response for all CVs studied. The degree of pollen tube elongation under $O_3$ stress resembles the parental response. Classes of pollen tube lengths were used to compare tube elongation for the different pollen populations at several $O_3$ levels. Unstressed pollen populations exhibited a normal distribution of tube lengths after 24hr incubation. Increasing $O_3$ stress narrowed the spread of tube lengths in $O_3$-sensitive pollen but much less so in $O_3$-tolerant pollen. (Figs.1&2)

## 4 Discussion and Conclusions

The data indicate a positive relationship between pollen and parental response to $O_3$ in the CVs studied. CVs were selected for study because the sporophytic $O_3$ response was known. The similarity in $O_3$ response of pollen and sporophyte confirms our earlier recognition of this relationship. Basis for the response is elucidated by the discovery of an 85% gene overlap between pollen and sporophyte. (Willing _et al_ 1984)

Based on this limited study with $O_3$ and 2 solanaceous genera, on the work of Willing _et al_, and on the work with _Lycopersicon esculentum_ (Tanksley _et al_ 1981) it seems that pollen response to $O_3$ could be used to predict sporophytic response to $O_3$. In our case, pollen response to $O_3$ clearly mimics sporophytic response to $O_3$. Using % germination, the response of the sporophyte to $O_3$ can be readily predicted by the pollen response to the same $O_3$ dosage. Also, by measuring pollen tube lengths after a standardized $O_3$ exposure and incubation period, we can define the shift in pollen tube length distribution due to the $O_3$ exposure. If further study with more plant genera confirms our data, we can then proceed to use pollen as a screen for predicting species response to ozone.

References

Berry, C.R.(1973) The differential sensitivity of eastern white pine to three types of air pollution. Can.J.Forest Res.3:543.

Cameron JW, Johnson H Jr, Taylor OC, Otto HW (1970) Differential susceptibility of sweet corn hybrids to field injury by air pollution. HortScience 5: 217-219.

Davis DD, Wood FA (1972) The relative susceptibility of eighteen coniferous species to ozone. Phytopathol. 62: 14-19.

Davis DD. Kress L (1974) The relative susceptibility of ten bean varieties to ozone. Plant Dis. Rept. 58: 14-16.

Feder WA (1968) Reduction in tobacco pollen germination and tube elongation by low levels of ozone. Science 160: 1122.

Feder WA (1970) Plant response to chronic exposure to low levels of oxidant type air pollution. Environ.Poll.1:73-76.

Feder WA (1981) Bioassaying for ozone with pollen systems. Environ. Health Perspectives 37: 117-123.

Feder WA, Sullivan F (1969) Differential susceptibility of pollen grains to ozone injury. Phytopathol. 59: 399.

Feder WA, Fox FL, Heck WW, Campbell FJ (1969) Varietal responses of petunia to several air pollutants. Plant Dis.Reptr. 53: 506-510.

Gentile AG, Feder WA, Young RE, Santer Z (1971) Susceptibility of _Lycopersicon_ spp to ozone injury. Proc. J.A.Soc.Hort.Sci. 96: 94-96.

Howell RK, Devine TE, Hanson CH (1971) Resistance of selected alfalfa strains to ozone. Crop Sci. 11: 114-115.

Jensen KF (1973) Response of nine forest tree species to chronic ozone fumigation. Plant Dis.Rptr. 57: 914-917.

Krause GHM, Riley WD, Feder WA (1975) Effects of ozone on petunia and tomato pollen tube elongation _in vivo_. Proc. Amer. Phytopathol. Soc. 2: 100.

Miller PR, Parmenter JR Jr, Flick BJ, Martinez CW (1969) Ozone dosage response of Ponderosa pine seedlings. J.Air Pollut. Control Assn. 19: 435-438.

Omrod DP, Adepipe HO, Ballantine DJ (1976) Air pollution injury to horticultural plants: A review. Hort.Abs. 46: 241-248.

Riley WD, Feder WA (1974) Pollen tube susceptibility to ozone as a function of tube extension. Proc.Amer.Phytopathol.Soc. 1:43.

Tanksley SD, Zamir D, Rick CM (1981) Evidence for extensive overlap of sporophytic and gametophytic gene expression in _Lycopersicon esculentum_. Science 213: 453-455.

Willing RP, Mascarenhas JP (1984) Analysis of the complexity and diversity of mRNAs from pollen and shoots of _Tradescantia_ Plant Physiol. 75: 865-868.

Willing RP, Eisenberg A, Mascarenhas JP (1984) Genes active during pollen development and the construction of cloned cDNA libraries to messenger RNAs from pollen. Plant Cell Incompatibility Newsletter 16: 11-12.

Table 1. Effect of ozone on the percent germination of pollen grains from ozone-sensitive and ozone-tolerant cultivars of tobacco and *Petunia*.

| | | Percent Germination | | | | |
|---|---|---|---|---|---|---|
| | | Ozone Concentration (PPM) | | | | |
| CV | Exposure Time (hr) | 0.00 | 0.04 | 0.10 | 0.25 | 0.50 |
| Tobacco[1/] Bel W-3 | 5.5 | 86.6* | 61.2 | 54.2 | 22.0 | 7.0 |
| Tobacco[2/] Bel B | 5.5 | 89.7 | 87.8 | 85.8 | 83.8 | 65.0 |
| Petunia[1/] White Bountiful | 5.5 | 87.1 | ---- | 86.1 | 71.7 | 53.7 |
| Petunia[1/] White Cascade | 5.5 | 70.9 | ---- | 65.3 | 58.5 | 38.6 |
| Petunia[1/] Tiffany | 5.5 | 82.8 | ---- | 80.2 | 73.5 | 52.9 |
| Petunia[1/] Pinwheel | 5.5 | 80.6 | ---- | 79.5 | 75.6 | 51.9 |
| Petunia[2/] Rose Perfection | 5.5 | 87.7 | ---- | 87.5 | 80.7 | 73.1 |
| Petunia[2/] Blue Lagoon | 5.5 | 89.0 | ---- | 82.3 | 83.1 | 85.3 |

* Mean of 300 pollen grains

1/ ozone-sensitive cultivar

2/ ozone-tolerant cultivar

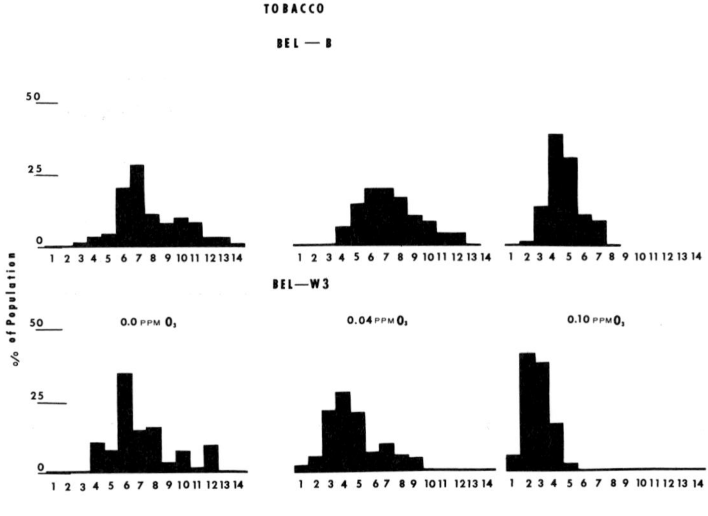


Fig.1 Effect of $O_3$ on growth of pollen tubes of $O_3$- sensitive tobacco CV Bel W-3 and $O_3$-tolerant CV Bel-B. Lengths are relative (mm). Lower numbers are shorter tubes.

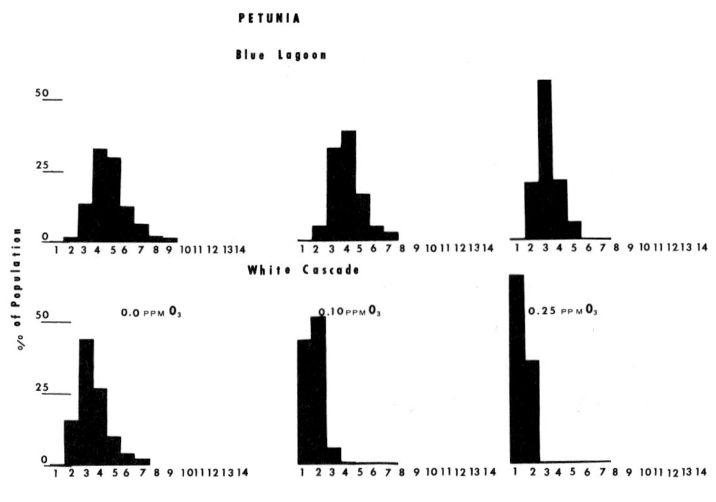


Fig.2 Effect of $O_3$ on growth of pollen tubes of $O_3$-sensitive Petunia CV White Cascade and $O_3$-tolerant CV Blue Lagoon. Lengths are relative (mm). Lower numbers are shorter tubes.

# In Vitro and In Vivo Effects of Acidity and Trace Elements on Pollen Function

R.M. Cox[1]

## INTRODUCTION

Pollen development and activity are known to be among the more sensitive botanical indicators of atmospheric pollution (Feder 1981). Atmospheric pollution may directly effect the pollen by reducing its viability prior to pollination or by affecting the chemical environment of the stigmatic surface. Marked reductions in cone dimensions, seed weight and viability together with reduced pollen viability have been observed near to sources of air pollution (Antipov 1970). Furthermore, these effects may occur at pollution levels lower than that required for foliar injury (Houston and Dochinger 1977). Acidity of $SO_2$ fumigations under wet or high relative humidities was taken for granted by Dopp (1931), who suggested that the increased *in vitro* effect of $SO_2$ fumigation over the *in vivo* fumigation effect was due to extra buffering of the stigmatic surface. The acidification of the pollen media by $SO_2$ fumigation was demonstrated by Karnosky and Stairs (1974). Murdy and Ragsdale (1980) working with *Geranium carolinianum* suggested that *in vitro* $SO_2$ effects on pollen may emulate *in vivo* effects under high relative humidity and the resultant reduced pollen germination and initial tube growth may reduce seed set. Such a reduction in *in vivo* pollen germination was related to reduced seed set in *G. carolinianum* by DuBay and Murdy (1983). Masaru *et al*. (1980), however, was to finally verify the importance of $H^+$ ion inhibition of pollen function *in vitro* and suggest the implications of acidic precipitation which was confirmed *in vivo* by Cox (1984). Trace element effects on pollen have also been investigated (Chaney and Strickland 1984), and trace elements in combination with acidity have been investigated by Cox (1983b) and will be further described here.

---

1 Maritimes Forest Research Centre, Fredericton, N.B., Canada E3B 5P7

## IN VITRO RESPONSE TO ACIDITY AND TRACE METALS

The pollen of twelve different species occupying various habitats in eastern Canada were sampled and assayed for response to pH in 50 µl standing drop cultures (Cox 1983a). This culture solution contained different sucrose concentrations depending on the osmotic requirements of the pollen, and was adjusted to pH 5.6, 4.6, 4.0, 3.6, 3.0, and 2.6 with dilute $H_2So_4$. The pollen was assayed according to Cox (1983a). When pollen tubes in the pH 5.6 treatment had grown to a suitable length for measurement, both % germination and the length of the longest tubes (90 percentile) of the first 50 pollen grains in random fields of view were recorded. Means and ranges of estimated dosage values of pH that caused a 50% probability of response (failure to germinate $LD_{50}$) were computed using a probit procedure (Cox 1983a) for each species, these are shown in order of sensitivity in Figure 1. It is of some concern that the threshold of response of all the pollens tested were within the range of pH experienced in ambient precipitation in eastern Canada. The broad-leaved species receiving direct precipitation were found to be the most pH sensitive. The next most sensitive pollen was sampled from the forest understorey and ground flora species while the least sensitive pollen was sampled from the conifers, which as a group, had significantly lower $LD_{50}$ values than the broad-leaved species ($P < .001$). These broad sensitivity groupings are in agreement with those described for foliar sensitivity by Evans (1980).

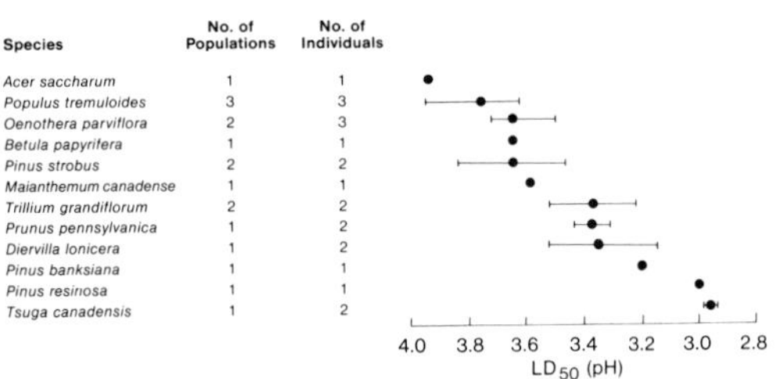


Fig. 1. Means and ranges of $LD_{50}$ pH for in vitro % pollen germination (arcsin).

A full understanding of the direct effects of acid precipitation on forest plant reproduction however can only be obtained by taking into account the species' location in the canopy strata due to canopy modification of the precipitation. The adaptive relationships between these strata in terms of foliar and pollen sensitivity are as yet unstudied. The combined effect of copper and pH on in vitro pollen germination of 11 species was assayed in factorial combination of four pH levels (5.6, 4.6, 3.6 and 2.6) with 5 levels of copper as chloride (0, 0.05, 0.1, 0.2

and 0.4 ppm). These copper/pH assays were carried out in a similar way as described by Cox (1983a). The conifer assays contained 15 ppm of chloramphenicol and Nystatin to prevent proliferation of fungi and bacteria. The summary statistics are shown in Table 1 and again indicate similar pH effects as those shown in Figure 1. The overall effect of copper on pollen germination, however, varied between species being significant in 5 of the 11 pollens tested.

Analysis of variance and range tests at each pH level indicated significant effects ($\alpha$ = 0.05) of copper on pollen germination at pH 5.6-4.6. At these particular pH's significant effects ($\alpha$ = 0.05) of copper were demonstrated in 8 of the pollens tested. However, the pollen germination response was either stimulatory inhibitory or a mixture of both (Table 1).

Table 1

Summary of effects of pH (2.6 - 5.6) and copper (0 - 0.4 mg $l^{-1}$) on *in vitro* pollen germination (arcsin) of 11 forest plant species.

| Species | pH effects: Overall | pH effects: Threshold of inhibition ($\alpha$ = .05) | Cu effects: Overall | Cu effects: pH of Cu effect ($\alpha$ = .05) | Cu effects: Response |
|---|---|---|---|---|---|
| *Populus tremuloides* | ** | 5.6 - 4.6 | ** | 4.6 | Inhibition |
| *Oenothera parviflora* | *** | 5.6 - 3.6 | N/S | 4.6 | Stimulation |
| *Acer saccharum* | *** | 4.6 - 3.6 | *** | 5.6 - 4.6 | Stimulation Inhibition |
| *Betula alleghaniensis* | *** | 4.6 - 3.6 | *** | 4.6 | Inhibition |
| *Betula papyrifera* | *** | 4.6 - 3.6 | N/S | None | None |
| *Diervilla lonicera* | *** | 4.6 - 3.6 | *** | 5.6 - 4.6 | Mixed |
| *Trillium grandiflorum* | *** | 4.6 - 3.6 | * | 5.6 | Stimulation |
| *Pinus strobus* | *** | 4.6 - 2.6 | N/S | None | None |
| *Tsuga canadensis* | *** | 4.6 - 2.6 | N/S | 5.6 | Stimulation |
| *Pinus resinosa* | *** | 3.6 - 2.6 | N/S | None | None |
| *Pinus banksiana* | *** | 3.6 - 2.6 | N/S | 5.6 | Mixed |

*** = P < .001    ** = P < .01    * = P < .05

Interactions between Cu and pH on pollen germination was investigated (Cox 1985) by determining the $LD_{50}$ pH dosages. These dosages were again computed using a probit regression model (Cox 1983a), however, the natural response rate used in the computation was that of the pH 5.6 treatment with no copper in order to relate changes in $LD_{50}$ pH with increase in copper concentration. This analysis revealed the pollen of *P. tremuloides* was made more sensitive to acidity by the addition of 0.2 ppm Cu to the liquid medium whereas pollen of *P. resinosa* and *O. parviflora* become more resistant to acidity by the addition of 0.1 ppm Cu for the former species and by 0.05 and 0.20 ppm Cu for the latter.

These changes in pollen dose response indicate that some species' sensitivity to pH is influenced by the presence of copper at concentrations currently occurring on occasion in ambient precipitation. However it is not known how these changes reflect the copper nutrient status of the pollen.

Variation in plant response to soil pH and differential adaptation of plants to various soil types within the species range are well known. The suggestion that sporophyte tolerances may be expressed in the pollen (Feder 1981) and the demonstrated genetic variability in tolerance to air pollutants in white pine (Houston and Stairs 1973) prompted an in vitro investigation of white pine pollen. The combined effects of pH and Al on white pine pollen sampled from individuals on adjacent sites in the Manitoulin region, Ontario, with soils varying in pH and base status was examined . Two comparisons of pollen response were made using pollen from different individuals from the contrasting soil types, similar assay procedures were used as that described above with Cu and pH. The same pH's were used in factorial combination with 0, 10, 50, 100 and 500 μM Al for the first comparison whereas the 500 μM Al treatments were dropped from the second. The response surface for % germination (arcsin) for the two comparisons are shown in Figure 2.

Analysis of variance indicated significant pH ($P<0.0001$) and Al effects ($P = 0.56 - 0.0001$). Significant interaction terms ($P = 0.05 - 0.0001$) were also detected due to synergism at pH 4.6 and Al concentrations greater than 50 μM. This interaction may be due to the change in Al speciation from the available $Al(OH)_2^+$ ion at pH 4.6 to the $Al^{+3}$ ion at pH 3.6 (Burrows 1977). The reduced toxicity at pH 3.6 may be caused by the chaelation of the $Al^{+3}$ ion by the organics in the culture solution thus reducing its availability. At pH 2.6. however, $Al^{+3}$ availability increases exerting a toxic effect. Apart from the synergistic interaction the general shape of the germination response surfaces of the pollen from the base-rich site indicates a greater sensitivity with higher linear component, whereas the response surface of the pollen sampled from the base-poor site was more resistant to pH with a higher quadratic component to the response surface. These pollen

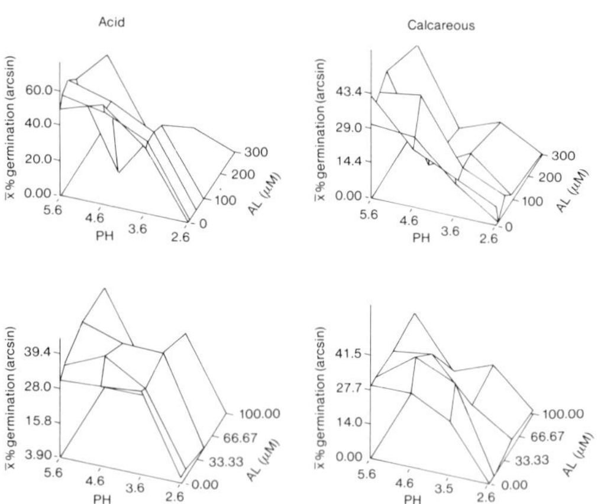


Fig. 2. Germination responses of pollen from two individuals of *Pinus strobus* growing on acid and a calcareous soil to Al and pH.

responses may reflect edaphic tolerance of their sporophytes which have been selected under the calcareous and acid soil conditions at the different sites.

## POLLEN RESPONSE TO ACIDITY *IN VIVO*

Simulations of precipitation events of depth 0.26 cm at various acidities on stigmatic receptivity of *Oenothera parviflora* was examined by Cox (1984). Virgin receptive stigmas of emascalated flowers of a Muskoka population whose petals had been removed in the bud to reduce natural pollination were treated with sprays of various pH's prior to controlled pollinations with known loads of pollen. After 5 h the stigma/ styles were excised and fixed independently for later microscopic examination. The percentage of potentially viable pollen (darkly stained contents) that a) germinated and b) produced germ tubes long enough to penetrate the stigma (i.e. 3 x diameter of the grain) were recorded (Figure 3). Analysis of variance indicated acid rain simulants caused a significant ($P < 0.01$) reduction in stigma receptivity using both criteria, and that the pH 3.6 and 2.6 simulants significantly ($\alpha = 0.05$) reduced receptivity compared to the pH 5.6 control spray and the unsprayed control treatment. The $LD_{50}$ dosages of pH computed for the two stigma receptivity criteria were pH 3.45 and pH 4.55 respectively indicating that germ tube growth and stigmatic penetration is more sensitive to acidity than pollen germination. In addition it must be noted that the 20% reduction in both criteria of stigmatic receptivity are additive effects and indicate that substantial reduction in stigmatic receptivity may result from acidic precipitation prior to pollination in this species.

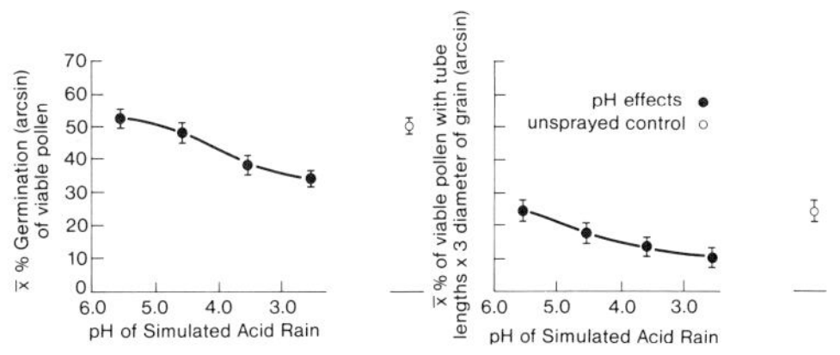


Fig. 3. The effects of field simulations of acid precipitation of *Oenothera parviflora* (Means 1 SE of 8 replicates).

Selection of pollen genotypes on the stigmatic surface by the direct effect of air pollution would effect the frequency of pollen genotypes contributing to the next generation. Furthermore such interference with the participants of gametic competition may effect progeny fitness

(Malcahy 1979). Such changes in gene frequency caused by acid rain may be adaptive in acidic habitats but may reduce fitness where adaptation to high pH's or calcareous habitats are important. This may imply subtle effects on populations on calcareous sites previously thought to be exempt from acid rain effects.

## REFERENCES


Antipov VG (1970) Ohrana priody naturale, Sverdlovak 7: 31-35 (Forest. Abstr., 32: (4) 752).

Burrows WD (1977) Crit Rev in Environ Control. 7: 167-216.

Chaney WR, Strickland RC (1984) J Environ Qual Vol 13: 391-394.

Cox RM (1983a) New Phytol 95: 269- 276.

Cox RM (1983b) Sensitivity of forest plant reporduction to acid rain, in: Proc. Int. Conf. Acid rain and forest resources. (Rennie, P.J. and Robitaille, G. eds.) Quebec City 1983 (in press).

Cox RM (1984) New Phytol 97: 63-70.

Cox RM (1985) The response of plant reproductive processes to air pollution. Proc NATO Advanced Res Workshop. Effects of Acidic Deposition and Air Pollutants on Forest, Wetlands and Agricultural Ecosystems. Toronto May 13-17 (In Press).

Dopp W (1931) Ber Dent Bot Ges 47: 173-221.

DuBay DT, Murdy WH (1983) Bot Gaz 144: 376-381.

Evans LS (1980) Foliar responses that may determine plant injury by simulated acid rain. In: Polluted Rain. pp 239-257.

Feder WA (1981) Health Perspectives. 37: 117-123.

Houston DB, Dochinger LS (1977) Environ Pollut 12: 1-5.

Houston DB, Stairs GR (1973) Forest Science 19: 267-271.

Karnosky DF, Stairs GR (1974) J Environ Qual 3: 406-409

Masaru N, Katsuhisa F, Sankichi T, Yukata W (1980) Environ Pollut 21: 51-57

Mulcahy DL (1979) Science. 206: 20-23

Murdy WH, Ragsdale HL (1980) J Environ Qual 9: 493-496

# Alfalfa Pollen and Callus Responses to Fusarium

D.E. ROWE, D.L. STORTZ, D.S. GILLETTE[1]

Fusarium wilt caused by Fusarium oxysporum Schlecht f. sp. medicaginis (Weimer) Snyd. and Hans. is an economically important disease affecting the root and crowns of alfalfa (Medicago sativa L.). The genetics for resistance appear to be a dominant gene at one locus and partially dominant gene at a second unlinked locus (Hijano et. al. 1983). Due to the dominance, many different genotypes can give a resistant phenotype and, obviously, some genotypes are of much greater value in advancing generations to a breeding program than others. Sexually produced progenies from either selfing or crossing may not accurately reflect differences in genotypes due to a naturally occuring phenomena in the autotetraploid called gametic disequilibrium (Rowe and Hill, 1983). Thus a search continues for phenotypic measures which more accurately reflect the parental genotypes and thus allow prediction or estimate of the value of a plant as a parent in a synthetic variety. The possibility of using measures on the diploid pollen of the autotetraploid alfalfa is not only of scientific interest, but potentially has value as a tool for plant breeding.

A crude extract of F. oxysporum has been used to develop resistant sectors in callus culture from which disease resistant plants were forthcoming (Hartman et. al. 1984). The effectiveness of this selection procedure indicated that the extract derived from F. oxysporium liquid cultures was involved in some manner, though not understood, in the dis-

---

[1] USDA-ARS, University of Nevada, Reno, NV 89557 USA

ease causing mechanism.

This current study was initiated to explore the possibility of using measurements made on pollen as better measures of the genotypic value of plants. Since susceptibility and resistance responses are not as easily defined at the pollen level as they are at the callus level and in the intact plant, comparisons were made with 25 clones for the in vitro responses of pollen to a F. oxysporum extract, the in vitro responses of callus to the F. oxysporum extract, and in the in vivo responses to the disease causing organism.

## Materials and Methods

Plants selected for this study are from three diverse sources and were initially selected for callus production. The evaluations were made on ten plants from the cultivar 'Moapa 69', (plants prefixed with Mo in Table 1) eight from the cultivar 'Narragansett' (prefixed Na), and seven from a plant introduction germplasm (prefixed I) from Hungary (PI 433639).

### Fusarium Filtrate Production

The procedure for filtrate production was a modification of the procedure described by Hartman et. al. (1984). Two isolates of F. oxysporum f. sp. medicaginis (GH7 and 983) were maintained in soil tubes. Liquid cultures of either isolate were grown in 1 L flasks inoculated with one inoculating loop of soil per flask. The cultures were grown for 14 days at $28^{o}C \pm 1^{o}C$. The media was then filtered sequentially through a No. 541 Whatman filter, a glass fiber filter, and a 1 μm polycarbonate membrane filter. The filtrate was autoclaved at $120^{o}C$ for 20 minutes.

### In vitro pollen measurements

The percent of pollen tube germination and average length of germinated tubes were determined for each plant media with and without Fusarium filtrate. The media used was a modification of that of Barnes and Cleveland (1963). A liter of media had 970 ml of distilled water, 120 g of sucrose, and either 30 ml of the Fusarium filtrate or as a control 30 ml of Czapek-Dox broth.

Plants were maintained in a controlled atmosphere for 24 hours prior to collection of pollen. The pollen was introduced to the plates by tripping florets over the plates. The plates were incubated for four hours at $24^{o}C \pm 1^{o}C$ and then the tubes were killed with acetocarmine stain. These plates were stored at 40% RH and $5^{o}C$ until measurements were completed. For each replicate the germination of 100 random spores and the length of 20 germinated spores, measured with an eyepiece micrometer, were determined. In each run there were two replicates with 3 to 7 runs per clone for each isolate and control depending on the availability of pollen.

## In vitro callus measurements

Alfalfa florets were surface sterilized for 2 minutes in 1.31% sodium hypochlorite and then rinsed twice in sterile water. The ovaries were aseptically dissected and placed on SHDN (Schenk and Hildebrandt 1972) media for 28 days. The callus was then partitioned into 50 mg pieces $\pm$ 10 mg and placed either on SHDN media or SHDN media modified by replacement of 50% of the water with the culture filtrate. Visual scores were made after seven days (for GH7) or weight of callus was measured after 28 days (for 983). There were five pieces of callus per plate with four replications.

## In vivo measurements

The in vivo responses to this pathogen were determined after the procedure of Frosheiser and Barnes (1978). The roots and tops of vegetative propogules of each plant were pruned to 4 inches. The roots were soaked for 20 minutes prior to planting in a homogenate of the 14 day old liquid culture of GH7 on Czapek-Dox broth. The propogules were planted in a sand bench and after six weeks the disease responses were evaluated after the visual criteria of Frosheiser and Barnes (1978). Roots were scored on a scale of 0 to 5 with a 0 being most resistant and 1 being highly susceptible. The average severity index (ASI) for each plant was determined as the average score of its progeny.

Table 1. In vitro pollen and callus responses to culture filtrate and average severity index (ASI) for 25 alfalfa plants. Percent measurements expressed as percents of controls.

| | Isolate 983 | | | Isolate GH7 | | | | |
|---|---|---|---|---|---|---|---|---|
| Entry | Tube Length | Germ | Callus | Tube Length | Germ | Callus | ASI | 28 Day Callus |
| | ----------------% | ---------------- | | | | | | (g) |
| I-1 | 40 | 82 | 46 | 45 | 87 | S | -- | 4.8 |
| I-3 | 38 | 106 | 31 | 38 | 82 | S | 1.2 | 6.3 |
| I-5 | 38 | 94 | 76 | 42 | 67 | S | 1.5 | 6.3 |
| I-7 | 40 | 93 | 21 | 37 | 88 | S | 1.6 | 6.5 |
| I-12 | 42 | 86 | 68 | 49 | 98 | S | 1.8 | 3.1 |
| I-13 | 37 | 77 | 61 | 49 | 101 | S | 1.8 | 8.0 |
| I-14 | 47 | 96 | 47 | 47 | 83 | R | 1.5 | 7.6 |
| MO-6 | 42 | 86 | -- | 44 | 98 | S | 1.0 | -- |
| MO-8 | 49 | 109 | 17 | 54 | 100 | S | 1.5 | 2.9 |
| MO-10 | 48 | 113 | -- | 44 | 84 | S | 1.3 | -- |
| MO-21 | 42 | 104 | 29 | -- | -- | S | 3.5 | 4.8 |
| MO-22 | 45 | 88 | 97 | 46 | 112 | R | -- | 5.9 |
| MO-32 | 49 | 97 | 28 | 48 | 83 | -- | -- | 11.2 |
| MO-35 | 41 | 77 | 57 | 48 | 106 | -- | 1.5 | 7.4 |
| MO-38 | 54 | 96 | 29 | 58 | 92 | R | 1.6 | 4.8 |
| MO-53 | 49 | 122 | 55 | 53 | 86 | I | 1.4 | 2.3 |
| MO-54 | 45 | 117 | 44 | -- | -- | S | -- | 5.1 |
| NA-23 | 44 | 102 | 13 | 51 | 104 | I | 4.0 | 6.1 |
| NA-32 | 42 | 96 | 54 | 42 | 92 | S | 3.8 | 7.2 |
| NA-34 | 40 | 100 | 44 | 39 | 105 | S | 3.4 | 7.1 |
| NA-35 | 40 | 98 | 32 | 42 | 98 | S | 4.0 | 4.9 |
| NA-38 | 43 | 125 | 50 | 42 | 113 | I | -- | 3.7 |
| NA-39 | 41 | 94 | 4 | 40 | 76 | S | 3.0 | 2.9 |
| NA-43 | 50 | 86 | 59 | 55 | 116 | S | 3.2 | 2.7 |
| NA-44 | 43 | 89 | 92 | 49 | 101 | R | 1.0 | 5.7 |
| Means | 43 | 97 | | 46 | 94 | | | |

## RESULTS

The germinations and tube lengths of pollen on media containing the filtrate from either isolate are expressed as percentages of their respective controls (Table 1). The over 100% ratios of germinations are explained as the consequence of both the variability in binomial variables and a slight inhibitory effect associated with Czapek-Dox broth media added to the controls.

The responses of the callus to the Fusarium filtrate are expressed as percentages of the growths on the control media for the 983 isolate but not for GH7. The responses to the GH7 isolate facilitated qualitative interpretations at seven days. When the callus died it was considered to be susceptible (designated S) and when the callus grew on the filtrate containing media apparently unaffected it was considered resistant (designated R). Those few cases where growth ocurred only in some sectors of the callus were considered to have an intermediate level of resistance (I). For calculation of product moment correlations (r) the callus exposed to GH7 isolate was assigned the following values: R=1.0, I=0.5, and S=0.0.

The product of the values for tube length and germ in Table 1 is interpreted as the average length of pollen tubes for 100 random pollen grains. This cross product (CP), though not in the Table 1, was used in calculations of correlations.

The correlations between responses of the plant materials to the two isolates of F. oxysporum were highly significant for tube length (r=0.73) and for callus growth (r=0.64). The correlation of germination responses was not statistically significant. An inspection of Table 1 suggests some interactions of responses over isolates.

The ASI scores are significantly correlated to the callus responses on both the 983 isolate (r=-0.43) and the GH7 isolate (r=-0.5). None of the values for pollen response, including the CP values, are significantly correlated to the ASI or the responses of the callus. But selection of 25% of the population on basis of minimum CP scores with either isolate will result in selection of only the resistant plants (ASI<2.0) with a single exception. For any trait under dominant gene control, the pollen of some heterozygous plants is expected to have a mean inferior to the parent. Interpretation of these results requires the comparison of six week old plants response to the Fusarium, 28 day old undifferentiated callus response to the filtrate, and a four hour response of n=2x pollen to the filtrate. A directional response between assays is not unexpected.

Acknowledgement

The authors thank Dr. T.J. McCoy for his input and the use of his genetic materials.

## References

Barnes DK, Cleveland RW (1963) Pollen tube growth of diploid alfalfa in vitro. Crop Sci 3:291-295.

Frosheiser FI, Barnes DK (1978) Field reaction of artificially inoculated alfalfa populations to the Fusarium and bacterial wilt pathogens alone and in combination. Phytopathology 68:943-946.

Hartman CL, McCoy TJ, Knous TR (1984) Selection of alfalfa (*Medicago sativa*) cell lines and regeneration of plants resistant to the toxin(s) produced by *Fusarium oxysporum* f. sp. *medicaginis*. Plant Sci Letters 34:183-194.

Rowe DE, Hill RR Jr (1984) Effect of gametic disequilibrium on means and on genetic variances of autotetraploid synthetic varieties. Theor Appl Genet 68:69-74.

Schenk RU, Hildebrandt AC (1972) Medium and techniques for induction and growth of monocotyledonous and dicotyledonous plant cell cultures. Can J Bot 50:199-204.

# Induction of Gametic Selection *in situ* by Stylar Application of Selective Agents

C.J. SIMON AND J.C. SANFORD[1]

## Introduction

Efforts to develop a pollen selection system have led to the development of a stylar selection method. This method involves the absorption of selective agent through the exterior of the style such that it can effect pollen as it grows through the style. Results of experiments about to be described indicate that selective agents can absorb into the style where they can affect pollen tube growth, and that progeny from such crosses reproducibly differ from progeny of control crosses.

## Measurement of Pollen Sensitivity to Selective Agents

Initial studies were performed to measure the effect of numerous selective agents on the germination of pollen. These tests were done *in vitro* in a liquid media based on that of Brewbaker and Kwack (1963). The base media consisted of 300 ppm $Ca(NO_3)_2.4H_2O$, 100 ppm $H_3BO_3$, and 10 % sucrose. The potential selective agent was incorporated into this media at a range of concentrations, and tobacco pollen was germinated in it for two hours on a rotating aeration wheel. Samples were then analyzed microscopically to determine the percentage of pollen that germinated. Results of these tests are shown in figure 1. Three general classes of selective agent response were found by this

[1]Graduate student and Assistant Professor, Cornell University, Department of Horticultural Sciences, New York State Agricultural Experiment Station, Geneva, New York, USA, 14456

Figure 1

Effects of numerous selective agents on *in vitro* pollen tube growth. The percent of pollen exhibiting a pollen tube in excess of one pollen tube diameter in length is shown by the Y-axis. Each line on the graphs corresponds to the agent with which it is labeled. The concentration of that agent is shown on the X-axis. Three general classes of agent were found. Fig. 1-A shows those agents that were relatively ineffective in inhibiting pollen germination. These included Glyphosate, Cyclocel, NaCl, Abscisic Acid, Methionine Sulfoximine, and an analog to proline. Fig. 1-B shows the agents having intermediate effect, including Simazine, Sinbar, Giberellic Acid and EDTA. Fig. 1-C shows those agents strongly inhibitory on pollen tube growth, including Picloram, $Al_2[SO_4]_3$, Paraquat and Fusaric Acid.

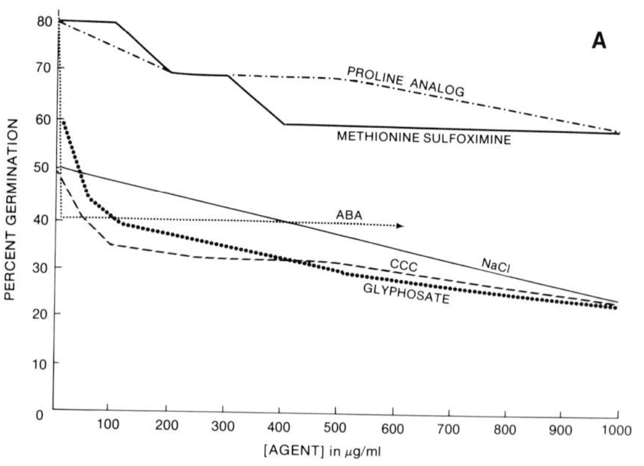


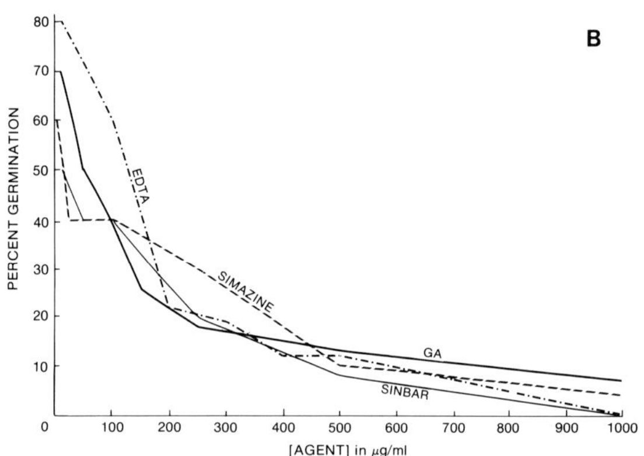


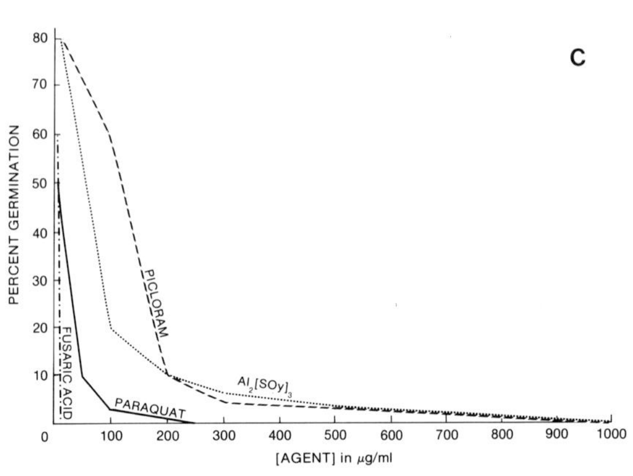

method. Figure 1-A shows those agents that were only weakly inhibitory when incorporated into the germination medium. Figure 1-B shows those agents that had intermediate effect, and figure 1-C showed those agents that were highly effective in inhibiting pollen tube growth *in vitro*. Fusaric acid is represented by the line nearly parallel to the y-axis on figure 1-C. This agent is a chemical associated with fusarium wilt disease (Kern 1972). At 6 ppm all pollen germination was inhibited by fusaric acid (FA). This agent seemed attractive for stylar selection experiments since it appeared that very little agent was necessary to provide strong selection pressure. Therefore, stylar selection tests were conducted with special focus on FA.

## Stylar Selection Experiments

Mature plants of a *Nicotiana langsdorffii* clone were used as females in these experiments. This accession of *N. langsdorffii* was the result of several generations of selfing, making it quite homozygous. Pollen For these experiments was taken from a population of highly heterozygous interspecific $F_1$ hybrids of *N. alata* x *N. sandrae*. FA, with or without a surfactant (Monsanto Corp., Adjuvant 08081) was applied to the entire length of the style with a small camel hair artist's brush. One hour after application of the treatment, fresh pollen collected from the hybrid plants was applied liberally to the stigma.

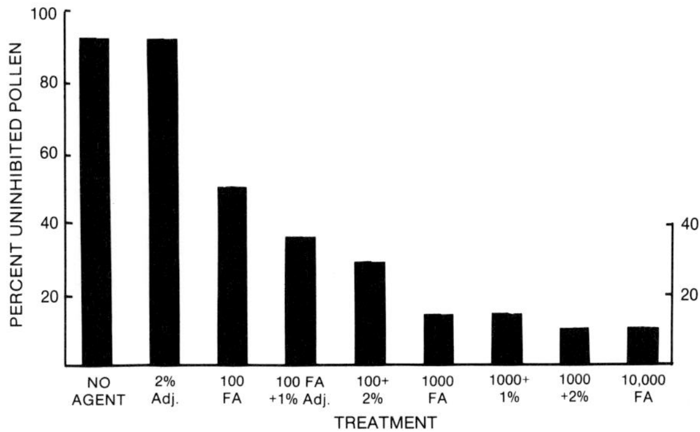


Figure 2. The effect of stylar applications of FA with or without surfactant (Adj) on *in situ* pollen tube growth. Pollen tubes were observed with fluorescent microscopy following the stylar treatment (FA given in ppm, Adj given as %).

Two days after the pollination, the styles were removed and placed in 1N NaOH. They were later stained with analine blue by the method of Kho and Baer (1968), and examined with fluorescent microscopy. The proportion of viable pollen that produced tubes extending the entire length of the style was measured and is shown in figure 2. It is seen that increasing concentrations of fusaric acid resulted in smaller proportions of pollen tubes of full length. It is also seen that the adjuvant enhanced the effect of FA, while the adjuvant alone did not affect pollen tube growth.

## Evaluation of Selected Progeny

Capsules resulting from two separate selection experiments (experiments 1 & 2) were collected and the progeny were grown in completely randomized designs in the greenhouse. A protocol was developed in which nine 9 mm leaf disks were cut from the blades of 6" long developing leaves that were growing from the basal rosette of the plant. The disks were then floated in 30 ml of 100 ppm FA in sealed glass petri plates. There were 3 disks per plate and three plates per plant, with ten plants per treatment. The plates were placed in a randomized complete block design (with three blocks) on a laboratory table top. Approximately one week after cutting the disks, they were scored visually for sensitivity to fusaric acid, based on a 1-5 scale for disk degradation, with '1' being not degraded and '5' being heavily degraded. Scoring was blind and proved to be highly reproducible, both within and across scorers. Analysis of variance showed that the replicates could be pooled. The pooled scores were then analyzed, and a summary of the results of experiments 1 & 2 are shown in figures 3 & 4. The results in figure 3 show that the populations resulting from FA treatments during pollination scored significantly higher than did the controls with this leaf disk assay. The F values for these individual tests were significant beyond the 99% level. The progeny test of experiment 2 (fig. 4) confirmed the findings of experiment 1. The most extreme treatment of FA with the adjuvant, in this test, resulted in progeny that differed significantly from the controls, while the less extreme treatments resulted in progeny that did not differ significantly from the controls. The strongest treatment in this experiment is comparable to the weakest treatment of experiment 1, and is thus interpreted as being the minimum treatment necessary to result in a significant difference under the conditions of this test.

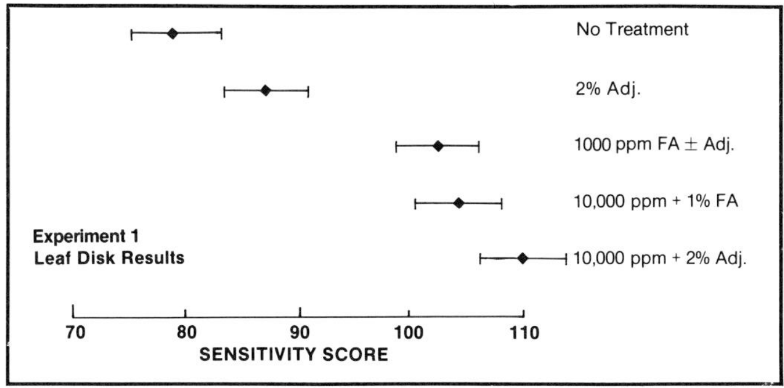


Figure 3. Results of leaf disk screening of progeny from experiment 1. A scale of 1-5 was used to evaluate the disks. Ten plants were screened, with nine disks per plant divided into three blocks. Sum scores for each block were averaged to give the value shown. The composite mean of three scoring tests is represented by the diamond and the error bars represent a 95% confidence interval based on the pooled standard deviation. F values were significant beyond the 99% level.

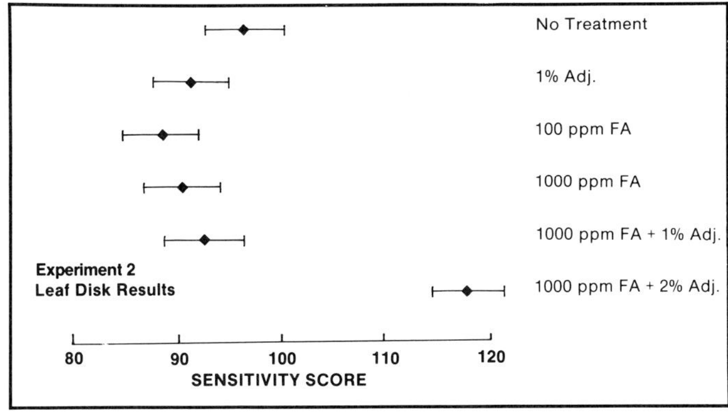


Figure 4. An independently generated population was tested in the manner described above. Scores were generated in precisely the same way. Again, the F value exceeded the 99% level of significance.

Surprisingly, the observed response of the progeny to selection pressure was opposite the expected response. We expected that growing pollen through a high FA environment would result in FA-resistant leaf disks. Instead we found that the disks from treated pollinations were significantly more susceptible to FA. This would suggest either a reverse selection mechanism, or an indirectly correlated response between leaf disk deterioration and pollen tube growth. Further investigation has shown that the leaf disk response of plants to EDTA parallel their response to FA, indicating that the leaf disk assay is probably based on the known chelating properties of FA (Kern 1972). EDTA is much less effective on _in vitro_ pollen tube growth than FA, as shown in figure 1, so the basis of the actual pollen selection process is probably not simply the chelation effect. This may suggest that the leaf disk reaction and the pollen tube response to FA may be physiologically distinct and negatively correlated. Whatever the case, it appears that stylar application of FA does result in a significant change in the progeny population, and this gives us hope that such a stylar selection method may someday be employed as a useful plant breeding method when used with other selection agents.

## Acknowledgments

We would like to thank Karen Knauerhase and John DeMarino for their technical help with this project.

## References

Brewbaker JL, Kwack BH (1963) The essential role of calcium ion in pollen germination and pollen tube growth. Am J Bot. 59:859-865.

Kern H (1972) Phytotoxins produced by Fusaria. In: Wood RKS, Ballio A, Graniti A (eds) Phytotoxins in Plant Diseases. Academic Press, pp 35-48.

Kho YO, Baer J (1968) Observing pollen tubes by means of fluorescence. Euphytica 17:298-302.

# Factors Influencing Artificial Gametophyte Selections using Synthetic Stigmas

ROBERT N. BOWMAN[1]

## Introduction

Demonstration that some genes expressed in sporophyte tissues are also expressed during growth of the male gametophyte (pollen), has led to the prospect of artificially selecting pollen genomes to influence succeeding sporophyte characteristics. As long ago as 1924, Haldane suggested that male gametophytes are cheap genetic experiments, multitudinously produced and each possessing a unique genetic complement. Inherent to this model is the necessity that the female stigma-style tract must be able to discriminate vigorous male genomes from the rest, and further must allow only these "desirable" males to achieve sexual fusion with the waiting eggs buried deep within the ovary. Much evidence (Mulcahy 1974, 1978, 1983) now suggests that the stigma-style tract of angiosperms serves precisely this function of male gametophyte selection.

To achieve artificial selection of pollen, the intact stigma-style tract must be deactivated or eliminated, and surplanted with a suitable replacement capable of sustaining pollen activity while accomplishing selection for desired genetic attributes. In _Clarkia unguiculata_ L. (Onagraceae), which possesses an inferior ovary, the stigma-style apparatus can be surgically decapitated and replaced with a synthetic stigma (Bowman 1984). Outcrossed $F_1$ progeny, ascertained by genetic markers in the parents, are the result of pollen grains having germinated and grown within the synthetic stigma capsule. Stressing agents, such as herbicides, pathogen toxins or salt, may conceivably be included within the capsule to suppress the growth of intolerant pollen genomes; resistant genotypes will achieve fertilization and resultant diploid seeds will be heterozygous for the resistant alleles (Bowman 1985).

---

[1]Botany Department, Colorado State University, Fort Collins, Colorado 80523 USA

**Characterization of Pollen Growth**

Considerable effort has been directed toward optimization of in vitro growth of pollen in both bi- and trinucleate species (Brewbaker and Kwack 1963; Roberts et al. 1983). Despite this effort, in vitro pollen growth in many species is typically much lower than that achieved in stigmatic pollinations (Heslop-Harrison, 1975). Pollen germination and growth involves complex recognition factors between the pollen grain and stigmatic environment including adhesion, surface recognition, and hydration (Dumas and Gaude 1981; Knox 1984). Osmoregulation of pollen growth is potentially influenced by contribution of solutes and water from the female tract. Undoubtedly, the failure of wide pollinations results most often from relatively crude maladjustments, as in the mismatch of water tensions in the stigma and alighted pollen which can either prevent inital hydration or bring about osmotic bursting of the grains (Heslop-Harrison 1975, 1979). To effect fertilization, synthetic stigmas must match the environmental parameters normally encountered in situ. Figure 1 illustrates the in vitro growth response of Clarkia unguiculata pollen to varying osmotic strengths, regulated by sucrose concentration. Table 1 indicates the relative success of synthetic stigma pollinations with pollination capsules containing similiar sucrose concentrations. Clearly, in vitro assays are, at best, only crude estimations of the pollen environment necessary to achieve fertilization. It is noteworthy that most standard in vitro assays for pollen growth (cf. Mulcahy and Mulcahy 1983) are

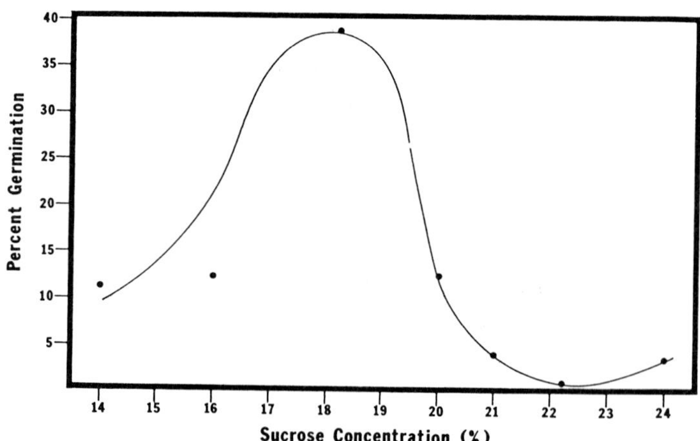


Fig. 1. Germination of Clarkia unguiculata pollen on media (Bowman 1984) with varying sucrose concentrations. Each point is an average of six trials from three different plants. Pollen was germinated at 23 C for 3 hours.

Table 1. Successful synthetic stigma pollinations in *Clarkia unguiculata* with media (Bowman 1984) of varying sucrose concentrations. Parentheses indicate number of attempted pollinations. A successful fruit was one that contained any outcrossed seeds.

| sucrose concentration | 16% | 18% | 19% | 20% | 21% |
|---|---|---|---|---|---|
| successful fruits | 0(16) | 0(33) | 0(180) | 23(104) | 22(64) |

made after a few minutes or hours, yet numerous angiosperms temporally separate pollination from fertilization by days, weeks or even months (Willson and Burley 1983). Thus, lower osmotic potentials, allowing for sustained pollen growth without bursting, may better approximate in situ environments than optimal growth rates indicated by in vitro assays.

Specificity of pollen growth requirements reflects the likelihood of successful synthetic stigma pollinations. Given the general difficulty of trinucleate pollen culture, binucleate species may be more amenable to artificial manipulations. Continued refinement of pollen growth media (Lafleur et al. 1981) favoring sperm cell formation will likewise facilitate successful artificial gametophyte selections. Though considerable investigations have centered on stigma receptivity and pollen-style interaction (Dumas et al. 1984), surprisingly little is known about regulation of pollen activity (if any) once the pollen tubes reach the ovary. Buchholz et al. (1932) have reported that pollen tubes within the ovary that fail to encounter an ovule may even reenter the style and grow backwards through the conductive tract. Synthetic stigma pollinations can circumvent all or part of the stigma-style tract, depending upon where the decapitation is made (Bowman 1984).

**Methodology and Fruit Development**

Pollen, by virtue of its meiotic origin, is genetically diverse. The essence of artificial selections via synthetic stigmas is to isolate ovules from pollen grains that are intolerant of the stressing agent; further, fertilization by tolerant male genomes must be accompanied by successful seed and fruit maturation.

Methods of decapitation directly influence successful seed set following synthetic stigma pollinations (Bowman 1984). The position and orientation of decapitation, inherent to the particular species involved, influences the amount of pollen that can be applied to the stump. Bacterial or fungal infection of the stump following post-fertilization removal of the cap may result in fruit abortion. After caps are removed, topical antiseptics can be applied, though best results seem to be obtained with natural callusing of the cut surface. Provided that stigma capsules are left in place for no more than 48 hours, infection does not seem to be a serious problem.

Fruit development in Clarkia unguiculata following synthetic stigma pollinations is only about 23 percent of that obtained with intact stigmatic pollinations. This value appears to be a function of the general procedure rather than deactivation of available pollen by stressing agents, since nonselection trials (Bowman 1984) were not significantly more precocious than selection trials (Bowman 1985). Fruit development is directly influenced by the number of maturing seeds within the ovary. Buchholz et al. (1932), working with Datura, reported that a minimum of about 20-30 ovules must be fertilized in order to ensure fruit maturation. In contrast, C. unguiculata will mature a fruit containing only one seed even though fruits from normal pollinations average 82 seeds. In those species which spontaneously abort fruits with low numbers of seeds (as in tomatoes), supplemental application of auxins may be used to enhance fruit development and seed maturation (Leopold 1964).

The duration of pollen activity is decidedly short in some species, including Clarkia unguiculata. Fertilization occurs approximately 36 hours after pollination. To be effective, pollen selections must occur within this window. Localization of stressing agents such as herbicides within the pollination capsule result in uniform immersion of the pollen grains. Little of the ovarian tissue is exposed to the stressing agent, and further, the source can be readily removed following fertilization. Selection for resistance to the herbicide chlorsulfuron (DuPont) in C. unguiculata (Bowman 1985) was accomplished without damage to the ovary, presumably because the compound is readily translocated (thus diluted) and little was available to the sporophyte. The capacity to uniformally and totally immerse the genetic dissemule within the selection medium is significant; a frequent criticism of in vitro tissue culture selections (except for suspension cultures) is that all cells are not equally exposed to the stressing agent (Handa et al. 1983). Synthetic stigmas make pollen genomes susceptible to exogenously supplied stressing agents yet interfere as little as possible with the normal course of seed development. Other techniques including direct in vitro pollination of ovules or ovary explants (Zenkteler 1980) must deal with regeneration problems in addition to achieving the desired selections.

## Traits Subject to Selection

Theoretically, a wide spectrum of traits can be selected in the male gametophyte and expressed in subsequent sporophyte generations, provided that the genes controlling the desired trait are active in both phases. Willing et al. (1984), Tanksley et al. (1981) and others have shown that the degree of genetic overlap between the two phases is extensive, encompassing 60 percent or more of the total genome. Fortunately, many desired traits including resistance to several herbicides, toxins or salt are probably controlled be genes regulating basic aspects of cell metabolism. The likelihood that these genes are expressed in both phases is high. With regard to salt tolerance, Sacher et al. (1983) were able to select for salt resistant pollen in Lycopersicon hybrids by immersing pollen in salt solutions prior to pollination. Resistance to salt was demonstrated in the succeeding sporophyte generation.

## Future Prospects

That gamete selection is possible is not surprising, given the fecundity of gamete production in sexually reproducing animals and plants. When precise single-target interference of basic metabolism can be achieved, as is often the case with herbicides, there is real possibility to accomplish, and demonstrate gamete (or in plants, gametophyte) selections. Resistance to multiple-target effectors such as toxins or salt can also be accomplished, though demonstration and quantification can be difficult. Synthetic stigma selection for salt tolerance is intriguing since pollen function appears to be governed by osmoregulation, itself highly regulated by exceptionally high sucrose concentrations. Possibly salt can be added as a stressing agent to the pollen growth medium, with a concurrent reduction in sucrose to maintain an optimal osmotic potential. Demonstration that these selections are possible remain to be experimentally documented.

## References

Bowman RN (1984) Experimental non-stigmatic pollinations in Clarkia unguiculata Lindl. (Onagraceae). Am J Bot 71:1338-1346

Bowman RN (1985) Synthetic stigmas and artificial selection for herbicide resistance (submitted for publication)

Brewbaker JL, Majumder SK (1961) Cultural studies of the pollen population effect and the self-incompatibility inhibition. Am J Bot 48:457-464

Brewbaker JL, Kwack BH (1963) The essential role of calcium ion in pollen germination and pollen tube growth. Am J Bot 50:859-865

Buchholz JT, Doak CC, Blakeslee AF (1932) Control of gametophytic selection in Datura through shortening and splicing of styles. Bull Torr Bot Club 59:109-118

Dumas CD, Gaude T (1981) Stigma-pollen recognition and pollen hydration. Phytomorph 31:191-201

Dumas CD, Knox RB, Gaude T (1984) Pollen-pistil recognition: New concepts from electron microscopy and cytochemistry. Int Rev Cytol 90:239-274

Handa S, Bressan RA, Handa AK, Carpita NC, Hasegawa PM (1983) Solutes contributing to osmotic adjustement in cultured plant cells adapted to water stress. Plant Physiol 73:834-843

Heslop-Harrison J (1975) Male gametophyte selection and the pollen stigma interaction. In: Mulcahy DL (ed) Competition in Plants and Animals. Elsevier, New York, pp 177-190

Knox RB (1984) Pollen-pistil interactions. In: Linskens HF, Heslop-Harrison J (eds) Cellular Interactions. Encyc Plant Physiol (new series) 17:508-608

Leopold AC (1964) Plant Growth and Development. McGraw-Hill, New York. pp. 259-281

Mulcahy DL (1974) Adaptive significance of gametic competition. In: Linskens HF (ed) Fertilization in Higher Plants. Elsevier, New York, pp 27-29

Mulcahy DL (1979) The rise of the angiosperms: a genecological factor. Science 206:20-23

Mulcahy DL, Mulcahy GB (1983) Pollen selection: an overview. In: Mulcahy DL, Ottaviano E (eds) Pollen: Biology and Implications for Plant Breeding. Elsevier, New York, pp xv-xvii

Roberts IN, Gaude TC, Harrod G, Dickinson HG (1983) Pollen-stigma interactions in Brassica oleracea: A new pollen germination medium and its use in elucidating the mechanism of self incompatibility. Theor Appl Genet 65:231-238

Sacher RF, Mulcahy DL, Staples RC (1983) Developmental selection during self pollination of Lycopersicon x Solanum $F_1$ for salt tolerance of $F_2$. In: Mulcahy DL and Ottaviano E (eds) Pollen: Biology and Implications for Plant Breeding. Elsevier, New York, pp 329-334

Tanksley SD, Zamir D, Rick CM (1981) Evidence for extensive overlap of sporophytic and gametophytic gene expression in Lycopersicon esculentum. Science 213:453-455

Willing RP, Eisenberg A, Mascarenhas JP (1984) Genes active during pollen development and the construction of cloned cDNA libraries to messenger RNAs from pollen. Incompat Newsl 16:11

Willson MF, Burley N (1983) Mate Choice in Plants. Princeton University Press, New Jersey. pp 93-107

Zenkteler M (1980) Intraovarian and in vitro pollination. Int Rev Cytol Suppl 11B:137-156

# In Vitro Reaction Between Apple Pollen and Apple Scab Fungus (*Venturia inaequalis* Cke. Wint.)

T. VISSER AND Q. VAN DER MEYS[1]

## INTRODUCTION

It would be interesting for the apple breeder aiming at cultivars resistant to scab (*Venturia inaequalis* (Cke.) Wint.), if pollen from scab susceptible or resistant donors reacted *in vitro* upon (diffusates of) scab spores or mycelium or, vice versa, if the scab fungus responded *in vitro* to the presence of pollen. Literature on this subject is scant. FOKKEMA (1971) found that the growth of *Helminthosporium sativum* on rye leaves was enhanced by the presence of rye pollen. *In vitro* trials showed that pollen diffusates did not affect the toxine production, but strongly stimulated the formation of two enzymes, which explained the increased fungus infection of leaves with pollen. CHAR & BHAT (1985) observed that *in vitro* growth of *Sclerospora graminicola* (Sacc.) Schroet. was inhibited by the addition of pollen of an Asterian plant, presumably on account of growth inhibiting chemicals. TRIPATHI et al. (1982), testing the *in vitro* effect of pollen of 54 plant species on *Helminthosporium oryzae* Breda de Haan, found that pollen of 11 plants stimulated and of one (*Xanthium strumarium*) inhibited the pathogen, possibly due to fungitoxic substances. Similarly, PANDEY et al. (1983) tested 45 pollens as to their *in vitro* action on the growth of *H. oryzae* and also of *Alternaria alternata*. Only the pollen of *Lycopersicon lycoperiscum* distinctly inhibited both pathogens, pollen of the other species produced a variable response.

As regards other *in vitro* reactions, specially with *Venturia inaequalis*, NOVEROSKE et al. (1964) showed that an enzyme mixture from apple leaves inhibited the germination of scab spores. On the other hand, RAA (1968), posing that resistance to scab is determined by the growth of the mycelium in the leaves rather than by the germination of spores, found no effect of extracts of leaves of scab resistant and susceptible apple cultivars on

[1]Institute for Horticultural Plant Breeding (IVT)
P. O. Box 16, 6700 AA Wageningen, The Netherlands

mycelium growth. Finally, it is noteworthy that ANDREWS (1983), investigating the antagonism between (50) microorganisms and apple scab, found that only Chactomium globusum reacted most clearly and consistently as an antagonist, resulting (in vitro) from nutrient competition and/or antibiosis.

As mentioned, our interest was in the possibility of an in vitro antagonism between apple pollen and the scab pathogen, specifically with respect to a differentiation between pollen of scab susceptible and scab resistant donors as this may provide the way to "pollen selection" for scab resistance.

## METHODS AND MATERIALS

The experimental scab material originated from a culture provided by the Laboratory of Phytopathology of the Agricultural University of Wageningen. It was also isolated from a spore suspension obtained by shaking scab infected leaves in water. The spores were grown on a "DPA medium", consisting of 20 g dextrose, 4 g potato extract and 15 g agar (oxoid L11) in 1:1 water.

The cultures were made in petri dishes or, occasionally, on object glasses -- coated with the medium -- in petri dishes; they sporulated (and the spores were germinable) after 6-7 wks incubation in the dark at 23°C. Spore germination was carried out under the same conditions and assessed after 2-5 days.

The apple pollen -- from scab resistant (underlined in the Tables) and susceptible cultivars -- used in the trials was 6 mths stored in the deep freeze and subsequently in a desiccator at 10 % RH and 2-3°C. The pollen was always rehydrated prior to the germination tests. Initially these were to be done in a hanging-drop culture of 10 % sucrose and 50 ppm $H_3BO_3$ (VISSER, 1955) but later it was found more practical to use an agar medium of 10 g agar (oxoid L11) per liter and 10 % sucrose and 50 ppm $H_3BO_3$ (MARCUCCI et al., 1982). Results were approximately the same, the mean % (of 9 apple cultivars) germination being 47.0 and 51.3 for the hanging-drop culture and agar medium, respectively. The pollen was either germinated in small petri dishes or on object glasses -- coated with the medium -- placed in petri dishes with wet filter paper to maintain humidity. The germination was assessed after 3-3 ½ hrs at ambient temperature (about 20°C).

Treatments were usually six, sometimes four times replicated; in each replicate the % germination was determined on the basis of 2 counts of 50 grains each.

RESULTS

Effect of scab spores on pollen. As regards the influence of a drop of spore suspension added to the pollen medium, it was first determined whether germination would be affected at all by just a drop of water spread over the surface either 2 hrs or 4 days previously.

As there was no difference between these two periods, only the results of the latter are presented in Table 1. It is seen that in the absence of

Table 1. Percentage germination of dry (-R) and rehydrated (+R) pollen on a medium untreated (-D) or wetted with a drop of water (+D) 4 days previously.

| Pollen Donor | -R | | +R | |
|---|---|---|---|---|
| | -D | +D | -D | +D |
| G. Delicious | 33 | 13 | 63 | 71 |
| Elstar | 36 | 13 | 72 | 62 |

rehydration, germination on the wetted medium was considerably poorer than on the untreated one, and still poorer when compared to the germination of rehydrated pollen which was equally good whether or not the medium's surface had been wetted. From the subsequent trials with a drop of spore suspension --

Table 2. Percentage germination on pollen medium wetted with a drop of water (-) or scab spore suspension (+) 7 days prior to pollen application.

| Pollen Donor | Trial 1 * | | Trial 2 ** | |
|---|---|---|---|---|
| | - | + | - | + |
| G. Delicious | 77 | 72 | 69 | 75 |
| D3 | 21 | 46 | 43 | 40 |
| Elstar | 66 | 54 | -- | -- |
| D11 | 83 | 83 | -- | -- |
| Priscilla | -- | -- | 35 | 34 |
| Mean | 61.8 | 63.8 | 49.0 | 49.7 |

* 250-500 spores/drop
**2000 spores/drop

added well in advance to allow spore germination -- it was observed that the presence or absence of scab spores made no difference to the germination of pollen, whether originating from scab susceptible or resistant (underlined) cultivars (Table 2).

Effect of scab mycelium on pollen. The possible influence of scab mycelium diffusates was ascertained by placing a circular piece (diam 8 mm) of DPA medium with mycelium on the pollen medium for 1 or 4 days. Subsequently, pollen was put on the spot where the mycelium had been (0) and at distances of 1, 2, 3 and 4 cm from this spot. A piece of DPA without mycelium served as a control. It was found that germination (and tube growth) were neither

Table 3. Percentage germination (and tube length x $10^{-2}$ mm) on pollen medium on which a piece of scab medium, without (-) or with (+) scab mycelium, was placed for 1 or 4 days prior to pollen application.

| Pollen Donor | 1 Day | | 4 Days | |
|---|---|---|---|---|
| | - | + | - | + |
| G. Delicious | 74(3.4) | 75(3.6) | 71(6.2) | 72(5.2) |
| Elstar | 47(5.2) | 39(5.5) | 47(5.2) | 44(5.6) |
| Priscilla | 44(1.2) | 47(1.8) | 32(1.3) | 36(1.2) |
| Mean | 55.0(3.3) | 53.7(3.6) | 50.0(4.2) | 50.7(4.0) |

significantly affected by the time that the mycelium was placed on the pollen medium (Table 3), nor by the distance between the pollen and the spot where the mycelium had lain (Fig. 1). There was no reliable difference between pollen of resistant (underlined) or susceptible cultivars.

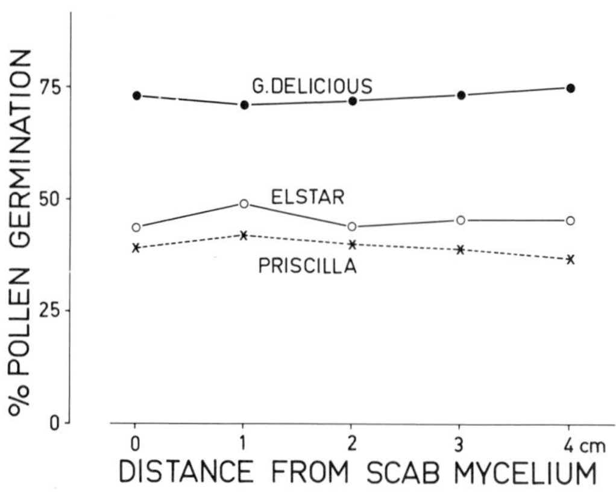


Fig. 1. Germination of pollen in relation to the distance to the spot where a piece of scab medium with mycelium had lain.

Table 4. Germination of spores in various media.

| Medium | Germination |
|---|---|
| Water | 67 % |
| DPA Medium | 60 % |
| Pollen medium | 77 % |

*Effect of pollen on scab spores.* After assessing that scab spores germinated about equally well in water, DPA and pollen medium (Table 4), it was found that the germination of spores on a pollen medium to which viable pollen had been applied 2 days earlier was no different from that of spores without pollen (Table 5). It made neither any difference whether the pollen was from a scab resistant or susceptible cultivar.

Table 5. Percentage germination of scab spores on pollen medium without (-) or with (+) germinated pollen.

| Pollen Donor | - | + |
|---|---|---|
| G. Delicious | 22 | 19 |
| D3 | 36 | 32 |

Effect of pollen on scab mycelium. As regards this effect, it was first investigated how scab and pollen grow on each other's media as well as on a scab medium adjusted to pollen by adding sucrose (10 %) and boron (50 ppm). It appeared (Table 6) that the mycelium had expanded equally well (measured after 7 weeks) on all three media, but it was on DPA much thicker than on the other two. The pollen which scarcely germinated on DPA did much better, though not yet optimally, if sucrose and boron had been added (B). On account of these observations mycelium growth as influenced by pollen was tested on DPA, 1 $cm^2$ of which was covered by 100 mg dead (sterilised) pollen (in this case only; also of pear) and thereafter inoculated (no pollen, served as a control). After 8 and 22 days mycelium growth was found to be equally good, irrespective of the presence of apple or pear pollen.

Table 6. The performance of the scab fungus and pollen in various media.

| Media | Mean diameter of scab mycelium after 7 weeks | Pollen germin. |
|---|---|---|
| A. DPA medium | 4.81 cm | 1 % |
| B. DPA + 10 % sucrose + $H_3BO_3$ | 4.38 cm | 28 % |
| C. Pollen medium | 4.62 cm | 56 % |

## CONCLUSIONS

The reviewed data (see Introduction), being few and diverse, lacked promise as to finding an interaction between a pathogen and the pollen of the same plant species. This was substantiated by the present findings insofar as apple pollen germination was not affected by diffusates of spores or mycelium of apple scab in the agar medium, while vice versa, growth of the scab mycelium and germination of the spores was not influenced by diffusates of pollen in the medium. It made no difference whether the pollen originated from scab resistant or scab susceptible apple cultivars.

Our study emphasized the importance of rehydration of the pollen, particularly when using a "wet" medium. It was further found that scab mycelium grew and spores germinated on the pollen medium; reversely, pollen did not germinate on the scab medium.

REFERENCES


Andrews, J. H., F. M. Berbee & E. V. Nordheim. 1983. Microbial antagonism to the imperfect stage of the apple scab pathogen, Venturia inaequalis. Phytopathology 73(2): 228-234.

Char, M. B. S. & S. S. Bhat. 1975. Antifungal activity of pollen. Naturwissenschaften 62(11): 536.

Fokkema, N. J. 1971. The effect of pollen in the phyllosphere of rye on colonization by saprophytic fungi and on infection by Helminthosporium sativum and other leaf pathogens. Neth. J. Plant. Pathol. 77, Suppl. no. 1: 60 pp.

Marcucci, M. Clara, T. Visser & J. M. van Tuyl. 1982. Pollen and pollination experiments. VI. Heat resistance of pollen. Euphytica 31: 287-290.

Noveroske, R. L., J. Kuc & F. B. Williams. 1964. Oxidation of phloridzin and phloretin related to resistance of Malus to Venturia inaequalis. Phytopathology 54: 92-97.

Pandey, D. K., R. N. Tripathi, R. D. Tripathi & S. N. Dixit. 1983. Fungitoxicity in pollen grains. Grana 22: 31-33.

Raa, J. 1968. Natural resistance of apple plants to Venturia inaequalis. A biochemical study of its mechanism. Dissertation Un. of Utrecht, 99 pp. (Schotanus & Jens, Utrecht N.V.).

Tripathi, R. N., D. N. Pandey, R. D. Tripathi & S. N. Dixit. 1982. Antifungal activity in pollens of some higher plants. Indian Phytopath. 35: 346-348.

Visser, T. 1955. Germination and storage of pollen. Mededelingen landbouwhogeschool 55: 1-68.

# Pollen Selection in Breeding Glasshouse Tomatoes for Low Energy Conditions

A. P. M. den Nijs,[1] B. Maisonneuve,[2] and N. G. Hogenboom[1]

## 1 Introduction

Pollen selection at low T for sporophyte vigour at low T could be valuable for accelerating breeding programs aimed at the development of new tomato cvs which grow and produce well under low temperature and low light intensity in the glasshouse in winter. Zamir et al. (1981) reported T-dependent selective fertilization in L. esculentum pollinated with a mixture of pollen from L. esculentum and low T tolerant L. hirsutum. Several genes of L. hirsutum were preferentially present in the L. esculentum x ( L. esculentum x L. hirsutum) backcross progeny produced after pollination at low T (Zamir et al., 1982).

To check whether this correlation between low T performance of pollen and progeny also holds up within L. esculentum itself, we examined the occurrence of pollen selection at low and normal T in the pollen of $F_1$-hybrids between cvs differing for growth at low T. We also studied in vitro pollen germination and tube growth in relation to T of four pairs of cvs which were used as parents of the $F_1$-hybrids (Maissoneuve and den Nijs, 1984). This report examines the pollination results of two such pairs of cvs with contrasting in vitro pollen germination. More detailed results will be reported elsewhere (Maisonneuve et al., 1985, in prep.).

---

1 Institute for Horticultural Plant Breeding (IVT), P.O. Box 16, 6700 AA Wageningen, the Netherlands

2 Present address: Station d'Amelioration des Plantes, INRA, Route de Saint-Cyr, 78000 Versailles, France

## 2 Materials and Experiments

The pure breeding cvs were chosen on the basis of their different vegetative growth in preliminary experiments. Fast growing cvs Marmande (Ma) and Immuna Prior Beta (I) were crossed with slow growing cv Monalbo (Mo) to produce $F_1$-hybrids (Maisonneuve and den Nijs, 1984). The $F_1$-hybrids were self pollinated at 22$^o$C day/15$^o$C night (normal T=N) and at 15$^o$C day/8$^o$C night (low T=L) in climate rooms under 10 hrs days at 24 W/m$^2$ light intensity. Furthermore, the pollen was collected from plants that had been growing for at least 3 weeks at either normal (N) or low (L) T. Thus 4 pollination treatments were created: NL (pollination at normal T with pollen formed at low T), LL, LN and NN. Some maternal plants were transferred from the normal T climate room to the low T room a few days before the pollinations started. Five days after the last pollination, all plants were set out in a glasshouse.

Vegetative growth of the parents and $F_2$-populations was assessed in a climate room under 8 hrs days at 24 W/m$^2$ light intensity. Temperatures were 20$^o$C day/14$^o$C night during the first 3 weeks, and 19$^o$C day/10$^o$C night during the next 4 weeks. The number of leaves of at least 3 cm length was counted 4 times at weekly intervals, the last date coinciding with the harvest date at 7 weeks from sowing.

The average dry weight per plant at harvest and the variance were compared by ANOVA between the populations within each $F_2$-progeny.

## 3 Results

Self pollinations usually succeeded, except for the LL treatment. Fruits usually weighed more than 60 g at harvest. Only fruits containing 60-150 seeds were used. The most common seed fraction was 2.5-3 mm, irrespective of the pollination treatment. The most common seed fraction was chosen for the growth analyses. The results for total dry weight are presented in Fig. 1

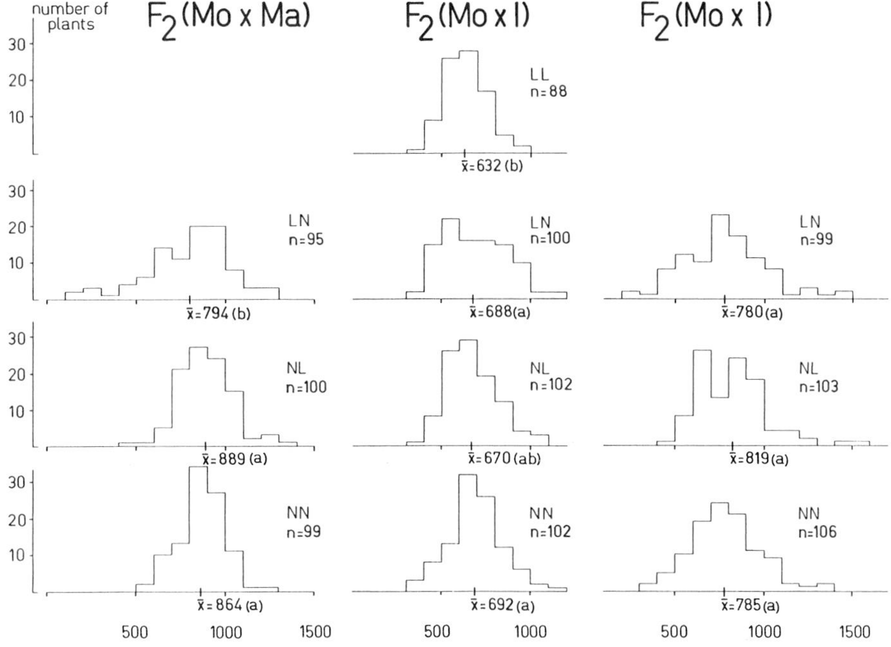


Fig. 1. Distributions over dry weight classes of $F_2$ plants originating from different pollination treatments. Within each column, averages followed by the same letter were not significantly different at p = 0.05

For the $F_2$ (Mo x Ma), average dry weight of the plants from the NL treatment was slightly, but not significantly, higher than that of the NN plants. The average of the LN plants was significantly lower. The average leaf initiation rate of NL plants was 1.98 leaves/week (l/w), slightly more than the 1.91 l/w of both NN and LN populations. The variance for dry weight per plant of the LN population was significantly greater than the variances of both other $F_2$s.

For the $F_2$ (Mo x I) two separate growth analyses were carried out. Average dry weight of NN, NL and LN populations did not significantly differ, but the LL plants had a lower average dry weight. The average leaf initiation rates of the $F_2$s ranged between 1.74 (LN) and 1.80 l/w (NN). The variance for dry weight was again significantly greater for the LN population than for the others in the first analysis. The second experiment was designed in such a way, that effects of pollination date on plant dry weight could be separately analysed. Pollination date significantly affected dry weight (Fig. 2).

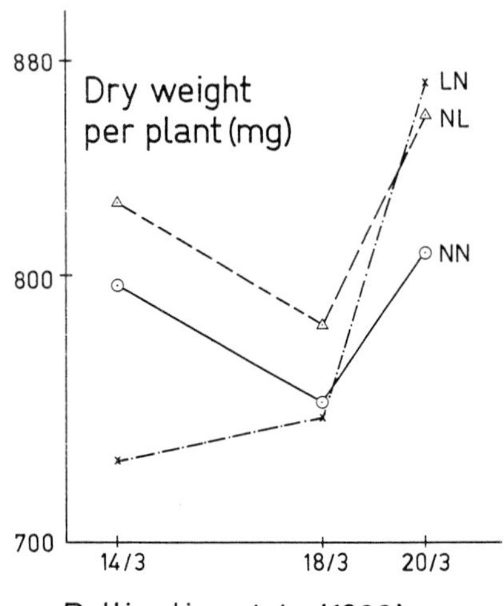


Fig. 2: Effect of pollination date on average of $F_2$ (Mo x I) originating from different pollination treatments.

Analysis of growth of the paired parental cvs showed that Marmande (Ma) had 7.6 leaves of at least 3 cm length after 7 weeks of growth at low T, significantly more than the 6.4 leaves of Monalbo (Mo). Average dry weights of the plants did not differ at this stage (0.57 and 0.58 g, respectively). Of the second pair Monalbo-Immuna Prior Beta, I had 7.9 leaves, significantly more than Mo (7.2 leaves). Dry weight, however, was lower for I (0.62 g) than for Mo (0.83 g).

## 4 Discussion

The average dry weights and leaf initiation rates of the different $F_2$-populations do not indicate differential pollen selection, neither in relation to pollen formation, nor to pollen function temperature. However, pollination date effects likely interfered with possible pollen selection effects. Due to the design, the maternal plants had to remain in the climate rooms during 15 days after the first pollinations. The significantly higher dry weight from the last pollination date in Fig. 2 may result from a shorter time span of the resulting embryos in the climate rooms. The low T is expected to have the most deleterious effect on development, and indeed the dry weight of LN plants increased most.

Seed size can be a complicating source of variation in plant dry weight. Although we mostly used only the most common seed size class, additional larger seed in the second growth analysis of $F_2$ (Mo x I) yielded heavier plants. There was no indication, however, for a systematic effect of pollination treatment on the distribution of seeds over size classes.

For pollen selection to be more effective at low T than at normal T, pollen behaviour must differ at the two Ts. If pollen germination would be an important factor in pollen selection, pollination at low T could provoke a more severe selection in the pollen population produced by the $F_1$ (Mo x Ma), because only at low T did the pollen of the cv with the faster vegetative development (Ma) germinate better than that of the other cv. Pollination T could not be expected to change selection in the pollen of $F_1$ (Mo x I), since at all Ts in vitro germination of I was superior to that of Mo (Maisonneuve and den Nijs, 1984). Indeed we did not find T dependent pollen selection in the latter hybrid. However, for $F_1$ (Mo x Ma) we obtained a significantly lower average dry weight of the LN plants, while leaf initiation rates were equal to that of NN plants.

The disagreement between our results and those of Zamir et al. (1982) could be due to three different factors. First, in Zamir's study the difference in low T tolerance between the parents was very large, indeed. The cvs used in our study differed much less for growth at low T, so differences due to pollen selection, if present, are more difficult to detect. Second, low T depressed pollen germination of _L. hirsutum_ much less than that of _L. esculentum_. This important prerequisite for differential pollen selection was met only in one pair of cvs in our study and even there much less pronounced. Third, the pollen selection temperature in our climate rooms was less extreme than in Zamir's study.

The greater variance for dry weight of the LN populations was at least for $F_2$ (Mo x M) due to rather more smaller than bigger plants. The depressed germination at low T may have left few pollen able to penetrate the ovules. The dry weight of the NN plants was more uniform, and higher than that of $F_1$-hybrid and the parental cvs (Maisonneuve et al., in prep.). This might indicate pollen selection at normal T. The absence of this at low T could be responsible for the more variable LN population (cf. Lewis, 1954). Alternatively, the low T may have led to more variable seed quality.

In conclusion, we have not found exploitable pollen selection at low T favouring sporophytic performance at low T of the resulting progeny, but we could not unambigously distinguish between pre- and post-fertilization events. From the plant breeding point of view, our results do not suggest, that making pollinations at high T for a breeding program designed to adapt the crop to low T may be counterproductive because of selection pressures in opposite directions. Using ample pollen for pollinations appears to be well advised to facilitate possible pollen selection effects.

## References


Lewis D (1954) Pollen competition and sporophytic characters. Ann Rep John Innes Hort Inst 45: 16-17

Maisonneuve B, den Nijs APM (1984) In vitro pollen germination and tube growth of tomato (Lycopersicon esculentum Mill.) and its relation with plant growth. Euphytica 33: 833-840

Maisonneuve B, Hogenboom NG, den Nijs APM, Nilwik HJM (1985) Pollen selection in breeding tomato (Lycopersicon esculentum Mill.) for adaptation to low temperature. Euphytica (in preparation).

Zamir D, Tanksley SD, Jones RA (1981) Low temperature effect on selective fertilization by pollen mixtures of wild and cultivated tomato species. T A G 59: 235-238

Zamir D, Tanksley SD, Jones RA (1982) Haploid selection for low temperature tolerance of tomato pollen. Genetics 101: 129-137

# Biochemical Alterations in the Sexual Partners Resulting from Environmental Conditions before Pollination Regulate Processes after Pollination

M.M.A. van Herpen[1]

## 1. The effects of environmental conditions during maturation of pollen and style on subsequent pollen tube growth

Environmental conditions during the growth of the *Petunia hybrida* plant have an influence on the condition of style and pollen. The time of the year in which the plant cuttings are prepared, the temperature during plant growth and the age of the plant all have an effect on pollen and style interaction during the progamic phase (van Herpen & Linskens 1981). More specifically, the length of the pollen tubes after 24 hours growth on the plant was found to depend for example on the temperature before the progamic phase. It is the kind of pollination (incompatible or compatible) which determines to which extent the conditions of pretreatment of pollen and style contribute to the ultimate length of the pollen tubes after 24 hours of growth on the plant. In the experiments, plants were grown under a day/night cycle of 16/8 h. The temperature at night was 18 $^{o}$C. During the day the temperature was either 25.5 $^{o}$C (high) or 19.5 $^{o}$C (low). The length of the pollen tubes of clones W166K ($S_1S_2$; self-incompatible) and $T_2U$ ($S_3S_3$ also self-incompatible) was measured after 24 h growth in $T_2U$ styles. Pollen developed at the day temperature of 25.5 $^{o}$C make longer pollen tubes than pollen developed at the lower temperature when the differently treated pollen are tested in an incompatible pollination. On the other hand, in a compatible pollination the longest pollen tubes can be found when the styles have developed at the high temperature. Even a single high temperature cycle given prior to the start of the progamic phase to a style matured at the low temperature, is so effective that its impact is vanished only when the time interval between the treatment and pollination exceeds four weeks. Thus, after an

[1] Botanisch Laboratorium, Katholieke Universiteit, Toernooiveld, 6525 ED Nijmegen, The Netherlands.

incompatible pollination the condition of the pollen has an effect on the ultimate pollen tube length; after a compatible pollination however, it is the condition of the style which is the determining factor. The fact that the kind of pollination determines whether the temperature before the progamic phase affects the ultimate length of the pollen tubes either via pollen or via the style, opens the possibility to explore the relation of the incompatible and compatible pollination with temperature as a main variable (van Herpen 1981, 1983, 1984a, b) in different ways.

## 2. The amount of certain compounds in pollen and style in relation to the temperature during development

The protein composition of pollen and style extracts, as studied with immuno-electrophoretic techniques, depends on the day temperature during plant growth. The total amount of protein in pollen was higher when the pollen developed at the high than at the low temperature; styles developed at different temperature cycles don't show a (quantitative) difference in the amount of protein.

The amount of fats is the same in pollen developed at the high or low temperature. Pollen developed at 25.5 $^{o}$C however contains more low-molecular weight carbohydrates and phytic acid than pollen developed at 19.5 $^{o}$C (van Herpen 1984b). A plausible interpretation is that pollen developed at the high temperature has more reserve material and because of that makes longer pollen tubes than pollen developed at the low temperature. The implicit assumption is that in an incompatible pollination no reserves are derived from the style.

The correlation between temperature during style development and the amount of lipids, low-molecular weight carbohydrates and intercellular substance was also investigated: styles developed at the low temperature possess more lipids and low-molecular weight carbohydrates but less intercellular substance than styles developed at the high temperature (van Herpen 1983, 1984b).

The next question was: whether the above mentioned changes in stylar compounds are related to the difference in length of the pollen tubes growing through those styles. An experiment was set up in which the effect of the substances extracted from those styles could be tested on growing pollen tubes (van Herpen 1984a).

## 3. The biochemical alterations in the style before pollination and their effect on processes after pollination

The growth of Petunia clone W166K pollen tubes is inhibited in a medium supplemented with style extracts from clone $T_2U$. After a compatible pollination, no differences could be detected between the length of the pollen tubes growing in extracts comprising the ethanol soluble compounds of unpollinated styles developed at the low and the high temperature (van Herpen 1984a). The length of these pollen tubes is probably not regulated by the amount of low-molecular weight carbohydrates.

High molecular weight extracts from styles developed at 25.5 °C (extracts A) had a less inhibitory effect than extracts from styles developed at 19.5 °C (extracts B). This means that pollen tubes growing in extract A have a greater length than pollen tubes submerged in extract B. After enzymatic breakdown of the proteins, the inhibition by extract A was as strong as that by extract B. Enzymatic proteolysis in extract B had no effect on the length of the pollen tubes.

From these results the conclusion seems justified that prior to anthesis pollen tube growth stimulating proteins are being synthesized in the style when the day temperature is sufficiently high (van Herpen 1984a). These proteins function probably only in a compatible pollination.

In incompatibility research much attention has been given to the malfunctioning of the pollen-pistil relationship. Investigations on the compatible pollination perhaps give more basic insight in the interaction of the two sexual partners.

## References

Herpen MMA van (1981) Effect of season, age and temperature on the protein pattern of pollen and styles in *Petunia hybrida*. Acta Bot Neerl 30: 277-287.

Herpen MMA van (1983) Temperature, *Petunia* stylar ultrastructure and free carbohydrates. Proceedings 7th International Symposium on Fertilization and Embryogenesis in Ovulated Plants. 14-17th June 1982. Račková Dolina (Czechoslovakia). Pp. 227-230.

Herpen MMA van (1984a) Extracts from styles, developed at different temperatures, and their effect on compatibility of *Petunia hybrida* in excised-style culture. Acta Bot Neerl 33:195-203.

Herpen MMA van (1984b) Environment and pollen-style interaction. Thesis, University Nijmegen.

Herpen MMA van, Linskens HF (1981) Effect of season, plant age and temperature during plant growth on compatible and incompatible pollen tube growth in *Petunia hybrida*. Acta Bot Neerl. 30:209-218.

# The Use of In Vitro Methods in the Production of Pollen

R.I. GREYSON, D.R. PAREDDY, V.R. BOMMINENI AND D.B. WALDEN[1]

In contrast to plant tissue culture procedures which promote callus proliferation, cell suspensions and plant propagation, the culture of flower and inflorescence primordia can be designed to favor normal growth and differentiation. The technique has revealed details of the nutritional requirements for development and the biochemical features of organ maturation and differentiation (Tepfer et al., 1962; Raman and Greyson, 1978; Bilderback, 1971). Prominent in these investigations has been the use of the plant growth regulators frequently associated with sexual differentiation (Galun et al. 1962).

The recent demonstration (Polowick and Greyson, 1981 and 1982), that corn tassels can acheive considerable normal growth and differentiation represents an opportunity to carry out some of these explorations on the inflorescence of an important crop plant but as well makes possible a number of potentially practical applications.

Immature tassels, 10-15 mm in length and bearing spikelet primordia in which some florets bear anther primordia, grow and mature in 20-25 days to produce 100 to 200 spikelets essentially normal in all respects. Thus, during the culture period in the flasks the anthers mature and meiosis, microsporogenesis and pollen formation take place (Polowick and Greyson, 1985).

The details of meiosis in cultured anthers resemble closely those of normal anthers, (Polowick and Greyson,

[1]Department of Plant Sciences, The University of Western Ontario, London, Ontario, Canada N6A 5B7

1985) and the pollen germinates on agar (Pareddy et al.1985) and on receptive silks. Fertilization and the production of germinable seed from this in vitro-produced pollen has yet to be demonstrated.

Surprisingly the ear can also serve as a source of pollen, though the details of its formation and its characterization are less well studied. The important requirements for significant anther production on cultured ears include a small (young) initial size and a cytokinin (kinetin) in addition to the usual media components. Though pollen is produced is some of these anthers, no data on the cytology of its formation or on its viability and germinability is available at present.

This demonstration that pollen can be produced from excised inflorescence primordia, under laboratory conditions allows us to contemplate a variety of basic and applied studies. Perhaps, it is now appropriate to return to those studies of a decade or so ago which attempted to unravel the biochemical events in meiosis. In addition, an exploration of the biochemistry and the nutritive requirements of deficient mutants such as the genic male steriles, morphological abnormalities such as the tassel seed mutants, anther ear (an-1), and tunicate (Tu-) mutants can be explored beyond the purely descriptive levels.

The most intriguing opportunities from this work however lie in the potential uses for biotechnology or genetic engineering. As has been suggested by Flavell and Mathias (1984) and others, pollen represents a potential place in the life cycle where heritable variability may be introduced. If so, crops could be modified along desired lines. So far this has not been accomplished and for many grass crops, including corn, no suitable method has been found by which to introduce genetic material.

Inflorescence cultures might represent an alternative vehicle by which genetic material and associated vectors might be introduced into developing microspores and thereby into the pollen, assuming, of course, that pollen from these cultures will fertilize ovules.

Acknowledgements: This work was supported by both Operating and Strategic Grants from the National Engineering and Scientific Research Council of Canada to both R.I. Greyson and D.B. Walden.

REFERENCES

Bilderback DE (1971) The effects of amino acids upon the development of excised floral buds of *Aquilegia*. Amer J Bot 58: 203-208

Flavell R, Mathias R (1984) Prospects for transforming monocot crop plants. Nature 307: 108-109

Galun E, Jung Y, Lang A (1962) Culture and sex modification of male cucumber *in vitro* Nature 194: 596-598

Pareddy DR, Greyson RI, Walden DB (1985) *In vitro* germination of pollen from cultured tassels. Maize Gen Coop Newslett 59: 73-74

Polowick PL, Raman K, Greyson RI (1981) *In vitro* liquid culture of corn tassels. Maize Gen Coop Newslett 55: 116

Polowick PL, Greyson RI (1982) Anther development, meiosis and pollen formation in *Zea* tassels cultured in defined liquid medium. Plant Sci Lett 26: 139-45

Polowick PL, Greyson RI (1985) Microsporogenesis and gametophyte maturation in cultured tassels of *Zea* mays L. Can J Bot (in press)

Raman K, Greyson RI (1978) Further observations on the differential sensitivites to plant growth regulators by cultured 'single' and 'double' flower buds of *Nigella damascena* (Rananculaceae). Amer J Bot 65: 180-191

# Maize Pollen Research: Preliminary Reports from Two Projects Investigating Gamete Selection

D.B. WALDEN AND R.I. GREYSON[1]

Conventional methods of plant breeding have led to variable success in horticultural and agronomically important crop plants during the last half century. The continued quest for further gains in productivity is leading to the testing and possible adoption of complementary and more stringent methods by the breeders.

Most conventional methods include various selection procedures directed towards - or utilizing - the sporophytic generation, which of course is the generation in which the gains are anticipated. With few exceptions, the gametophytic generation is 'passive' in breeding strategies - required for the tranfer of genes and or cytoplasm, but otherwise not involved in the strategies. The challenge remains, however, to determine whether or not the gametophytes could be employed in selection procedures. Gamete selection - processes through which individual or a small sub-population of gametes may be selected and recovered - represents a potentially powerful tool, the dimensions of which are not well understood. Selection of a gametophyte, compared with our earlier demonstration of recovery in the FI (Raman et al. 1980), suggest the feasiability of such procedures.

Central to the proposal of gamete selection is the hypothesis that a gamete or gametophyte may express its own genes rather the phenotype of the parent. Preliminary data

[1]Department of Plant Sciences, The University of Western Ontario, London, Ontario, Canada N6A 5B7

are reported below for two studies which focus on this hypothesis. Both studies utilize pollen from *Zea mays* L.

In the first study, we employ a not well recognized normal cytogenetic event to demonstrate that the size of a pollen grain is a product of its genotype. Many chromosomal observations are available in maize such that the chromosomal site of a specific aberration is at the choice of the investigator. Reciprocal translocations in the heteromorphic - but not homomorphic - state normally produce three segregational patterns: alternate (approximately 50%) and adjacent (50%) I and II. The alternate segregants contain a balanced chromosomal complement and are normal. The adjacent segregants contain one of two kinds of duplication/deficiency (dp/df) complements and are 'unbalanced'. The dp/df condition in maize routinely manifests itself in the male gametophyte by a reduction in size of the pollen grain from a virtually undetectable

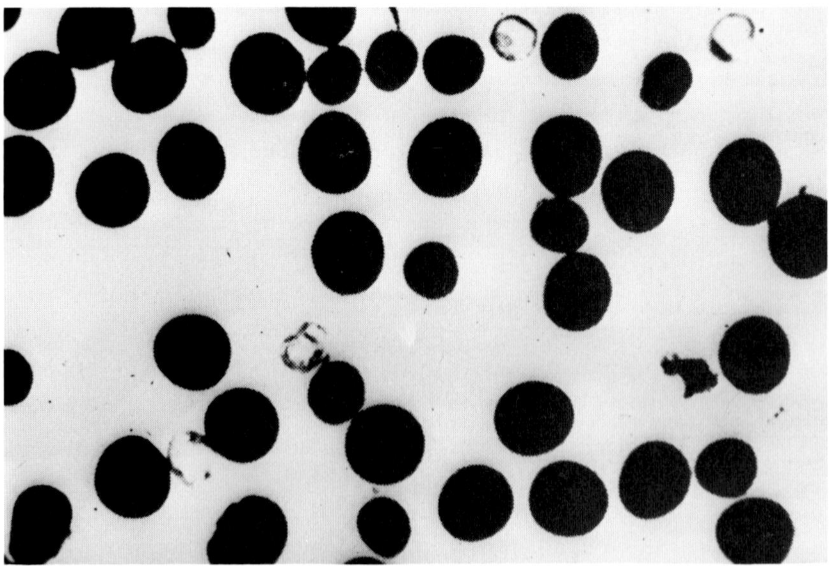

Fig. 1. Photomicrograph of propriocarmine stained maize pollen from a translocation heteromorph.

diminuition to the loss of contents and formation of 'ghosts' (Figure 1). Choice of the specific translocation conditions the size of the adjacent segregants. Very small non-viable pollen grains and ghosts are obtained whenever the site of translocation is not distal. This observation is attributable to the fact that a long interchange segment (the segment lost in an adjacent I segregation) apparently carries genes required for normal male gametophyte formation and function. In other words, such genes appear to be many in number and are distributed around the chromosome complement. However, reduction of the length of the interchange segment (by choice of a translocation site more distal) will yield from some but not all translocations pollen grains of near normal size and functional capacity. In bulk pollinations, such male gametophyte adjacent segregants rarely compete with alternate segregants.

Employment of sparse pollination or enrichment procedures can be introduced to test the fecundability of appropriate adjacent segregants. In our study, we have been testing the distal arms of most chromosomes; the most intensive study has been on arms where distal markers are available. Such recessive markers are incorporated in the female parent, such that if uncovered by the pollination of an adjacent segregant deficient in that segment, the marker phenotype will be expressed in the next generation. such marker plants are then tested for the absence of the specified chromosome segment.

To enrich, we have attempted to take advantage of the pollen size differences. Normal maize pollen (from a diploid) is approximately 100-105 u in diameter. Pollen from a tranlocation heteromorph can be passed through a graded size set of sieves. Pollen collected on each sieve can be employed in sparse polinations, thereby maximizing the likelihood of completing fertilization in the absence of competing gametophytes.

Data from four tests are presented in Table 1. Two features appear in these data: first, no marker progeny have been found in the chromosome IX material, suggesting that an essential factor exists in the telomanic region of

IXs; second, the frequencies of successful transmission are clearly pollen size and translocation specific.

Table 1. The Transmission Frequencies of Selected DP/DF Male Gametophytes of Maize

| SIEVE SIZE | | N | FREQUENCY | |
|---|---|---|---|---|
| CHROMOSOME VI | S.92[a] | | | |
| Control | | 5634 | 0 | |
| 63 u | | 296 | 0 | |
| 76 u | | 1701 | (11) | .0073 |
| 90 u | | 2388 | (4) | .0018 |
| 105 u | | 410 | 0 | |
| 125 u | | 85 | 0 | |
| CHROMOSOME IX | S.96[a] | | | |
| Control | | 7810 | 0 | |
| 63 u | | 1216 | 0 | |
| 76 u | | 11019 | 0 | |
| 90 u | | 12115 | 0 | |
| 105 u | | 697 | 0 | |
| 125 u | | 711 | 0 | |
| CHROMOSOME X | L.89[a] | | | |
| Control | | 1221 | 0 | |
| 63 u | | 110 | 0 | |
| 76 u | | 431 | (34) | .08 |
| 90 u | | 579 | (6) | .01 |
| 105 u | | 128 | (1) | .01 |
| 125 u | | 92 | 0 | |
| CHROMOSOME X | L.94[a] | | | |
| Control | | 2319 | (1) | |
| 63 u | | 191 | (140) | .73 |
| 76 u | | 476 | (45) | .10 |
| 90 u | | 1211 | (15) | .01 |
| 105 u | | 876 | 0 | |
| 125 u | | 110 | 0 | |

[a]Cytological position of the translocation, S = Short Arm, L = Long Arm

In the second study, we have estimated the _in vitro_ response of the male gametophyte from 51 maize genotypes to three concentrations of ten agrichemicals, chosen on the basis of their widespread use on maize or maize fields in

North America. Diffential response of the pollen may indicate: a) unique aspects of the male gametophyte generation; and b) the potential for selection within populations of pollen.

Our (Cook and Walden, 1965; 1967) in vitro culture techniques were employed in an experimental design that included three replicates from each of which were drawn five samples per treatment. An agrichemical was applied to the surface of a 15 cm petri dish 15 minutes before innoculation with pollen. Thirty minutes elapsed at ambient temperature before the replicates were sampled. Five fields per dish were selected at random and photographed. Two operators conducted the experiments over a 30-day period. Two sets of innoculations were made beginning at 1:30 and 3:30 p.m. All data were recorded from the film as numbers of grains within a field (between 25 and 50), the numbers of germinated grains and the length of ten tubes. The percent germination was calculated and submitted to an arc sine transformation for further analysis. An extensive report of this project will appear later.

To permit comparisons among treatments, sets and days, a standard source of pollen (inbred Oh43) was employed at all times. All data were standardized against this control. In each agrichemical genotype treatment, a control and three concentrations of agrichemical were employed: FD = manufacturer's recommended field dose, 0.1 FD and 0.01 FD.

The performance of Oh43 across the experimental period is shown in Figure 2.

All agrichemical/genotype interaction responses were assigned into one of four classes:

Class I Decreased germination at all concentrations of agrichemical

Class II Decreased germination at 0.1 FD and F.D. but not 0.01 FD

Class III Decreased germination at FD but not at 0.1 or 0.01 FD

Class IV no decreased germination

No genotype or agrichemical response was invariate.

The most consistent response across genotypes were to the

insecticide Lannate (89% of the responses were class IV, ie and to the herbicide Banvel (75% class IV).

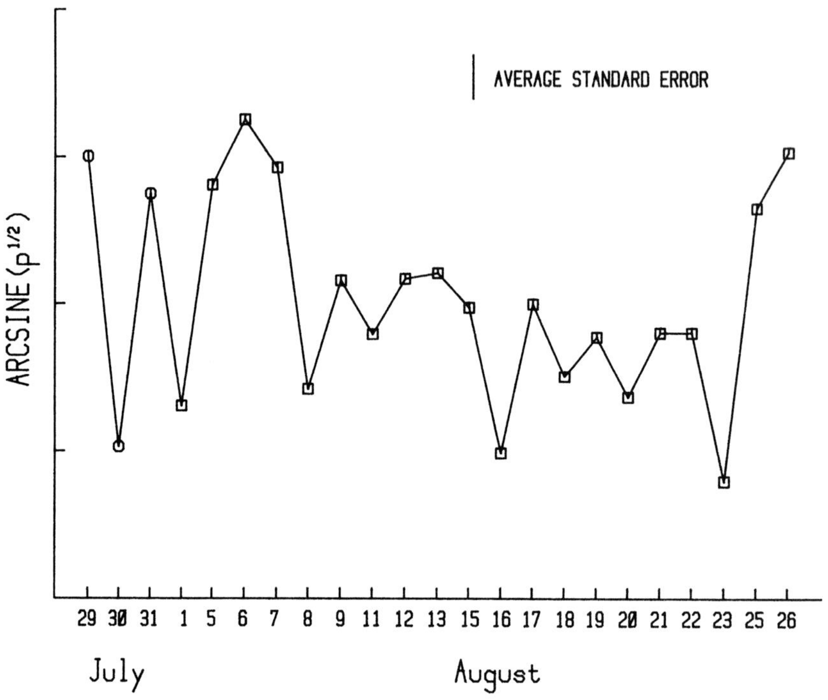


Fig. 2. Performance (arc sine percent germination) of Oh43 pollen during the period July 19 through August 26, 1981.

Approximately 25% of the several hundred genotype/ agrichemical interactions measured were class IV responses. No one genotype was insensitive to all agrichemicals. The remaining 75% of the interactions all demonstrated a dose response (reduced geminations). The insecticide Malathion produced 85% of the responses in Class I or II. The herbicide 2,4-D was unique in that there were no class I interactions and most responses were class III. There was some indication of enhanced germination in the 0.01 FD 2,4-D treatment. Genotype response ranged from those with 90% class I response to all agrichemicals to one yielding 85% class IV responses.

In summary, our second project illustrates the sensitivity of the maize male gametophyte to various chemical stresses. The sensitivity to certin physical and environmental stresses has been demonstrated earlier. The recovery of successful 'stress tolerant' gametophytes has been initiated. We rely on conventional breeding procedures to illustrate the future use for gamete selection.

Acknowledgement: NSERC Strategic and Operating Grant support (D.B.W.) is gratefully acknowledged.

REFERENCES

Cook FS, Walden DB (1965) The male gametophyte of *Zea mays* L. II *In vitro* germinations. Can J Bot 43: 779-786

Cook FS, Walden DB (1967) The male gametophyte of *Zea mays* L. III The influence of temperature and calcium on germination and tube growth. Can J Bot 45: 605-613

Raman K, Walden DB, Greyson RI (1980) Fertilization of *Zea mays* by cultured gametophytes. J Hered 71: 311-314

# Pollen Storage Effects on Early Seedling Growth in Maize

P. L. Pfahler[1]

## 1 Introduction

Maize (*Zea mays* L.) has been subjected to many generations of deliberate sporophytic selection followed, in each generation, by pollination with excess quantities of fresh pollen. Under this pollination procedure, extreme pollen genotype selection and variance reduction for genetic factors affecting pollen competitive ability could occur. Changes in this pollination scheme may amplify pollen transmission differences in these heterogeneous pollen populations previously exposed to pollen genotype selection. Extended pollen storage at 2°C prior to pollination has produced differential pollen transmission of alleles at various qualitative loci in maize (Pfahler 1974b). This storage effect has not been examined for quantitative characters associated with vigor and heterosis. Early seedling growth has been positively correlated with heterozygosity level and vigor expressed later in the life cycle (Whaley 1952).

This study was conducted to determine the effect of pollen storage at 2°C prior to pollination on various seedling growth characters in maize.

## 2 Materials and Methods

On each of two dates, large quantities of pollen grains from at least 30 plants of three single cross hybrids (FR9xH55=F,B73x

---

[1]Agronomy Department, University of Florida, Gainesville, FL 32611, USA

Oh545=B,K64xK55=K) were collected before 900 hours. Then, eight ears of each hybrid were pollinated with an excess amount of pollen grains to produce a $F_2$ population. The remaining pollen grains of each hybrid were then stored at 2°C in an open container. At 1,2,3,4, and 5 days of storage, the pollen grains of each hybrid were allowed to reach room temperature (23°C) and mixed thoroughly. A pollen sample from each hybrid was then used to pollinate eight ears of the same hybrid with an excess of pollen. Seeds of intermediate size and weight from each $F_2$ population were selected and treated with a fungicide (Captain).

These seeds were soaked in deionized water for 24 hours at 2°C and then put in plastic growth pouches (containing 20 ml deionized water) which were placed in closed containers at 25°C. The embryos were positioned vertically so that the shoot and root would grow normally. After 5 days, various measurements (shoot length, primary root length, seedling length=shoot + root length) from the scutellar node were made. At least 85 seedlings from each hybrid-pollen storage-date combination were measured.

Minimum differences for significance (MDS) values were obtained by means of the revised Duncan's ranges using for $P$ only the maximum number of means to be compared (Harter 1960). Variance homogeneity among pollen storage days within each hybrid was tested using Bartlett's test (Snedecor 1956).

## 3 Results

*Shoot length.* No consistent pattern was found with increased pollen storage in any hybrid (Table 1). In F, storage from 1 to 4 days increased the length in comparison to 0 day, but the length at 5 days was slightly lower than 0 day. On the other extreme, increased pollen storage in B decreased the length in no consistent pattern in relation to 0 day.

The variances of the hybrids were very inconsistent (Table 2). In F and K, no significant differences among variances resulting from pollen storge were found. In B, a highly significant difference was present with the variance (1981) at 5 days considerably higher than any of the other 5 storage days. However, the variance (962) at 4 days was considerably lower than any of the other 5 storage days.

Table 1. Mean length (mm) of each hybrid at each storage day

| Character[1] | Hybrid | Pollen storage (day) 0 | 1 | 2 | 3 | 4 | 5 |
|---|---|---|---|---|---|---|---|
| Shoot length | F | 81 | 95 | 89 | 86 | 89 | 78 |
| | B | 106 | 96 | 93 | 95 | 85 | 98 |
| | K | 98 | 100 | 99 | 102 | 105 | 92 |
| Root length | F | 107 | 123 | 120 | 106 | 117 | 89 |
| | B | 146 | 133 | 140 | 134 | 122 | 130 |
| | K | 131 | 118 | 124 | 122 | 122 | 131 |
| Seedling length | F | 189 | 218 | 209 | 193 | 207 | 167 |
| | B | 252 | 229 | 234 | 229 | 208 | 228 |
| | K | 229 | 218 | 223 | 224 | 227 | 223 |

[1]MDS values among any combination of hybrid and pollen storage means at the 5 and 1% level, respectively; shoot length = 15 and 19; root length = 21 and 27; and seedling length = 32 and 42

Table 2. Variance of each hybrid at each storage day

| Character[1] | Hybrid | Pollen storage (day) 0 | 1 | 2 | 3 | 4 | 5 |
|---|---|---|---|---|---|---|---|
| Shoot length | F | 1970 | 2530 | 2124 | 2747 | 2410 | 2839 |
| | B | 1350 | 1276 | 1449 | 1332 | 962 | 1981 |
| | K | 986 | 888 | 1034 | 905 | 1000 | 1093 |
| Root length | F | 4021 | 4185 | 3711 | 5102 | 4518 | 4253 |
| | B | 3164 | 3550 | 3662 | 3283 | 3733 | 3670 |
| | K | 1951 | 2121 | 2558 | 2263 | 1861 | 1763 |
| Seedling length | F | 9823 | 10754 | 9492 | 13771 | 11777 | 12566 |
| | B | 6768 | 7399 | 7829 | 6682 | 6818 | 8722 |
| | K | 4170 | 4451 | 5382 | 4531 | 4038 | 4500 |

[1]Chi square values for variance homogeneity among pollen storage days within each hybrid. Shoot length: F = 5.03, $P$ = 0.05-0.25; B = 17.44, $P$ = less than 0.01; and K = 1.58, $P$ = 0.90-0.90. Root length: F = 2.86, $P$ = 0.75-0.50; B = 3.67, $P$ = 0.75-0.50; and K = 5.40, $P$ = 0.50-0.25. Seedling length: F = 9.41, $P$ = 0.10-0.05; B = 5.57, $P$ = 0.50-0.25; and K = 2.93, $P$ = 0.75-0.50

*Root length*. Pollen storage did not produce a consistent pattern (Table 1). In comparison to 0 day, the length of F was higher at 1, 2, 3, and 4 days of storage, but, at 5 days, the length was considerably lower. In B, pollen storage over 0 day decreased length.

No significant differences among the variances resulting from pollen storage were present within any hybrid (Table 2).

*Seedling length*. Pollen storage exceeding 0 day produced no change or a decreased length in B and K (Table 1). In F, intermediate levels (1-4 days) of pollen storage were longer than 0 day. However, the length of F at 5 days was considerably shorter than 0 day.

No significant differences among variances as a result of pollen storage were present within any hybrid (Table 3).

## 4 Discussion

In this study, pollen storage did not consistently influence the length and variance of shoots, roots, and seedlings of the resulting $F_2$ populations in any hybrid. Apparently, pollen genotype selection was not present for early seedling growth characters.

Our understanding of the male transmission of genetic elements and its relationship to the resulting sporophyte is very limited. With only a few exceptions, the Mendelian assumption that male transmission is independent of the alleles being transmitted was found to be valid. However, at some qualitative loci in maize, differential male transmission was intensified by pollen storage prior to pollination (Pfahler 1974b). At these loci, the competitive ability of pollen grains containing the dominant allele was generally increased with extended pollen storage. Recently, a major factor (relative pollen tube growth rate) of the competitive ability of fresh maize pollen grains was positively correlated with certain quantitative traits presumably associated with heterosis in the resulting sporophyte (Mulcahy 1971; Ottaviano et al. 1980). Since the accumulation of dominant alleles at many loci is now accepted as the primary cause of heterosis (Sprague and Eberhart 1977), differential male transmission of dominant alleles is the most logical explanation. In the study reported here, the resulting sporophyte should have had increased length and reduced variance if increased male transmission of the dominant alleles was occurring and early seedling growth exhibited heterosis. The mode of inheritance of seedling growth characters in maize has not been closely examined. A positive relationship between early growth and heterozygosity level was found when inbred parents were compared to their $F_1$ hybrids (Whaley 1952). However, in rye which closely parallels maize in the degree of inbreeding depression and heterotic response, coleoptile length was not related to vigor expressed later in the life cycle (Pfahler 1974c) and was responsive to both positive and negative selection (Pfahler, 1974a). Thus, heterosis *per se* had limited influence on coleoptile length and this probably was the case with the seedling growth characters measured in this study with maize.

The relationship between the gametophytic and sporophytic generations on the physiological, biochemical, and ontogenetic level is not clear. If pollen genotype selection is to be effective in genetic improvement programs, differences among competing pollen grains resulting from their individual haploid genotype must be present. No obvious differences in $F_1$ variances were found for *in vitro* pollen tube growth in maize even when pronounced differences among the parental inbreds were present (Pfahler 1970). This suggests that the pollen source or diploid effects rather than the haploid genotype of the pollen grain was the prime factor in controlling pollen tube growth. The question arises as to how pollen tube growth *per se* is related to sporophytic processes and responses. For example, drought tolerance in the sporophyte is known to involve numerous diverse and possibly unrelated factors including: stomatal number, size, and response; epidermal condition; root density, efficiency, and depth, and osmotic pressure gradients (Hurd 1971). All these factors are, in turn, influenced by genetic and environmental factors. What aspects of pollen tube growth rate *per se* could be associated with this complex interaction in the sporophyte is difficult to visualize. At this time, pollen storage is not associated with any sporophytic trait. Since the reproductive process in known to reduce the transmission of those pollen grains defective in development because of deletions, duplications and chromosomoal abnormalities (Swanson and Stadler 1955), pollen storage may enhance this process. However, this effect of pollen storage would probably have minimal impact on the mean and variance of the resulting sporophytic population. Before any generalizations as to the extent of differential pollen transmission and its value to genetic improvement programs can be made, this relationship between the gametophytic and sporophytic generations must be more fully understood especially in those species such as maize which have possibly been exposed to intense pollen genotype selection during their development.

*Acknowledgement.* Appreciation is expressed to Beverly Gillis for her skill and diligence in typing and organizing this manuscript and to W. T. Mixon for his outstanding technical assistance.

References

Harter HL (1960) Critical values for Duncan's multiple range test. Biometrics 16:671-685.

Hurd EA (1971) Can we breed for drought resistance? In: Larson KL and Eastin JD (eds.) Drought Injury and Resistance in Crops. Crop Science of America Special Publ. 2, pp. 77-88.

Mulcahy DL (1971) A correlation between gametophytic and sporophytic characteristics in *Zea mays* L. Science 171:1155-1156.

Ottaviano E, Sari-Gorla M, Mulcahy DL (1980) Pollen tube growth rates in *Zea mays:* Implications for genetic improvement in crops. Science 210:437-438.

Pfahler PL (1970) *In vitro* germination and pollen tube growth of maize (*Zea mays*) pollen. III. The effect of pollen genotype and pollen source vigor. Can. J. Bot. 48:111-115.

Pfahler PL (1974a) Effect of selection for coleoptile length in rye, *Secale cereale* L. Euphytica 23:515-520.

Pfahler PL (1974b) Fertilization abilty of maize (*Zea mays* L.) pollen grains. IV. Influence of storage and the alleles at the shrunken, sugary and waxy loci. In Linskens HF (ed) Fertilization in Higher Plants. North Holland, pp. 15-25.

Pfahler PL (1974c) Relationships between coleoptile length and forage production and grain yield in rye, *Secale cereale* L. Euphytica 23:405-410.

Sprague, GF, Eberhart SA (1977) Corn breeding. In Sprague GF (ed) Corn and Corn Improvement. American Society of Agronomy Series 18, pp 305-362.

Snedecor GW (1956) Statistical Methods, Fifth edition. Iowa State Univ. Press, Ames, Iowa.

Swanson CP, Stadler LJ (1955) The effect of ultraviolet radiation on the genes and chromosomes of the higher organisms. In Hollaender A (ed) Radiation Biolgoy Vol. 2 McGraw-Hill, pp 249-284.

Whaley WG (1952) Physiology of gene action in hybrids. In Gowen JW (ed) Heterosis. Iowa State Univ. Press, pp. 98-113.

# Pollen Selection Through Storage: A Tool for Improving True Potato Seed Quality?

N. PALLAIS, P. MALAGAMBA, N. FONG, R. GARCIA, AND P. SCHMIEDICHE[1]

## 1 Introduction

Selection pressure applied to pollen may considerably alter the gene frequency in the following generation of the species in question. The principles of the Hardy-Weinberg Law do not hold true if there is selection pressure at the gametophytic level. Studies on pollen selection mechanisms have increased our understanding of Angiosperm evolution (Baker et al. 1983; Hoeckstra, 1983), and gametophytic selection has facilitated the achievement of breeding objectives (Ottaviano et al. 1982; Zamir et al. 1983).

Pollen selection may become a useful technique for the production of high quality true potato seed (TPS). The use of TPS appears a viable alternative to low quality seed tubers in many regions of the torrid zone (Malagamba, 1984). However, potato seedlings are slow to develop and lack uniformity. Such characteristics need improvement in order to increase the utilization of TPS under sub-optimal agro-ecological conditions. The high level of sporophytic heterogeneity in the tetraploid potato usually results in many undesirable phenotypes in a TPS progeny. Modification of genotype frequencies through pollen selection would benefit progeny performance significantly.

## 2 Materials and Methods

The only pollen source used for all experiments was CIP 376999.6, hereafter referred to as *R*. Female progenitors: CIP 720045, hereafter referred to as *A* and CIP 720125, hereafter referred to as *M*. Two experiments were conducted at CIP (Lima) during 1984. Pollen was collected, mixed thoroughly and stored, or utilized with minimum delay. All pollen was placed in gelatine capsules over silica gel inside sealed black film containers. In the first experiment, pollen stored for 28 days at −12° C (± 4° C) was used to pollinate clones A and M once. The pollen had undergone two different pre-storage treatments: (1) FD (frozen dry) pollen dehydrated 3 hours at +5° C (± 2° C) before freezing and (2) FW (frozen wet) pollen frozen without delay. The pollen thus stored was used for two pollination treatments that corresponded to the two pre-storage treatments. Randomly selected flowers were tagged and emasculated one day before pollination and anthesis. Younger buds and open flowers were pruned. In the second experiment, three lots of pollen were used; 1) pollen stored 1 1/2-2 months at +5° C (± 2° C), 2) the same pollen stored 1 1/2-2 months at − 12° C (± 4° C) and 3) fresh pollen from a later planting. Flowers were emasculated daily and pollinated every other day for two weeks at peak flowering.

Berries were harvested the same day at proper maturity. All TPS was extracted, dried, stored, and evaluated 5-6 months after harvest. Seed size separations were made by bulking the total population of seeds, using standard round screens for width (W) and rectangular wire screens for thickness (T) stacked on a vibrator. Total number of TPS from each sample was counted by hand and weighed on an analytical scale. Before the evaluations, TPS was soaked in gibberellic acid (GA-1500 ppm) for 24 hours to break dormancy unless otherwise indicated. Seed vigor tests were conducted in dark, controlled environments at 15° C and 25° C (± 1° C). For the first experiment, berry and seed weight (wt) was evaluated from 10 single berry samples and from bulked groups of approximately 10 berries each. Seed vigor was evaluated by the slant test method (Smith et al. 1973) five days after start of germination. The germination media were modified to 1/50M in one evaluation (Table 4). For the second experiment, all TPS harvested was bulked, and samples were selected at random. Seed vigor was measured

[1]International Potato Center (CIP) Lima-Peru

daily and expressed as speed of germination (Pinthus et al. 1979; Tekrony et al. 1977). Tests for speed of germination were performed on moist filter paper in closed petri dishes (watered as needed).

## 3 Results and Discussion

The effects of FD and FW pollen are summarized in Table 1. The use of FW pollen resulted in a significant reduction of berry wt and number of seeds/berry when clone M was used as a female. The significant increase in 100-seed wt appeared to be related to the significant decrease in the berry wt and number of seeds/berry. However, calculating the ratio of berry wt to number of seeds/berry (Pallais, unpublished), 100-seed wt increased as this ratio decreased within each treatment indicating that increased availability of maternal tissue does not explain higher seed wt. The correlation coefficients in Table 2 also suggest that a negative effect on seed wt may result from increased berry wt. These results are important since TPS quality has been found to be correlated with increased seed wt (Dayal et al. 1984).

Since there is no conclusive physiological explanation for these results, the observed effects suggest that there has been a screening effect on the pollen genotypes due to the FW treatment. However, there must also have been a maternal effect in the expression of the characters investigated judging from the different response of clone A to the pollen that had received the same FW treatment. The correlation coefficients presented in Table 2 demonstrate significant differences between the response of clone A and the response of clone M to FW pollen also suggesting genotypic effects.

Table 1. Berry weight (g), seed number/berry and 100-seed weight (mg) in two TPS hybrids produced using stored pollen from two pre-storage treatments; dried before freezing (FD) and freezing without drying (FW).

| | A x R | | M x R | |
|---|---|---|---|---|
| | FD | FW | FD | FW |
| Average berry wt. | 11.1 | 9.5 | 15.7** | 9.3 |
| Seed number/berry | 172 | 113 | 179* | 74 |
| 100-seed wt. | 75.4 | 76.0 | 81.0 | 88.8* |

* = $P < 0.05$; ** = $P < 0.01$ (mean separations within treatments)

Table 2. Correlation coefficients for three seed attributes of two TPS hybrids produced using stored pollen from two pre-storage treatments; dried before freezing (FD) and freezing without drying (FW).

| | Average berry weight | | | | 100-seed weight | | | |
|---|---|---|---|---|---|---|---|---|
| | A x R | | M x R | | A x R | | M x R | |
| | FD | FW | FD | FW | FD | FW | FD | FW |
| Seed number/berry | .78 | .55 | .78 | .83 | −.33 | −.39 | −.51 | −.35 |
| Seed wt./berry | .76 | .46 | .82 | .75 | – | – | – | – |
| 100-seed wt. | −.47 | −.20 | −.20 | −.57 | – | – | – | – |

The data of the 100-seed wt, obtained from seed size separations (W; W x T) (Fig. 1a), strongly support the association between FW pollen and increased seed wt for clone M. The data, depicted as the distribution of different seed sizes (Fig. 1a), show that increased seed size (W and T) is the result of using FW pollen. A striking and uniform increase in the proportion of seeds of middle widths ($W_M$) is evident for FW seed obtained from clone M (Fig. 1a). In contrast, the effect on the 100-seed wt of clone A is unclear, and the effect on seed size distribution for FW seed from clone A (Fig. 1b) is less distinct, yet similar to clone M. These observations, however, are insufficient evidence to prove that R pollen genotypes were selected for their ability to withstand freezing in hydrated condition, and a physiological explanation for these results may still be found. Therefore, our hypothesis of a genotype effect requires further discussion. The distribution of seed size depicted on Figs. 1 a-b also suggests that clone A exerts a stronger genetic control over seed size than does clone M. The maternal influence on seed size is powerful in crosses between inbred lines (Harper et al. 1970). However, the regulation of seed size may also be influenced by the paternal contribution to the endosperm (1/3) as well as to the embryo (1/2). Both the magnitude of the paternal influence and the variation in seed size distribution is increased in seeds produced from interspecific and other hybrid combinations (Valentine, 1956). In crosses between sweet and starchy corn, for example, the contribution of pollen to the endosperm determines seed size (Kiesselbach, 1926). The high degree of heterozygosity of many TPS progenitors, which are often the result of interspecific crosses, suggests that

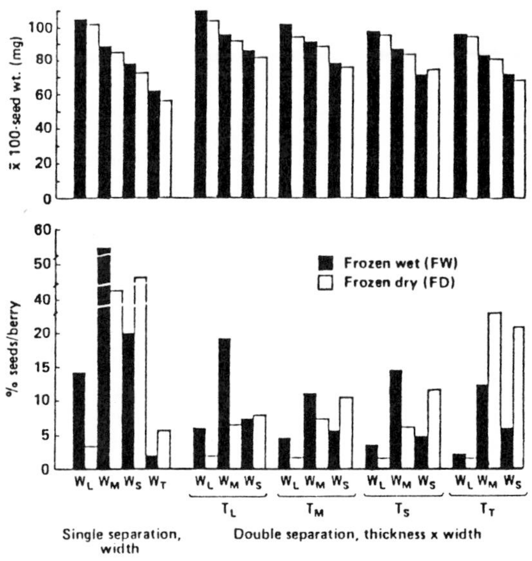


Figure 1a. Effect of 100-seed weight and percent of seeds per berry, of two pollen-storage treatments on seed-size frequency distribution, separated by width (W, mm) and by thickness (T, mm) of a TPS progeny from Mex705383 x R128.6; (M x R). $W_L$: >1.8; $W_M$: 1.8-1.6; $W_S$: 1.6-1.4; $W_T$: <1.4. $T_L$: >0.85; $T_M$: 0.85-0.79; $T_S$: 0.79-0.75; $T_T$: <0.75.

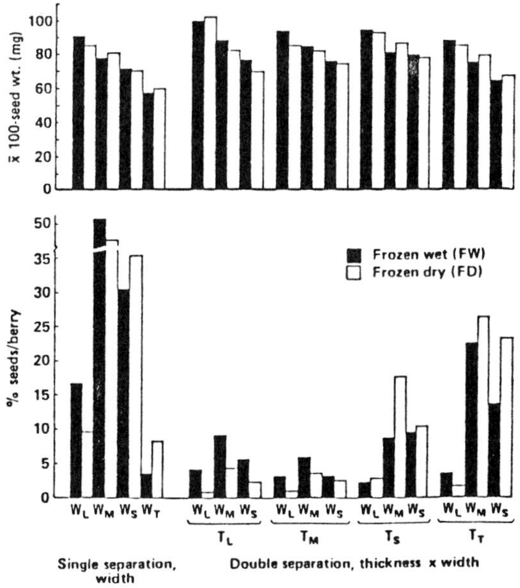


Figure 1b. Effect of 100-seed weight and percent of seeds per berry, of two pollen-storage treatments on seed-size frequency distribution, separated by width (W, mm) and by thickness (T, mm) of a TPS progeny from Atzimba x R128.6; (A x R). $W_L$: >1.8; $W_M$: 1.8-1.6; $W_S$: 1.6-1.4; $W_T$: <1.4. $T_L$: >0.85; $T_M$: 0.85-0.79; $T_S$: 0.79-0.75; $T_T$: <0.75.

specific components of seed size (e.g. width, thickness and length) would be highly heritable as has been shown in oats (Murphy, 1962).

The superior performance of some small seeded progeny selections (CIP, unpublished) indicates that TPS size per se may not be the most important quality factor to be selected for. Uniform and vigorous seedling emergence under sub-optimal conditions is not necessarily a function of increased seed size and weight. Field emergence and vigor, however, are important components of TPS performance and have to be improved. Therefore, increased importance will have to be placed on the genotype—of the sporophyte as well as of the gametophyte—as the ultimate determinant of TPS quality. The data presented in Tables 3 and 4 show a general significant increase in germination and radicle length of FW seed produced on clone M. In contrast, the performance of FD seeds produced on the same clone is generally lower than that of the FW seeds and also less uniform within the sizes tested (Table 3). The data presented in Table 3 were obtained with a germination media that contained salt in a concentration of 1/15M which is high for potatoes. These results are of particular importance as it is known that sub-optimal germination conditions sharply increase the difference between vigorous and non-vigorous seeds (Harrington, 1974). Data in Table 4 show a significant but less dramatic difference between the performance of FW over FD seed in clone M. The data in Table 4 for clone A, however, show a significant increase in germination and radicle growth of the small ($W_S$) (FD) seed. This supports our thesis that the genotype which results from pollen selection is the primary factor in TPS quality improvement, and that seed size associations play a lesser role. In addition, these data suggest that pollen handling techniques for TPS improvements will differ according to the genotype of the progenitor.

Table 3. Effect of two pollen pre-storage treatments on germination (%) and radicle length (mm) in TPS of large, medium, small, and thin width (mm) after 5 days in a slant test under high salt concentration.[1/]

| Pollen pre-storage treatments | $W_L$ = Large (> 1.8 mm) | | $W_M$ = Medium (1.8-1.6 mm) | | $W_S$ = Small (1.6-1.4 mm) | | $W_T$ = Thin (<1.4 mm) | |
|---|---|---|---|---|---|---|---|---|
| | Germ. | Radicle length | Germ. | Radicle length | Germ. | Radicle length | Germ. | Radicle length |
| Frozen but not dried | 85 | 9.6 | 80* | 9.6** | 88 | 8.0** | 83* | 9.4** |
| Frozen, previously dried | 65 | 8.5 | 55 | 9.0 | 75 | 5.0 | 61 | 6.6 |

* = $P < 0.05$; ** = $P < 0.01$ (mean separation within columns)

[1/] TPS progeny used M x R

Table 4. Effect of two pre-storage pollen treatments; dried before freezing (FD) and freezing without drying (FW) on germination of medium ($W_M$) and small ($W_S$) width TPS of equal thickness (> .85 mm) in two progenies (A x R and M x R).

| | A x R | | | | M x R | | | |
|---|---|---|---|---|---|---|---|---|
| | $W_M^a$ 1.8-1.6 | | $W_S$ 1.6-1.4 | | $W_M$ 1.8-1.4 | | $W_S$ 1.6-1.4 | |
| | FD | FW | FD | FW | FD | FW | FD | FW |
| Germination | 80 | 72.5 | 77.5* | 52.5 | 98.75 | 97.5 | 96.25 | 97.5 |
| Radicle length | 12.6 | 10.6 | 14.3* | 10.7 | 11.6 | 18.2* | 9.6 | 15.8* |

* = $P < 0.05$ (mean separations between columns)

[a] See definitions in Table 3

The data in Fig. 2 (second experiment) further support the hypothesis that specific methods of storing pollen have considerable potential as an instrument to manipulate the mean performance of the resulting progeny. The results show the presence of a low level of dormancy as demonstrated by the reaction of the seeds treated with GA. The faster germination rate of the TPS obtained from stored pollen is obviously unaffected by dormancy indicating once more that a screening effect on the genotype of the male gametophyte may have occurred. Still caution has to be excercised when comparing data resulting from the performance of TPS obtained from stored pollen with data from TPS obtained from non-stored pollen. This is due to the fact that the determining factors during pollen formation are difficult to control. The concept of gametophytic screening in pollen should, however, not be limited to different types of storage. Various screening techniques such as the modification of the environment during the formation or function of potato pollen should also be explored.

Additional gains for the improvement of TPS quality may be made by manipulating the competitive, selective interactions during pollen tube growth (Mulcahy, 1974; Yamada, 1983). Our research may result not only in TPS of higher quality but also of higher quantities. Finally, it may be concluded that the gametophyte is not merely a carrier of the haploid genotype but that it plays a significant role in the establishment of the gene frequency of the next generation. Pollen selection, as related to the production of high quality TPS of potato is an ideal subject for further investigation because of its potential for immediate practical application for food production in lesser developed nations.

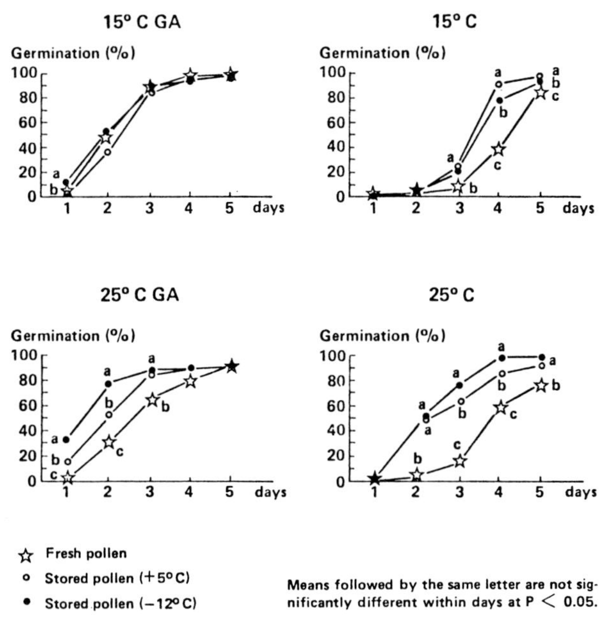


Figure 2. Speed of germination of a TPS progeny (A x R) under two temperatures with or without GA.

*Acknowledgement.* The authors wish to express their gratitude to Dr. Peter Gregory for editing this manuscript.

## REFERENCES

Baker HG, Baker I (1983) Some evolutionary and taxonomic implications of variation in the chemical reserves of pollen. In: Pollen: Biology and Implications for Plant Breeding. Mulcahy DL, Ottaviano E (Editors) Elsevier Biomedical, New York 43-51

Dayal TR, Upadhya MD, Chaturvedi SN (1984) Correlation studies on 1000-true-seed weight, tuber yield and other morphological traits in potato (*Solanum tuberosum* L.) Potato Res 27:185-188

Harper JL, Lovel PH, Moore KG (1970) The shapes and sizes of seeds. Ann Rev of Ecology and Systematics 1:327-356

Harrington JF (1971) The necessity for high quality vegetable seed. Hort Sci 6:2-3

Hoekstra FA (1982) Physiological evolution in Angiosperm pollen: possible role of pollen vigor. In: Pollen: Biology and Implications for Plant Breeding. Mulcahy DL, Ottaviano E (Editors) Elsevier Biomedical, New York 35-41

Kiesselbach TA (1936) The immediate effect of genetic relationship and of parental type upon kernel weight of corn. Nebr Agr Exp Sta Res Bull 33:1-69

Malagamba P (1984) Design and evaluation of different systems of potato production from true seed, Proc of 9th Trien Conf of the Eur Assoc for Potato Res (EAPR). Winiger FA, Stockli A (Editors) Interlaken (Switzerland) 412 pp

Mulcahy DL (1974) Adaptive significance of gametic competition. In: Fertilization of Higher Plants. Linskens HF (Editor) North Holland Publishing Company, Amsterdam 27-30

Murphy CF, Frey KJ (1962) Inheritance and heritability of seed weight and its components in oats. Crop Sci 2:509-512

Ottaviano E, Sari Gorla M, Pe E (1982) Male gametophytic selection in maize. Theor Appl Genet 63:249-254

Pinthus MJ, Kimel U (1979) Speed of germination as a criterion of seed vigor in soybeans. Crop Sci 19:291-292

Smith OE, Welch NC, Little TM (1973) Studies on lettuce seed quality: I. Effect of seed size and weight on vigor. J Amer Soc Hort Sci 98:529-533

Tekrony DM, Egli DB (1977) Relationship between laboratory indices of soybean seed vigor and field emergence. Crop Sci 17:573-577

Valentine DH (1956) Studies in British Primular. Part 5. The inheritance of seed compatibility. New Phytol 55:303-318

Yamada M, Murakami (1983) Superiority in gamete competition of pollen derived from F. plants in maize. In Pollen: Biology and Implications for Plant Breeding. Mulcahy DL, Ottaviano E (Editors) Elsevier Biomedical, New York, 389-395

Zamir, D, Vallejos EG (1983) Temperature effects on haploid selection of tomato microspores and pollen grains. In Pollen: Biology and Implications for Plant Breeding. Mulcahy DL, Ottaviano E (Editors) Elsevier Biomedical, New York, 335-342

# Gametophytic Expression of Heavy Metal Tolerance

KAREN B. SEARCY AND DAVID L. MULCAHY[1]

Pollen selection based on genes expressed by both the gametophyte and sporophyte may be useful for breeding plants adapted to specific environments, and may also be important in the evolution of natural populations. Metal tolerance in pollen was investigated using closely related species, and individuals of the same species tolerant or sensitive to zinc or copper. The study was directed toward three questions. 1. Is tolerance to copper or zinc expressed in pollen? 2. Is it possible to select for metal tolerance in pollen? 3. Is metal tolerance in pollen determined by the haploid pollen or by the sporophytic genome?

## Materials and Methods

The species studied were: *Viola calaminaria* spp. *westfalica* (Lej.) Ernst, *Viola arvensis* Murr, *Silene dioica* (L.) Clairv., *Silene alba* (Mill.) Krause, and *Mimulus guttatus* DC.

*Metal tolerance of pollen*. Pollen was collected shortly after dehiscence from plants grown in standard potting soil. Pollen was hydrated (Shivanna and Heslop-Harrison, 1981) and tested at 6 concentrations of $ZnSO_4$ or

[1]Botany Department, University of Massachusetts, Amherst, Massachusetts 01003 USA

$CuSO_4$ in modified Brewbaker's medium (Brewbaker and Kwack, 1963) containing 1.62 mM $H_3BO_3$, 1.27 mM $Ca(NO_3)_2$, 15 mM Mes, pH 5.5, and either 0.4 M sucrose (*Silene*) or 0.8 M sucrose (*Mimulus* and *Viola*). Pollen germination at 20° (production of pollen tubes 2x the diameter of the pollen grain) was scored for 200-400 grains and tube lengths measured for 50 pollen grains at each concentration. Each experimental series was repeated 3-6 times.

*Selection experiments*. Plants of *Silene* and *Mimulus* were cloned and controls were grown in sand culture in 1/4 strength Hoagland's solution. Treated clones were grown in the same solution amended with additional copper (2 ppm) or zinc (5 or 10 ppm). After 4-6 wks and at the end of the experiments, 2-4 flowers from each clone (3 plants/treatment) were collected, dried at 60° and metal concentration determined by atomic absorption spectrophotometry.

To determine if pollen selection could occur through pollen competition, that is, differential pollen tube growth rates, flowers of the same age from treated and control plants were pollinated using pollen from untreated plants homozygous for metal tolerance or sensitivity. Flowers were placed in an incubator at 20°, removed at intervals and fixed in 70 % ethanol. Styles were cleared in NaOH and length of pollen tubes measured for 2-8 flowers from each combination after staining with analine blue (Martin, 1959).

To test selection during pollen development in the anther, a tolerant individual, heterozygous for tolerance to copper or zinc, was used (Searcy and Mulcahy, in press). Pollen quality in treated and control plants was determined shortly after dehiscence by fluorescein diacetate (Heslop-Harrison and Heslop-Harrison, 1970) or by scoring micro-pollen (*S. dioica*). Crosses to untreated metal sensitive plants were also made using pollen from the treated and control plants. The number of metal tolerant progeny in a sample from two crosses of each type was determined by comparing the ability to produce roots in solutions with 0.5 ppm copper or 2.5 ppm zinc and solutions without metals (Searcy and Mulcahy, in press).

## Results and Discussion

*Expression of metal tolerance in pollen.* For *Silene dioica*, *S. alba* and *Mimulus guttatus*, metal tolerance of the pollen was correlated with that of the pollen source. Germination of pollen from metal sensitive plants was reduced 50 % at concentrations of zinc or copper that were lower than those needed for a similar reduction in germination of pollen from metal tolerant plants (Table 1). In addition, germination of pollen from metal sensitive plants was completely inhibited at concentrations which pollen from tolerant plants could germinate and grow (Table 1). Pollen from zinc tolerant *V. calaminaria* ssp. *westfalica* appeared to require zinc for maximum germination. However, there seemed to be no other relation to the metal tolerance of the pollen source. Pollen germination was reduced by the same concentration of zinc, and some pollen from the zinc tolerant and sensitive species of violet germinated at all concentrations of zinc tested (Table 1).

Table 1. Metal tolerance of the pollen source and pollen germination

| Species | Pollen Source | Pollen Germination concentration (μM/l) Cu or Zn max germ | 50 % germ | 0 % germ |
|---|---|---|---|---|
| *Silene dioica* | Zn tol[1] | 0.01-100 Zn | 200 Zn | 500 Zn |
| *S. alba* | Zn sen | 0 | 100 Zn | 200 Zn |
| *Mimulus guttatus* | Cu tol[2,3] | 0.1 Cu | 100 Cu | 200 Cu |
| *M. guttatus* | Cu sen | 0 | 50 Cu | 100 Cu |
| *Viola calaminaria* | Zn tol[1] | 1-100 Zn | 200 Zn | >1000 Zn |
| *V. arvensis* | Zn sen | 0.01-100 Zn | 200 Zn | >1000 Zn |

Tolerant = tol
Sensitive = sen

1. Kakes (1981)
2. Searcy and Mulcahy (1985)
3. Macnair (1983)

*Selection during pollen tube growth*. An increase from 6 ± 1 to 48 ± 6 ppm Cu in flowers of *M. guttatus* and from 37 ± 4 to 175 ± 35 ppm zinc in styles of *S. dioica* did not affect the relative growth rates of pollen tubes. Pollen tubes from tolerant and sensitive individuals reached the ovaries of treated and control pistils within 5 hrs in *S. dioica* (avg style length 6.7 mm) and by 10 hrs in *M. guttatus* (avg style length 10.2 mm). Thus, selection for metal tolerance through differential pollen tube growth does not appear likely. Since pollen tube growth is important in pollen selection (Mulcahy, 1979), these results were unexpected. It may be that metal concentrations were not sufficiently high or that metals were present in an inactive form or in a location (Thurman, 1981), which would not affect the growth of pollen tubes. However, irradiation of mature pollen has little effect on the growth rate of pollen tubes (Brown and Cave, 1954; Pfahler, 1967), so that pollen growth may not be sensitive to some kinds of stress.

*Selection during pollen development*. The average metal concentration in the flowers increased from 5 ppm in the control to 54 ± 4 ppm copper in treated plants of *M. guttatus* and from 40 to as much as 600 ppm zinc in *S. dioica*. (Note. Concentrations in these segregating clones are higher than in the tolerant homozygotes referred to above.) In both species the increase in metals was accompanied by a 40-45 % reduction in viable pollen. The data from test crosses to metal sensitive females indicated that the reductions were largely selective. That is, pollen from treated plants produced a higher proportion of metal tolerant progeny than did pollen from the untreated plants (Fig. 1).

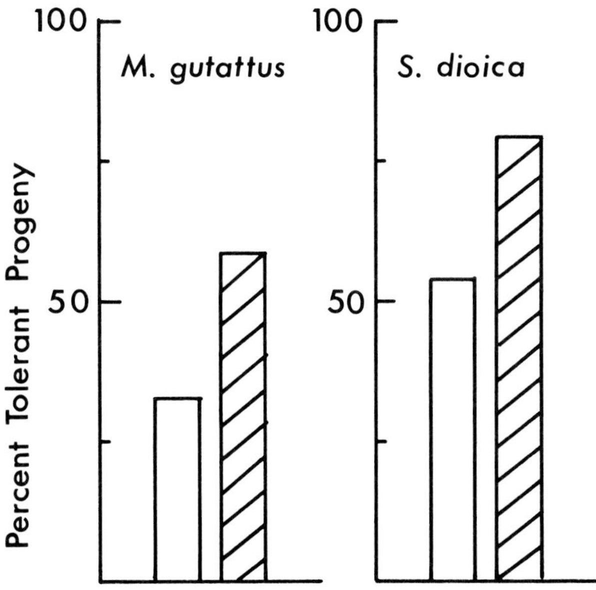


Figure 1. Percent metal tolerant progeny from crosses between a metal sensitive female plant and pollen from untreated (open bars) or treated (hashed bars) pollen sources. Metal tolerance was measured for 82 progeny of M. guttatus and 24 from S. dioica for each type of pollination. The increase in the number of tolerant plants produced by pollen from the treated source is significant for both M. guttatus ($X^2$ = 26.1, $p < 0.001$), and S. dioica ($X^2$ = 5.08, $p < 0.05$).

However, measurements of root lengths (not given here) showed that the maximal level of tolerance did not increase significantly as a result of selection. This, too, indicates that a sensitive subpopulation was eliminated by the treatment.

Since the pollen sources in this experiment were single individuals, heterozygous for metal tolerance genes, the response to selection also indicated that metal tolerance in the pollen of the S. dioica and M. guttatus is largely determined by the pollen genes. Otherwise there could be no phenotypic variation on which selection could act. Thus, in some species, pollen selection for copper or zinc tolerance may be useful in plant breeding and may also contribute to the development and maintenance of metal tolerant plants in natural populations.

Acknowledgment. We thank P. Kakes for seeds of *Viola calaminaria* ssp. *westfalica* and M. R. Macnair for seeds of copper tolerant and sensitive *Mimulus guttatus*.

References

Brewbaker JL, Kwack BH (1963) The essential role of calcium in pollen germination and pollen tube growth. Amer J Bot 50: 859-865

Brown SW, Cave MS (1954) The detection and nature of dominant lethals in *Lilium*. I. Effects of x-rays on the heritable component and functional ability of the pollen grain. Amer J Bot 41: 455-469

Heslop-Harrison J, Heslop-Harrison Y (1970) Evaluation of pollen viability by enzymatically induced fluorescence: intracellular hydrolysis of fluorescein diacetate. Stain Technol 45: 115-120

Kakes P (1981) Genecological investigations of zinc plants. IV. Zinc tolerance of *Viola calaminaria* spp. *westfalica* (Lej.) Ernst, *Viola arvensis* Murr and their hybrids. Acta Oecologica Oecol Plant 2(16): 305-317

Macnair MR (1983) The genetic control of copper tolerance in yellow monkey flower, *Mimulus guttatus*. Heredity 50: 283-293

Martin FW (1959) Staining and observing pollen tubes in the styles by means of fluorescence. Stain Technol 34: 125-128

Mulcahy DL (1979) The rise of the angiosperms: a genecological factor. Science 206: 20-23

Pfahler, PL (1967) Fertilization ability of maize pollen grains. III. Gamma irradiation of mature pollen. Genetics 57: 523-530

Searcy KB, Mulcahy DL (1985) Pollen selection and the gametophytic expression of metal tolerance in *Silene dioica* (L.) Clairv. (Caryophyllaceae) and *Mimulus guttatus* DC. (Scrophulariaceae). Amer J Bot (*in press*)

Shivanna KR, Heslop-Harrison J (1981) Membrane state and pollen viability. Ann Bot 47: 759-770

Thurman DA (1981) Mechanism of metal tolerance in higher plants. In: Lepp NW (ed) Effects of Heavy Metal Pollution on Plants. Vol. II, Applied Sci Publ, Barking, England, pp. 239-249

# Self-Incompatibility And Pollen-Style Interactions

# The Interaction Between Compatible or Compatible and Self-incompatible Pollen of Apple and Pear as Influenced by Pollination Interval and Orchard Temperature

T. VISSER[1]

## INTRODUCTION

The results of mentor or pioneer pollen trials, applying compatible and self-incompatible pollen mixed or in succession, are inevitably influenced by the ambient temperature, though to what extent is unknown. Temperature directly affects flower senescence -- in particular that of the stigmas and embryos -- and the growth rate of the pollen tubes and possibly also the functioning of the style barrier for self-pollen. Irrespective of the temperature, the growth rate of compatible pollen may be faster than that of self-pollen, as has been found by e.g., MODLIBOWSKA (1945) and MARCUCCI & VISSER (1986) for apple and pear.

The picture is more complicated in double pollinations, results of which for apple and pear indicate that the first pollen stimulates the performance of the second (VISSER & VERHAEGH, 1980b). This effect is diminished, reducing the contribution of the second pollen accordingly, by increasing temperature and pollination interval (VISSER & MARCUCCI, 1983). In order to see how differences in temperature affect seed yield of double pollinations with compatible pollen only or with compatible and self-incompatible pollen, the data from dozens of experiments carried out over an eight-year-period have been evaluated.

## METHODS AND MATERIALS

In the period 1975-1983 a total of 56 experiments were carried out in the orchard involving double pollinations, either with compatible pollen applied both time (C/C) or with compatible and self-incompatible pollen applied first or second (C/S or S/C) at an interval of one or two days; a single pollination with compatible pollen served as a control (C). The 56 experiments can be categorised as follows:

[1] Institute for Horticultural Plant Breeding (IVT)
P. O. Box 16, 6700 AA Wageningen, the Netherlands

- 20 with C/C, C/S, S/C (6 on pear, 14 on apple);
- 20 with C/C only (7 on pear, 13 on apple);
- 16 with C/S and/or S/C only (7 on pear, 9 on apple).

Thus there were 40 trials in which C/C and 36 in which C/S and/or S/C figured, comprising together some 25,000 treated flowers. Each treatment/cultivar/trial was represented by 5-10 replicates (shoots) totalling 100-200 flowers (4-5 flowers/cluster). The flowers were depetalled in the balloon stage, soon thereafter pollinated for the first time with freshly collected pollen and subsequently bagged. The second pollination was carried out one or two days later. For each trial the maximum orchard temperature of the first and second day of pollination was recorded, but it is to be noted that on clear days temperatures in the bag were likely to be higher in the bags than outside.

Results were measured by the % fruit set and seed set per fruit, the product of which was called the pollination index (PI = number of seeds produced/pollinated flower; VISSER & VERHAEGH, 1980a) serving as an overall parameter of the efficiency of a pollination treatment. In order to compare the trials, the treatment PI was expressed as a percentage of the PI of the control (= 100 %). The percentages were arranged in temperature classes, based on the mean two-day-maximum temperature per trial, and averaged per class (see Appendix). In a limited number of trials "marker pollen" (VISSER & VERHAEGH, 1980b; VISSER et al., 1983) was used, which made it possible to determine the relative contribution of the first and second pollen in a double pollination. The abbreviations of cultivars and locations used in the Appendix were the following:

Pears---AF = Abate Fetel, BH = Beurré Hardy, BL = Bonne Louise, Con = Condo, CO = Conference, DdC = Doyenne de Comice, GW = Gieser Wildeman.

Apples--Alk = Alkmene, Cox = Cox's Orange Pippin, GD = Golden Delicious, Ida = Idared, JGr = James Grieve, Man = Mantet, Pri = Priscilla, Sta = Starkrimson, Sum = Summerred.

Locations-B = Bologna (Italy), E = Elst, G = Geldermalsen, N = Numansdorp, S = St. Anna Parochie, Wa = Wageningen, W = Wilhelminadorp.

RESULTS, DISCUSSION AND CONCLUSIONS

The general trends for apple and pear were similar so that the data in the Appendix were averaged irrespective of the species. As regards the double pollinations with compatible pollen (C/C), the mean PIs are shown as a function of temperature in Fig. 1. The proportion of seed attributable to the first and second pollen is illustrated in Fig. 2 (data from marker gene trials: VISSER & MARCUCCI, 1983).

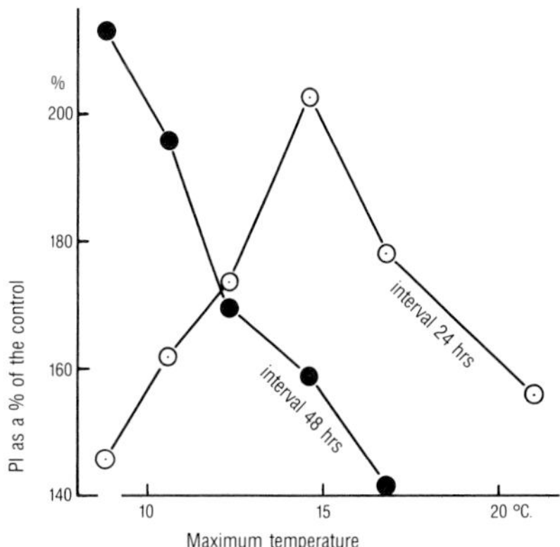


Fig. 1. The relative pollination index (PI = seed set/pollinated flower) of double pollinations with compatible pollen at intervals of one or two days in relation to the mean maximum orchard temperature for the first and second day of pollination.

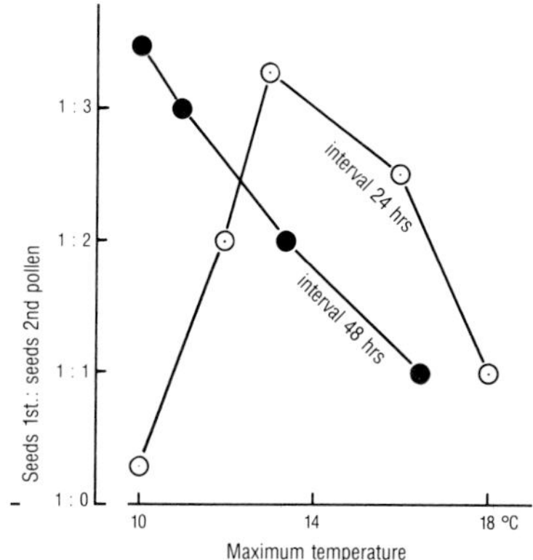


Fig. 2. The ratios between seed set of the first and second pollination of double pollinations with compatible pollen at intervals of one or two days in relation to mean maximum orchard temperature for the first and second day of pollination.

The two figures show very similar pictures. At a pollination interval of two days both the seed production, as expressed by the PI, and the contribution of the second pollen towards this production decreases linearly (significant for P = 0.01). Where there is an interval of one day between pollinations, an optimum curve is obtained, indicating that seed production and the second pollen's contribution to it are maximal around 14°-15°C.

The similarity of the trends presented in these figures suggests that the explanation as to their configuration is the same. It is based on the reasoning that the first applied pollen stimulates ("paves the way for") the second pollen (VISSER & VERHAEGH, 1980b), probably doing so optimally when the first pollen has germinated on the stigma and its tubes have grown no further than about 1/4th of the style at the time of the second pollination (VISSER & MARCUCCI, 1983; VISSER et al., 1983). With an increasing temperature or pollination interval, the first pollen grows increasingly ahead of the second, or, reversely, at a low temperature or short interval it has not yet made much of a start. In both cases this would result in a reduction of the seed production because the contribution of the second pollen is reduced, as is indeed shown in Figs. 1 and 2. Presumably, the seed production would eventually level off to that obtained by a single pollination, but it is noteworthy that this happened rarely; in most trials (3 out of 40) the double pollinations distinctly scored over the single ones (see Appendix).

With respect to the double pollinations with compatible and self

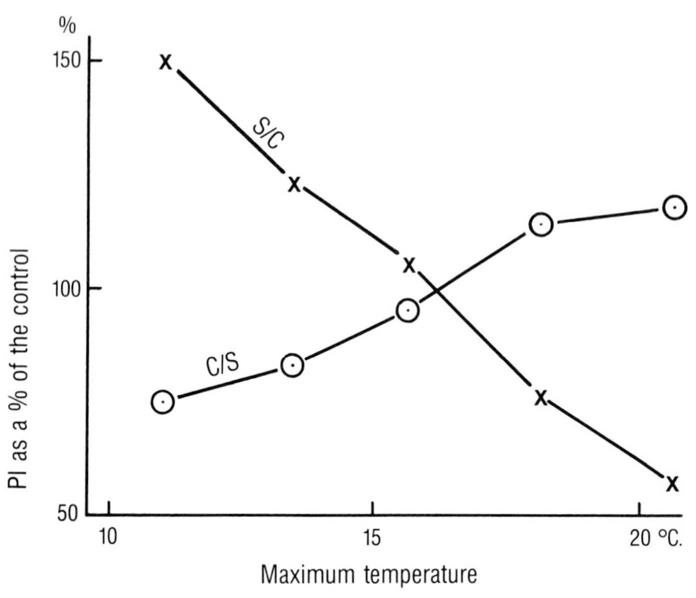


Fig. 3. The relative pollination index (PI = seed set/pollinated flower) of successive cross- and self-pollination (C/S) and the reverse combination (S/C) at a one day interval in relation to mean orchard temperature for the first and second day of pollination (see Appendix).

(incompatible) pollen, the mean PIs of S/C and C/S show reverse trends, the one decreasing, the other increasing with orchard temperature (Fig. 3; see also Appendix). Essentially, they illustrate the same temperature effect as in the aforegoing, taking into account the different behaviour of compatible and self pollen. In S/C the self pollen reverses from stimulating (VISSER et al., 1983) to hindering the compatible pollen (PI above resp. below control), as its "head start" over the latter increases with increasing temperature. Presumably, at relatively low temperature (< 15°C) self-pollen tubes stagnate in the style, while at a higher temperature (15°-20°C) a number of tubes reach the ovules (LOMBARD et al., 1972; BASINA, 1973; WILLIAMS & MAIER, 1977). However, it is a reasonable assumption that a proportion of the selfed embryos abort (VISSER et al., 1983). Due to the resulting lower number of "unoccupied" ovules, it is unlikely that this loss is fully compensated for by the second, compatible, pollinations. Eventually, the linear decrease (significat at P = 0.01) of the PI of S/C is likely to stop at a low level of self-seed only, when at a still higher temperature (or longer interval) the compatible pollen cannot reach the ovules in time due to flower (stigma) senescence. This was also observed in S/C pollinations on *Petunia* of which the seed production decreased (by 33 %) by increasing the interval between self- and cross-pollination (SASTRI & SHIVANNA, 1978). The reverse trend for C/S is similarly explained on the basis of the larger part played by the first, compatible, pollen because with increasing temperature (or interval) its tubes grow increasingly ahead of the self pollen and thereby also lose their "stimulative" capacity. Although the trend can be presented linearly (significant for P = 0.01), logically it is an optimum curve which tops at the level of a single compatible pollination (control) when at a high orchard temperature the self pollen is "outdistanced."

Although differences are small due to the low temperature (11°C), it is in accordance with the above observations that the mean C/S is lower and the mean S/C is higher at a two-day than at a one-day interval (see Appendix).

In conclusion it may be said that the reviewing of all available data has led to a finer point of view than before (VISSER et al., 1983; VISSER, 1984). Whereas it was previously contended that, judging by fruit and seed set, "S/C usually exceeds C/S, the former being often better, the latter poorer than the control" (VISSER, 1984), this now appears to be true only for orchard temperatures below 15°C. In other words, the degree to which these combinations differ from a single pollination largely depends on pollination temperature and interval. This also applies to the outcome of double pollinations with compatible pollen and the contributions of the first and second pollen.

Taking into account the situation in the orchard, it is evident that yield results from many pollen interactions (see also VISSER & MARCUCCI, 1984) in which the self-incompatible pollen plays a distinct role.

REFERENCES


Basina, I. G. 1973. Hort. Abstr. 1974, 4: 7340.

Lombard, P. D., R. R. Williams, K. G. Stott & C. J. Jefferies. 1973. Compt. Ren. Symp. 'Culture due Poirier' 4-8 Sept. 1972. Angers: 266-279.

Marcucci, M. Clara & T. Visser. 1986. Euphytica: in press.

Modlibowska, I. 1945. J. Pom. 21: 57-89.

Sastri, D. C. & K. R. Shivanna. 1978. Incomp. Newsletter 9: 91-93.

Visser, T. 1984. Acta Hort. 109-116.

Visser, T. & M. Clara Marcucci. 1983. Euphytica 32: 703-709.

Visser, T. & M. Clara Marcucci. 1984. Euphytica 33: 699-704.

Visser, T. & J. J. Verhaegh. 1980a. Euphytica 29: 379-383.

Visser, T. & J. J. Verhaegh. 1980b. Euphytica 29: 385-390.

Visser, T., J. J. Verhaegh, M. Clara Marcucci & B. A. Uijtewaal. 1983. Euphytica 32: 57-64.

Williams, R. R. & M. Maier. 1977. J. Hort. Sci. 52: 475-483.

Appendix. The PI (seeds/pollinated flower) of double pollinations at an one or two day interval with compatible pollen (C/C-left) or with compatible and self-incompatible pollen (S/C, C/S-right) in relation to the maximum temperature of the first and second day of pollination; results are expressed as a % of that of a single pollination with compatible pollen, serving as a control.

| Cultivar trial year (location) | Temp. $^{o}$C | Interval 1 day C/C | Interval 2 days C/C |
|---|---|---|---|
| BH '80 (Wa) | 8 | – | 250 |
| Co '80 (E) | 9 | 146 | – |
| DdC '80 (N) | " | – | 256 |
| " '81 (S) | " | – | 100 |
| GD '81 (G) | " | – | 247 |
| Mean | 8.8 | 146 | 213 |
| DdC '80 (E) | 10 | 140 | – |
| " '82 (E) | " | 92 | 77 |
| Sum '82 (E) | " | – | 283 |
| DdC '82 (N) | 11 | 200 | 100 |
| Sum '77 (N) | " | 121 | – |
| " '82 (N) | " | 259 | 306 |
| Cox '81 (G) | " | – | 246 |
| " '81 (W) | " | – | 107 |
| Mean | 10.6 | 162 | 196 |
| BL '79 (E) | 12 | 198 | – |
| Man '76 (E) | " | 162 | – |
| " '76 (E) | " | 125 | – |
| Pri '77 (W) | " | – | 200 |
| JGr '80 (E) | " | 165 | – |
| " '83 (E) | 13 | 174 | – |
| Alk '80 (E) | " | 270 | 170 |
| Sum '80 (E) | " | 123 | 141 |
| Mean | 12.3 | 174 | 170 |
| BL '82 (N) | 14 | 250 | 178 |
| Pri '83 (E) | " | 255 | – |
| GD '80 (E) | " | 118 | – |
| Ida '82 (E) | " | 128 | 107 |
| BL '82 (E) | 15 | 190 | 255 |
| DdC '83 (E) | " | – | 85 |
| Cox '75 (E) | " | – | 169 |
| " '76 (E) | " | 330 | – |
| Sum '83 (E) | " | 151 | – |
| Mean | 14.6 | 203 | 159 |
| Alk '80 (Wa) | 16 | 146 | – |
| Cox '82 (G) | " | 177 | 142 |
| " '80 (E) | 17 | 191 | 138 |
| " '82 (E) | 18 | 200 | – |
| Mean | 16.8 | 178 | 140 |
| GD '79 (E) | 19 | 181 | – |
| DdC '79 (E) | 20 | 138 | – |
| Cox '81 (S) | 22 | 122 | – |
| GD '81 (S) | " | 212 | – |
| " '76 (W) | " | 156 | – |
| " '76 (W) | " | 127 | – |
| Mean | 21.1 | 156 | – |

| Cultivar trial year (location) | Temp. $^{o}$C | Interval 2 days C/S | Interval 2 days S/C |
|---|---|---|---|
| DdC '80 (N) | 9 | 70 | 130 |
| " '81 (S) | " | 101 | – |
| GW '83 (E) | " | 60 | 125 |
| Cox '81 (G) | " | 56 | – |
| " '81 (W) | " | 82 | – |
| GD '81 (W) | " | 55 | – |
| " '81 (G) | 12 | 117 | – |
| " '83 (E) | " | 139 | 152 |
| Mean | 10.8 | 85 | 136 |

| Cultivar trial year (location) | Temp. $^{o}$C | Interval 1 day C/S | Interval 1 day S/C |
|---|---|---|---|
| Co '80 (E) | 9 | 86 | 150 |
| DdC '80 (E) | 10 | 114 | 138 |
| BL '79 (E) | 12 | 49 | – |
| JGr '80 (E) | " | 90 | 172 |
| Ida '81 (E) | " | 43 | 143 |
| Mean | 11.0 | 76 | 151 |
| Alk '80 (E) | 13 | 108 | 107 |
| Sum '80 (E) | " | 86 | 149 |
| GD '80 (E) | 14 | 56 | 104 |
| " '83 (B) | " | 48 | 145 |
| Con '83 (E) | " | 119 | 108 |
| Mean | 13.6 | 83 | 123 |
| Alk '83 (E) | 15 | 58 | 72 |
| Ida '83 (E) | " | 123 | 98 |
| Alk '80 (Wa) | 16 | 100 | 145 |
| Cox '82 (G) | " | 135 | 221 |
| DdC '83 (E) | " | 68 | 35 |
| " '83 (S) | " | 83 | 36 |
| Mean | 15.7 | 95 | 105 |
| Cox '80 (E) | 17 | 106 | 111 |
| " '82 (E) | 18 | 115 | 38 |
| Sta '82 (B) | " | 82 | – |
| DdC '82 (B) | " | 103 | – |
| " '83 (B) | " | 165 | 80 |
| Mean | 17.8 | 114 | 76 |
| GD '79 (E) | 19 | 158 | 116 |
| AF '83 (B) | 20 | 102 | 38 |
| DdC '79 (E) | " | 55 | – |
| GD '82 (E) | " | 103 | – |
| Sta '83 (B) | 22 | 135 | 15 |
| Cox '81 (S) | " | 127 | – |
| GD '81 (S) | " | 146 | – |
| Mean | 20.7 | 118 | 56 |

# Pollen-Pistil Interaction

GABRIELLA BERGAMINI MULCAHY AND DAVID L. MULCAHY[1]

The need, site, and eventually, mode of pollen-pistil interaction; in other words--the extent of compatibility--is the focus of my work. In the case of the angiosperm flower, stigma, style and ovary are the sites where interaction occurs. In binucleate species, no active interaction seems to occur at the stigma site. Konar and Linskens (1966) suggested the stigma acts just as a site for germination of pollen without being involved in its nutrition. Let's skip the style for now and consider the ovary.

The activation of the ovary has long been a subject of interest to researchers: White (1907), Tupy (1961), Linskens (1974), Jensen (1983), Deurenberg (1977), and van der Donk (1974) just to name a few of them.

All the work done so far points to the fact that, well before the pollen reaches it, the ovary responds not only to the fact that pollination has occurred, but also to the type of pollination. The difference in reponse to cross versus self pollination seems to be a quantitative rather than a qualitative one (in many cases anyway). Recognition signals must travel from stigma-style to the ovary. There have been different suggestions on the nature of such a stimulus: an electrical signal was proposed by Linskens and Spanjers (1973) and a hormonal one by Jensen et al. (1983). The pathway, in the case of fast diffusing substances, could

[1]Department of Botany, University of Massachusetts, Amherst, Massachusetts 01003 USA

be the vascular bundles in the style or the vast intercellular connections of the transmitting tissues where plasmodesmata, parallel to the length of the style, are much more numerous than the transverse ones (Gunning, pers. comm.) Rapid intercellular communication in plants is not unique to the pollen ovary interaction: Davis and Shuster (1981) give evidence for a rapid traveling wound signal generating as a response the formation of polysomes and increased protein synthesis.

The activation of the ovary, before actual involvement in the development of the embryo must be responsible, at least in part, in promoting the pollen tube growth. In effect, in a previous work, that is just what we noticed (Mulcahy and Mulcahy, 1985). We cross-pollinated styles of Petunia hybrida with and without ovary attached, and incubated them at 25$^{\circ}$C in a saturated atmosphere. Twenty hrs later we prepared the styles to collect pollen tubes in semivivo. Both sets of styles were cut just ahead of the front of the pollen tubes and the cut end dipped in Brewbaker media. The difference in tube length (calculated as the average of the 10 longest tubes extruding from the cut end of each style, 30 styles for each treatment) was very significant.

Table 1. Pollen tube lengths beyond the cut stylar ends when the first part of growth occurred in pistils with ovaries removed and in pistils with ovaries attached.

| | Mean | S. E. |
|---|---|---|
| Ovary Removed | 1,014 | ± 29 µm |
| Ovary Present | 1,801 | ± 42 µm |

One possible sequence of events, therefore, may be that following pollination: the ovary becomes activated and the activated ovary, in turn, stimulates pollen tube growth. Van der Donk (1975) explained the lack of growth in incompatible tubes in terms of an inability to activate the stylar genome. The question is now, if the activation of the ovary stimulates the pollen tube growth, could an ovary activated by compatible pollination stimulate the growth of the incompatible pollen? I found the application of vital dye very informative in this regard.

The fluorochrome we chose was Bisbenzimidazole dye 33258 Hoechst which binds tightly to DNA (Latt and Wohlleb, 1975). The staining property of such a dye is due to A-T specific binding. Upon binding to DNA, the dye fluorescence intensity is practically independent of pH variation from 4-8 (while free dye fluorescence will decrease 10 fold going from low to high pH values). The dye absorption spectrum peaks were at 260 and 338 nm. The observations were conducted with an American Optic Scope equipped with a mercury vapor lamp, 50 watt, BG 12 and KU 418 exciter filters, 500 nm dichromic beam splitter and OG 515 barrier filters. Dye-germination media contained 0.5 mg dye/ml. Germination media was a simplified Brewbaker and Kwack (1963): 300 ppm $Ca(NO_3)_2$, 100 ppm $H_3BO_3$, 10 % sucrose (12 % being optimum for in vitro pollen germination).

Two clones of _Nicotiana alata_, strictly self-incompatible (selfpollen tubes stopping at 1/3 of the style) were used for the experiments. Pollen to be used for pollination was presoaked in media with and without the dye for 10 min (in a rotary shaker), spun down and resuspended, both stained and control, in control media to remove the free dye. The pollen was then spun down again and used for pollinations.

Styles were pollinated with: compatible pollen presoaked in media without dye; with compatible pollen prestained; with incompatible pollen prestained; and some with 1/2 stigma with unstained compatible pollen and 1/2 stigma with prestained incompatible, simultaneously.

Pollinated styles were incubated for 24 hr at 25$^{o}$C in saturated atmosphere. The styles were then cut just ahead of the pollen tube front and dipped in control media, with 12 % sucrose. Twenty hr later the pollen tubes had grown out of the above cut styles into the germination media. The Hoechst stain, as soon as the cytoplasmic streaming subsided, was clearly visible in the brightly fluorescent nuclei (2 sperm nuclei and a less bright vegetative nucleus).

No tubes protruded from the styles that had been

selfed. No flourescent nuclei were detectable in tubes deriving from the compatible pollination with unstained pollen. In the following tables is reported the percentage of the tubes exhibiting fluorescent nuclei, as seen in two experiments.

In the first experiment there are 8 styles per treatment, and unstained, fresh pollen was used (as control). In the second experiment 5 styles per treatment were used, the control pollen was presoaked in Brewbaker media without dye. In both experiments, more than 50 % of the tubes deriving from mixed pollination exhibited fluorescent nuclei.

Table 2.

| Pollen Source | % Fluorescent Nuclei | S. E. |
|---|---|---|
| 21 May | | |
| Compat, not stained, not soaked | zero | --- |
| Incompat, prestained, presoaked | no tubes | --- |
| Compat, prestained, presoaked | 82.55 | ± 5.815 |
| 1/2 Incompat, prestained, presoaked and 1/2 Compat, unstained, not soaked | 55.127 | ± 6.471 |
| 27 May | | |
| Compat, not stained, presoaked | zero | --- |
| Incompat, prestained, presoaked | no tubes | --- |
| Compat, prestained, presoaked | 39.34 | ± 3.711 |
| 1/2 Incompat, prestained, presoaked and 1/2 Compat, unstained, presoaked | 57.108 | ± 4.745 |

The discrepancy from the optimum 100 % tubes with stained nuclei in the case of pollination with all prestained compatible pollen are due to the fact that even in 10 min of presoaking in a rotary shaker, not all pollen grains become hydrated: air bubbles, clumping of a naturally sticky pollen, adhesion to the vial walls are few of the many factors that will interfere with an homogenous presoaking and prestaining.

To be sure that the dye would not move from one pollen type to the other, in another experiment, we used only compatible pollen: 1/2 prestained and 1/2 just soaked in control media, on 5 styles; and all prestained pollen on other 5.

Table 3.

| Pollen Source | % Stained Tubes | S. E. |
|---|---|---|
| Compat, unstained, presoaked | zero | --- |
| Compat, prestained, presoaked | 73.0477 | ± 3.1311 |
| 1/2 Compat, prestained, presoaked and 1/2 Compat, not prestained, presoaked | 43.0298 | ± 17.2366 |

The proportion of tubes containing stained nuclei was reduced approximately in half. To further check the results, this time prestained pollen was rinsed twice. We pollinated 5 more styles: 1/2 stigma with compatible unstained pollen and 1/2 stigma with incompatible (self) stained. Again, the percentage of tubes with fluorescent nuclei, deriving from mixed pollination, was quite constant.

Table 4.

| Pollen Source | % Fluorescent Nuclei | S. E. |
|---|---|---|
| Double rinse of prestained pollen 1/2 Incompat, prestained and 1/2 Compat, unstained | 53.924 | ± 4.936 |

These are only preliminary results; the simultaneous employment of a second vital stain will probably give us more definite results.

However, it appears that the mentor pollen had a positive effect on the pollen tube length of the incompatible pollen. How far this effect will carry them is still a question.

Jensen et al. (1983) coined the term of "cross-talk" between pollen and ovary. The question is: once the compatible pollen established the conversation, will the incompatible, always around, join in?

To quote Visser (1984): "Our data have shown that self-incompatibility might be less pronounced than is suggested by the outcome of single selfpollination."

## References


Brewbaker JL, Kwack BH (1963) The essential role of calcium in pollen germination and pollen tube growth. Am J Bot 50: 859-865

Davis E, Shuster A (1981) Intercellular communication in plants; evidence for a rapid generated, bidirectionally transmitted wound signal. Proc Natl Acad Sci USA 78: 2422-2426

Donk, JA van der (1974) Synthesis of RNA and protein as a function of time and type of pollen tube-style interaction in _Petunia hybrida_ L. Molec Gen Genet 133: 93-98

Donk, JA van der (1975) Recognition and gene expression during the incompatibility reaction in _Petunia hybrida_ L. Molec Gen Genet 141: 305-317

Deureberg JJM (1975) In vitro protein synthesis with polysomes from unpollinated cross and selfpollinated _Petunia_ ovaries. Planta 128: 29-33

Jensen WA, Ashton ME, Beasley CA (1983) Pollen tube-embryo sac interaction in cotton. In: Mulcahy DL, Ottaviano, E (ed) Pollen: Biology and Implication for Plant Breeding, Elsevier, pp. 67-72

Konar RN, Linskens HF (1966) Physiology and biochemistry of the stigma fluid of _Petunia hybrida_. Planta 71: 372-387

Latt SA, Wohlleb JC (1975) Optical studies of interaction of Hoechst 33298 with DNA, chromatin and metaphase. chromosomes. Choromosoma 52: 297-316

Linskens HF, Spanjers AW (1973) Changes of the electrical potential in the transmitting tissue of _Petunia_ styles after cross and self pollination. Incomp News Let 3: 81-85

Linskens HF (1974) Translocation phenomena in the _Petunia_ flower after cross and self pollination. In: Linskens HF (ed) Fertilization in Higher Plants, Elsevier, pp. 285-292

Mulcahy GB, Mulcahy DL (1985) Ovaries influence on pollen tube growth as indicated by the semivivo technique. J Amer Bot: in press

Tupy J (1961) Changes in hydrocarbons in ovaries after self or cross pollination. Biologica Plantarum 3: 1-14

Visser T, Marcucci MC (1984) The interaction between compatible and selfincompatible pollen of apples and pears as influenced by their ratio in pollen cloud. Euphytica 33: 699-704

White J (1907) The influence of pollination on the respiratory activity of the gynoecium. Ann Bot 21: 489-499

# Self-incompatibility Recognition and Inhibition in Nicotiana alata

NEELAM SHARMA AND K.R. SHIVANNA[1]

## 1 Introduction

Self-incompatibility is a clear manifestation of cellular recognition between the pollen and the sporophytic tissues of the pistil. However, there have been no experimental studies on the biochemical basis of self-incompatibility recognition (see Shivanna and Johri 1985). Our recent investigations on *Petunia*, using an in vitro bioassay in which self-pollen grains are selectively recognised and inhibited, have indicated the involvement of lectin-like components of the pollen and specific sugar moiety of the pistil in self-incompatibility recognition (Sharma and Shivanna 1982, 1983; Sharma et al. 1985). We have extended these studies to *Nicotiana alata* and this paper presents the details of the in vitro bioassay and the biochemical basis of self-incompatibility recognition.

## 2 Material and Methods

Field grown plants of *Nicotiana alata* Linn., which were strictly self-incompatible, were used in our studies. The methodologies followed were similar to those used for *Petunia* (Sharma and Shivanna 1983). The extract (crude as well as dialysed) from the stigma and style from unpollinated flowers were incorporated (0.5-2.5 pistils $ml^{-1}$) in the pollen culture medium (150 g $l^{-1}$ of sucrose + 0.1 g $l^{-1}$ of boric acid in 0.015 M phosphate buffer pH 5.9). Pollen grains were germinated in sitting drop cultures on microscope glass slides kept in petri plates lined with moist filter paper. To study the role of lectins and sugars self-incompatibility recognition,

------------

1 Department of Botany, University of Delhi, Delhi-110007, India.

in one set of experiments a lectin (1 mg $ml^{-1}$) was incorporated in the culture medium; in another set pollen grains were treated with a sugar (100 mM) before culture (Sharma and Shivanna 1983). Per cent pollen germination and tube growth were scored 6 h after culture. For each treatment, pollen grains were cultured in two replicates and the experiment was repeated 3-5 times. In each replicate over 200 pollen grains were scored for germination and 50 tubes for length.

## 3 Results

**The In Vitro Bioassay.** Figure 1 presents the response of pollen in the presence of self and cross-extracts*. Self-extract markedly inhibited pollen germination as well as tube growth. Cross-extract had no effect on the response of pollen. These differential responses of self and cross-extracts were consistently observed over a range of concentrations (0.5-2.5 pistils $ml^{-1}$). Even the dialysed extract was effective in bringing about selective inhibition of self-pollen, indicating that the recognition factors are macromolecules. The differences in the responses

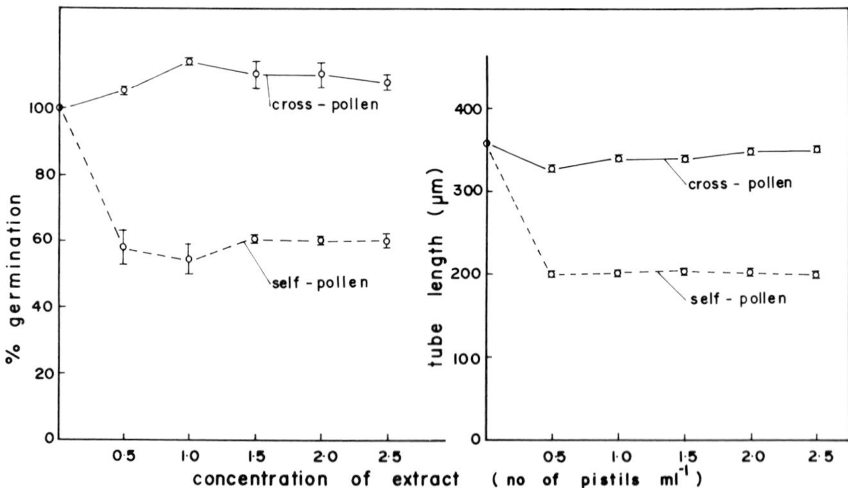


Fig.1. Effect of self- and cross-pistil extract on the responses of pollen. Culture period: 6h.

-------------

* Self-pollen/self-extract -- pollen (used for culture) and pistils (used for extract) obtained from the same plant. Cross-pollen/cross-extract -- pollen and pistils obtained from different plants which are cross-compatible.

of self and cross-pollen were highly significant ($P \leqslant 0.05$). The in vitro bioassay, in which self-pollen grains are selectively recognised and inhibited, has been confirmed for a number of plants during the last four years. Addition of calcium ($Ca(NO_3)_2$, 300 $\mu$g $ml^{-1}$) in the culture medium did not affect the differential responses of self and cross-extracts.

**Location of Incompatibility Factors.** Self-pollen was selectively inhibited even when the stigmatic exudate alone was incorporated in the culture medium. When the extracts from the stigma and the style were incorporated separately in the culture medium, self-pollen grains were inhibited in both the extracts. However, when the extracts from the stigma along with 2-3 mm of the subjacent style, and the remaining part of the style were incorporated separately in the medium, inhibition of self-pollen was observed only in the former and not in the latter (Fig. 2). These results clearly show that the incompatibility factors are present in the stigmatic exudate as well as the tissues of the stigma and upper 2-3 mm of the style.

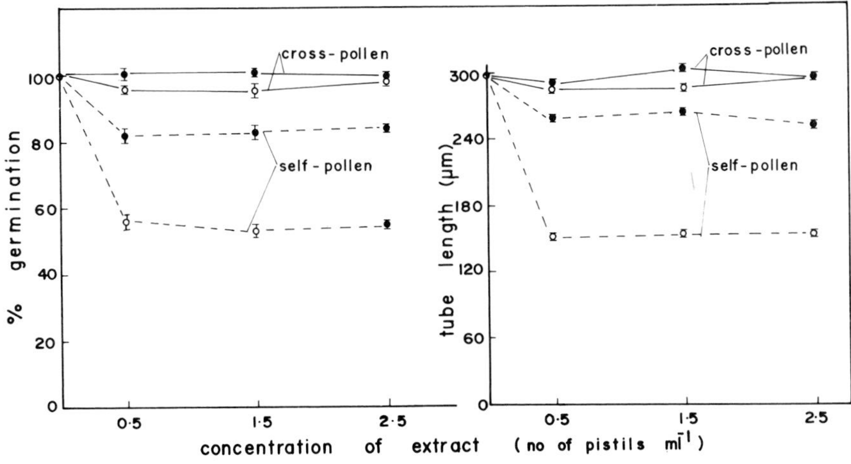


Fig. 2. Effect of the extract from the stigma along with 2-3 mm of the subjacent style (O) and the remaining part of the style (●), incorporated separately in the culture, medium on the responses of pollen. Culture period : 6h.

**Role of Lectins and Sugars in Self-incompatibility Recognition.** Effect of four lectins -- concanavalin A (Con A), phytohemagglutinin (PHA), lectin from Phytolacca americana (PHY) and lectin from Solanum tuberosum (SOL) -- and seven sugars -- D-glucose, D-galactose, D-lactose, N-acetyl-D-glucosamine, N-acetyl-D-galactosamine, D-fucose and D-mannose -- were studied on the response of five self-incompatible but cross-compatible plants. The logic of these experiments was that if a lectin/sugar is effective in blocking recognition molecules, self-pollen grains are not recognized and hence are not inhibited in the presence of the pistil extract.

Figure 3 presents the effects of Con A (incorporated in the culture medium) on the responses of pollen of two of the plants tested. Con A was effective in overcoming inhibition of both germination and tube growth of self-pollen in plant 6 but not in plant 2; it did not affect the responses of cross and control-pollen. The other three lectins were also effective in overcoming inhibition of self-pollen in some of plants but ineffective in others.

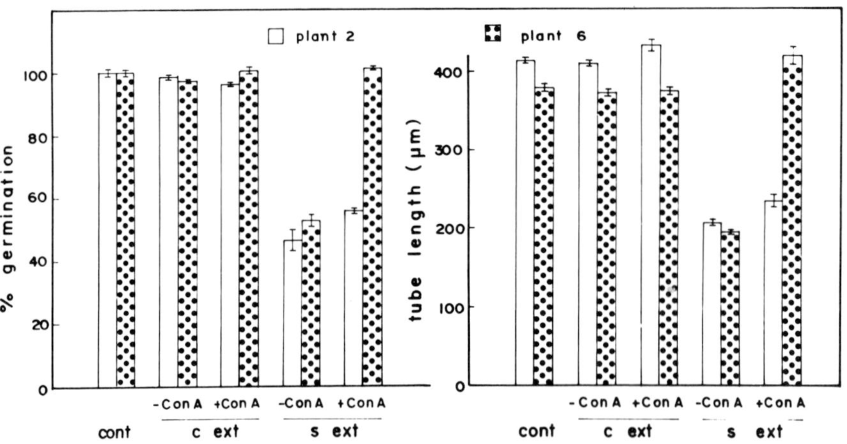


Fig. 3. Effect of Con A in overcoming inhibition of self-pollen in two of the plants. Culture period : 6 h. (cont - Control, c ext - cross-extract, s ext - self-extract)

Treatment of pollen with D-glucose (Fig. 4) or D-galatose was also effective in overcoming inhibition of self-pollen. D-mannose inhibited pollen germination in the control medium itself, and hence its effect on self-compatibility inhibition could not be studied. The other four sugars were not effective in overcoming inhibition of self-pollen. Unlike lectins the efffect of sugars was uniform on all the plants tested.

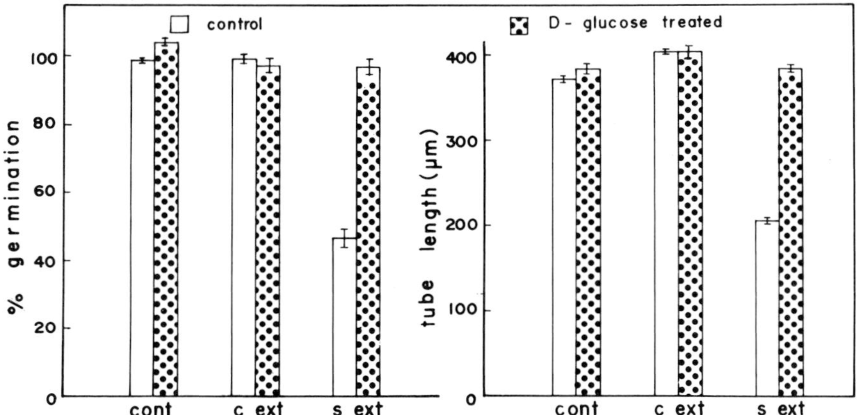


Fig. 4. Effect of treatment of pollen with D-glucose in overcoming inhibition of self-pollen. Culture period : 6h.

**Role of Protein and RNA Syntheses in Pollen Inhibition.** Recognition of self-pollen initiates metabolic processes to bring about pollen inhibition. To investigate whether the inhibitory processes are dependent on protein and RNA syntheses in the pollen, cycloheximide ( a translation inhibitor) or actinomycin D/cordycepin (transcription inhibitors) was incorporated in the culture medium. As cycloheximide inhibited pollen germination even in the control medium, its role in inhibition of self-pollen could not be studied. Actinomycin D and cordycepin did not inhibit

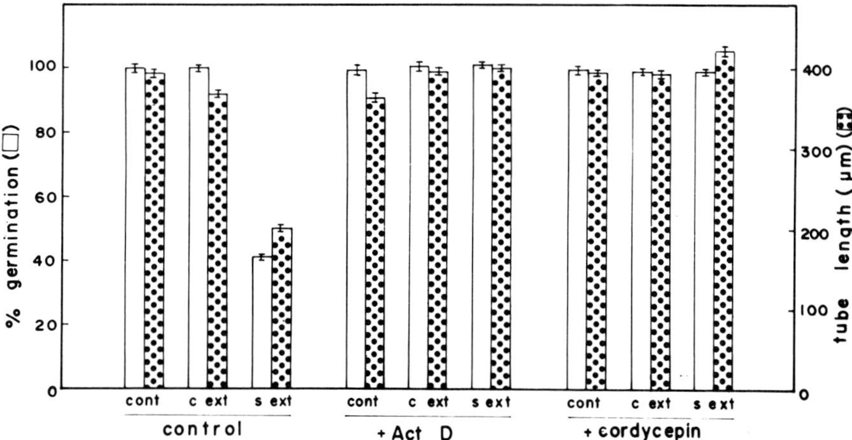


Fig. 5. Effect of act D and cordycepin in overcoming inhibition of self-pollen. Culture period : 6h.

pollen germination (Fig.5). Both of them were effective in overcoming inhibition of self-pollen.

## 4 Conclusions

Our results clearly indicate that in Nicotiana alata self-incompatibility recognition is established as a result of complementation between lectin-like components of the pollen and specific sugar moieties, presumably of glycoproteins, in the pistil. Recognition molecules can be blocked by either treating the pollen with a suitable sugar or incorporating a suitable lectin in the culture medium used in the bioassay. Neither the lectins nor the sugars show S-allele specificity. Apparently, the subtle S-allele-specific differences in the structure of the recognition molecules cannot distinguish related sugars/lectins. The inhibition of self-pollen, following recognition, is dependent on fresh transcription and translation in the pollen.

## References


Sharma N, Shivanna KR (1982) Effects of pistil extracts on in vitro responses of compatible and incompatible pollen. Indian J. exptl Biol. 20:255-256.

Sharma N, Shivanna KR (1983) Lectin-like components of pollen and complementary saccharide moeity of the pistil are involved in self-incompatibility recognition. Curr. Sci. 52:913-916.

Sharma N, Bajaj M, Shivanna KR (1985) Overcoming self-incompatibility through the use of lectins and sugars in Petunia and Eruca. Ann. Bot. 55:139-141.

Shivanna KR, Johri BM (1985) The Angiosperm Pollen: Structure and Function. Wiley Eastern, New Delhi.

# Role of Stigma in the Expression of Self-incompatibility in Crucifers in View of Genetic Analysis

K. HINATA AND K. OKAZAKI[1]

1. Introduction

A number of contributions on the genetics of self-compatibility vs. -incompatibility in Crucifers may suggest that there are two cases for the appearance of self-compatible plants in self-incompatible ones. The first is due to the changes or the interactions of _S_ alleles (Bateman 1954, Sampson 1957, Thompson and Taylor 1966, Zuberi et al 1981), and the second is that self-compatibility is ascribed to genes different from _S_ alleles (Murakami 1965, Thompson and Taylor 1971, Nasrallah 1974, Hinata et al 1983).

In the previous study on the genetics of self-compatibility in _Brassica campestris_ var. _yellow sarson_ with the aid of S-glycoprotein analysis, we have postulated that (1) the self-compatibility is controlled by a recessive epistatic modifier (_m_) independent of the _S_, (2) the _m_ homozygotes suppress the expression of self-incompatibility in stigma but not in pollen, (3) the activity of the _S_ allele in _yellow sarson_ is generally weak enough to be neglected (Hinata et al 1983). The present experiment was undertaken to learn whether such genetic factors are found in other self-compatible plants.

2. Materials and methods

A self-compatible strain (Q) in _Brassica campestris_ L. isolated from a cultivar 'Mizuna' was provided by Prof.

---

1. Faculty of Agriculture, Tohoku University, Sendai, Japan

S. Tokumasu, Ehime Univ. Two self-incompatible strains ($\underline{S}^{i}$) used here were isolated by us from a naturalized population of the same species in Japan. They were $\underline{S}^{8}$ and $\underline{S}^{12}$ homozygotes having S-glycoproteins of pI 8.5 and 6.0, respectively. Hybrids of $F_1$, $F_2$ and backcrossed progenies obtained by bud pollination between self-compatible and -incompatible strains were subjected to the observation of compatibility in selfing and test-crosses with the parents. The compatibility was determined by the pollen-tube penetration into stigmas under fluorescent microscopy after squashing in decolorized aniline blue. The pollen-tube penetration was scored for each stigma by 0, 1, 2 and 3 for the penetration of no pollen-tubes, of less than 5, of less than half and of abundant pollen-tubes, respectively. An average value of 5-10 stigma observations in each combination was calculated and compatibility was presented by -, ±, + and ++ for the values of 0-0.4, 0.5-1.0, 1.1-2.0 and 2.1-3.0, respectively. Analysis of S-glycoprotein bands was carried out by cellulose acetate electrofocusing (Hinata and Nishio 1981).

## 3 Results

In the test-crosses data varied more often than in the case of _yellow sarson_ (Hinata et al 1983), and the results were presented by the average values. Table 1 shows the incompatibility behavior of the hybrid progenies, Q x $\underline{S}^{8}$. The $F_1$ plants were incompatible in selfing as well as in the test-crosses with $\underline{S}^{8}$. In the backcross with Q, the progenies segregated into four types (I - IV), of which the types I - III were identical to the types in the _yellow sarson_ analysis, ie. the type I had no $\underline{S}^{8}$-band and was compatible in all test-crosses as well as selfing, the type II having the $\underline{S}^{8}$-band was compatible in all test-crosses except between its pollen and $\underline{S}^{8}$-stigma, and the type III having the $\underline{S}^{8}$-band showed the same incompatibility relations as the $F_1$. The type IV, that was found in the current experiment, had no $\underline{S}^{8}$-band and was incompatible in selfing and in the test-cross between its stigma and Q-pollen. The backcrossed progenies with the $\underline{S}^{8}$ produced only the type III. The $F_2$ segregated into the four types. Similar results were obtained in the cross, Q x $\underline{S}^{12}$.

The genetic behavior of compatibility in the strain Q

was assumed as follows; (1) there are M vs. m alleles whose locus is independent of S, (2) the m homozygotes suppress the S action in stigmas but not in pollen, and (3) the S in the Q has still its activity, though its S-glycoprotein has not been determined. The last is different from the case of yellow sarson (Hinata et al 1983). The expected incompatibility behavior under the current assumption shown in Table 2 agreed with the data in Table 1. Segregation of the four types in the $F_2$ and backcrossed progenies in the two hybrids was tested as shown in Table 3. The segregation ratio fitted to the expectation in all cases but one, the $F_2$ of Q x $S^{12}$, where heterotic type III was abundant.

Incompatibility behavior was tested in a diallel cross-pollination between the four types in the backcrossed progeny of (Q x $S^{12}$)x Q. The results and expected relations are shown in Table 4. The present scheme could be accepted as a general rule, though some irregular cases were also found. The penetrance of the m gene was considered to be low.

4 Discussion

The present results together with the former ones (Hinata et al 1983) may add one case to the genetic compatibility. Self-incompatibility is a totally integrated metabolic system that involves, at least, recognition and pollen-tube disturbance sub-systems. Self-compatibility is expected to appear by a defect of a part of the total SI systtem; ie. some compatibility is due to the defect of recognition sub-system (Nasrallah 1974) and there must be some others due to the defect of pollen-tube disturbance sub-system. We have an interest in the present genetic scheme because (1) the M allele is independent of S allele, (2) the M may be a necessary factor for the SI system, (3) the mm may be assigned to a defect of the pollen-tube disturbance at the pistil side and (4) if such scheme plays in the interspecific incompatibility, this could be applied to the unilateral incompatibility proposed by Lewis and Crowe (1958).

There are a number of reports about the physiological breakdown of self-incompatibility by treatments; ie. treatments of stigma by physical stress, electric current, high temperature (Matsubara 1980), organic solvents, organic acids, cycloheximide (Roberts et al 1984), etc(see Hinata and Nishio for review). It is curious that all these treatments were applied to stigma, except two cases. The $CO_2$ treatment was effective when pollen-tubes attached to stigmas (Nakanishi and Hinata 1973) and the other was cycloheximide dust treatment to pollen (Ferrari and Wallace 1977). This might be due to the difficulty of the treatment of pollen. However, these results also seem to indicate the importance of stigma which may involve a pollen-tube disturbance sub-system in self-incompatibility.

It is tempting to speculate that the genotypic information from pollen may be received in stigmas by a receptor, that is probably S-glycoprotein, and the following event is controlled by the *M* metabolic sub-system in stigmas.

References

Bateman AJ (1954) Self-incompatibility system in Angiosperms II. *Iberis amara*. Heredity 8:305-332.

Ferrari TE Wallace DH (1977) Incompatibility on *Brassica* stigmas is overcome by treating pollen with cycloheximide. Science 196:436-438.

Hinata K Nishio T (1980) Self-incompatibility in Crucifers. In Tsunoda S et al (eds) *Brassica* Crops and Wild Allies. Japan Scientific Societies Press, Tokyo, pp223-234.

Hinata K Nishio T (1981) ConA-peroxidase method; an improved procedure for staining S-glycoproteins in cellulose-acetate electrofocusing in Crucifers.Theor Appl Genet 60:281-283.

Hinata K Okazaki K Nishio T (1983) Analysis of self-compatibility in *Brassica campestris* var *yellow sarson* (A case of recessive epistatic modifier). In Proc 6th Intern Rapeseed Conf, Paris, I:354-359.

Lewis D Crowe LK (1958) Unilateral interspecific incompatibility in flowering plants. Heredity 12:233-256.

Matsubara S (1980) Overcoming self-incompatibility in *Raphanus sativus* L. with high temperature. J Amer Soc Hort Sci 105: 842-846.

Murakami K (1965) Selective fertilization in relation to plant breeding I. Chinese cabbage (*Brassica pekinensis* Rupr) 3. Inheritance of self- and cross-incompatibility. Japan J Breed 15:97-109.

Nakanishi T Hinata K (1973) An effective time for $CO_2$ gas treatment in overcoming self-incompatibility in *Brassica*. Plant & Cell Physiol 14:873-879.

Nasrallah ME (1974) Genetic control of quantitative variation in self-incompatibility proteins detected by immunodiffusion. Genetics 76:45-50.

Roberts IN Harrod G Dickinson HG (1984) Pollen stigma interactions in *Brassica oleracea* II. The fate of stigma surface proteins following pollination and their role in the self-incompatibility response. J Cell Sci 66:255-264.

Thompson KF Taylor JP (1966) The breakdown of self-incompatibility in cultivars of *Brassica oleracea*.Heredity21:637-648.

Thompson KF Taylor JP (1971) Self-compatibility in kale. Heredity 27:459-471.
Sampson DR (1957) The genetics of self- and cross-incompatibility in *Brassica oleracea*. Genetics 42:253-263.
Zuberi MI Zuberi S Lewis D (1981) The genetics of incompatibility in *Brassica* I. Inheritance of self-compatibility in *Brassica campestris* L. var. *Toria*. Heredity 46:175-190.

Table 1 Incompatibility behavior of the hybrid progenies of Q x $\underline{S}^8$

| Progeny | No. of plants | Test cross | | | | | $\underline{S}^8$band | Type |
|---|---|---|---|---|---|---|---|---|
| | | x$\underline{S}^8$ | $\underline{S}^8$x | Self | xQ | Qx | | |
| $F_1$ | 9 | - | - | - | | | + | III |
| $F_1$ x Q | 3 | ++ | ++ | + | + | + | - | I |
| | 2 | ++ | - | ++ | ++ | + | + | II |
| | 6 | - | - | - | - | ± | + | III |
| | 6 | + | ++ | - | - | + | - | IV |
| $F_1$ x $\underline{S}^8$ | 23 | - | - | - | | + | + | III |
| $F_2$ | 1 | ++ | ++ | + | | + | - | I |
| | 2 | + | - | + | | + | + | II |
| | 11 | - | - | - | | + | + | III |
| | 5 | ++ | ++ | - | | + | - | IV |

x$S^8$ (xQ): Progeny plants x $\underline{S}^8$ pollen (Q pollen).
$\underline{S}^8$x (Qx): $\underline{S}^8$stigma (Q stigma) x progeny plant pollen.
Symbols + and - in testcrosses indicate compatible and incompatible, respectively. + and - in $\underline{S}^8$band show presence or absence of it, respectively.

Table 2 Expected incompatibility behavior of hybrid progenies under the assumption of recessive epistataic modifier ($\underline{M}$ vs. $\underline{m}$)

| Progeny | Type | Genotype | Test cross | | | | | $\underline{S}^i$band |
|---|---|---|---|---|---|---|---|---|
| | | | x$\underline{S}^i$ | $\underline{S}^i$x | Self | xQ | Qx | |
| $F_1$ | III | $\underline{S^iS' \ Mm}$ | - | - | - | - | + | + |
| $F_1$ x Q | I | $\underline{S'S' \ mm}$ | + | + | + | + | + | - |
| | II | $\underline{S^iS' \ mm}$ | + | - | + | + | + | + |
| | III | $\underline{S^iS' \ Mm}$ | - | - | - | - | + | + |
| | IV | $\underline{S'S' \ Mm}$ | + | + | - | - | + | - |

$\underline{S'}$ represents the $\underline{S}$ allele of the Q strain.

Table 3 $\chi^2$ test for segregation of backcross and $F_2$ to the recessive epistatic modifier assumption

| Progeny | Test cross type I | II | III | IV | Total | $\chi^2$ | P |
|---|---|---|---|---|---|---|---|
| Backcross: | | | | | | | |
| Expect. ratio | 1 | 1 | 1 | 1 | | | |
| (Q x $\underline{S}^8$) x Q | 3 | 2 | 6 | 6 | 17 | 3.00 | .50-.30 |
| (Q x $\underline{S}^{12}$) x Q | 7 | 6 | 4 | 7 | 24 | 1.00 | .90-.80 |
| Total | 10 | 8 | 10 | 13 | 41 | 1.24 | .80-.70 |
| $F_2$: | | | | | | | |
| Expect. ratio | 1 | 3 | 9 | 3 | | | |
| Q x $\underline{S}^8$ | 1 | 2 | 11 | 5 | 19 | 1.30 | .80-.70 |
| Q x $\underline{S}^{12}$ | 1 | 1 | 21 | 2 | 25 | 8.07* | .05-.02 |
| Total | 2 | 3 | 32 | 7 | 44 | 5.86 | .20-.10 |

Table 4 Compatibility relations in the diallel cross between the four types of the backcrossed progenies, (Q x $\underline{S}^{12}$)x Q, and the expectation

Data:

| Type | Male / Female | I: 5 | 8 | 17 | II: 7 | 11 | III: 9 | 14 | 21 | IV: 15 | 20 | $\underline{S}^{12}$band |
|---|---|---|---|---|---|---|---|---|---|---|---|---|
| I | 5 | ± | + | | ± | ++ | ++ | − | | + | − | − |
| | 8 | − | + | | ++ | | | | | − | + | − |
| | 17 | ++ | | ± | ± | | ++ | + | ++ | + | ++ | − |
| II | 7 | ± | − | ++ | ± | − | − | ++ | ± | ± | ± | + |
| | 11 | ++ | ++ | | ± | + | ++ | ++ | | ++ | + | + |
| III | 9 | ± | − | − | − | − | − | − | − | − | − | + |
| | 14 | ± | + | | − | − | − | − | − | + | − | + |
| | 21 | − | | − | − | | − | − | − | − | − | + |
| IV | 15 | − | − | | − | − | − | − | ± | − | − | − |
| | 20 | − | − | | − | − | − | − | − | − | − | − |

Expectation:

| Type | Expected genotype | Type I | II | III | IV | $\underline{S}^i$band |
|---|---|---|---|---|---|---|
| I | $\underline{S'S'}$ $\underline{mm}$ | + | + | + | + | − |
| II | $\underline{S^iS'}$ $\underline{mm}$ | + | + | + | + | + |
| III | $\underline{S^iS'}$ $\underline{Mm}$ | − | − | − | − | + |
| IV | $\underline{S'S'}$ $\underline{Mm}$ | − | − | − | − | − |

# A New S-Allele and Specific S-Proteins Associated with Two S-Alleles in Nicotiana Alata

A. KHEYR-POUR AND J. PERNES[1]

## 1 Introduction

It has been proposed that the S-locus, in gametophytic, monofactorial, poly-allelic systems of self-incompatibility, has a tripartite structure: two parts controlling the pollen and the style specificities, and one common part, named the specificity segment (Lewis 1949, 1960). Support for this concept of an integrated S-locus has been provided by the occurrence of pollen-part mutations (in many species) and the style-part mutations (in few cases) as well as revertible mutations (Nettancourt 1977). Genetics and some recent biochemical data appeared to provide a rather satisfactory and consistant explanation of experimental results. However, in a recent paper, Mulcahy and Mulcahy (1983) pointed out several "shortcomings" for the oppositional model and proposed an alternative, heterosis, model which created quite a bit of interest (Lawrence et al. 1985, Mulcahy et al. 1985).

Our previous data in N. alata, concerning the self-incompatibility behaviour of dihaploid plants (Khey-Pour et al. 1983) and the mapping of the S-locus (Labrouche et al. 1983) were two pieces of evidence, supporting the oppositional model. In this paper, we discuss the occurrence of a new functional S-allele and the genetics of two S-specific proteins for two S-alleles.

## 2 The Origin Of New S-Alleles

Extensive studies with radiation and chemical mutagens were unable to generate new functional S-alleles (Nettancourt 1977). Thus it was suggested

---

[1]Laboratoire de Genetique et Physiologie du Developpement des Plantes, C.N.R.S. - 91190 Gif Sur Yvette, FRANCE

that new S-specificities probably do not result from point mutations at the S-locus. Inbreeding techniques seemed to be the only method which generated new S-alleles (Denward 1963, Nettancourt et al. 1971, Pandey 1970a, Hogenboom 1972, Anderson et al. 1974, Gastel et al. 1975). However, Sree Ramulu (1982a) was unable to confirm the results of Nettancourt et al. (1971) and Gastel et al. (1975) when the $S_1$ and $S_2$-alleles in L. peruvianum were put into different genetic inbred backgrounds. On the other hand, he detected a new $S_2*$ allele in L. peruvianum plants (1982b) regenerated from the somatic tissues in anther culture. In this case the new $S_2*$ was first expressed in the pollen while, in inbreeding cases, the new S-specificities were expressed first in the stylar-part. These data showed that there was not a single mechanism in the occurrence of such new S-specificities.

TABLE 1 - Genetics of new $S_z$-allele

| ♀ $S_{F11}S_{F11}$ x ♂ $S_3S_{F11}$ (D.H) * → | Selfs | $S_3S_{F11}$ ♂ | $S_3S_{F11}$ ♀ | $S_3S_3$ ♂ | $S_3S_3$ ♀ | $S_{F11}S_{F11}$ ♂ | $S_{F11}S_{F11}$ ♀ |
|---|---|---|---|---|---|---|---|
| 21 plants ( $S_3S_{F11}$) | - | - | - | - | + | - | + |
| Plant 18 ($S_zS_{F11}$) | 25-/25 ** | 10+/10 | 11+/12 | 17+/18 | 5+/5 | 30-/30 | 5+/5 |

Crosses with plant 18

| | Progeny | |
|---|---|---|
| 2 - Plant 18 : $S_zS_{F11}$ x♂ $S_3S_{F11}$ gave : | 11 $S_zS_3$ | 10 $S_3S_{F11}$ |
| 3 - Plant 18 : " x♂ $S_3S_3$ " : | 10 $S_zS_3$ | 10 $S_3S_{F11}$ |
| 4 - ♀ $S_3S_{F11}$ x♂ $S_zS_{F11}$ (plant 18) " : | 9 $S_zS_3$ | 11 $S_zSF_{11}$ |
| Total | 30 | 31 |

Bud-selled Progeny

| | Progeny | | |
|---|---|---|---|
| | $S_zS_z$ | $S_zS_{F11}$ | $S_{F11}S_{F11}$ |
| 5 - $S_zS_{F11}$(plant 18) B.P. | 10 | 24 | 10 |
| 6 - $S_zS_{F11}$ (from cross n°2) B.P. | 14 | 10 | 4 |
| 7 - $S_zS_{F11}$(from cross n°2) B.P. | 22 | 10 | 4 |
| | $S_zS_z$ | $S_zS_3$ | $S_3S_3$ |
| 8 - $S_zS_3$ (from cross n°2) B.P. | 14 | 10 | 0 |
| 9 - $S_zS_3$ (from cross n°2) B.P. | 8 | 22 | 4 |

* DH = dihaploid

** = number of fruits / number of flowers, (+) = compatible, (-) = incompatible

3 The Origin Of New $S_2$-Allele In N. *alata* (Table 1):

The progeny of one of the dihaploid (D.H.) $S_{F11}S_{F11}$ plants (discussed in Khey-Pour et al. 1983) crossed by o$S_3S_{F11}$, produced 22 $S_3S_{F11}$ progeny (as expected) and one unexpected plant (no 18). This latter plant was self-incompatible, but almost all the flowers crossed by $S_3S_{F11}$ (o and o) and o$S_3S_3$ set seeds in large numbers. However, this plant (no 18) was incompatible with $S_{F11}S_{F11}$ pollen. Such results were in favor of a new functional S-allele. The genotype of the unexpected plant 18 was designated $S_zS_{F11}$.

The unexpected $S_zS_{F11}$ plant 18 and its cross progeny with o$S_3S_{F11}$, which gave 11 $S_zS_3$ and 10 $S_3S_{F11}$, (cross no 2 in Table 1) were reciprocally crossed with our S-allele stocks ($S_1S_1$, $S_3S_3$, $S_1S_3$, $S_eS_a$, $S_{F10}S_{F10}$, $S_{F11}S_{F11}$, and $S_3S_{F11}$) in a diallele model. The new S-allele was compatible with all our genotypes under study; it has a new specificity which was absent in our S-allele bank. The cross progeny of the plant 18 ($S_zS_{F11}$) with o$S_3S_3$, and with o and o$S_3S_{F11}$ segregated in 1:1 ratios. Such results indicated that in the original plant 18, both style and pollen specificities of the new $S_z$-allele were exhibited.

Bud-selfed progeny of plant 18, two other $S_zS_{F11}$ and two $S_zS_3$ (Table 1), yielded the three expected genotypes in classical proportions with, however, sometimes a significant excess of the $S_zS_z$ genotype. Such excess should indicate that the $S_z$-allele could be more competitive in the bud-pollination of $S_zS_{F11}$ and $S_zS_3$ genotypes over the sister alleles ($S_{F11}$ and $S_3$).

The results of Table 1 indicated that the new $S_z$-allele specificities were stable over three generations of genetic studies. Later, two other plants were found which exhibited the $S_z$ specificities. They were among the progeny of the third generation studies of two other $S_{F11}S_{F11}$ dihaploid plants (distinct from the first D.H. plant discussed in Table 1), crossed by o$S_3S_{F11}$ genotype. The progeny of these two new cases of $S_z$-allele occurrence is under study.

It seems most likely that the new $S_z$-allele in three independent pedigrees originated from the mutation of the $S_3$-allele of the same $S_3S_{F11}$ clone used as male and female tester in our genetic studies of $S_{F11}S_{F11}$ D.H. plants. Apparently, the female backgrounds of the D.H. plants do not seem to be directly involved in the occurrence of the new $S_z$-allele.

The $S_zS_z$ styles accept the pollen of N. *Langsdorfii* ($S_f$-allele, Pandey 1964) as does the $S_3S_3$ genotype, but the $S_f$ pollen tubes are inhibited in the stigmatic zone of $S_{F11}S_{F11}$ plants. Other genetic markers are being used to obtain a better explanation of the effective origin of the new $S_z$-allele.

Our data seem to fit with classical mutational events of $S_3$ to $S_z$ allele. At least that is the simplest interpretation of the observed facts. These three independent mutational events are comparable to the results fo Sree Ramulu (1982b) in L. peruvianum : he detected a new $S_z$*-allele which was absent in his genetic stock as is the case for our new $S_z$-allele. This $S_z$-allele is very stable. We did not detect any unexpected fruit and seed sets in three generations of genetic study (Table 1).

## 4 Specific S-Proteins Associated With the New $S_z$ and $S_{F11}$ Alleles

The basic question concerning the S-proteins is to demonstrate whether they are coded by the S-locus itself or by genes closely linked to the S. This question needs a careful genetic study. Pandey (1967) proposed isoenzymes as the S-genes product in N. alata, but Bredmeijer et al. (1975, 1980) did not confirm this hypothesis. Labroche et al. (1983) by a careful genetic analysis of leaf peroxidase isoenzymes demonstrated that two cathodic peroxidases were controlled by separate genes ($P_I$ and $P_{II}$) which are linked to the S-locus. The distances between S-$P_I$ and S-$P_{II}$ are respectively 3 and 34 centimorgans. Two other genes which coded for anodic peroxidase isoenzymes were independent of the S-locus. Bredmeijer et al. (1981) have detected specific proteins for $S_2$ and $S_6$ alleles in N. alata. Harris et al. (1984) study the same material. The $S_2$-protein seems to be a glycoprotein (arabinogalactan protein). The glycoproteins were proposed as the S-specific proteins in Brassica species (Nishio et al. 1977, 1982, Hinata et al. 1982, Ferrari et al. 1981, Nasrallah et al. 1983).

The genetic analysis of two specific proteins for the new $S_z$ and $S_{F11}$-alleles is discussed here. Our first aim was to be sure that these proteins were associated with these two S-alleles.

## 5 Materials and Results (Table 2)

The supernatant of crude extracts (by phosphate buffer, 0.01M, pH 7) of the stigmatic zone of advanced mature buds and open flowers were electrofocused, in horizontal ultrathin (0.5 mm) polyacrylamide gel following L.K.B. instructions. Specific S-associated proteins were found only in the pistil extracts of advanced buds and open flowers for $S_{F11}$-allele ($B_1$ Protein, iP = 9.5, MM = 27,000) and for $S_z$-allele ($B_3$ Protein iP = 9, MM = 30,000). For other S-alleles studied ($S_1$, $S_3$, $S_{F10}$, $S_a$ and $S_e$) we did not detect S-associated bands (Table 2).

Despite elaborate genetic study we did not find any recombinant plants between $S_{F11}$ and $B_1$ protein in the more than 231 plants bearing the $S_{F11}$-allele. 210 plants without the $S_{F11}$-allele did not have this protein. Also 188 plants possessing the $S_z$-allele did have the $B_3$ protein; once again, there were no recombinant plants between $S_z$-allele and the $B_3$ protein.

These two bands ($B_1$ and $B_3$) were quantitatively the most abundant proteins observed by electrofocusing. In most cases. extracts from each plant were repeated several times.

TABLE 2 - GENETICS OF $B_1$ AND $B_3$ PROTEINS

| GENEOTYPES | Numbers of plants | $B_1$ | $B_3$ | O S B* |
|---|---|---|---|---|
| $S_{F11}S_{F11}$ | 35 | + | - | - |
| $S_{F11}S_{F11}$ (dihaploid plants) | 3 | + | - | - |
| $S_{F11}S_{F11}S_{F11}S_{F11}$ (Audrogenetic) | 2 | + | - | - |
| $S_{F11}S_{F11}S_{S3}$ (3N) | 30 | + | - | - |
| $S_{F11}S_{F11}S_{F10}S_{F10}$ (somatic anther culture) | 12 | + | - | - |
| $S_{F11}S_3$ | 91 | + | - | - |
| $S_{F11}S_z$ | 58 | + | + | - |
| $S_zS_z$ | 68 | - | + | - |
| $S_zS_3$ | 62 | - | + | - |
| $S_1S_1$ | 5 | - | - | - |
| $S_3S_3$ | 35 | - | - | - |
| $S_1S_3$ | 12 | - | - | - |
| $S_{F10}S_{F10}$ | 4 | - | - | - |
| $S_3S_{F10}$ | 6 | - | - | - |
| $S_aS_a$ | 3 | - | - | - |
| $S_eS_e$ | 2 | - | - | - |
| $S_eS_a$ | 10 | - | - | - |
| N. Langsdorfii ($S_fS_f$) | 3 | - | - | - |
| Total of plants | 441 | | | |

(+) = presence of S-allele associated bands , (-) = absence of S-allele associated bands
* OSB = other specific bands.

The plants possessing $S_{F11}$ and $S_z$-alleles, analyzed in Table 2, were parts of many crosses or bud-selfed progeny. The genetic backgrounds of some of them are indicated in Table 1. Recombinant rate studies for closely S-linked genes were performed. The probability of having recombinant phenotypes of $S_{F11}B_1^-$, $S_3B_1^+$, or $S_{F11}B_3^+$, is more than 99%, even in 1% recombination rates between the S-locus and $B_1$ or $B_3$ supposed to be coded by linked genes. We did not detect any such recombined phenotypes. Such results seem to be in favor of the assumption that the $B_1$ and $B_3$ proteins are likely to be coded by $S_{F11}$ and $S_z$-alleles respectively, or at least by the genes very closely linked to the S-locus.

The biochemistry and purification of these proteins ($B_1$ and $B_3$) are under study in connection with R. Remy (unpublished data). These proteins did not have peroxidasic activities. Thus they are distinct from the two cathodic peroxidases coded by two linked genes ($P_I$ and $P_{II}$) to S (Labroche et al. 1983). We study the carbohydrate contents and other components of these proteins in collaboration with A. Clarke in comparison with that of $S_2$-allele of _N. alata_ studied by this laboratory.

We thank Jacqueline Knight and Almas Sadr for their valuable technical assistance.

REFERENCES

Anderson MK,Taylor N,and Ducan IF,(1974) Euphytica 23. 140-148
Bredmeijer GMM,and Blass J (1975) Acta Bot. Neerl.24 : 37-48
Bredmeijer GMM, and Blass J (1980) Theor.Appl.Genet. 57 : 119-123
Bredmeijer GMM,and Blass J (1981) Theor.Appl.Genet. 59 :185-190
Denward T (1963) Hereditas 49 : 189-334
Ferrari TE, Bruns D, and Wallace DH, (1981) plant.Physiol. 67 : 270-277
Gastel AJG Van and Nettancourt D (De) (1975) Incompatibility Newsletter 6 66-69
Harris PJ, Anderson AM, Basic A, and Clarke AE (1985) Oxford surveys of plant molecular and cell biology (BJ Milfin ed.) Oxford University Press (in press)
Hogenboom NG (1972 Euphytica 21 : 228-243
Hinata K, Nishio T, and Kimura J (1982) Genetics 100 : 649-657
Kheyr Pour A, Bui Dang Ha D, and Pernès J (1983) ; pollen : biology and implacations for Plant Breeding DL Mulcahy and E Ottaviano, eds, pp 303-309
Labroche Ph, Poirier-Hamon S, and Pernès J (1983) Theor.Appl.Genet. 65 : 163-170
Lawrence MJ Marshall DF, Curtis VE, and Fearon CH (1985) Heredity 54 : 131-138
Lewis D (1949) Heredity 3 : 339-355
Lewis D (1960) Roy.Sco.London B 151 : 468-477
Mulcahy DL, and Mulcahy GB, (1983) Science 220 : 1247-1251
Mulcahy DL, and Mulcahy GB (1985) Heredity 54 : 139-144
Nettancourt D, de (1977) Incompatibility in Angiospermes. Springer Verlag Berlin
Nettancourt D, de, Ecochard R, Perquin MDG, Drift T Van der and Westerhof M (1971) Theor.Appl.Genet. 41 : 120-129
Nasrallah ME, Doney RC, and Nasrallah JB (1983) Pollen : biology and implications for plant breeding p 251-257, edited by DL Mulcahy and E Ottavio, Elsevier Biochemical
Nishio T, and Hinata K (1977) Heridity 38 : 391-396
Nishio T, and Hinata K (1982) genetics 100 : 641-647
Pandey KK (1964) Elements of the S-gen complex. Genet. Res. Comb. 2 : 397-409
Pandey KK (1967) Nature (London) 213 : 669-672
Pandey KK (1970 b) nature (London) 227 : 689-690
Sree Ramulu K (1982 a) Incompatibility Newletter 14 : 103-110
Sree Ramulu K (1982 b) Heredity 49 : 319-330

# Immunodetection of S-gene Products on Nitrocellulose Electroblots

M.E. NASRALLAH AND J.B. NASRALLAH[1]

## 1 Introduction

Genetic, immunochemical and electrophoretic approaches to the study of the self-incompatibility mechanism of higher plants have identified some of the molecular components involved in the pollen-stigma interaction of incompatibility (Nasrallah & Wallace, 1967; Nishio & Hinata, 1977; Bredemeijer & Blass, 1981; Mau et al. 1982). In general, self-incompatibility is believed to be mediated by the expression of identical S alleles carried in the anther (or pollen in the case of gametophytic systems) and in the pistil (Heslop-Harrison, 1975). In Brassicaceae, the detection, in stigmas, of S-locus specific glycoproteins (SLSG) which are believed to play a role in the pollen/stigma recognition events, has been reported by several groups. However, no reports of the corresponding pollen determinants of incompatibility have as yet been made in Brassica. In this chapter, we describe the detection of S-gene products following electrophoretic separation of protein extracts and transfer to nitrocellulose filters, by the so-called "Western" technique. The sensitivity of this technique has greatly enhanced our ability to visualize small quantities of SLSG, so that it is now possible to detect these and antigenically related molecules in extracts of pollen as well

[1]Section of Plant Biology, Cornell University, Ithaca, New York 14853

as of single stigmas. The application of this technique has allowed the localization of SLSG in the surface papillar cells of the stigma. Our results further suggest that contact with pollen causes release of the previously sequestered glycoproteins.

**2 Immunodetection of S-glycoproteins on electroblots**

Single stigmas were homogenized in 0.1 M Tris buffer, pH 7.2 and centrifuged at 15,000 x g for 15 minutes. One-dimensional sodium dodecyl sulfate-polyacrylamide gel electrophoresis was carried out on whole soluble extracts from single stigmas in 10% (w/v) gel slabs according to Laemmli (1970), and as previously described (Nasrallah & Nasrallah, 1984). The separated proteins were electrophoretically transferred to nitrocellulose paper according to Towbin et al (1979). Immunolabeling of nitrocellulose blots was done essentially as described by Whitehouse and Putt (1983). Briefly, the blots were first washed in phosphate-buffered saline (PBS)-0.15% (v/v) Tween 20, then reacted for four hours with rabbit serum raised against S-specific glycoproteins purified by preparative isoelectric focusing of stigma extracts in granulated beds of Sepharose. Excess rabbit antibodies were removed by washing in several changes of PBS-Tween, before treatment with peroxidase-conjugated protein A for two hours. After washing in PBS-Tween, the complex was visualized by the diaminobenzidine method.

**3 S-gene expression in the stigma and pollen**

SLSG have been functionally correlated to the self-incompatibility reaction in stigmas harvested at various stages of development (Nasrallah, 1974; Nasrallah & Nasrallah, 1984). In these earlier studies, extracts from several stigmas, typically 25, and by necessity collected from several matched inflorescences were required for isoelectric focusing and double diffusion analyses. The immunodetection method described here allows not only the analysis of single stigmas, but in addition the monitoring during stigma development of the

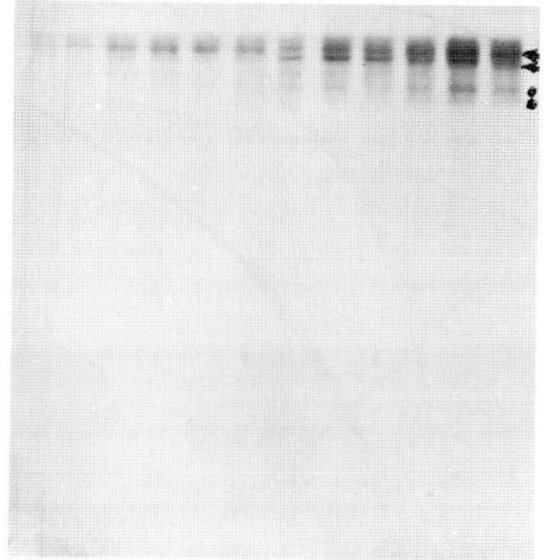

Fig. 1. Developmental analysis of S-gene expression in stigmas. Individual stigmas at various stages of development were harvested in sequence from one inflorescence. The immunoblot shows increasing levels of SLSG (arrows) during stigma development from the immature stage (left lane) to the mature open flower stage (right lane). Dots mark the immunologically related lower molecular weight molecules.

various molecular weight species of SLSG (Nasrallah & Nasrallah, 1984) and of their related antigens. As shown in Fig. 1, the developmental profile of single stigmas harvested along one inflorescence, reveals trace levels of the SLSG molecular weight species (arrows) in immature self-compatible stigmas and an increase in their concentration during flower maturation. Significantly, pollen-tube development as monitored by fluorescence microscopy is inhibited coincidently with the accumulation of these molecules. A further observation can be made from an inspection of the immunoblots. In addition to the SLSG bands, lower molecular weight bands (dots in Fig. 1) that do not correspond to S-specific isoelectric focusing bands are stained. Although the identity of these cross-reactive molecules is not known, their reaction with antisera specific to SLSG suggests their structural and perhaps functional relatedness to SLSG.

The usefulness of the "Western" blotting technique is also demonstrated in Fig. 2. It can be seen in this figure that different S-allele homozygotes can be discriminated from one another since they exhibit different molecular weight patterns of S-antigens, and heterozygous plants can be identified by their hybrid patterns (Fig. 2).

The antiserum which was developed against SLSG from stigmas also reacts with pollen components as shown in the righthand most channel in Fig. 2. Interestingly, the pollen antigens differ from their stigma counterparts in apparent

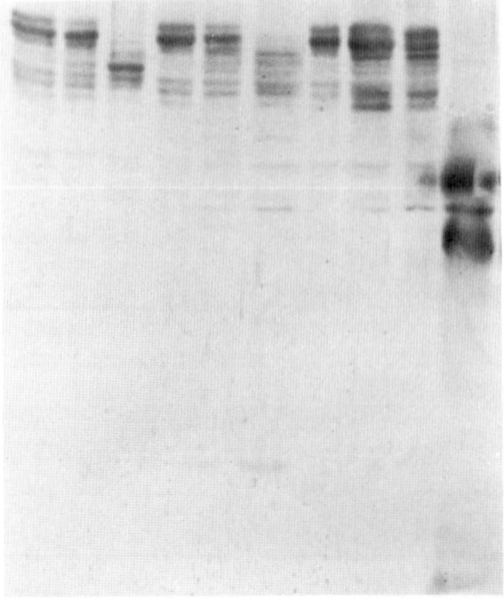

Fig. 2. Detection of SLSG on "Western" blots. The stigma patterns for four S-homozygotes and three heterozygous combinations, and the pollen (P) antigens are shown. Starting at the left, lane 1: $S_6S_6$; lane 2: $S_6S_{11}$; lane 3: $S_{11}S_{11}$; lane 4: $S_6S_6$; lane 5: $S_6S_{13}$; lane 6: $S_{13}S_{13}$; lane 7: $S_6S_6$; lane 8: $S_6S_{14}$; lane 9: $S_{14}S_{14}$; lane 10: pollen.

molecular weight. The basis of these observed differences is unknown, but could conceivably be due to differential glycosylation in the two tissues.

## 4 Localization of SLSG in the stigma

A central question concerning SLSG is the site of their synthesis and localization in the stigma and pollen. As a first step in answering this question, papillar cells were dissected away from the underlying tissues of the stigma under a stereoscope. The papillae, numbering about three to five thousand cells and the underlying stigma tissues were extracted separately and subjected to "Western" analysis. As shown in Fig. 3, the majority if not all of SLSG and related molecules are found in the papillar fraction (right lane). We interpret the less intense bands visible in the underlying tissues (left lane) to be due to contamination from the disruption of papillar cells during dissection.

In conclusion, we have found the "Western" technique to be very useful for the analysis of various developmental and physiological aspects of pollen-stigma interaction. Applications range from efficient screening for S-genotypes to

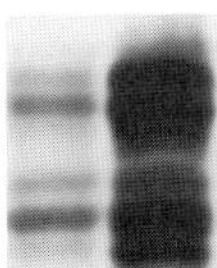

Fig. 3. Localization of SLSG to the papillar cells of the stigma. Papillar cells (right) were dissected away from the underlying tissues of one stigma (left). The two fractions were extracted separately and subjected to the "Western" technique. The majority of SLSG is seen in the papillar cell fraction.

investigations in which only small amounts of tissue can be practically handled.

This work was supported by NSF grant No PCM 8118035.

References

Nasrallah, ME, Wallace DH (1967) Immunogenetics of self-incompatibility in Brassica oleracea L. Heredity 22: 519-527

Nishio T, Hinata K (1977) Analysis of S-specific proteins in stigma of Brassica oleracea L. by isoelectric focusing. Heredity 38: 391-396

Bredemeijer GMM, Blass J (1981) S-specific proteins in styles of self-incompatible Nicotiana alata. Theor. Appl. Genet. 59: 185-190

Mau SL, Raff J, Clarke AE (1982) Isolation and partial characterization of components of Prunus avium L. styles, including an antigenic glycoprotein associated with a self-incompatibility genotype. Planta 156: 505-516

Heslop-Harrison J (1975) Incompatibility and the pollen-stigma interaction. Ann. Rev. Plant Physiol. 26: 403-425

Laemmli UK (1970) Cleavage of structural proteins during the assembly of the head of bacteriophage T4. Nature 227: 680-685

Nasrallah JB, Nasrallah ME (1984) Electrophoretic heterogeneity exhibited by the S-allele specific glycoproteins of Brassica. Experientia 40: 279-281

Towbin H, Staehelin T, Gordon J (1979) Electrophoretic transfer of proteins from polyacrylamide gels to nitrocellulose sheets: procedure and some applications. Proc. Natl. Acad. Sci. USA 76: 4350-4354

Whitehouse DB, Putt W (1983) Immunological detection of the sixth complement component (C6) following flat bed polyacrylamide gel isoelectric focusing and electrophoretic transfer to nitrocellulose filters. Ann. Hum. Genet. 47: 1-8

Nasrallah ME (1974) Genetic control of quantitative variation in self-incompatibility proteins detected by immunodiffusion. Genetics 76: 45-50

# Applications of a New Membrane Print Technique in Biotechnology

P.M. O'NEILL, M.B. SINGH AND R.B. KNOX[1]

## Introduction

In exploring the biotechnology of pollen, nitrocellulose membrane is used extensively for electro-blotting of proteins from polyacrylamide gels, e.g. dot-blotting (Hawkes et al., 1982). The proteins bind electrostatically to the membrane surface and can then be stained with various probes, e.g. general protein stains, chromophore substrates for enzymes, lectin or lectin-like molecules for glyco-conjugates and monoclonal or monospecific polyclonal antibodies for antigens. Nitrocellulose membrane has been utilized to make pollen prints or micro-dot blots of surface proteins of pollen of oilseed rape, *Brassica campestris* and ryegrass *Lolium perenne*.

Pollen surface proteins are of particular interest in the sporophytic self-incompatibility system of *Brassica* where the self rejection response occurs on the stigma surface. Recognition events are triggered by a surface component from the pollen grains. Several studies (see Heslop-Harrison et al., 1975; Roberts et al., 1980; Kerhoas et al., 1983) suggest that wall-held protein fractions are implicated in eliciting the self-incompatibility response. Pollen surface proteins are also involved in inducing allergic reactions in susceptible humans to pollen of ryegrass (Howlett et al., 1981).

Pollen prints were first developed by Heslop-Harrison et al (1973). Pollen grains were applied either directly or on a

[1] Plant Cell Biology Research Centre, School of Botany, University of Melbourne, Parkville, Victoria 3052, Australia.

piece of adhesive tape to an agarose film. They were allowed to release their surface components, removed and the resulting prints were dried, fixed and stained. An initial release of surface components was observed followed by a greater release through the germinal apertures of the pollen grains.

Our study was concerned with investigating the nature of pollen surface proteins by capturing them on nitrocellulose membrane from individual grains for subsequent microanalysis. A technique has been developed where these surface proteins are immobilized without fixation, and exposed to various cytochemical probes without disrupting the integrity or viability of the pollen. This means that screened intact grains are available for pollination experiments.

## Materials and Methods

A monolayer of pollen was applied within a defined area on a clean microscope slide. Strips of nitrocellulose membrane (Schleider and Schuell, Membranfilter 50, pore size 0.1 µm) were soaked in 0.05 M Tris buffered saline solution. Each paper was blotted to remove excess buffer, secured to the slide at one end with a piece of adhesive tape and gently laid over the pollen. Another microscope slide was laid over the paper, light pressure applied to ensure pollen-paper contact and a 100 g weight placed on top for the required printing time.

After this time the weight and slide were removed, the paper cut from the adhesive tape and any attached grains removed with a camel hair brush. The papers were left to air dry and were either processed immediately or were stored desiccated at -20°C until required. Papers were handled throughout this procedure with fine forceps.

## Results

Amido black was used as a general protein stain to assess the efficiency of the technique with various printing times. After only 10 s, *Brassica* pollen protein could be detected. However, the quantity and resolution of these pollen prints was much improved after 30 s (Fig. 1A). Prints made after 1 min and 2 min showed a marked increase in accumulated surface proteins.

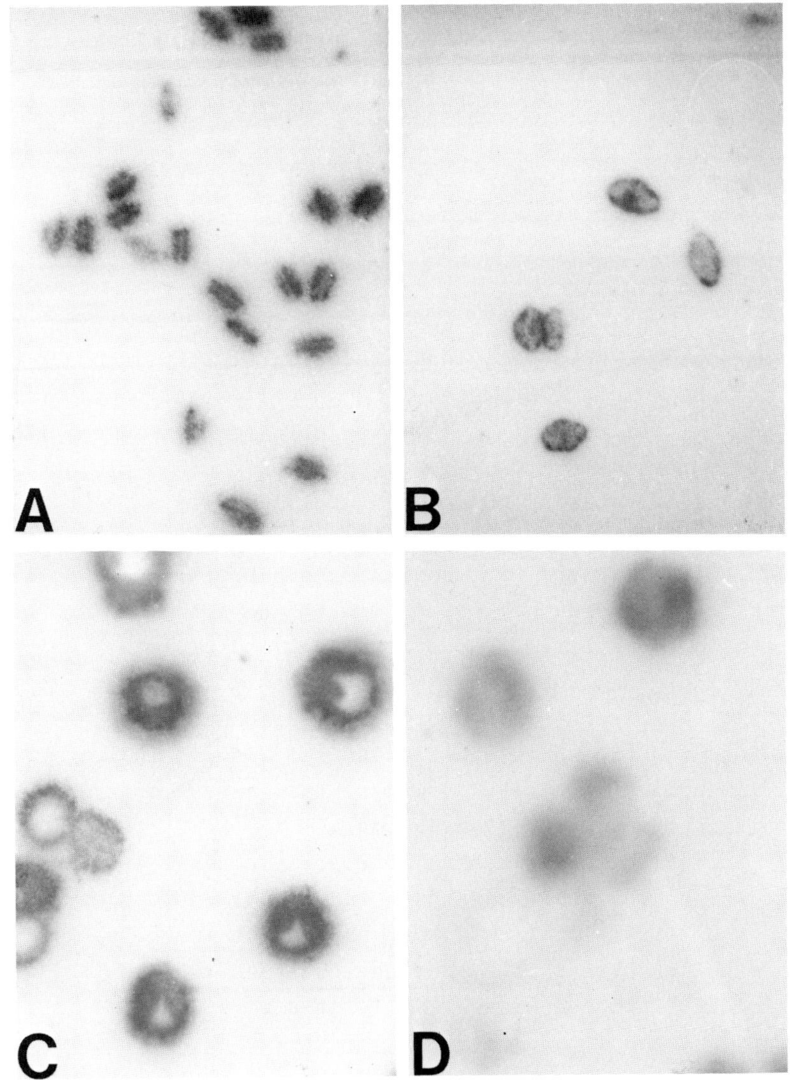


Fig. 1. Pollen Prints x 330
A. 30 sec Brassica prints stained with amido black.
B. 30 sec Brassica prints stained with Con A peroxidase.
C. 2 min L. perenne prints treated with mab 1.
D. 2 min L. perenne prints treated with mab 2.

Pollen quality was monitored continually throughout these studies using the fluorochromatic reaction (FCR) test. Pollen quality was retained at a maximum up to 2 min printing time. Pre-hydrated pollen, kept at 100% RH for 1 h, released large amounts of protein from its surface after only 10 s printing time.

Amido black (see Fig. 1A) and the India ink staining of pollen prints show the precision with which pollen protein is released onto the paper. The germinal apertures are obvious as slits between areas of dense staining reflecting intimate wall contact. These indicate the orientation of the pollen grain touching the paper.

30 s _Brassica_ prints were probed for glycoproteins using Con A labelled with peroxidase (see Fig. 1B). Glucoside and/ or mannoside residues are present in pollen surface prints. In control prints stained only for peroxidase, pollen prints remained unstained. Thus, the peroxidase method of labelling can be utilized as an indirect staining method.

Immunofluorescent studies by Howlett et al (1981) have shown that polyclonal antibodies raised to allergens of rye-grass (_Lolium perenne_) pollen bind to the cytoplasm and wall of sectioned material. Monoclonal antibodies (mabs) raised to these allergenic components bind to specific bands of western blots of _L. perenne_ pollen proteins (Singh and Knox, 1985). Homologous antigens in other grass pollens are indicated by a dot immunobinding assay. However, this assay requires significant mass of pollen (50 mg).

The pollen print method offers the potential to analyse antigens of individual grains. Monoclonal antibodies selected from the library available were applied to 10s, 30s, 1 min and 2 min pollen prints of _L. perenne_. They were then labelled indirectly with peroxidase. Allergens were detected in the prints after only 10s. After 2 min printing, different amounts of each antigen could be distinguished (see Fig. 1C and 1D) illustrating that the allergens may be present in different concentrations in the surface proteins of _L. perenne_ or that they are present in different sites. The greater staining intensity of the reaction with mab $A_1$ as opposed to mab $A_2$ in the pollen wall could indicate greater allergenic activity as it is the pollen surface which first contacts human nasal passages. Human serum from a patient sensitive to ryegrass has also been shown, by indirect labelling with peroxidase anti-IgE, to react with pollen prints of _L. perenne_.

After localizing β-galactosidase histochemically using 5-bromo- 4-chloro- 3-indoxyl- β-D-galactoside as substrate in semithin sections of *Brassica* pollen, the enzyme was detected in the cytoplasm and possibly the intine wall (Singh and Knox 1984). Any surface enzyme, however, may have been lost during processing. To oetermine if the enzyme is present in surface components, pollen prints of *Brassica* were made where half the prints were surface prints of intact grains and half were prints of crushed pollen. These prints allowed a direct comparison of total protein, known to have β-galactosidase activity, with surface protein of the pollen. When the prints were stained with the cytochemical method, only the total protein prints stained, and only very lightly. A more sensitive method of β-gal detection was developed using monospecific polyclonal antibodies. Polyclonal antibodies to Jack bean β-galactosidase were raised in rabbits and rendered monospecific using Western blotting techniques (see Singh and Knox, 1985). 30 s pollen prints were processed by the indirect immuno-peroxidase method. Only crushed pollen prints showed β-galactosidase activity. When these prints were post-stained with India ink the surface prints were revealed. This study confirmed that β-galactosidase is not a component of the surface protein of *Brassica campestris* pollen.

## Conclusions

As shown from these results, the pollen print or micro-dot blot technique is a useful method with which to investigate pollen proteins. We conclude that:

- Pollen surface proteins can be immobilized on nitrocellulose membrane as they are released from the grain.
- Printing is carried out without protein being lost through fixation, freezing and other procedures disruptive to pollen integrity. Pollen quality is retained through the printing process.
- Protein bound to the membrane retains its antigenicity, lectin-binding ability and enzymic activity after storing at -20°C for up to six months.
- Microanalysis and screening of pollen samples can be carried out simultaneously on many individual living pollen grains.

Further, the screened grains can be recovered for pollination and fertilization experiments.

## References

Hawkes R, Niday E, Gordon J (1982) A dot-immunobinding assay for monoclonal and other antibodies. Analyt Biochem 119: 142-147.

Heslop-Harrison J, Heslop-Harrison Y, Knox RB, Howlett B (1973) Pollen wall proteins: 'Gametophytic and sporophytic' fractions in the pollen walls of the Malvaceae. Ann Bot 37: 403-412.

Heslop-Harrison J, Knox RB, Heslop-Harrison Y, Mattsson O (1975) Pollen wall proteins: emission and role in incompatibility responses. In: Duckett JG, Racey PA (eds) The Biology of the Male Gamete. Academic Press, New York, pp 189-202.

Howlett BJ, Vithanage HIMV, Knox RB (1981) Immunofluorescence localization of two water soluble glycoproteins, including the major allergen of ryegrass, *Lolium perenne*. Histochem J 13: 461-480.

Kerhoas C, Knox RB, Dumas C (1983) Specificity of the callose response in stigmas of *Brassica* Ann Bot 52: 597-602.

Roberts IN, Stead AD, Ockendon DJ, Dickinson HG (1980) Pollen-stigma interactions in *Brassica oleracea* Theor Appl Genet 58: 241-246.

Singh MB, Knox RB (1984) Quantitative cytochemistry of β-galactosidase in normal (Gal) and enzyme deficient (gal) pollen of *Brassica campestris*: an application of the indigogenic method. Histochem J 16: 1273-1296.

Singh MB, Knox RB (1985) Immunofluorescence applications in plant cells. In: Robards AW (ed) Botanical Microscopy 1985. Oxford University Press, Oxford (in press).

# Pollen-Stigma Interactions and S-Products in <u>Brassica</u>

T. GAUDE AND C. DUMAS[1]

## 1 Introduction

The cellular events that occur during pollen-stigma interactions and the mechanisms by which self-incompatibility is expressed have been the subjects of much research and hypotheses these past ten years (Heslop-Harrison 1975; Dumas et al. 1984). Recently, the initial interaction between male and female partners have been interpreted in terms of new concepts based on cellular and molecular recognition (Dumas et al. 1984). When pollen landed on the stigma, an information, specific of the species and genotypes, is read-out by the stigma or other pistil tissues. Both information and read-out system are encoded in a complex genetic system of which the commonest and simplest type is the mono-factorial S locus with many alleles (see De Nettancourt 1984). Pollen acceptance or rejection is thus based on the results of a dialogue between the S-gene products. As a consequence, if information and read-out are compatible, fertilization ensues. If not, the interactions lead to the incompatibility response. The <u>Brassica</u> model, controlled by a sporophytically inherited S gene system, with recognition reaction at the stigma surface, provides an ideal experimental model. In this article we will review the data available on stigma read-out system and pollen information in <u>Brassica</u> that constitutes certainly one of the best-defined processes of cell recognition found in higher plants.

## 2 Stigma read-out system

Components of the stigmatic papillae that may be involved in the recognition reaction between pollen and stigma have been investigated by a number of workers. The pioneer work is certainly that of Wallace's

1 Université Cl. Bernard-LYON I, R.C.A.P., UM CNRS 380024, 69 622 Villeurbanne Cedex, FRANCE.

group which by immunological and electrophoretic methods identified components that presented S-allele specificity in _Brassica oleracea_ (Nasrallah and Wallace 1967; Nasrallah et al. 1970). Sedgley (1974) confirmed these results by serological techniques. In Japan, Nishio and Hinata (1977) demonstrated the presence of S-specific glycoproteins in homogenates of _Brassica_ stigmas by using isoelectric focusing (IEF) Since this time, isoelectric focusing was revealed as a very appropriate technique in the study of stigma proteins. By this method, Roberts et al. (1979) detected a single glycoprotein whose appearance concides with the acquisition of self-incompatibility in _Brassica_. As bud stigmas do not express the self-incompatible response, it has been proposed that a new component is excreted on the stigma papillae surface during the pistil maturation (Shivanna et al. 1978). This glycoprotein may be a candidate for such a role, enabling the recognition and rejection of self-pollen. More recently, Ferrari et al. (1981) isolated and partly characterized a glycoprotein specific for the $S_2$ allele. It has been found to regulate pollen germination _in vitro_ and to modify the behaviour of compatible pollen at the stigma surface. Most recently, a very interesting work has been realized by Nasrallah's group at Cornell University (Nasrallah and Nasrallah 1984). A number of self-incompatible genotypes of _Brassica oleracea_ have been analyzed by IEF and by electrophoresis. Among the nine S genotypes tested, each exibits a unique basic glycoprotein pattern with one or more differential bands binding the lectin concanavalin A (Con A). These seemingly S-specific bands, only detected in the stigmatic tissue of the flower, were functionally correlated to the self-incompatibility reaction. Although differing in their isoelectric point, the S allele specific molecules are all resolved into several glycoprotein components of similar molecular mass on SDS-gels (between 57-65 kilodaltons molecular mass). The question arises as to whether the S locus codes for the core protein or controls the modifying enzymes involved in glycosylation. All the data available on the molecular size and nature of the S-specific molecules of _Brassica_ stigmas are grouped in Table 1.

In conclusion, the putative S molecules of _Brassica_ stigmas are exclusive to the stigma, correlate with the stigma receptivity, bind the lectin Con A and possess related molecular structure.

| S allele | Nature and molecular mass (daltons) | Reference |
|---|---|---|
| $S_2$ | Glycoprotein; MM 54,000 | Ferrari et al.(1981) |
| $S_6$ | Con A-binding glycoproteins MM 63,000; MM 65,000 | Nasrallah and Nasrallah (1984) |
| $S_7$ | Con A-binding glycoproteins MM 57,000; MM 59,000 | " |
| $S_{13}$ | Con A-binding glycoproteins MM 61,000; MM 63,000 | " |
| $S_{14}$ | Con A-binding glycoproteins MM 62,000; MM 64,000 | " |
| $S_{22}$ | Con A-binding glycoproteins MM 60,000; MM 65,000 | Nishio and Hinata (1982) |
| $S_{39}$ | Con A-binding glycoprotein MM 57,000 | " |

Table 1: Nature and molecular mass of putative S specific molecules of Brassica oleracea stigmas.

What is the location of the read-out system?

Brassica stigma belongs to the dry-type stigma with a surface covered by a thin proteinaceous layer termed the pellicle (Mattsson et al. 1974). The pellicle is not visible by transmission electron microscopy with heavy metal salts and needs specific cytochemical methods to be visualized (Gaude and Dumas, in press). An interesting feature of these techniques is that the Con A is able to bind to the pellicle suggesting that the S-specific glycoproteins may be components of the pellicle. Moreover , it has been shown that the efficiency of the stigmatic barrier depends on the physical and chemical integrity of the stigma surface (see Gaude et al. 1985). All this data allows to consider that the read-out system of Brassica is pellicle located.

3-Pollen information

A considerable number of proteins and glycoproteins are present in the pollen and pollen wall (see review in Knox 1984). By isoelectric focusing of homogenates of Brassica pollen, Nishio and Hinata (1978) detected some thirty bands. However, neither S gene specific differences nor S gene specific antigens have been determined in Brassica pollen. In the literature, the existence of S-specific molecules has been reported only in 2 species: Petunia and Oenothera, which present a gametophytic self-incompatibility system (see review in Dumas et al. 1984). Unfortunately, no further progress has been made with this system in identifying and characterizing the active proteins.

In order to elucidate the possible presence of S-specific molecules in Brassica pollen, we have carried out an electrophoretic study with the aim of comparing the pollen protein patterns of several self-incompatible lines. 4 homozygous self-incompatible genotypes $S_3$,$S_{11}$,$S_{16}$ $S_{17}$ of Brassica oleracea var. acephala were employed (kindly supplied by Dr. D.Ockendon, NVRS, Wellesbourne, UK). Our analysis by sodium dodecyl sulfate polyacrylamide gel electrophoresis (SDS-PAGE) combined with a highly sensitive silver stain allowed the characterization of a few particular bands that are not common to all genotypes (see Fig.1)

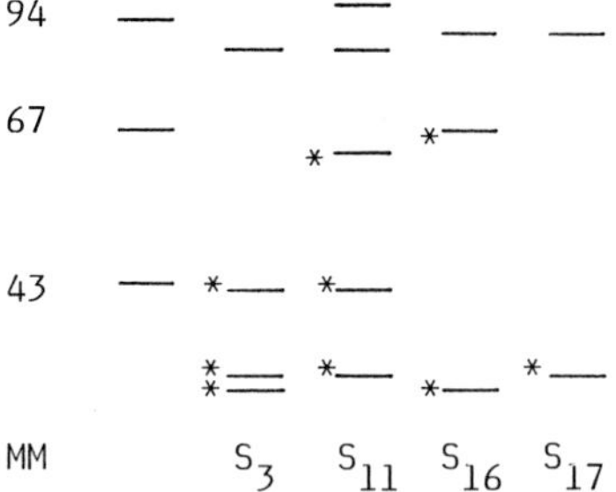


Figure 1: Particular bands of protein patterns from S different pollen extracts (after SDS-PAGE: *:Con A-binding glycoprotein
MM: molecular mass markers x $10^3$ daltons)

Following electrophoretic transfer of proteins from SDS-gels to nitrocellulose membranes (electroblotting) and staining of glycoproteins by the Con A-peroxydase method (Hawkes 1982), we demonstrated that some of these particular bands are glycoproteins. For $S_{16}$ and $S_{17}$ pollen extracts, the genotype-specific bands can be only detected in mature anthers (a few hours prior to anthesis) or mature viable pollens but never in immature anthers harvested from buds at about 5 days prior to anthesis. After electroblotting of equivalent amounts of protein from these three pollen developmental stages, Con A-peroxidase staining revealed that a striking increase in the glycoprotein concentration occurs with maturation. Consequently, it appears that the few hours prior to pollen maturation are concerned with glycosylation. Whether glycosylation implicates sporophytic or gametophytic originated proteins is yet not known? The acquisition of new glycoproteins during pollen development has been confirmed by IEF which allows the characterization of two new diffuse bands with basic isoelectric points.

These new findings are very exciting since it is the first time that proteins (or glycoproteins) of Brassica pollen are reported to be possibly related to the S gene expression. However, the S-specificity of these molecules remains to be more clearly demonstrated. The use of the callose reaction as a biological test for discriminating S-specific molecules may be very helpful in this perspective. The callose response of stigmas, which is a consequence of self-incompatibility, has effectively been proved to be induced by specific pollen components, probably associated with expression of the S-gene (Kerhoas et al. 1983). Moreover, a question arises as to whether all the putative S-specific bands constitute subunits of a native oligomeric S-molecule.

What is the location of the pollen information?

The proteins housed in the exine are first to make contact with the pistil while those in the intine may be released more gradually during germination (Heslop-Harrison et al. 1975). There is thus evidence for a considerable number of potential informational molecules at the surface of the male partner. Recently, we have detected a surface exinic layer which presents some of the characteristics of a membrane (Gaude and Dumas 1984). This layer, designated the exinic outer layer (EOL) could participate in the attachment of the pollen grain to the stigma and constitutes a site of choice for the location of S-specific molecules. Immunocytochemical methods using monoclonal antibodies raised against the putative S-proteins may permit a precise location of these compounds in the pollen.

We may conclude that the molecular bases of the pollen-stigma recognition in Brassica seem to be based on interactions between glycoproteins located on cell surfaces of both partners. A greater knowledge of the molecular nature of S-products will considerably improve our understanding of this recognition process that regulates fertilization and seed-setting in flowering plants.

Acknowledgments: We thank Miss Anne-Marie Thierry for skilled technical assistance.

References

Dumas C, Knox RB, Gaude T (1984) Pollen-pistil recognition: new concepts from electron microscopy and cytochemistry. Intern Rev Cytol 90: 239-271.

Ferrari TE, Bruns D, Wallace DH (1981) Isolation of a plant glycoprotein involved with control of intercellular recognition. Plant Physiol 67: 270-277.

Gaude T, Dumas C (1984) A membrane-like structure on the pollen wall surface in *Brassica*. Ann Bot 54: 821-825.

Gaude T, Dumas C (1985) Organization of stigma surface components in *Brassica*. J Cell Sci (in press).

Gaude T, Palloix A, Hervé Y, Dumas C (1985) Molecular interpretation of overcoming self-incompatibility in *Brassica*. *In*: Willemse MTM, Went Van JL (eds) Sexual reproduction in seed plants, ferns and mosses. Purdoc Wageningen, pp 102-104.

Hawkes R (1982) Identification of Concanavalin A-binding proteins after sodium dodecyl sulfate gel electrophoresis and protein blotting. Anal Biochem 123: 143-146.

Heslop-Harrison J (1975) Incompatibility and the pollen stigma interaction. Ann Rev Plant Physiol 26: 403-425.

Heslop-Harrison J, Knox RB, Heslop-Harrison Y, Mattsson O (1975) Pollen wall proteins: emission and role in incompatibility responses. *In*: Duckett JG, Racey PA (eds) The biology of the male gamete. Academic Press, pp 189-202.

Kerhoas C, Knox RB, Dumas C (1983) Specificity of the callose response in stigmas of *Brassica*. Ann Bot 52: 597-602.

Knox RB (1984) Pollen-pistil interactions. *In*: Linskens HF, Heslop-Harrison J (eds) Cellular interactions. Springer Verlag, pp 508-608.

Mattsson O, Knox RB, Heslop-Harrison J, Heslop-Harrison Y (1974) Protein pellicle of stigmatic papillae as a probable recognition site in incompatibility reactions. Nature 247: 298-300.

Nasrallah ME, Wallace DH (1967) Immunogenetics of self-incompatibility in *Brassica oleracea* L. Heredity 22: 519-527.

Nasrallah JB, Nasrallah ME (1984) Electrophoretic heterogeneity exhibited by the S-allele specific glycoproteins of *Brassica*. Experientia 40: 279-281.

Nasrallah ME, Barber JT, Wallace DH (1970) Self-incompatibility proteins in plants: detection, genetics and possible mode of action. Heredity 25: 23-27.

Nettancourt D de (1984) Incompatibility. *In*: Linskens HF, Heslop-Harrison J (eds) Cellular interactions. Springer Verlag, pp 624-639.

Nishio T, Hinata K (1977) Analysis of S-specific proteins in stigma of *Brassica oleracea* L. by isoelectric focusing. Heredity 38: 391-396.

Nishio T, Hinata K (1978) Stigma proteins in self-incompatible *Brassica campestris* L. and self-compatible relatives, with special reference to S-allele specificity. Jap J Genet 53: 197-205.

Roberts IN, Stead AD, Ockendon DJ, Dickinson HG (1979) A glycoprotein associated with the acquisition of the self-incompatibility system by maturing stigmas of *Brassica oleracea*. Planta 146: 179-183.

Sedgley M (1974) Assessment of serological techniques for S-allele identification in *Brassica oleracea*. Euphytica 23: 543-551.

Shivanna KR, Heslop-Harrison Y, Heslop-Harrison J (1978) The pollen-stigma interaction: bud pollination in the Cruciferae. Acta Bot Neerl 27: 107-119.

# Isozyme Markers for the Incompatibility Loci in Rye

G. WRICKE[1]

## Introduction

Genetic incompatibility systems are already used for the production of hybrid seed in some species. *Brassica oleracea* is the best known example in which this method has reached economic importance.

Recently it has been proposed (Wricke, 1984a, b) to use the two factor gametophytic incompatibility system in rye for breeding hybrid varieties. A presupposition for the application of this system is that enough selfed seeds can be produced from single plants. In rye it is possible to obtain enough seed set after selfing by heat treatment (Wricke, 1978). Among the plants selfed in this way are genotypes $S_1S_2Z_3Z_4$, i.e. heterozygous at both loci (het het). Lines developed from such genotypes can be propagated permanently under normal environmental conditions when the incompatibility system is active. England (1974) proposed this method for breeding hybrid varieties for forage grasses. He was the first to suggest that by simple interplanting of such lines, heterozygous at both incompatibility loci, 83.3% crossing between lines should occur. Without incompatibility only 50% crossing between lines would be expected. Thus, by the proposed method, a high amount of heterosis between two lines can be exploited.

In a species with a two factor gametophytic incompatibility system genotypes occur also which are heterozygous at one locus and homozygous at the other, for example $S_1S_1Z_3Z_4$ or $S_1S_2Z_3Z_3$,

1 Inst. f. Angewandte Genetik, Universität Hannover, Germany F.R.

so called ho het types. Lines developed from ho het genotypes can not be propagated under normal environmental conditions because only one genotype is formed. The question arises in which proportions het het and ho het genotypes do occur in random mating populations. The relation of these types depends on the number of alleles at the two loci. General formulae are given by Weber et al. (1982). Trang et al. (1982) estimated the number of alleles in the German variety 'Halo' and found 7 alleles at the first and 12 alleles at the second locus. In this case 12.5% of the genotypes in the population are of the type $S_iS_iZ_kZ_l$, 6.3% $S_iS_jZ_kZ_k$ and 81.1% of the het het type $S_iS_jZ_kZ_l$. Therefore one might conclude that for the range of the observed numbers of alleles the probability of het het genotypes in a population is sufficiently high for practical purposes.

First experimental results (Wricke, 1984a) with hybrids generated by the proposed method point towards such a high rate of intermating between lines as predicted by the theory. These first results suggest that it should be possible to breed hybrid varieties by exploiting the self-incompatibility system. However, there are still further questions which have to be investigated experimentally to confirm the practicability of this method. These concern the propagation of the lines before the production of hybrid seed. The first question refers to the spontaneous occurrence of mutations leading to new incompatibility alleles. Furthermore it is important to know if there are functional differences between the two incompatibility loci and differences in the efficiency of incompatibility alleles. Such questions can be more conveniently investigated if the two incompatibility loci can be marked by other genes. Therefore, it was tried to find isozyme loci which are linked to the incompatibility loci. Preliminary experiments had suggested, that one incompatibility locus is linked to the peroxidase locus Prx 7.

## Materials and Methods

Peroxidase isoenzymes were separated by isoelectric focussing in polyacrylamide flat gels having a pH 3.5 - 10 gradient (LKB ampholytes). For a more detailed description of separation and staining techniques, see Wehling et al. (1985). Heterozygous plants at the peroxidase isozyme locus Prx 7 (P) of the cultivars 'Dankowskie' and 'Animo' were selfed in the growth chamber under a constant temperature of 30°C during flowering time (Wricke, 1978). Within each selfed progeny, plants homozygous for one of the electrophoretically distinguishable Prx 7 alleles (hom 1 or hom 2) were crossed to heterozygous individuals (het). If P is closely linked to the incompatibility locus Z, with $P_1$ being linked to $Z_3$ and $P_2$ to $Z_4$, three types of crosses can occur. In the first type, for example $S_1S_2Z_3Z_3P_1P_1$ x $S_1S_1Z_3Z_4P_1P_2$, a large excess of genotypes heterozygous for the P-locus is expected. Another class of crosses, for example $S_1S_1Z_3Z_3P_1P_1$ x $S_1S_2Z_3Z_4P_1P_2$, would yield a segregation of about 1 homozygous to 2 heterozygous and in a third class, e.g. $S_1S_1Z_3Z_3P_1P_1$ x $S_2S_2Z_3Z_4P_1P_2$, the typical 1 : 1 segregation for backcrosses is expected.

If P and Z segregate independently, a 1 : 1 ratio would be expected in all crosses. Reciprocal het x hom crosses should, if they yield any seed at all, display a 1 : 1 segregation of the P-locus in all cases.

Analysis of segregational data were made by means of $\chi^2$ using Yates' correction for continuity where necessary.

## Results

Table 1 gives the results of 22 crosses between heterozygous and homozygous peroxidase genotypes (Wricke and Wehling, 1985). Most of the crosses (No. 1, 3, 5, 7-11, 14, and 15) show either zero or only few homozygous genotypes and belong to the first type of described testcrosses. The second class is represented by the crosses No. 12, 16, 20 and 22. The third class comprises crosses 18 and 21.

All progenies of het x hom crosses display a 1 : 1 segregation and thus the results of all crosses in Table 1 confirm a close linkage between Prx 7 and one locus of the incompatibility system.

From the progenies of the first type of cross a recombination value of p = 1.9% is obtained. From the second type a zero value is estimated. We may conclude, therefore, that the true value of p lies between 1.9% and zero.

Table 1: Segregation at the Prx 7 locus after crossing Prx 7 homozygous and heterozygous plants.

| Cross No.: | Prx 7 genotype of parents | Segregation of progeny | | |
|---|---|---|---|---|
| | | Prx 7 hom : het | $\chi^2$ (1 : 1) | $\chi^2$ (1 : 2) |
| 1 | hom 2 x het | 1 : 13 | 8,64+ | - |
| 2 | het x hom 2 | 24 : 25 | 0,00n.s. | - |
| 3 | hom 2 x het | 0 : 71 | 69,01+ | - |
| 4 | het x hom 2 | 9 : 10 | 0,00n.s. | - |
| 5 | hom 2 x het | 0 : 53 | 51,02+ | - |
| 6 | het x hom 2 | 45 : 49 | 0,10n.s. | - |
| 7 | hom 1 x het | 0 : 42 | 40,02+ | - |
| 8 | hom 1 x het | 3 : 70 | 59,67+ | - |
| 9 | hom 2 x het | 0 : 49 | 47,02+ | - |
| 10 | hom 2 x het | 1 : 41 | 36,21+ | - |
| 11 | hom 2 x het | 1 : 49 | 44,18+ | - |
| 12 | hom 2 x het | 35 : 69 | 10,47+ | 0,001 n.s. |
| 13 | het x hom 2 | 11 : 10 | 0,00n.s. | - |
| 14 | hom 2 x het | 1 : 40 | 35,22+ | - |
| 15 | hom 1 x het | 2 : 34 | 26,69+ | - |
| 16 | hom 1 x het | 5 : 18 | 6,26+ | 0,92 n.s. |
| 17 | het x hom 1 | 13 : 18 | 0,52n.s. | - |
| 18 | hom 1 x het | 39 : 47 | 0,57n.s. | 5,06+ |
| 19 | het x hom 1 | 7 : 12 | 0,84n.s. | - |
| 20 | hom 1 x het | 12 : 25 | 3,89+ | 0,003 n.s. |
| 21 | hom 1 x het | 48 : 64 | 2,01n.s. | 4,15+ |
| 22 | hom 1 x het | 9 : 19 | 2.89n.s. | 0,004 n.s. |

+ : significant for $\alpha = 0,05$

## Discussion

This system should enable us to study the questions outlined above. It is also tried to map the second incompatibility locus via an isozyme marker. Preliminary data indicate that the other incompatibility locus is linked to a 6-phosphogluconate-DH locus. The progeny of one cross of homozygous x heterozygous for the isozyme gave a segregation of 103 heterozygous : 1 homozygous.

The proposed method of breeding hybrid varieties by means of self-incompatibility in rye depends on a well isolated propagation of the parental lines. Migration of foreign pollen genotypes with other incompatibility alleles leads to a reduction of the percentage of intercrossing between the two lines. The same effect would be observed if a mutation to a new self-incompatibility allele occurs within the lines. Generally the probability of losing a new mutation by drift is relatively high. A new incompatibility allele, however, has a great selective advantage and will spread very fast through the line.

There are only few results published in the literature concerning the spontaneous occurrence of new incompatibility alleles in other species. In connection with the fact that small natural populations maintain polyallelic series of unexpectedly large size, S. Wright discussed the question from the population genetics point of view already in 1939. But until now no clear answer can be given what are the mutation rates of S-loci. In spite of intensive research efforts our understanding of the origin and nature of genetic polymorphism at the self-incompatibility locus is still very limited.

With the two-factor gametophytic incompatibility system lines stemming only from one plant can be established and propagated permanently. These lines contain two alleles at each locus. If complete isolation from other rye plants during propagation is guaranteed third alleles can originate only by mutation. Lines which are propagated during the breeding process represent a suitable material for studying the occurrence of new alleles. For such investigations marker genes can be very valuable to distinguish between spontaneous mutations and fertilization by foreign pollen. Thus, such investigations of determining the spontaneous mutation rate are not only important for the practical plant breeder but should give also valuable information on the nature of incompatibility loci.

Isozyme markers for the incompatibility loci should be useful to investigate whether there are functional differences between the two incompatibility loci and differences in the efficiency of different S-alleles. Differences in levels of self-incompatibility have been found in many species. In _Festuca pratensis_ self-incompatibility is governed by the same two factor incompatibility mechanism as in rye. Here, Lundqvist (1964) observed, that in selfed families (obtained by pseudo-compatibility) deviations from the expected segregation, both at the level of S - Z pairs and among the individual alleles, occur. Similarily in rye marker genes should help to investigate whether pseudo-compatibility leads preferentially to certain genotypes and how this is related to the degree of self-incompatibility under normal environmental conditions.

## References

ENGLAND, F.J.W., 1974: The use of incompatibility for the production of $F_1$ hybrids in forage grasses. Heredity 32, 183-188.

LUNDQVIST, A., 1964: The nature of the two-loci incompatibility system in grasses. IV. Interaction between the loci in relation to pseudo-compatibility in _Festuca pratensis Huds_. Hereditas 52, 221-234.

TRANG, Q.S., G. WRICKE and W.E. WEBER, 1982: Number of alleles at the incompatibility loci in _Secale cereale L_. Theor. Appl. Genet. 63, 245-248.

WEBER, W.E., G. WRICKE and Q.S. TRANG, 1982: Genotypic frequencies at equilibrium for a multi-locus gametophytic incompatibility system. Heredity 48, 377-381.

WEHLING, P., G. SCHMIDT-STOHN and G. WRICKE, 1985: Chromosomal location of esterase, peroxidase and phosphoglucomutase isozyme structural genes in cultivated rye (_Secale cereale L_.). Theor. Appl. Genet. (in press).

WRICKE, G., 1978: Pseudo-Selbstinkompatibilität beim Roggen und ihre Ausnutzung in der Züchtung. Z. Pflanzenzüchtg. 81, 140-148.

WRICKE, G., 1984a: Hybridzüchtung beim Roggen mit Hilfe der Inkompatibilität. Vortr. Pflanzenzüchtg. 5, 43-54.

WRICKE, G., 1984b: Incompatibility and hybrid breeding in rye. Vortr. Pflanzenzüchtg. 7, 50-60.

WRICKE, G., and P. WEHLING, 1985: Linkage between an incompatibility locus and a peroxidase isozyme locus (Prx 7) in rye. Theor. Appl. Genet. 69 (in press).

WRIGHT, S., 1939: The distribution of self-sterility alleles in populations. Genetics 24, 538-552.

# Regulation of Pollen Germination and Overcoming the Incompatibility Mechanism in *Brassica campestris* by $CO_2$

CHANDER P. MALIK[1] AND AKOMKAR S. DHALIWAL[1]

## 1 Introduction

The auto-incompatibility in *Brassica* is sporophytic and at least 50 different alleles at the S locus regulate the specificity of reaction. The rejection of incompatible pollen is at 3 levels: failure of germination; barrier to pollen tube penetration and inhibition of tube growth in the stigmatic surface. Role of $CO_2$ in metabolic regulation and cellular control has been discussed(Mitz, 1979 and Basra & Malik, 1985). It is shown to affect many plant physiological processes. $CO_2$ reportedly stimulates *in vitro* pollen tube growth in several species. Its role in overcoming incompatibility is recognized (Nakanishi & Hinata, 1975; Dhaliwal *et al.*, 1981 and O'Neill *et al.*, 1984). The present communication reviews some observations on the self-incompatibility responses in *Brassica campestris* var. toria following $CO_2$ enrichment (4-5%).

## 2 $CO_2$ Enrichment and Overcoming Self-Incompatibility

Self pollination encounters several barriers to successful fertilization and seed set; pollen adhesin, hydration, germination, tube penetration and its growth through stigma-style tissue. High RH, temperature and $CO_2$ overcome incompatibility leading to self seeding. $CO_2$(4%) with high RH enhanced volume, fresh wt. and shape of pollen in *B. campestris*. $CO_2$ increases the flow of water molecules from the stigmatic papillae to the pollen by affecting the lipid protein of the membrane. It may influence the osmoregulation by affecting plasma-membrane. It also affects adhesion of pollen to stigmatic papillae firmly. How $CO_2$ affects the formation of "biologically-active-molecule" causing pollen tube growth is not clear even though it is reported to affect protein synthesis in several systems(see Basra

[1]Department of Botany, College of Basic Sciences & Humanities, Punjab Agricultural University, Ludhiana - 141004, India.

& Malik, 1985)! We have already reported stimulation of germination and tube growth and release of esterases when $CO_2$ was increased from 0.03% (in air) to 4% (Dhaliwal et al., 1981 ; Dhaliwal & Malik, 1982). Non-specific esterases are reportedly involved in the active cutinase complex essential for the dissolution of cuticle. There is a good scope to examine the effect of $CO_2$ on G-inh or G-act (Ferrari & Wallace, 1977).

Our studies also suggest that the fastest self-pollen tube(in the presence of $CO_2$) grew nearly as fast as the fastest cross-pollen tube($-CO_2$) (Dhaliwal & Malik, 1982). We recognize 5 check points for self pollen to germinate and we believe that $CO_2$ does not affect one step but is pleiotropic in action.

3 Non-photosynthetic $CO_2$ Fixation

Dhaliwal et al.(1981) have demonstrated that self- and cross-pollinated pistils of *B. campestris* could fix $CO_2$ non-photosynthetically via PEP carboxylase(PEPc); label was significantly high in cross-pollinated pistils. PEPc activity was more in pollen germinated in $CO_2$ atmosphere. Most label from $^{14}CO_2$ was in water-soluble fraction, with lesser amounts in proteins, starch, lipids and insoluble fraction. Ion exchange chromatography of water-soluble components into acidic, basic and neutral fractions revealed 57 : 12 : 10 distribution, respectively. High % of fixed $^{14}CO_2$ among the acid and basic fractions was in malate and glutamic acid, respectively. It seems that $CO_2$ stimulates tube growth by osmoregulation via synthesis of organic anions to balance cation uptake. We also estimated higher malate content in cross-pollinated pistils compared with self pistils.

4 Effect of $CO_2$ on Metabolic Status of Stigma-Style Tissue

Following $CO_2$ treatment we have studied changes in some enzymes and metabolites in selfed- and crossed stigma-style tissues :

5 Esterase Activity

Cross-pollinated stigmas had esterase activity higher than self-pollinated ones. With $CO_2$ the activity doubled in the latter.

6 Stigma Callose Response

The solubility properties of labelled glucan and lipid fractions formed from UDP-glucose-$^{14}C$ indicated that $^{14}C$-glucan soluble in hot water was predominant in selfed stigmas. In self-pollinated stigmas $^{14}C$ incorporation in different fractions was more than crossed ones. With $CO_2$, no noticeable change was recorded.

7 Glycolytic and Pentose Phosphate Pathway

$C_1/C_6$ ratio was calculated from $^{14}CO_2$ evolution from glucose-(1-$^{14}C$) or glucose-(6-$^{14}C$) which reflected the relative contribution of the two pathways. Predominance of PPP in cross-pollinated pistils is noticed. Glucose-6-P DH activity was nearly double in cross-pollinated pistils. The rejection of incompatible male in *B. campestris* is characterized by occlusion of pollen tube and stigmatic papillae by callose(B-1,3-glucan) which prevents tube entry into the papillae and is the first visible manifestation of incompatibility reaction(O'Neill *et al.*, 1984). Elevated $CO_2$ blocked the formation of callose in the stigmatic papillae. Pre-pollination treatment of pollen or stigma with $CO_2$ did not alter SI(Dhaliwal & Malik, 1981). High $CO_2$ levels must be present at a time of intercellular communication between pollen and stigma. O'Neill *et al.*(1984) have suggested that $CO_2$ primarily prevents physical blockage of self pollen-tube entry into papillae by removing the callose rejection response. $CO_2$ might disturb molecular communication between pollen and stigmatic papillae at a time when S-gene product acts. There is a possibility of $CO_2$ reaction involved in conformational changes associated with carbamate formation affecting S-specific glycoprotein in order that they fail to form biologically active molecules resulting in normal growth of self-pollen tubes through stigma papillae. Activity of synthases as indicated by high incorporation of $^{14}C$ into different fractions, was more following illegitimate pollination compared with the compatible one. $CO_2$ (4%) apparently did not affect B-1,3-glucan synthase activity indicating that $CO_2$ enables the tube penetration by activating callase. The data for PPP and G-6-PDH imply that following callose degradation, glucose moiety released is diverted to pentose phosphate and glycolytic pathways. The activities of these pathways were high in compatible mating or following $CO_2$ treatment of selfed stigmas. In cross-pollinated stigma supplied with $CO_2$, the

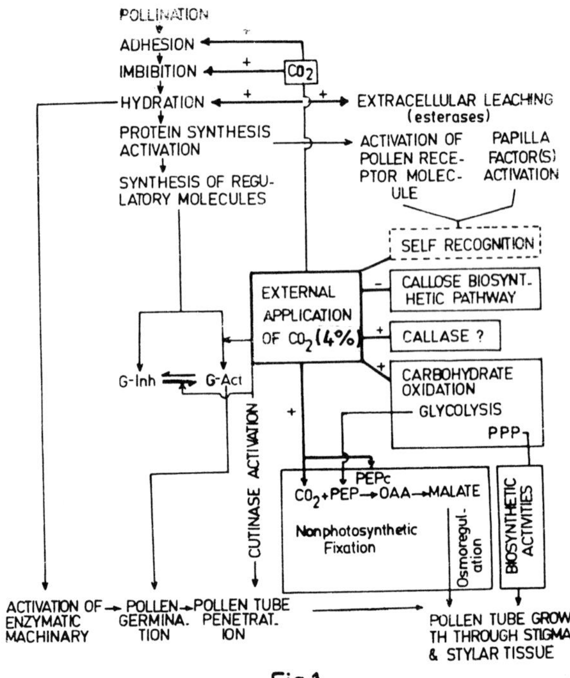


Fig.1

Table 1. Solubility of radioactive products of self- and cross pollinated stigma-style with UDP-glucose-$^{14}C$ as substrate. 1 hr assay with enzyme equivalent to one mg fresh wt of tissue

| Fraction | Pollination | | |
|---|---|---|---|
| | Self | Cross | Self + $CO_2$ |
| Chloroform-methanol soluble | 1018 | 654 | 481 |
| Hot water soluble | 518 | 3725 | 4591 |
| Alkali soluble | 50 | 30 | 46 |
| Alkali insoluble | 348 | 380 | 207 |
| Pellet | 7262 | 5482 | 7227 |

Table 2. $^{14}CO_2$ evolution from glucose-(1-$^{14}C$) or G-(6-$^{14}C$)by self- and cross-pollinated pistils. G-6-PDH activity expressed as 340 $mg^{-1}$ protein $min^{-1}$.

| Pollination | G-6-PDH activity | | cpm/pistil | | |
|---|---|---|---|---|---|
| | *in vitro* | *in vivo* | $C_1$ | $C_6$ | $C_1/C_6$ |
| Self | 6.5 | 3.4 | 15635 | 6099 | 1:0.39 |
| Cross | 12.5 | 5.7 | 19870 | 10681 | 1:0.54 |
| Self + $CO_2$ | 12.0 | - | - | - | - |

enzyme activity increased and became almost equal to compatible mating. Non-specific esterases, are suggestively involved in the active cutinase complex essential for the breakdown of the cuticle. Compatible pollination and also $CO_2$(4%) increased the activity of some esterases after selfing. $CO_2$ also increased leaching of <u>in vitro</u> esterases from incompatible pollen. Possibility of $CO_2$ affecting membrane permeability is suggested.

The role of different metabolic pathways in SI pollinations with elevated $CO_2$ levels is presented(Fig.1). It is adopted from Ferrari & Wallace(1977) model. Elevated $CO_2$ levels possibly mobilize and stimulate the utilization of respiratory intermediates; seemingly stimulates initially glycolysis and TCA cycles, etc., and causes starch or stored glucan breakdown to glucose-6-phosphate. Glucose-6-phosphate in the incompatible mating enters callose biosynthesis while in compatible mating it is diverted to PPP or glycolytic pathway. In SI pollination, normal carbohydrate metabolism is disturbed and a shift towards callose biosynthesis occurs. One of the $CO_2$ effects include disturbance of callose biosynthesis in some way. In cross-pollination, PP and glycolytic pathways were efficient, the former is the source of pentoses and NADPH while the latter supplies PEP. In cross-pollination, as a consequence of non-photosynthetic fixation, there is efficient malate production and high operation of PPP. During tube growth, high malate, availability of more reducing power and organic acids, are utilized in divergent biosynthetic activities. Following $CO_2$ application, increase in PPP, PEP carboxylase, malate dehydrogenase and malate is observed. Clearly, elevated $CO_2$ affects these metabolic pathways in self-pollinated pistils in some way. The role of $CO_2$ at the recognition level is obscure though our studies bring out its role in pollen adhesion, hydration and some phases of germination, tube growth and penetration of stigmatic papillae. The study of direct effects of $CO_2$ on the pellicle may provide information on events prior to pollen germination. The direct reaction of $CO_2$ with amino acids, peptides and proteins to form carbamate which causes structural changes in the molecule following the temporary change of charge from (+) to (-) resulting in attraction of opposite or repulsion of similar charge to open or close the holes in the membrane is envisaged. $CO_2$ may increase the flow of water from papilla to pollen by affecting lipid portion of the membrane which softens under $CO_2$ tension and becomes thinner and thinner.

References

Basra AS, Malik CP (1985) Non-photosynthetic fixation of carbon dioxide and possible biological roles in higher plants. Biol Rev 22: (in press).

Dhaliwal AS, Malik CP (1980) Effect of relative humidity and $CO_2$ on the shape, volume and fresh weight of *Brassica campestris* L Pollen *in vitro*. Ind J Expt Biol 18 : 1522-23

Dhaliwal AS, Malik CP (1981) Differences in the pentose phosphate pathway and glycolytic pathway in the self-incompatible and compatible genotypes of *Brassica campestris*. Newslett No. 13: 47-51

Dhaliwal AS, Malik CP (1982) Localization of some hydrolases in pollen and stigma of *Brassica campestris* following compatible and incompatible pollination. Phytomorphology 32: 37-41

Dhaliwal AS, Malik CP, Singh MB (1981) Overcoming incompatibility in *Brassica* by carbon dioxide and dark fixation of the gas by self and cross pollinated pistils. Ann Bot 48: 227-233.

Ferrari TE, Wallace DH (1977) A model for self-recognition and regulation of the incompatibility response of pollen. Theor Appl Genet 50: 211-225

Mitz MA (1979) $CO_2$ biodynamics: A new concept of cellular control. J Theor Biol 80: 537-551

O'Neill P, Singh MB, Neales TF, Knox RB, William EG (1984) Carbon dioxide blocks the stigma callose response following incompatible pollinations in *Brassica*. Plant Cell and Environment 7: 285-288

# A Cytoembryological Analysis of the Results of Different Types of Pollinations in Sugar Beet

N.J. WEISMAN[1]

The morphological aspect of incompatibility in sugar beet has not been studied. The little we know comes from comparative morphological studies of the embryo sac which demonstrated that fertilization does not occur after self-pollination (Charetschko-Sawizkaja 1940). Fluorescence microscopy has made it possible to observe the growing pollen tube in pistil tissue and its entering the ovule. With this method, it was established that pollen grain either failed to germinate or produced short tubes mainly at the stigma surface after self-pollination of a single self-incompatible plant (Zaikovskaja, Jujalova 1976).

Using standard methods (Weisman et al. 1984), we analysed the results of self-pollination of plants from four inbred lines of sugar beet and crosses of wild Beta procumbens with plants from two sugar beet lines. The inbreeding coefficient (F) of these lines was 0.67-0.75.

The lines differed in self-pollination (Table 1). The stigmata had abundant quantities of pollen which grew well. Pollen fertility was high and varied from 43.4 to 100%. Two lines, SOAN-26 and SOAN-27, were completely self-incompatible and formed only short pollen tubes. These tubes remained on the surface of the stigma or penetrated superficially and ceased to grow (Fig.1a). The different number of pollen tubes within the lines was probably due to the different onset of

1 Institute of Cytology and Genetics, Siberian Department of the USSR Academy of Sciences, 630090, Novosibirsk, USSR

Table 1. Results of cytoembryological analysis.

| Inbred lines | No. of plants involved | No. of flowers isolated | Pollen tube growth | | | Embryo development | | | Seed set | | |
|---|---|---|---|---|---|---|---|---|---|---|---|
| | | | No. of ovules examined | Of these tube-containing No. | Of these tube-containing % | No. of ovules examined | Of these embryo-containing No. | Of these embryo-containing % | No. of glomerulaes examined | Of these seed-containing No. | Of these seed-containing % |
| SELF-POLLINATION | | | | | | | | | | | |
| SOAN-23 | 10 | 1134 | 240 | 33 | 13.8 | 462 | 89 | 19.3 | 432 | 92 | 21.3 |
| SOAN-24 | 5 | 527 | 97 | 81 | 83.5 | 186 | 170 | 91.4 | 244 | 156 | 63.9 |
| | 5 | 417 | - | - | - | 165 | 96 | 58.2 | 252 | 104 | 41.3 |
| SOAN-26 | 7 | 947 | 220 | 0 | 0.0 | 294 | 0 | 0.0 | 233 | 0 | 0.0 |
| SOAN-27 | 7 | 612 | 138 | 1 | 0.7 | 223 | 2 | 0.9 | 251 | 3 | 1.2 |
| CROSS WITH B.PROCUMBENS | | | | | | | | | | | |
| SOAN-23 | 1 | 83 | 52 | 22 | 42.3 | 31 | 8 | 26.1 | - | - | - |
| SOAN-25 | 2 | 89 | 36 | 21 | 58.2 | 53 | 12 | 22.6 | 201 | 46 | 22.3 |
| SOAN-25(4x) | 1 | 66 | 30 | 12 | 40.0 | 36 | 4 | 11.1 | - | - | - |

germination. The ovules appeared unfertilized on 12 day after beginning of flowering - they were small and shriveled. Analysis at the stage of mature seeds revealed the presence of seeds only in 3 of 484 fruits.

Line SOAN-23 was partly self-compatible. There were many short pollen tubes penetrating into the stigmata at a short distance. Howeever, some tubes were capable of growing into pistil tissue, reach the ovary and enter the ovule. Most plants had several flowers with pollen tubes entering the ovule. And, accordingly, on 12 day from the beginning of flowering a portion of the ovules in the majority of the plants had embryos. Along with normally developing 12-day embryos, there also occurred embryos lagging in development. The data on seed set agreed well with those cytologically obtained. There were variations in the values for sed set, pollen tubes growth and embryo development. The maximum value was 57.4%.

All plants of line SOAN-24 showed high degree of self-compatibility. After germination, the pollen tubes grew in bundles towards the ovary. Coiling, they entered the micropyle (Fig.1b). There was a high percentage of 12-day old embryos and mature seeds in plants of this line. The values obtained for this line varied from 3.4% to 100%.

The results of pollination in ♀ sugar beet x ♂ B.procumbens crosses were as follows . In some of the flowers pollen either failed to germinate or produced a few superficial pollen tubes. In other flowers, pollen tubes developed like in the case of compatible intraspecific pollination in sugar beet. After entering pistil tissue, the bundle of the tubes were orientated towards the ovary. Fig.1c shows how pollen tube coils upon entrance into the micropyle portion of the ovule. In the two lines studied, the percentage of flowers with pollen tubes penetrated into the ovule was 42.3 - 58.2% of the total number of flowers examined. In the tetraploid counterpart SOAN-25, there were 40% flowers compatible with the pollen of B.procumbens .

The reciprocal pollination was incompatible. The pollen grew, produced short tubes either scattered at the surface or at stigma upper layer. The pattern was similar to the one observed in incompatibility pollinations in sugar beet.

The growth of pollen tubes in self-pollinated B.procumbens

and compatible pollinated sugar beet was quite comparable. The general appearance of pollen tubes under the fluorescent microscope: brightness of fluorescence, callose plugs, thickenings and bendings in the tube like in sugar beet pollen tubes.

The data for 12-day old ovules are in accord with those for pollen tube growth. Sugar beet x B.procumbens crosses produced embryos in 22.6-26.1% ovules. The tetraploid line, the counterpart SOAN-25, produced 11.1% of embryos. Most of the ovaries were normally developed, they had cotyledons, but these embryos were much shorter than those of 12-day old ones from intraspecific pollinations. The difference between the number of compatible pollination and the number of observed embryos was due to the death of some of the zygotes of the interspecific hybrid at the early stage of development. The 12-day old ovules from ♀ B.procumbens x ♂ sugar beet crosses were not studied.

It may be concluded that incompatibility in sugar beet is manifested in the inhibition of the growth of pollen tubes immediately below the stigmatic surface ragion. The majority of tubes do not grow beyond it. However, depending on random factors and the genotype, some tubes can penetrate deeply into the pistil and, reaching the ovule, accomplish fertilization. As a result, plans become, to some extent, pseudo-

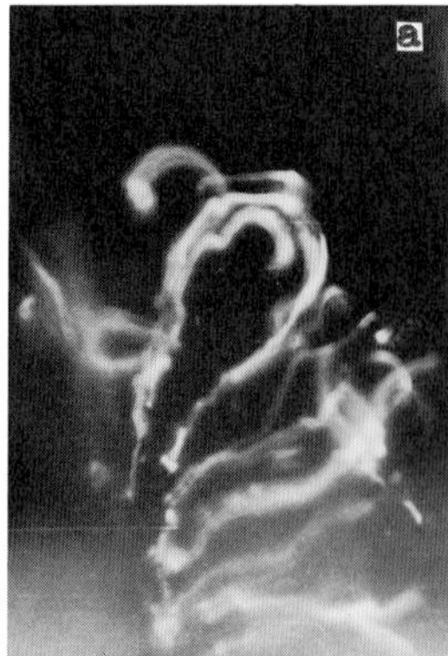


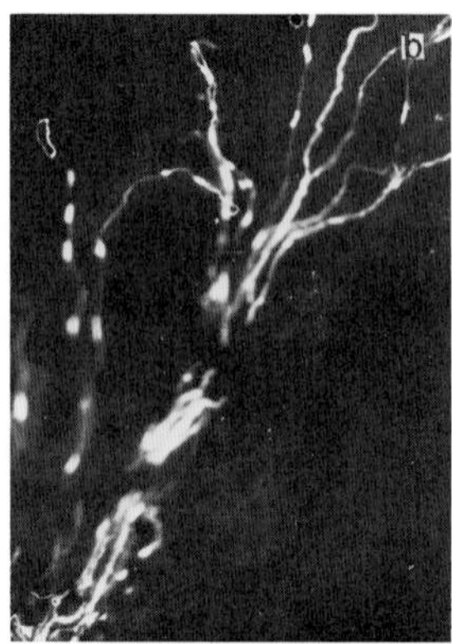


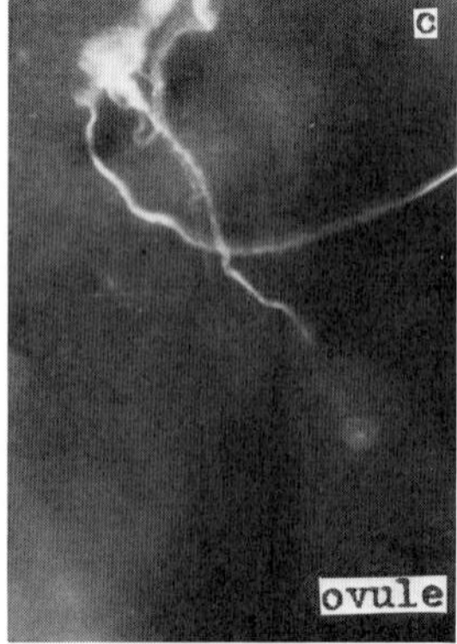


Fig. 1 a-c. Cytological phenomena connected with different types of pollinations. a: inhibition of pollen tubes growth in self-incompatible stigma tissue; b: the bundle of compatible tubes; c: pollen tube entered the ovule.

self-compatible. The incompatibility between sugar beet and B.procumbens was of unilateral kind. The inhibited growth of alien pollen tubes in some of the pollinated flowers and death of hybrid embryos were the causes of low seed set in the interspecific crosses. In the compatible pollinations, the growth patterns of B.procumbens pollen tubes from pollen germination to penetration into the ovule was similar to the one established for sugar beet.

## References

Charetschko-Sawizkaja EI (1940) Beta vulgaris species as auto-sterile plant. In: Sugar beet breeding. Kiev, pp 501.

Zaikovskaja ME, Jujalova TP (1976) Development of pollen tubes in self-sterile and self-fertile lines of sugar beet in isolation. Cytology and Genetics 10: 57-60.

Weisman NJ, Jujalova TP, Agaphonov HS (1984) The cytoembryology of incompatibility in sugar beet. In: The genetics of sugar beet. Nauka, pp 121-129.

# Selective Fertilization and Pseudocompatibility in Sugar Beet

S. I. MALETSKY[1]

1 Introduction. It is known that the S-gene system prevents self-fertilization in sugar beet. The incompatibility reaction occurs near the surface of the stigma, and it is manifest in inhibited growth of the pollen tubes (Zaikovskaya and Jujalova, 1976; Weisman et al., 1984). The incompatibility reaction can be significantly modified by environmental factors, and this made it possible to obtain seeds from plants which otherwise do not self-pollinate (pseudocompatibility). It was shown in the thirties that self-pollination carried out at an average ambient air temperature of $12^{o}$-$14^{o}$C sharply increases the level of pseudocompatibility (Kharechko-Savitskaya, 1938). In the sixties, we started studies on pseudocompatibility in sugar beet populations grown in different ecological conditions (Maletsky et al., 1970; Denisova et al., 1971). A noteworthy result was a 5-10-fold increase in pseudocompatibility of sugar beet plants grown in high altitude regions (Tien-Shan, Kirghizia). The results of these studies gave a general idea of the effects of ecological conditions on pseudocompatibility. However, no estimates of the contribution of genetic factors to pseudocompatible self-pollination were possible without comparisons of the fertilization frequencies with known genotypes.

It has been detected that one of the sugar beet S-loci is linked to the Adh-locus controlling the synthesis of alcohol dehydrogenase (Maletsky and Konovalov, 1985). This offered a novel approach for studying pseudocompatibility to base analysis of self- and cross-pollination on segregation patterns for the marker locus Adh. Two alleles at the Adh-locus were identified, one controlling the synthesis of the fast (FF), and the other the slow (SS) variant of the enzyme. Homozygotes have either the FF or SS variant while heterozygotes possess three of its variants (FF, SS and FS). The . locus

[1] Institute of Cytology and Genetics, Siberian Department of the U.S.S.R. Academy of Sciences, Novosibirsk (U.S.S.R.)

Adh is linked to the S- and l-loci. The genes are arranged in the following linear order: 1 (0.18) Adh (0.35) S; the number in parentheses is the recombination coefficient. The recombination coefficient between Adh- and S-loci was calculated under the assumption that only pollen tubes with one of the two alleles at the S-locus were involved in fertilization.

Herein, we present the results of studies on selective fertilization in the case of pseudocompatible self-pollination and compatible cross-pollination of sugar beet. In analyses of self- and cross-pollination, judgements were based on the segregation ratios estimated for the marker locus Adh.

2 Materials and Methods. An inbred sugar beet G-49-c originated from a single plant of population 140361 (Klein Wanzleben, DDR) after two generations of inbreeding (self-pollination and sibs-mating) was used. Consequently, the number of alleles at the S-loci in G-49-c could not exceed two. In all, there were 8 plants heterozygous for the Adh-locus. Two plants from a male sterile (ms) line homozygous for Adh-F allele were also used. They were grown in high altitude regions. They were self-pollinated under cotton bags. The level of pseudocompatibility was estimated. For this purpose, pieces 10 cm long (counting from the base of a stem) were taken from branches of the second and third orders, and the number of flowers and fruits set was counted on them. In the test crosses, pollen was applied to male sterile plants with shoots protected by grease-proof paper bags. To estimate self-incompatibility, we made observations on the growth of pollen tubes on the stigma and on the penetration of pollen tubes into ovary and ovule after 1 day and 7-8 days after application of pollen to the stigma. The outcome of competition between the pollen tubes was estimated on the basis of the Adh phenotype of the seeds. The seeds were taken out of the pericarp by hand, ground with pincers on an object glass in a drop of Tris-citrate buffer (pH 7.4) and then a piece of chromotographic paper, Whatman 3 MM, was moistened with the suspension. Electrophoresis and staining of the preparations were carried out according to the standard technique.

3 Results and Discussion. Analysis of the growth of pollen tubes demonstrated that all the 8 plants from G-49-c are self-incompatible; 16-24 flowers from each plant were examined on the first day after pollen application, and no pollen tubes penetrated into the ovaries and ovules. Only on days 7-8, did single pollen tubes penetrate in 10 of the 165 preparations (in the case of compatible pollination, penetration of the pollen tubes into the embryonic sac and fertilization was usually observed on the first day). The number of fruits of the self-pollinated plants varied from 1 %- 17 % of the total number of

flowers.

The segregation data for Adh in offspring of self-pollinated plants and the test crosses data are summarized in Table 1. The segregation observed differs significantly from monogenic (1 FF:2 FS:1 SS). The 8 plants are assigned to two groups: Group 1 -- in the first 3 plants the proportion of SS phenotypes is close to 14 %; Group 2 -- in the remaining plants it is about 3 %. Their genotypes may be written as: 1 -- + Adh-F $S_1$/+ Adh-S $S_2$; 2 -- + Adh-F $S_1$/l Adh-S $S_2$. In Group 1 the deviation in the segregation is due to the selective advantage of pollen tubes with $S_1$-allele over those with the $S_2$-allele. In Group 2, the deviation is also due to linkage of the Adh with a lethal gene. The results of the test crosses exclude differential viability of pollen grains in self-pollinated plants. In 5 of the 6 test crosses, the ratio of FF to FS phenotypes was close to the expected 1:1, i.e., the pollen tubes carrying alleles Adh-F and Adh-S have equal probability of being involved in fertilization.

TABLE 1.

Segregation and test cross data for the Adh gene in an inbred sugar beet population G-49-c.

| No. plant | self-pollination: number of seeds | phenotype frequency (%) FF | FS | SS | $\chi^2$ (1:2:1) | test cross: number of seeds | phenotype frequency(%) FF | FS | $\chi^2$ (1:1) |
|---|---|---|---|---|---|---|---|---|---|
| **Group 1** | | | | | | | | | |
| 9 | 105 | 32.4 | 52.4 | 15.2 | 6.12 | 69 | 46.38 | 53.6 | 0.36 |
| 16 | 77 | 39.0 | 49.3 | 11.7 | 11.46 | 261 | 52.5 | 47.5 | 0.64 |
| 27 | 105 | 40.0 | 45.7 | 14.3 | 18.15 | | | | |
| Total | 287 | 36.9 | 49.2 | 13.9 | 30.44 | 330 | 51.2 | 48.8 | 0.19 |
| **Group 2** | | | | | | | | | |
| 3 | 48 | 39.5 | 54.2 | 6.3 | 12.58 | 217 | 48.4 | 51.6 | 0.22 |
| 4 | 104 | 47.1 | 48.1 | 4.8 | 37.36 | 179 | 53.6 | 46.4 | 0.94 |
| 6 | 20 | 45.0 | 55.0 | 0.0 | 8.21 | 97 | 47.4 | 52.6 | 0.29 |
| 10 | 101 | 46.5 | 51.5 | 2.0 | 40.19 | | | | |
| Total | 273 | 45.4 | 50.9 | 3.7 | 95.30 | 493 | 50.1 | 49.9 | 0.00 |
| 24 | 60 | 45.0 | 53.3 | 1.7 | 22.8 | 206 | 27.2 | 72.8 | 87.2 |

The test cross and self-pollination data for plant No. 24 (bottom line, Table 1) are of interest. Judging by the self-pollination data, the pollen with Adh-F had selective advantage. However, judgement based on test cross data indicated that those with Adh-S also had a selective advantage. Obviously, the lethal gene affecting the segregation of the marker gene in self-pollinated plants did not affect segregation in the test crosses. When pollen from plant No. 24 is applied to ms-tester, it may be suggested that half of the grains are not involved in fertilization (pollen grain with the $S_1$-allele), and this is evidence that plants No. 24 and ms are partially cross-incompatible. If so, the differences in the results of the test crosses between sibs plants of population G-49-c (genotypically identical with respect to the S-locus linked to the Adh-locus) may be explained by their differences in other loci controlling self- and cross-incompatibility.

The first digenic model is that developed for self- and cross-incompatibility with complementary type of interaction of the S-alleles of two S-loci in the pollen and pistil tissue (Owen, 1942). Let the S-genotype of the ms-tester be written as 13.12 (13 designates the alleles at the first S-locus linked to Adh-locus, 12 that of the second S-locus with independent type segregation). The genotype of plant No. 24 is then written as 12.11, that of the other plants of population G-49-c as 12.33. This means that progenitor inbred population G-49-c had genotype as 12.13. In crosses of plant No. 24 to the tester line (13.12 x 12.11) one half of the pollen grain would be incompatible (1.1), the other half would be compatible (2.1). For this reason segregation for Adh is related to the recombination value between the Adh- and S-loci, on the one hand, and to partial cross-incompatibility in the test crosses, on the other hand. In crosses of the other plants from population G-49-c (13.12 x 12.33) with the testers, all the pollen grain would be compatible. The expected ratio for Adh phenotypes would be 1 FF:1 SS.

The rf between Adh- and S-loci is 0.27-0.28, i.e., equal to the proportion of FF homozygotes in the test cross with plant No. 24. Granting that rf = 0.28, the ratio of compatible pollen grains marked with Adh in the test crosses (ms-tester x No. 24) would be 0.72 for Adh-S and 0.28 for Adh-F.

The segregation for the marker locus is the same when considering the three loci models of inheritance of self- and cross-incompatibilty (for example, 13.12.12 female x 12.11.12 male cross, where the first locus is linked to the marker and the other segregates independently). The model can be expanded to accommodate a larger number of S-loci. The estimated recombination values between the Adh- and S-loci agree well with the present and previous values (Maletsky and Konovalov, 1985). Our present results agree with: 1/ those of the test crosses (in case, when the effect of other S-loci on formation of pollen specificity is considered); 2/ those of self-pollination in Group 1 (in case of pseudocompatible self-pollination only pollen tubes with the $S_1$-alleles are involved, but not those with $S_2$-alleles), the expected proportion of SS phenotype would be close to 13.5 %-14.0 %, the observed value is 13.9 %; 3/ the calculated rf explains also the results of self-pollination in Group 2, when taking into account that the marker locus is affected by two loci linked to it (S- and l-loci).

In 1984, the rf between Adh- and S-locus was about 0.28; in 1983, it was about 0.35-0.36 in sibs. The results obtained in 1983 were affected not only by the distance between the loci, but also by the differential competititve capacity of pollen tubes with various alleles at the linked S-locus. The difference in rf in the two experiments allowed us to estimate the contribution of pollination conditions to the competitive capacity of pollen tubes with different S-genotypes. The value of 0.35-0.36 for rf is an overestimate because there was the probability of pollen tubes with $S_1$- and $S_2$-alleles being involved. Those with the $S_2$-allele were not involved in the self-pollination in 1984. Support for this suggestion came from data on pseudocompatibility for 2 years: pseudocompatibility level was about 4 times higher per plant in 1983 than in 1984 (the variation ranges were 5.0 %-47.1 % and 1.0 %-17.0 % respectively). The proportion of SS homozygotes in self-pollination of Adh-F $S_1$/Adh-S $S_2$ was about 17.5 % in 1983 and it was only 13.9 % in 1984. The difference was possible only in the case when all pollen tubes were involved in self-pollination in 1983; however, the pollination frequency pollen tubes with $S_1$-alleles exceeds 7-8 times that with the $S_2$-alleles. In 1984, environmental factors excluded involvement of pollen tubes with $S_2$-alleles.

Thus, when self-pollination of self-incompatible plants of sugar beet occurs, pollen tubes with different S-genotypes show different degrees of incompatibility. The conditions during flowering significantly affect pseudocompatibility level in some particular S-genotypes of the pollen tubes and of the whole plant. The contribution of single environmental factors to pseudocompabilitity and competitive capacity of pollen tubes with various S-genotypes have not been, so far, studied.

Acknowlegement. Studies on sugar beet incompatibility with the use of the marker Adh-gene were carried out with the participation of graduate student A.A. Konovalov. The author is grateful to Miss A. Fadeeva for translating this paper from Russian into English.

References

Denisova EV, Maletsky SI, Lutkov AN (1971) Genetika (USSR) 8(7): 36-41

Kharechko-Savitskaya EI (1938) Proc Acad Sci USSR 18: 469-474

Maletsky SI, Denisova EV, Lutkov AN (1970) Genetika (USSR) 6(6): 180-183

Maletsky SI, Konovalov AA (1985) Genetika (USSR): in press

Owen FW (1942) J Agr Res 64: 679-698

Weisman NJa, Jujalova TP, Agafonov NS (1984) Genetics of Sugar Beet. Nauka, pp. 121-129 (USSR)

Zaikovskaya NE, Jujalova TP (1976) Cytology and Genetika (USSR) 10: 57-60

# Self-incompatibility in Light of Population Structure and Inbreeding

R.G. Olmstead[1]

## 1 Introduction

Self-incompatibility has been viewed traditionally as a mechanism to promote outcrossing and minimize inbreeding in flowering plants (Darwin 1876; de Nettancourt 1977). This has led some authors to equate self-compatibility (SC) with inbreeding and self-incompatibility (SI) with outbreeding in natural populations (Ruiz and Arroyo 1978; Mcleod et al. 1983; Wiens 1984). It is true that high levels of inbreeding exist in certain self-compatible, self-fertilizing species. However, it is not necessary that self-compatible species be highly inbred, nor that self-incompatible species be highly outbred (Bawa 1974, Kress 1983). The breeding system of a plant and the level of inbreeding in a population are related, but often in such a way that one may have little effect on the other. SC and SI are terms that relate to the breeding system of an individual plant, as are the terms selfing and outcrossing. In contrast, inbreeding and outbreeding refer to the degree of identity by descent of genes within a population and depend primarily on population size, being essentially independent of breeding system.

In the discussion that follows, the origin of SI is shown to be largely independent of the level of inbreeding in the population as a whole, and an explanation is presented based on the difference in fitness among individuals that may be sufficient to account for the evolution of SI. The foundation for this argument is the reduction in fitness exhibited by individuals that carry deleterious or lethal recessive alleles and that self-pollinate in a finite population. This reduction in fitness due to selfing may be enough to account for the maintenance of any system that can avoid selfing, regardless of the degree of inbreeding in the population.

The basis for developing this approach to the evolution of SI rests on the observation that most flowering plant species are characterized by small effective population sizes (Ehrlich and Raven 1969; Levin and Kerster 1974;

[1]Department of Botany, University of Washington, Seattle, WA 98195

Levin 1981). An important consequence of this observation is that most flowering plant populations maintain a certain level of inbreeding. The weight of empirical evidence documenting this observation (Kerster and Levin 1968; Price and Waser 1979; Schaal 1981) lends support to the primarily theoretical argument that the ensuing level of inbreeding is beneficial to plant populations (Shields 1982).

An important advantage of inbreeding is the ability that inbreeding confers on the population to reproduce a co-adapted genome (Shields 1982; Ohta 1980). Outbreeding will tend to disrupt associations of alleles that function well together by randomly recombining them into new associations. Inbreeding also can reduce the "cost of meiosis" that results from each offspring receiving only one-half of the male or female parent's genes, unless both halves come from the same parent (ie., self-fertilization) or unless some alleles share identity by descent from a common ancestor (Williams 1975). Inbreeding, with more alleles sharing identity by descent, obviates this cost of meiosis, approaching zero as the level of inbreeding increases. Both theoretical considerations (Shields 1982, 1983; Partridge 1983) and empirical research (Price and Waser 1979) suggest that an optimal level of inbreeding may exist for many, or most, flowering plant populations.

Inbreeding depression has been cited most often as the phenomenon responsible for the evolution of SI (de Nettancourt 1977; Charlesworth and Charlesworth 1979). At the population level, inbreeding depression is a reduction in the mean of individual fitnesses over the entire population. This can be manifested in more than one way. The gradual decrease in heterozygosity as a population becomes increasingly inbred may cause a general reduction in fitness throughout the population. This would be expected to occur concurrently with an increase in the benefits of inbreeding. A balance in these two factors will be expected to approach an optimal level of inbreeding (Shields 1983; Partridge 1983) without necessarily affecting the breeding system. Another manifestation of inbreeding depression, however, is the dramatic reduction in fitness of some individuals in the population, but not necessarily other members of the population, resulting in a reduction in the mean of individual fitnesses. This may come about through the greater probability of the expression of recessive lethal or deleterious alleles in the homozygous state through selfing than through outcrossing.

Whether SI has arisen once or many times in the angiosperms, it must have been derived at some time from primitively self-compatible stock. The absence of SI in extant gymnosperms and "primitive" angiosperm families supports this hypothesis. The population structure most likely to give rise to a SI system is one in which the population size is relatively small, mating is more or less at random with some self-fertilization, and reproductive capacity is moderate (Charlesworth and Charlesworth 1979; Shields 1982). Species composed of large populations with widespread pollen and/or seed dispersal and high fecundity will maintain an adequate amount of gene flow to tolerate a low level of selfing (Lande and Schemske 1985). Species with high levels of selfing will most likely maintain very low levels of deleterious or lethal alleles (Lande and Schemske 1985). Species with small population sizes and limited gene flow that exhibit very low levels of selfing as a result of effective structural or temporal outcrossing mechanisms will not be affected strongly enough by the fitness differential between the rare selfing events and predominant outcrossing to necessitate a SI system (Kress 1983).

## 2 Eliminating Selfing from Plant Populations

The avoidance of selfing is the outcome common to all SI systems. The effect of eliminating selfing from a population may be viewed in two contexts: as it relates to the reduction in the level of inbreeding in the population as a whole, and as it relates to fitness of individual members of the population. The level of inbreeding in a population may be represented by the inbreeding coefficient, F, a measure of the probability that two alleles share identity by descent from a common ancestor (Wright 1922). F will increase over time at a rate dependent upon the effective population size, $N_e$, such that: $\Delta F = 1/(2N_e)$ per generation (Falconer 1960). An equilibrium level of inbreeding that is somewhat lower than the level suggested by population size alone may be maintained by the effects of mutation, gene flow from remote populations, and selection for genetic polymorphism. Small populations will, therefore, become inbred more rapidly and reach a higher equilibrium level (fig. 1).

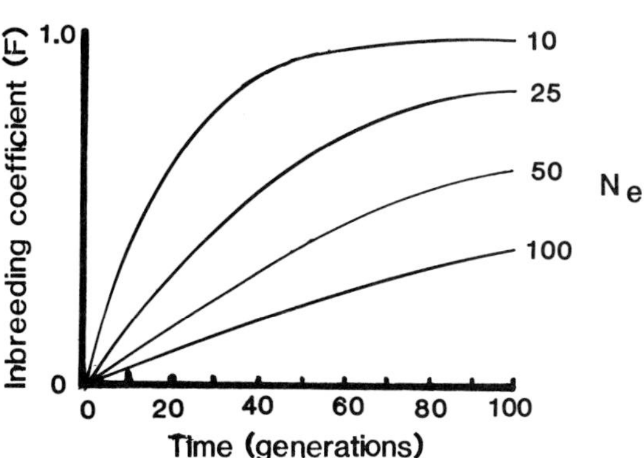


Fig. 1. The relationship between effective population size, $N_e$, and the level of inbreeding, represented by the inbreeding coefficient, F, over time, in generations (after Parkin, 1979).

Avoidance of selfing in a random mating population would have an effect on the level of inbreeding that closely approximates an increase in effective population size of one-half individual (Wright 1931; Crow and Kimura 1970). $N_e$ (outcrossing) = $N_e$ (random) + 1/2. Another way of illustrating this effect on the level of inbreeding is to consider the increase in level of inbreeding over time, assuming that an equilibrium level does not already exist. Consider the hypothetical case of two populations equivalent in all respects except that in one population random mating including selfing occurs and in the other population random mating occurs, but selfing is not allowed. A short delay in the time required by the second population to reach the same level of inbreeding as the first results from the greater effective population size of the second population and equals approximately $1 + 1/(2N_e)$ generations. Both populations will reach similar equilibrium levels of inbreeding with the second population perhaps equilibrating at a slightly lower, but not substantially different, level.

The primary effect of inbreeding is the increase in homozygosity in the population (Wright 1931). This can cause a reduction in fitness in some individuals resulting from the increased occurrence of deleterious or lethal recessive alleles in the homozygous state. Most deleterious alleles exist as recessives and occur in a population at low frequencies and primarily as heterozygotes. By selfing, a greater proportion of those alleles will recombine in a homozygous state than through cross-pollination.

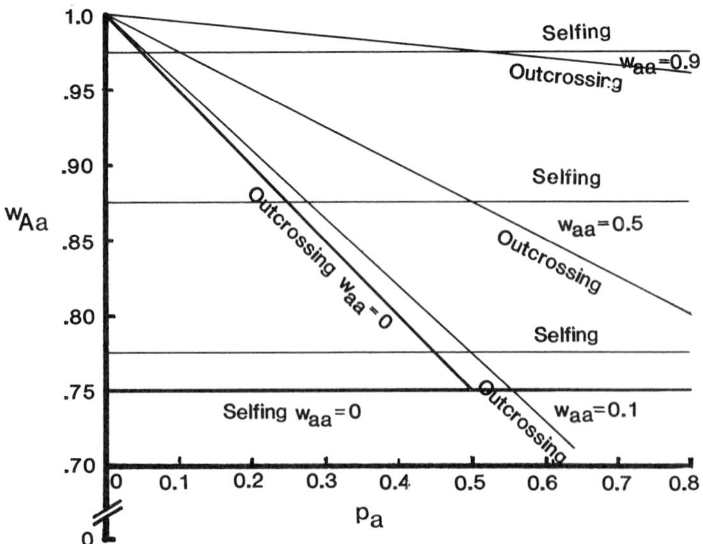


Fig. 2. Mean fitness of individuals heterozygous for lethal ($w_{aa}$ = 0) and deleterious alleles (with homozygote fitnesses, $w_{aa}$, equal to 0.1, 0.5, and 0.9) when selfing and when outcrossing.

In the case of a lethal recessive, a, the relative fitness (compared to AA) of an individual heterozygous at the locus in question, Aa, will differ depending on whether it is self- or cross-pollinated. The relative fitness will be 0.75 when self-pollinating, regardless of the frequency of the allele in the population (fig. 2). When outcrossing, the mean fitness of individuals heterozygous at the locus in question will be dependent on the frequency of the recessive allele in the population. The difference in mean fitness between heterozygous individuals that are outcrossed and those that are selfed is most pronounced at the low frequency with which a lethal recessive would be expected to occur in a population, as in the case of a recently arisen mutation. This reasoning can be extended to include cases of deleterious recessive alleles. The effective reduction in relative fitness of the heterozygote, Aa, compared to the homozygote, AA, is less than in the case of the lethal recessive, but a similar pattern emerges (fig. 4). As before, the presence of deleterious alleles at low frequencies offers a relatively greater advantage to outcrossers than to selfers.

In summary, the selective advantage gained by the avoidance of selfing can be measured by the increase in the mean of individual fitnesses in the population in the immediate generation. At the same time the avoidance of selfing has an insignificant effect on the level of inbreeding in the population. Thus, SI effectively decouples the beneficial effects of inbreeding (reduced cost of meiosis and a co-adapted genome) from the negative effects of selfing on individual fitness.

## 3 Conclusions

The relationship between breeding system and level of inbreeding in populations needs to be assessed more clearly to allow a better understanding of population dynamics in natural ecosystems. Most research into the genetics of SI systems uses cultivated plants, often highly inbred lines, where the effects of natural plant population dynamics cannot be considered in concert with the effect of the breeding system. More research is needed on breeding systems in natural populations where the effective population size can be estimated and taken into consideration when assessing the effects of breeding system.

What is the evolutionary response in plant populations when levels of inbreeding become too high? Selection controlling the level of inbreeding is most likely to affect those characteristics most directly associated with gene flow in flowering plants: pollen dispersal and seed dispersal. The role that SI plays in the avoidance of selfing contributes to the ability of flowering plant populations to maintain high levels of inbreeding while minimizing the ill effects of deleterious recessive alleles. This view of SI is compatible with the concept of optimal inbreeding (Shields 1983; Partridge 1983), which may be responsible for maintaining the evolutionary flexibility that has made angiosperms the dominant group of land plants.

Acknowledgements. I thank the following people for beneficial discussions and critical comments: William DiMichele, Jerrold Davis, Joe Felsenstein, Steven Seavey, David Mulcahy, Robyn Burnham, Ruth Shaw, and Melinda Denton.

References

Bawa KS (1974) Breeding systems of tree species of a lowland tropical community. Evolution 28:85-92.
Charlesworth D, Charlesworth B (1979) The evolutionary genetics of sexual systems in flowering plants. Proc Roy Soc London, Ser B, Biol Sci 205:513-530.
Crow JF, Kimura M (1970) An introduction to population genetics theory. Harper and Row.
Darwin C (1876) The effects of cross and self-fertilization in the vegetable kingdom. John Murray.
Ehrlich PR, Raven P (1969) Differentiation of populations. Science 165:1228-1232.
Falconer DS (1960) Introduction to quantitative genetics. Oliver and Boyd.
Kerster HW, Levin DA (1968) Neighborhood size in *Lithospermum caroliniense*. Genetics 60:577-587.
Kress WJ (1983) Self-incompatibility in Central American *Heliconia*. Evolution 37:735-744.
Lande R, Schemske D (1985) The evolution of self-fertilization and inbreeding depression in plants. I. Genetic models. Evolution 39:24-40.
Levin D (1981) Dispersal vs. gene flow in plants. Ann Missouri Bot Gard 68:233-253.
Levin D, Kerster HW (1974) Gene flow in plants. Evol Biol 7:139-220.
McLeod MJ, Guttman SI, Eshbaugh WH, Rayle RE (1983) An electrophoretic study of evolution in *Capsicum* (Solanaceae). Evolution 37:562-574.
de Nettancourt D (1977) Incompatibility in angiosperms. Springer-Verlag.
Ohta A (1980) Coadaptive gene complexes in incipient species of Hawaiian *Drosophila*. Amer Naturalist 115:121-132.
Parkin DT (1979) An introduction to evolutionary genetics. Edward Arnold.
Partridge L (1983) Non-random mating and offspring fitness. In: Bateson (ed) Mate choice. Cambridge Univ Press, pp 227-253.
Price MV, Waser N (1979) Pollen dispersal and optimal outcrossing in *Delphinium nelsonii*. Nature 277:294-297.
Ruiz T, Arroyo MTK (1978) Plant reproductive ecology of a secondary deciduous tropical forest in Venezuela. Biotropica 10:221-230.
Schaal BA (1980) Measurement of gene flow in *Lupinus texensis*. Nature 284:450-451.
Shields WM (1982) Philopatry, inbreeding, and the evolution of sex. State Univ NY Press.
Shields WM (1983) Optimal inbreeding and the evolution of philopatry. In Swingland JR, Greenwood PJ (ed) The ecology of animal movement. Clarendon Press, pp132-159.
Wiens D (1984) Ovule survivorship, brood size, life history, breeding systems, and reproductive success in plants. Oecologia 64:47-53.
Williams GC (1975) Sex and evolution. Princeton Univ Press.
Wright S (1922) Coefficients of inbreeding and relationship. Amer Naturalist 56:330-338.
Wright S (1931) Evolution in Mendelian populations. Genetics 16:97-159.

# The Heterosis Model: A Progress Report

DAVID L. MULCAHY, GABRIELLA BERGAMINI MULCAHY, AND DOUGLAS MACMILLAN[1]

The classical model of gametophytic self-incompatibility is based on a single multi-allelic locus, the S-locus. However, exceptions may involve as many as 4 different loci. According to this interpretation, the S-locus (or loci) control the active inhibition of incompatible pollen types. Although widely accepted, this model leaves several observations unexplained, three of which will be considered in the present paper.

In attempting to resolve some of these anomalies, we (Mulcahy and Mulcahy, 1983) have suggested an alternative to the classical model. It is termed the heterosis model because it implies that interactions between pollen and style mimic the heterotic interactions between genomes. This alternative interpretation assumes that self-incompatibility is determined by many loci and that failure of incompatible pollen tubes is passive rather than active. More specifically, it suggests that self-incompatibility results when pollen carries the same deleterious recessive alleles for which the style is homozygous recessive. The greater the number of such loci common to pollen and style, the stronger will be the self-incompatibility reaction. This interpretation has been criticized (Lawrence, et al., 1985) and, hopefully, clarified (Mulcahy and Mulcahy, 1985). In the present paper we provide further evidence from the literature and from our own ongoing studies.

The first topic to be considered is Larsen's observed association, in <u>Beta vulgaris</u>, between homozygosity of S-loci and the strength of the incompatibility reaction. Incorporated into the heterosis model, this concept suggests that inbreeding, by increasing the proportion of

[1] Botany Department, University of Massachusetts, Amherst, Ma. 01003 USA

homozygous loci, should generally strengthen the degree of self-incompatibility. Conversely, if a high degree of self-incompatiblity is indicative of a high frequency of heterozygous loci, hybridizations between highly self-incompatible individuals should produce individuals which are less self-incompatible than are their parents. We scored the degree of self-incompatibility within a mass-sib population of *Lycopersicon peruvianum* from Lima, Peru. Self-incompatibility was expressed as the proportion of the style which is penetrated by self pollen tubes, 48 hours after pollination at $17^{o}$C. Clones identified as M, A, and E, were the three most self-incompatible of 18 measured and clone F was one of the least. With clone M, the proportion of the style penetrated by self pollen was 50.48% (standard error = 3.89). In clone A, the figure was 64.96 $\pm$ 2.77, while in the hybrid, M x A, the figure was 81.19 $\pm$ 2.32. In clone E, self pollen penetrated 55.70% $\pm$ 3.38 of the style while 69.74% $\pm$ 1.70 of the hybrid (A x E) style was penetrated on selfing. Contary to the increased penetration exhibited by the above hybrids, one generation of selfing clone F resulted in progeny which showed 51.08% $\pm$ 1.56 stylar penetration on selfing, compared to 100% in the parental clone F.

The next issue to be considered is the source of new incompatibility specificities. The heterosis interpretation suggests that this question may be related to the apparent relationship between the strength of self-incompatibility and the level of heterozygosity. It is thus significant that the literature contains substantial support for the hypothesis that inbreeding somehow generates new S-alleles (Denward, 1963; McGuire, 1950; Gastel and de Nettancourt, 1975). These last results have been questioned (Sree Ramalu, 1983; Lawrence et al. 1985; but see Mulcahy and Mulcahy, 1985). Therefore, it seems appropriate to reexamine at least two studies which also concluded that inbreeding seems to generate new specificities. The first of these, by Denward (1963), analyzed the progeny resulting from a rare selfing in *Trifolium pratense*. Family 0111a consisted of 8 plants which, if the seed source was S1S2, should have segregated into three incompatibility classes: 1 S1S1 : 2 S1S2 : 1 S2S2. Six of the 8 plants fell into two intrasterile, interfertile groups (see Table 1). The fact that plants 3, 4, and 8 accept pollen from their parent but fail in the reciprocal crosses, indicates that these plants should be an S-allele homozygote. Denward also concluded that plants 1, 2, and 6, represent that other homozygote. However, the fact that 1, 2, and 6 can backcross to the seed source indicates the presence of a different (or silent) S-allele.

Of course, contamination is a quite possible explanation

Table 1. Partial diallele among inbred progeny of <u>Trifolium pratense</u> (after Denward, 1963).

| | Diallele Crosses | | Parent used as | |
|---|---|---|---|---|
| | 1, 2, 6 | 3, 4, 8 | Female | Male |
| 1, 2, 6 | - | + | + | + |
| 3, 4, 8 | + | - | - | + |

for these results, and Denward tested for this by analyzing the backcross of plant number 1 to the parent. The maternal parent, if S1S2, will classically produce four possible compatibility classes upon outcrossing. Backcrossing any one of these to the parent should classically exclude any S1 or S2 gametes from the pollen source, resulting in two, and only two, incompatibility classes from backcrossing. However, Denward found that one backcross progeny apparently consisted of four incompatibility classes. Thus the observed results cannot be explained by assuming that contamination has taken place. One possible explanation is that a new incompatibility locus had been activated by the inbreeding. An alternative explanation is that the parent S1S2 was outcrossed and plant number 1 was actually S1S3, for example. Backcrossing this would still produce only two compatibility classes, S1S3 and S2S3. The only way to obtain 4 classes in this backcross would be to invoke both contamination and self-compatibility induced by the mentor effect (see Visser, 1985 and Mulcahy and Mulcahy, 1985, both in this volume). In that case, S1S2 x S1S3 would produce S1S1, S1S3, S2S1, and S2S3. Although other explanations are possible, this interpretation would explain the observed results.

The second study which also gives strong evidence that new specificities can be generated by inbreeding comes from McGuire (1950) who hybridized an accession of <u>Lycopersicon peruvianum</u> from Lima (B and T no. 1263) with another from Tacna (B and T no. 1243). This cross produced three intrasterile, interfertile groups, plants 410-1, 410-2, 410-5, 410-7, and 410-9, in one group, 410-3, 410-4, 410-6, 410-10, and 410-12 in another, and plants 410-8 and 410-11 in the third.

Forced selfing of plant number 410-9, presumed to be S1S3, resulted in 6 inbred offspring. Each of these could be S1S1, S1S3, or S3S3. Plant 25-2 rejected pollen from its parent, 410-9, and must therefore have also been S1S3. As S1S3, 25-2 should accept pollen from the group containing 410-3, 410-4, 410-6, 410-10, and also that containing 410-8. Plant 25-2 did accept pollen from 410-3, 410-4, and 410-10. However, contrary to expectations, it rejected pollen from 410-6 and 410-8.

Table 2. Compatibility reactions of the I1 (plant 25-2) from plant 410-9 to pollen from three incompatibility classes (after McGuire, 1950).

| | Pollen Source | | | |
|---|---|---|---|---|
| | S1S3<br>410-1, -2, -5, -7, -9 | S1S4<br>-3, -4, -6, -10 | S2S3<br>-8 | S2S4<br>none |
| S1S3?<br>(25-2) | − − − + − | + + − + | − | |

This indicates the apparent activation, in 25-2, of S4 and S2 incompatibility specificities, respectively. However, the fact that pollen from 410-3, -4, and -10 are not rejected indicates that even this explanation is inadequate. It would seem that several loci, each possibly represented by different alleles in different plants, must be involved. Furthermore, the acceptance of 410-7 pollen indicates that either the S1 or the S3 specificity has been inactivated. Alternatively, multiple locus systems might be involved.

The reciprocal crosses, 410-6 x 25-2 and 410-8 x 25-2 also failed, contrary to classical expectations, although the involvement of additional specificities is compatible with these results. As always, the possibility of contamination must be considered. However, the fact that 25-2 retains its reciprocal incompatibility qwith its parent 410-9 rules out contamination as an explanation.

Another difficulty with the classical model of gametophytic self-incompatibility is that the S-locus is apparently difficult to map. That this may not be impossible to do is suggested by recent work by Kheyr-pour and colleagues, reported in this volume. However, the problem is apparently a difficult one. Data obtained by backcrossing hybrids of L. esculentum x L. chilense into marker stocks of L. esculentum suggested that the S-locus was apparently located on chromosome number 2, somewhere between the loci dl (dwarf) and Wo (Wooly) (Martin, 1961). These backcrosses to L. esculentum indicated also that at least 2, 3, or even 4 loci were required to produce the incompatibility reaction. It is, of course, known that the S-alleles are subject to modifiers, but, as the term, "modifiers" implies, these are generally believed to play a secondary role. Martin was the first to demonstrate that more than one gene is actually required to produce the self-incompatibility reaction. Furthermore, it is highly significant that, using Martin's backcross material, Rick (1963) concluded that, "...the segregation for the S alleles... was independent of the chromosome 2 markers", (Rick, 1963), and later that what was assumed to be the S-locus was instead, "another SI regulator" Rick (1982).

Why should the S-locus be so difficult to map? We (Mulcahy and Mulcahy, 1983) have suggested that the self-incompatibility reaction represents the combined effect of several loci, each of which is an SI regulator. If all but one of these regulators are fixed by inbreeding, the remaining regulator will be termed the S-locus and easily mapped. However, each investigator may map a different regulator as the S-locus.

Where does this leave us? At this point, it seems that the relationship between inbreeding and the strength of self-incompatibility, first reported by Larsen (1977) and incorporated into the heterosis model, may be a general one. Furthermore, inbreeding and interpopulational crosses may indeed generate or nullify incompatibility specificities, respectively. This too is in accord with the heterosis model. An additional point is that several loci may be involved in the incompatibility reaction, but none of these is definitively established as the S-locus.

Finally, we present a syllogism which, we feel, strongly suggests that the classical interpretation of self-incompatibility, that is, a system controlled by a single, multiallelic locus, must be reexamined:

Premise 1. When F1 and later generations between genotypes are vigorous and fertile, single locus, mendelian factors are routinely transferred from one genotype to another.

Premise 2. Self-compatible species, pollinated by self-incompatible relatives, give rise to vigorous, fertile, and self-incompatible hybrids. The self-incompatibility is, however, lost in subsequent backcrosses to the self-compatible parent.

Conclusion. Self-incompatibility is not controlled by a single, multiallelic locus.

Acknowledgements: We wish to express sincere thanks to Professor C. Rick, University of California, Davis, for providing samples of Lycopersicon peruvianum and to the National Science Foundation for support in the form of NSF Grant No. BSR 8407472.

## References

Denward, T. 1963. The function of the incompatibility alleles in red clover (Trifolium pratense). Hereditas 49, 189-234.

Gastel, A.J.G., Nettancourt, D. de, 1975. The generation of new incompatibility alleles. Incompatibility Newsletter 6, 66-69.

Larsen, K. 1977. Self-incompatibility in Beta vulgaris L. I. Four gametophytic, complementary S-loci in sugar beet. Hereditas 85, 227-248.

Lawrence, M.J., Marshall, D.F., Curtis, V.E., Fearon, C.H. 1985. Gametophytic self-incompatibility re-examined: a reply. Heredity 54, 131-143.

Martin, F.W. 1961. The inheritance of self-incompatibility in hybrids of Lycopersicon esculentum Mill. x L. chilense DUN. Genetics 46, 1443-1454.

McGuire, D.C., 1950. Self-incompatibility in Lycopersicon peruvianum and its hybrids with L. esculentum. PhD. thesis, U. California, Davis.

Mulcahy, D.L. and Mulcahy, G.B. 1983. Gametophytic self-incompatibility reexamined. Science 220, 1247-1251.

_______________. 1985. Gametophytic self-incompatibility or the more things change... Heredity 54, 139-144.

Rick, C.M. 1963. Search for the S locus. Tomato Genetics Newsletter 13, 22-23.

_______________. 1982. Genetic relationships between self-incompatibility and floral traits in the tomato species. Biol. Zbl. 101, 185-198.

Sree Ramulu, K., 1982. Failure of obligate inbreeding to produce new S-alleles in Lycopersicon peruvianum Mill. Incompatibility Newsletter 14, 103-110.

# Incompatibility and Incongruity: Two Mechanisms Preventing Gene Transfer Between Taxa

PETER D. ASCHER[1]

Pollen-pistil incompatibility generally refers to the ubiquitous self-incompatibility (SI) phenomenon characteristic of angiosperms. Interpopulation pollen-pistil interactions resulting in malfunction also occur and often have been ascribed to the genetic system controlling SI (Lewis and Crowe, 1958; Pandey, 1969). A generally held concept unifying theoretical treatments of SI and intertaxa incompatibility is that pollen-pistil incompatibility is an active process, evolving as the result of selection. In 1973, Hoganboom proposed that failure in pollen-pistil interactions can be passive and termed this phenomenon incongruity.

Whitehouse (1950) was of the opinion that the gametophytic form of SI was, in fact, the primary trait enabling angiosperms to diversify and so dominate the plant kingdom. SI controls mating within populations so as to minimize selfing by preventing normal post-pollination development of male gametophytes when they and pistil express matched self-incompatibility (S) alleles (East and Mangelsdorf, 1925; Gerstel, 1950; Hughes and Babcock, 1950). As a consequence, affected male gametophytes fail to deliver sperm to females.

Unilateral interspecific incompatibility (UI) describes intertaxa combinations in which reciprocal pollinations differ, succeeding when a given species is used as female but failing when that species is used as male with the same partner (Mather, 1943; McGuire and Rick, 1954; Harrison and Dar-

[1] Department of Horticultural Science & Landscape Architecture, University of Minnesota, St. Paul, MN 55108, USA. Paper Number 14,543 of the Minnesota Agricultural Experiment Station Scientific Journal Article Series.

by, 1955; Lewis and Crowe, 1958; Pandey, 1969). Because many intertaxa combinations exhibiting UI involved self-compatible (SC) taxa paired with SI taxa, Lewis and Crowe (1958) and Pandey (1969) attributed UI to the self-incompatibility locus (S). However, data reported by Mather (1943) and Martin (1964) suggest independent segregation of SI and UI. While SI imparts an identity such that individuals within a population differ, UI patterns suggest that all individuals within a population (Grun and Radlow, 1961; Martin, 1964, 1967; Grun and Aubertin, 1966) or within a species (Lewis and Crowe, 1958; Pandey, 1969) bear the same UI phenotype.

Hoganboom (1973, 1975, 1984) proposed that failure of pollen tubes to develop normally and fertilize female gametophytes in interpopulation crosses could result from incongruity, a lack of critical information in one partner about aspects of the other. He reasoned that changes in pistils due to mutation, selection, and genetic drift would select for compensation among male gametophytes from sympatric individuals. Similarly, changes in male gametophytes would select for compensation among pistils. Lack of compensation would preclude reproduction, dooming the particular genotype to extinction. When populations are isolated, Hoganboom continued, there would be no opportunity for compensatory selection between pistil and male gametophytes. Therefore, taxa with a history of isolation could fail to intermate because of a passive feature of isolation--lack of opportunity for co-evolution.

SI in *Lilium longiflorum* is clearly evident in pistils incubated at 23$\pm$1$^{o}$C (room temp) for 48 h after pollination: incompatible pollen tubes average 0.9 mm/h, while compatible tubes grow about 1.7 mm/h (Ascher and Campbell, 1983). Pollen tubes of other lily taxa in *L. longiflorum* pistils incubated 48 h at room temp appear to grow at 2 different velocities, neither of which resembles compatible tube growth (Ascher and Peloquin, 1968). Tubes of species and interspecific hybrids involving taxa from the section Leucolirion, of which *L. longiflorum* is a member, appear strongly inhibited just within the style proper. These pollen tubes were often branched, grew in a cork-screw fashion, stained abnormally, and did not appear to grow longer with longer incubation. On the other hand, pollen tubes of species and hybrids from other sections of the genus grew at rates up to and including those typical for SI tubes and, like SI tubes, continued growing at decreasing velocities with continued incubation. Neither type of intertaxa pollen tube was affected by incubation at 39°C, a treatment destroying SI in the *L. longiflorum* cultivars used.

Reciprocal pollination of combinations which produced the strongly-inhibited tubes in *L. longiflorum* pistils revealed UI, as tubes of *L. longiflorum* grew at compatible rates, penetrating 80% or more of the style (Ascher and Peloquin, 1968; Ascher and Drewlow, 1971a). However, reciprocal pollination of individuals producing slow-growing pollen tubes in *L. longiflorum* styles resulted in strongly inhibited tubes (Ascher and Drewlow, 1971a). While these tubes resembled UI-inhibited tubes, the response cannot be considered UI. Neither pollination resulted in compatible pollen tube growth. UI in *Lilium* exhibited a pollen tube phenotype differing from that of SI tubes, as has been reported for other species (McGuire and Rick, 1954; Harrison and Darby, 1955; Lewis and Crowe, 1958).

*L. longiflorum* stigmatic exudate, injected into the stylar canal at pollination, released pollen tubes from UI (Ascher and Drewlow, 1975). Similar treatment before self pollination has no effect on SI in *L. longiflorum* (Ascher and Drewlow, 1971b). Released UI tubes, however, did not grow as compatible, as do those from the reciprocal pollinations, but at rates typical of SI tubes (Ascher and Drewlow, 1975). Slow-growing pollen tubes from other sections of the genus were not affected by the presence of *L. longiflorum* stigmatic exudate in the stylar canal. These data, along with the insensitivity of intertaxa pollen tubes to high-temperature incubation which destroys SI (Ascher and Peloquin, 1968), suggest that slow-growing SI, slow-growing intertaxa, and sharply-inhibited intertaxa (UI) pollen tube phenotypes in *L. longiflorum* pistils are caused by fundamentally different phenomena.

Injecting 1 x $10^{-3}$M 6-methylpurine or 1 x $10^{-3}$M puromycin in water into pistils 6 or 12 h before intertaxa pollination indicated that blockage of RNA and protein synthesis in styles of buds from 2 days before to the day of anthesis alleviated the UI reaction (Ascher, unpublished). Treatment of styles 1 day after anthesis had no effect. Again, release of the UI inhibition resulted in pollen tube velocities typical of SI rather than compatible tubes. Introducing the same inhibitors in water into L. longiflorum styles before self pollination produced a different pattern (Ascher, 1974). Though the SI reaction was diminished by puromycin in styles of buds from 2 days before anthesis up to the day of anthesis, 6-methylpurine caused SI tubes to grow as compatible in styles injected 1 day after anthesis as well. While sensitivity to transcription and translation inhibitors for both SI and UI reactions suggests a protein basis for each, differences in time of sensitivity during maturation of the pistil would indicate different synthetic events for SI and UI.

A potential explanation for the slow-growing intertaxa pollen tube phenotype comes from data generated while attempting to investigate the effect of transcription and translation inhibitors on heat inactivation of SI. Submerging L. longiflorum pistils in water at 50°C for 5 minutes before pollination destroys the SI reaction (Hopper, et al, 1967). The intent of the experiment was to determine whether metabolic inhibitors in water, injected into heat-treated styles, interfered with the high-temperature response. However, controls (heat-treated pistils injected with distilled water before pollination) revealed that the protocol destroyed stylar capacity to support compatible pollen tube growth. Flushing the style with 10 drops of distilled water after heat treatment caused both compatible and incompatible pollen tubes to grow at rates typical of incompatible tubes (Ascher, 1975).

Hot-water treatment seems to enhance stylar secretion such that subsequent flushing removes all of the stylar secretory product. Without this substance, pollen tubes grow at incompatible rates, regardless of genotype. This slow growth is not due to activity of the S locus, but rather, to lack of a stylar product necessary for normal pollen tube growth. In this sense, intrataxon pollen tubes and heat-treated, flushed L. longiflorum pistils are incongruous: normal pollen tube growth is prevented by a lack of substances in the pistil required by pollen tubes. Slow growth of intrataxa pollen tubes in L. longiflorum styles may be caused by the same phenomenon. Since the response appears only between relatively unrelated, certainly long isolated taxa, it seems reasonable to suppose that these pollen tubes fail to recognize or, perhaps, are unable to use the stylar secretion stimulating increased growth rate of compatible pollen tubes. Failure, then, would not be an active rejection or incompatibility, but rather, a passive malfunction best described as incongruity.

The fact that intertaxa pollen tubes released from interspecific incompatibility (UI) by the presence of stigmatic exudate, 6-methylpurine, or puromycin in the style failed to grow at compatible velocities deserves comment. Since stylar secretion products of taxa producing UI in *L. longiflorum* are recognized as appropriate by *L. longiflorum* pollen tubes, which grow at compatible rates in these styles, why do the tubes grow as if incongruous when the UI inhibition is relieved? It may be that UI was selected in *L. longiflorum* to prevent fertilization by a specific sympatric taxon and the response selected also affects related taxa which have been isolated from *L. longiflorum* long enough to develop unilateral incongruity. In other words, the mechanisms resulting in UI are sufficiently general to recognize and control pollen tubes of taxa other than the selection agent.

*L. longiflorum* is extremely specialized with regard to habitat and occurs in a very small natural range. The only lily known to produce fertile hybrids with *L. longiflorum* is the more generalized and much more widely distributed *L. formosanum*. Perhaps UI was selected to protect *L. longiflorum* from *L. formosanum* but the nature of UI recognition is such that all members of the trumpet section, Leucolirion, are affected. Thus, when looking at pollen-pistil interactions involving *L. longiflorum*, it is possible to have active interspecific incompatibility and passive incongruity in the same pollen-pistil combination.

REFERENCES

Ascher PD (1974) The self-incompatibility reaction in detached styles of *Lilium longiflorum* Thunb. injected before pollination with 6-methylpurine or puromycin. Incompatibility News 4:57-60

Ascher PD (1975) Special stylar property required for compatible pollen-tube growth in *Lilium longiflorum* Thunb. Bot Gaz 136:317-321

Ascher PD, Campbell RJ (1983) Irreversible differentiation of pollen tubes by the self-incompatibility reaction in *Lilium longiflorum* Thunb. J Palyn 19:143-151

Ascher PD, Drewlow LW (1971a) Unilateral interspecific incom-

patibility in _Lilium_. Ybk N Am Lily Soc 24:70-74
Ascher PD, Drewlow LW (1971b) Effect of stigmatic exudate injected into the stylar canal on compatible and incompatible pollen tube growth in _Lilium_ _longiflorum_ Thunb. In: Heslop-Harrison J (ed) Pollen: Development and physiology. Butterworths, London, pp 267-272
Ascher PD, Drewlow LW (1975) The effect of prepollination injection with stigmatic exudate on interspecific pollen tube growth in _Lilium_ _longiflorum_ styles. Plant Sci Let 4:401-405
Ascher PD, Peloquin SJ (1968) Pollen tube growth and incompatibility following intra- and interspecific pollinations in _Lilium_ _longiflorum_. Am J Bot 55:1230-1234
East EM, Mangelsdorf AJ (1925) A new interpretation of the behavior of self-sterile plants. Proc Natl Acad Sci USA 11:166-171
Gerstel DU (1950) Self-incompatibility studies in guayule. II. Inheritance. Genetics 35:482-506
Grun P, Aubertin M (1966) The inheritance and expression of unilateral incompatibility in _Solanum_. Heredity 21:131-138
Grun P, Radlow A (1961) Evolution of barriers to crossing of self incompatible with self compatible species of _Solanum_. Heredity 16:137-143
Harrison BJ, Darby LA (1955) Unilateral hybridization. Nature 176:982
Hoganboom NG (1973) A model for incongruity in intimate partner relationships. Euphytica 22:219-233
Hoganboom NG (1975) Incompatibility and incongruity: Two different mechanisms for the non-functioning of intimate partner relationships. Proc R Soc Lond B 188:361-375
Hoganboom NG (1984) Incongruity: Non-functioning of intercellular and intracellular partner relationships through non-matching information. In: Linskens HF, Heslop-Harrison J (eds) Cellular interactions. Springer-Verlag, pp 640-654.
Hughes MB, Babcock EB (1950) Self incompatibility in _Crepis_ _foetida_ L. subspecies _roeadifolia_ (Biel) Schniz et Keller. Genetics 35:570-588
Lewis D, Crowe LK (1958) Unilateral interspecific incompatibility in flowering plants. Heredity 12:233-256
Martin FW (1964) The inheritance of unilateral incompatibility in _Lycopersicon_ _hirsutum_. Genetics 50:459-469
Martin FW (1967) The genetic control of unilateral incompatibility between two tomato species. Genetics 56:391-398
Mather K (1943) Specific differences in petunia. I. Incompatibility. J Genet 45:215-325
McGuire DC, Rick CM (1954) Self-incompatibility in species of _Lycopersicon_ sect. Eriopersicon and hybrids with _L_. _esculentum_. Hilgardia 23:101-124
Pandey KK (1969) Elements of the _S_-gene complex. V. Interspecific cross-compatibility relationships and theory of the evolution of the _S_ complex. Genetica 40:447-474
Whitehouse HLK (1950) Multiple allelomorph incompatibility of pollen and style in the evolution of angiosperms. Ann Bot 14:199-216

# Intra- and Interspecific Incompatibility in Brachiaria ruziziensis Germain et Evrard (Panicoideae)[1]

G. Coppens d'Eeckenbrugge, M. Ngendahayo, B.P. Louant[2]

## Introduction

An introgression program aiming at transfering apomixis from tetraploid species Brachiaria decumbens (B.d) and Brachiaria brizantha (B.b) into the genome of the sexuate diploid Brachiaria ruziziensis (B.r), an interesting forage crop, started in our laboratory (Gobbe and al., 1983). Its first step was the obtention of autotetraploid plants of B.r. The second step was to cross them with the apomictic tetraploid species. As self-seed-set is generally very low in B.r, about 1-7 % in both diploids and tetraploids, these interspecific crosses were not preceded by emasculation (a very uneasy operation in Brachiaria). A few seeds were obtained, but most of these were self-seeds (Ndikumana, 1985). To investigate the reason of this and to localize precisely the interspecific barrier in Brachiaria, self-incompatibility and interspecific incompatibility in the progamic phase were studied and compared. For this purpose, we observed pollentube growth after self- and cross-intraspecific pollinations and interspecific pollinations of emasculated flowers of diploid and tetraploid B.r.

---

1.This work is part of a doctorate thesis to be presented by M. Ngendahayo

2.Laboratoire de Phytotechnie Tropicale et Subtropicale
Université Catholique de Louvain, 1348 Louvain-la-Neuve, Belgique

## Material and methods

Following pollinations were realized using diploids and tetraploids :

intraspecific pollinations : - self-pollination
- intraploidic crosses (2nx2n,4nx4n)
- interploidic crosses (2nx4n,4nx2n)

interspecific pollinations : - B.r. (2n) x B.d. (4n)
- B.r. (2n) x B.b. (4n)
- B.r. (4n) x B.d. (4n)
- B.r. (4n) x B.b. (4n)

Spikelets were cropped and fixed at varying time intervals after pollination (30', 1h, 2h, 3h, 6h, 9h, 12h, 24h) and the whole pistil observed in U.V. microscopy after staining with decolourized aniline blue.

Before cross-pollinations, the anthers were removed. However, as they often come out quite at the same time as the stigmas, contamination by self-pollen could rarely be totally avoided.

## Results and discussion

We shall first describe pollen tube growth after compatible cross between diploids and secondly compare it with the results of the other types of pollination.

### Pollen tube behaviour after compatible cross between diploid B.r plants

Pollen germination and tube growth are very fast : after 30 minutes, pollen tubes may be as long as 10 times the pollen diameter. Generally, they penetrate the stigma by piercing the cuticle at the junction of two cells and are oriented towards the papillae base (fig.1). They then grow intercellularly, following the stigma cells outlines, till the stigma axis,where they quite always take the ovary direction (fig.2), because of the stigma morphological polarization (Heslop-Harrison and al., 1984). Some pollen tubes do not achieve penetration and curl around the stigma papillae (fig.3).

Although the stigma constitutes the normal germination site, pollen germination may also take place upon the style or the ovary. When germinating upon the style, pollen tubes take either the ovary direction or that of stigma (fig. 4 and 5). Pollen tubes entering directly the ovary are unable to reach the transmitting tissue (fig.6).

The earliest pollen tubes reach the style one hour after pollination and the ovary one hour later, but they constitute a very low proportion of the germinated pollen grains. From the flowers cropped later than 2 hours after pollination, we could estimate that 92 % of the pollen tubes are stopped at any level in the stigmas and 47 % of the remaining ones are stopped in the style. The tips of some of these blocked tubes swell and become fluorescent. Subsequently, they may burst (fig.7). Callose occlusion often occurs also in the pollen grain itself (fig.8). In other cases, the tube tip is filled with callose and a new branch emerges just before the tip ; the succession of branching-and-plugging sometimes gives a serrated appearance to the tube. More marked branchings also occur. However, most of the arrested pollen tubes present no morphological abnormalities.

Callose deposition in the pollen tube is uniform in the stigma. In the style, many tubes are less fluorescent and present regularly callose plugs (fig.9). Since pollen tubes need the same time to run along the stigma and the style, which is longer, the mean growth rate is higher in the latter than in the former.

Entering the ovary, the pollen tubes become again uniformly fluorescent (fig.9 and 10). Another consequence of their entry consists in the formation of brightly fluorescent granules at the junction of style and ovary (fig.9 and 11). Their accumulation increases in proportion to the pollen tubes number. Pollen tube growth abnormalities in the ovary only occur at this level (growth arrest and tip swelling : fig.11) or near the micropyle (branching of pollen tubes or growth arrest and tip swelling). Since pollen tube inhibitions near the micropyle only occured when other tubes had already reached the embryo sac, they are probably a post-fertilization event, similar to those described in _Spinacea oleracea_ by Wilms (1974).

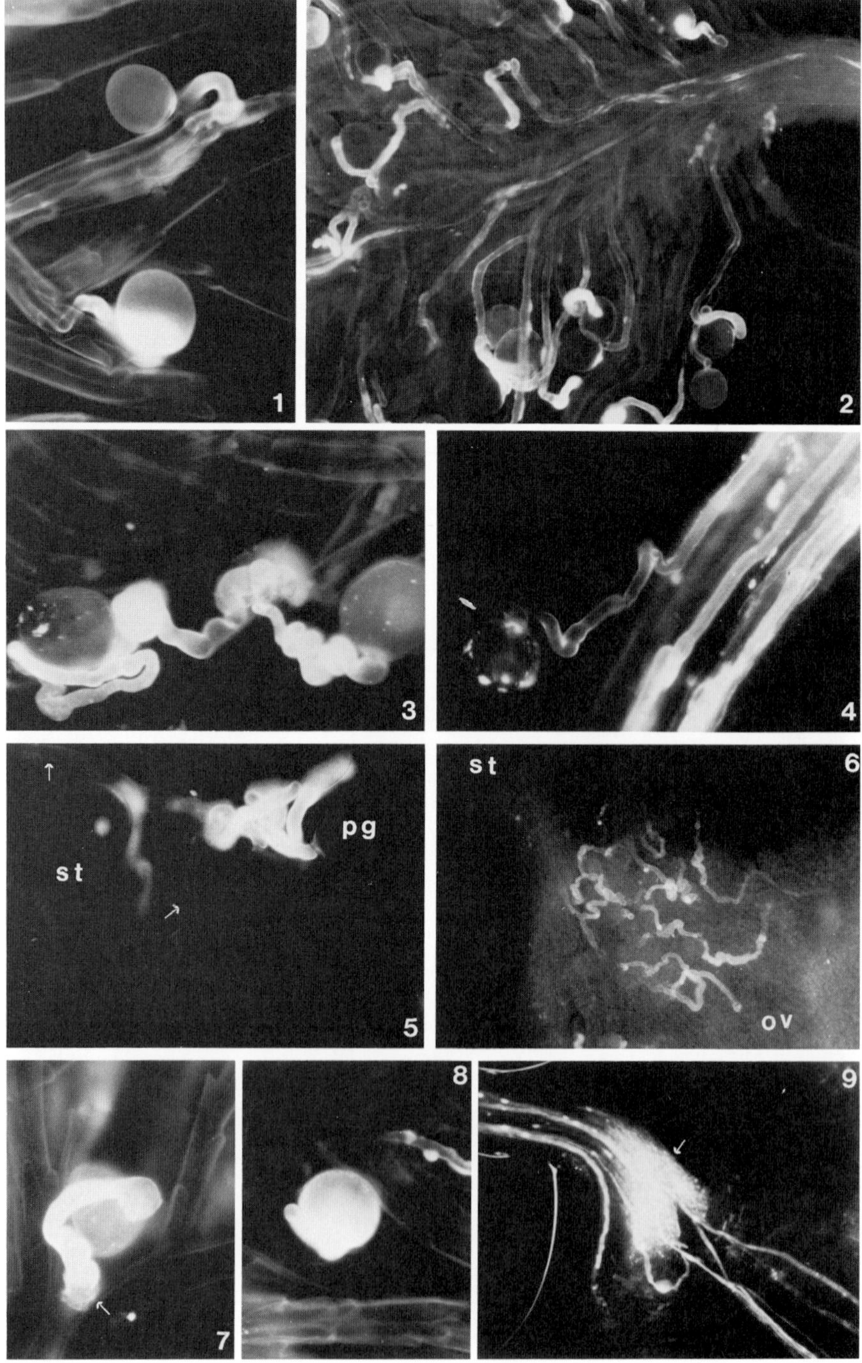
1
2
3
4
st
pg
5
st
6
ov
7
8
9

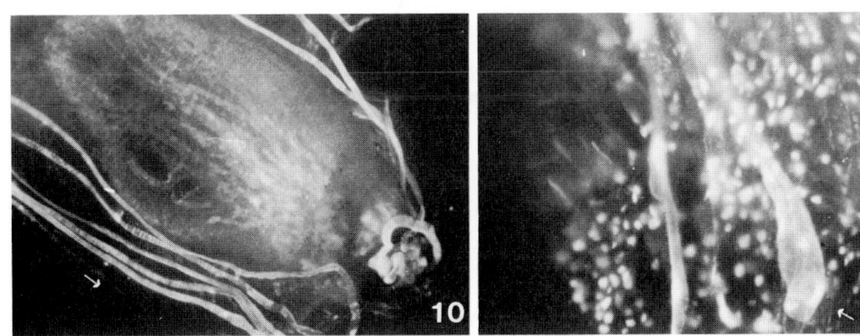


Abbreviations : pg : pollen grains ; st : style ; ov : ovary.

fig. 1 and 2. Pollen tube growth in the stigma ; some pollen grains are partially filled with callose, (fig. 1 : x 300 ; fig. 2 : x 100).

fig. 3. Pollen tubes coiling on the stigma after failing to penetrate it (x 300).

fig. 4. A pollen grain having germinated on the style (x 150).

fig. 5. Pollen grains having germinated on the style ; two pollen tubes have penetrated and grown in the style, one in each direction (arrows) (x 150).

fig. 6. Pollen germination on the ovary, near its junction with the style ( x 90).

fig. 7. Callose accumulation in a pollen tube and its subsequent bursting (arrow) (x 300).

fig. 8. Callose accumulation in an early inhibited pollen grain (x 300).

fig. 9. Pollen tube growth in the style and the ovary : note the callose plugs in the pollen tubes at the style level and the accumulation of fluorescent granules (arrow) at the junction between style and ovary (x 75).

fig. 10. Pollen tubes in the ovary. The arrow indicates the direction of growth (x 75).

fig. 11. Accumulation of fluorescent granules at the junction of style and ovary : note the swollen pollen tube tip (arrow) (x 300).

## Results of the other types of pollination

The proportion of inhibited pollen tubes varies from one type of pollination to another, but no difference in individual tube behaviour could be observed.

Results obtained from flowers cropped more than 2 hours after pollination are synthetized in tables 1 and 2. Table 1 gives the proportion of pollen tubes having penetrated the style and the ovary, allowing to compare the strength of the mechanisms controlling the pollen from different origins. Table 2 gives the proportion of flowers where tubes have been observed in the style and the ovary, allowing to compare the chances of fertilization of their unique ovule depending on the type of pollination.

The occurence after cross-pollination of typical pollen tube inhibitions such as the accumulation of callose at their tip, or their bursting, probably concerns self-pollen since contamination by this could not always be avoided. More generally, self-pollen might constitute a significant part of all the inhibited pollen tubes ; therefore, the proportions of non-inhibited cross-pollen tubes presented in table 1 are probably underestimated. However, the comparison between these data remain valid since they agree very well with table 2 data, which are not biased.

In both diploid and autotetraploid _B.r_, the percentage of selfed flowers containing pollen tubes in their ovary is slightly lower than self-seed set observed on the same plants. This indicates that there is no incompatibility reaction in the ovary. Selfings are characterized by a greater proportion of tubes stopped in the stigma (99%) and in the style (98% of the remaining). This inhibition may be manifested very far in the style : some tubes are stopped after they ran 4/5 of its length. Such a delay of the incompatibility reaction is not related to its strength. So, both style and stigma play an important role in incompatible pollen tubes arrest. The situation in _Brachiaria_ could be an exception to the generally admitted early-stigma-localized incompatibility reaction (Heslop-Harrison and Heslop-Harrison, 1982). However, this rule was deduced from studies of grasses most of which pertain to genera of the _Festucoideae_, e.g. _Alopecurus_, _Secale_, _Dactylis_, _Hordeum_, _Phalaris_, _Lolium_, _Gaudinia_, _Festuca_, _Briza_ (see Thomas and Murray, 1975 and Heslop-Harrison and Heslop-Harrison, 1982). Another exception is a _Chloridoideae_, _Cynodon dactylon_, where the incompatibility reaction takes place in the style (Thomas and Murray, 1975). So one may admit that the rule is not general or depends on the subfamily. But it is also possible that a strong incompatibility reaction in the _Festucoideae_ prevents totally pollen tubes from passing the stigma trichomes, so rendering unobservable an eventual reaction in the transmitting tract. Indeed, in self-fertile _Gaudinia fragilis_

Table 1. Proportion of pollen tubes having grown till the style or the ovary, more than 2 h after pollination (germinated pollen grains = 100 %).

| mother plant | B.r 2n | | | | | B.r 4n | | | | |
|---|---|---|---|---|---|---|---|---|---|---|
| pollinator | B.r2n(self) | B.r2n(cross) | B.r4n | B.d | B.b | B.r4n(self) | B.r4n(cross) | B.r2n | B.d | B.b |
| number of flowers | 231 | 124 | 78 | 79 | 70 | 227 | 78 | 54 | 103 | 61 |
| % pollen reaching the style | 1,01 | 7,94 | 5,68 | 4,44 | 1,35 | 1,11 | 5,00 | 6,46 | 6,95 | 4,93 |
| % pollen reaching the ovary | 0,02 | 4,17 | 4,56 | 2,54 | 0,49 | 0,10 | 2,47 | 4,49 | 3,78 | 0,93 |

Table 2. Proportion of flowers with pollen tubes in the style or the ovary, more than 2 h after pollination (flowers with germinated pollen upon the stigma = 100 %).

| mother plant | B.r 2n | | | | | B.r 4n | | | | |
|---|---|---|---|---|---|---|---|---|---|---|
| pollinator | B.r2n(self) | B.r2n(cross) | B.r4n | B.d | B.b | B.r4n(self) | B.r4n(cross) | B.r2n | B.d | B.b |
| number of flowers | 231 | 124 | 78 | 79 | 70 | 227 | 78 | 54 | 103 | 61 |
| % styles with pollen tubes | 32,03 | 69,36 | 55,13 | 56,96 | 34,29 | 9,69 | 79,46 | 92,59 | 39,81 | 36,07 |
| % ovaries with pollen tubes | 1,30 | 52,42 | 55,13 | 35,44 | 15,71 | 1,76 | 48,72 | 81,48 | 21,36 | 8,20 |
| mean number of tubes/ovary (when present) | 1,00 | 3,28 | 10,77 | 2,64 | 1,82 | 1,00 | 2,90 | 7,43 | 3,14 | 1,20 |

plants, most of self-pollen tubes are arrested in the secondary stigma branches or in the stylodia (Shivanna and al., 1982).

The whole pistil of B.r exerts a control over the number of male gametophytes reaching the embryo sac. Very probably, part of this control is passive : after compatible pollinations, many tubes seem blocked because of crowding effects. But an active control may not be excluded : callosic granules formation and inhibitions of pollen tubes at the junction of ovary and style and near the micropyle strongly suggest such a regulation.

Ploidy differences between parents have no evident effect upon pollen tube growth.

Interspecific pollinations give results intermediary between self- and intraspecific cross-pollinations. Pollen tubes from B.b are more frequently inhibited than those from B.d and their growth is markedly slowed down : the earliest ones penetrate the ovary only 3 hours after pollination. This could be related to a greater phylogenetic distance between B.r and B.b than between B.r and B.d as shown from cytogenetic studies (Ndikumana, 1985). But on the whole, these results do not explain the interspecific barrier between B.r and B.d or B.b. Therefore, it must be expressed at a later stage, fertilization or embryogenesis.

## References

Gobbe J, Longly B, Louant B-P (1983) Apomixie, sexualité et amélioration des graminées tropicales. Tropicultura 1(1) : 5-9.

Heslop-Harrison J, Heslop-Harrison Y (1982) The pollen-stigma interaction in the grasses. 4. An interpretation of the self-incompatibility response. Acta Bot Neerl 31(5/6) : 429-439.

Heslop-Harrison Y, Reger BJ, Heslop-Harrison J (1984) The pollen-stigma interaction in the grasses. 6. The stigma ('silk') of Zea mays L. as host to the pollens of Sorghum bicolor (L.) Moench and Pennisetum americanum (L.) Leeke. Acta Bot Neerl 33(2) : 205-227.

Ndikumana J (1985) Transfert du caractère "Reproduction apomictique" d'une espèce apomictique, naturellement tétraploïde, à une espèce sexuée, naturellement diploïde, dans le genre Brachiaria. Doctorate thesis, Université Catholique de Louvain.

Shivanna KR, Heslop-Harrison Y, Heslop-Harrison J (1982) The pollen-stigma interaction in the grasses. 3. Features of the self-incompatibility response. Acta Bot Neerl 31 : 307-319.

Thomas SM, Murray BG (1975) A new site for the self-incompatibility reaction in the Gramineae. Incomp Newslett 6 : 22-23.

Wilms HJ (1974) Branching of pollen tubes in Spinach. In : Linskens HF (ed) Fertilization in Higher Plants. North Holland, American Elsevier. pp 155-160.

# Incompatibility Relationships in Intra- and Interspecific Crosses of Zinnia elegans Jacq. and Z. angustifolia HBK (Compositae)

THOMAS H. BOYLE AND DENNIS P. STIMART[1]

## Introduction

Zinnia (Compositae-tribe Heliantheae) is comprised of approximately 17 species native to the New World (Torres 1963). Two annual species, Z. angustifolia HBK (formerly Z. linearis Benth.) and Z. elegans Jacq., have long been in cultivation as ornamentals. The two species are distinct, both morphologically and cytologically, with n=11 or 12 for Z. angustifolia and n=12 for Z. elegans (Torres 1963; Terry-Lewandowski et al. 1984).

Previous studies indicated hybridization barriers existed in Z. elegans x Z. angustifolia but not in reciprocal matings (Boyle and Stimart 1982). Zinnia pistils have papillate stigmas with little or no secretory exudate and pollen is trinucleate (Pulliah 1981). These are characteristics of species expressing sporophytic self incompatibility (SI) (Heslop-Harrison 1975). However, self incompatibility has been reported only in Z. angustifolia (Olorode 1970). The objectives of this research were to elucidate the nature of barriers in intraspecific crosses of Z. angustifolia and Z. elegans and determine if interspecific incompatibility acts as a prezygotic barrier to gene flow between these species.

## Materials and Methods

Five lines of Z. elegans and four clones of Z. angustifolia were crossed according to procedures reported previously (Boyle and Stimart 1982).

1 Department of Horticulture, University of Maryland, College Park, MD 20742 USA

Following controlled pollinations, barriers were assessed from: 1) pollen germination (PG), pollen tube (PT) growth, and callose (C) formation; 2) presence or absence of embryos 14 days following pollination; or 3) seed set. For PG and PT analysis, a modified Martin's (1959) procedure was used. Pistillate ray florets were pollinated and removed at 10 minute intervals for the first hour, and at 24 hr, and fixed for 24 hr in Carnoy's fluid (6:3:1 of 95% ethanol, chloroform and glacial acetic acid, respectively). Florets were washed in water, soaked in 2N NaOH for 4-6 hr, and re-washed in water. Softened pistils were stained with 0.1% solution of aniline blue dissolved in 0.1M $K_3PO_4$ for two or more hours. Callose resolution was optimum in pistils refrigerated for 24 hr or more in 0.1% aniline blue solution. Styles were removed from ovaries, mounted in a drop of 0.1% aniline blue solution, and squashed under a cover slip. To estimate the number of ungerminated pollen grains and short PT, styles were removed 24 hr after pollination and mounted fresh in 0.1% aniline blue solution. Fluorescence microscopy was performed using Leitz SM-LUX microscope equipped with a Phillips 100W high pressure Hg lamp.

## Results and Discussion

*Z. angustifolia*. Little or no C in either PT or stigmatic papillae was observed following cross compatible pollinations. As reported for *Helianthus annuus* (Vithanage and Knox 1977) and *Cosmos bipinnatus* (Knox 1973) compatible crosses were characterized by weak fluorescence of PT and the tube tip penetrated the stigmatic surface at the base, entering the style. Generally, PT could not be traced after penetration of the stigmatic surface. Seed set following compatible pollinations ranged from 55-87%, and ranged from 0-31% for self pollinations. Three clones averaged 0-4 pollen grains per pollinated pistil attempting PT penetration of the stigma before arrest and C accumulation in the PT while other pollen grains were arrested as a result of failing to germinate or producing a short PT occluded with C. Intense C fluorescence was evident in papillae adjacent to grains and PT. Self seed set in these clones was less than 8%. In contrast, one clone set 31% self seed and up to 50 grains per pollinated pistil were observed attempting penetration. Some C deposition occurred in adjacent papillae but fluorescence was weaker than in other clones. Seed set following SI and cross compatible matings was correlated positively with PT penetration of the stigmatic surface. Callose rejection to self pollinations showed that a sporophytic SI system was present in _Z. angustifolia_ (Heslop-Harrison 1975) and reconfirms an earlier report (Olorode 1970).

_Z. elegans_. Similarily, cross compatible pollinations were characterized by little or no C in either PT or papillae, whereas for incompatible pollinations C was prominent in both PT and papillae. Pollen germinated and PT penetrated the stigmatic surface 10-20 minutes after cross compatible matings. Two lines produced extensive C in papillae and in PT after self pollinations, and set less than 1% self seed. In contrast, one line regularly set 60-80% self seed with little or no C deposition in papillae of PT. Variability in SI response has been reported for other Compositae genera: _Ageratum_ (Stephens et al. 1982), _Chrysanthemum_ (Ronald and Ascher 1975), and _Helianthus_ (Frankel and Galun 1977). Self compatibility (SC) in _Z. elegans_ has resulted probably from selection for maximum seed set during domestication. Despite the occurrence of SC, the presence of a stigmatic C response following incompatible matings indicates a sporophytic SI system in _Z. elegans_.

_Z. angustifolia_ x _Z. elegans_. Two types of PT growth were observed: 1) inhibited PT (curled or deformed) failing at stigmatic penetration and accumulating C in the PT and adjacent papillae (typical of intraspecific incompatible matings) or 2) uninhibited stigmatic penetration of PT without C in PT or papillae (typical of intraspecific compatible matings). Some pollen grains elicited a weak C reaction in adjacent papillae although PT penetrated the stigma without C deposition. Other pollen grains failed to germinate and produced C in adjacent papillae. The C rejection reaction in inhibited PT was usually weaker than in intraspecific incompatible matings. Both inhibited and uninhibited PT were observed on each pistil and no apparent relationship existed between interspecific PT growth and the self seed setting capacity of _Z. angustifolia_ clones or _Z. elegans_ lines. Embryos observed 14 days after pollination confirmed interspecific fertilization and 67-81% of pollinated florets contained embryos. This was comparable to compatible matings within _Z. angustifolia_ and suggests that PT growth in transmitting tissue is not inhibited from reaching the embryo sac. The only evident barrier is inhibition of PG and PT penetration of some pollen at the stigmatic surface.

_Z. elegans_ x _Z. angustifolia_. Pollen tubes showed either inhibited or uninhibited penetration of the stigmatic surface, as observed for the reciprocal cross. Lines of _Z. elegans_ displaying high self seed set contained more uninhibited PT whereas _Z. elegans_ lines with low self seed set had more ungerminated pollen grains and inhibited PT. The stigmatic response of PT was not affected by *Z. angustifolia* clones used as pollen parents. Additionally, aberrant PT growth in the style was observed when one *Z. elegans* line was used as a female. No relation existed between aberrant PT in the style and self seed set of *Z. elegans* lines. Examination of florets 14 days after pollination indicated less than 1% embryos. This does not conclusively indicate that postzygotic breakdown is not a factor, but reciprocal differences in visible embryo development suggests prezygotic barriers do act to unilaterally limit gene flow between these species.

Interspecific incompatibility has been defined as post-pollination processes preventing formation of hybrid zygotes combining genomes of two fertile species (de Nettancourt 1977). Inhibition of PG and/or abnormal PT growth are manifestations of this interaction. When pollen of one species is inhibited from functioning on pistils of another, while no inhibition occurs in the reciprocal cross, unilateral incompatibility is operating (Lewis and Crowe 1958). Lewis and Crowe (1958) observed that interspecific and intergeneric incompatibility usually occurred unilaterally in which zygote formation was prevented when SI species were pollinated by SC species. Intergeneric crosses among Cruciferae (sporophytic SI) species supported the SI x SC incompatibility hypothesis, with SI x SI crosses reported as compatible (Lewis and Crowe 1958). In contrast, Sampson (1962) found the majority of crosses among 10 SI Cruciferae species incompatible, suggesting that SI x SI inhibition may operate in other families with sporophytic SI systems. Although stigmatic inhibition of PG and PT occurred for some pollen in *Z. angustifolia* x *Z. elegans*, our embryological observations suggest unilateral incompatibility operates among interspecific crosses of SI species.

Interspecific matings in families with sporophytic SI systems display rejection responses in pollen and stigma associated with C accumulation in papillae and PT (Sampson 1962; Vithanage and Knox 1977). The response phenotype in inhibited PT is similar to the SI reaction, suggesting that S-gene expression is involved in interspecific incompatibility (Dumas and Knox 1983). However, interspecific incompatibility is usually not S-allele specific but rather occurs between pistils of all plants of 1 species and pollen of all plants of another (Hogenboom 1975). Our data support the contention that interspecific incompatibility lacks S-allele specificity: inhibited PG and PT were observed on all stigmas following reciprocal crosses between *Z. elegans* and *Z. angustifolia*. These observations do not conclusively indicate that SI and interspecific incompatibility are not related, and evidence from interspecific matings among *Populus* species (sporophytic SI) suggest that the sites of recognition in SI and interspecific incompatibility are the same (Knox et al. 1972).

In *Zinnia*, inhibited PT growth following intraspecific SI matings and reciprocal interspecific crosses indicates an active stigmatic rejection process for both pollen types. The large amount of uninhibited pollen in reciprocal interspecific crosses of *Zinnia* suggests that interspecific pollen recognition by receptors on the stigmatic papillae is not as efficient as intraspecific recognition. An alternative possiblity is that a gametophytic factor in the pollen plays a role in discrimination and recognition of foreign pollen by the pistil. Since one of the sites of expression of interspecific incompatibility in *Zinnia* conforms to the site at which SI is expressed, the former hypothesis would be most likely. Thus, if S-gene expression on papillae plays a role in recognition of foreign pollen, it is less efficient than in recognition of intraspecific pollen. The nature of S-gene expression in interspecific matings remains to be resolved and requires further experimentation.

Other barriers, however, appear to operate unilaterally in interspecific crosses with *Z. elegans* as the female parent. This may be concluded from the aberrant PT growth observed in the style in some crosses and the absence of embryos following pollinations in which stigmatic penetration of PT was observed. Hogenboom (1975) used 'incongruity' to describe interspecific incompatiblity not due to S-gene expression but caused by differentiation of genes relevant to PG and PT growth. An alternative hypothesis, suggested from our data, is that loci involved in regulation and coordination of events from pollination to fertilization and other loci involved with SI response interact to prevent or hinder zygote formation. Thus, the presence of incongruity to the exclusion of S-gene expression (Hogenboom 1975) appears doubtful, based on the phenotypic observations in the study. Characterization of these stylar barriers and genetical studies of the SI and interspecific incompatibility system may resolve these questions.

Scientific Article No. A-4193, Contribution No. 7178 of the Maryland Agricultural Experiment Station.

References

Boyle TH, Stimart DP (1982) Interspecific hybrids of Z. elegans Jacq. and Z. angustifolia HBK: Embryology, morphology and powdery mildew resistance. Euphytica 31:857-867
Dumas C, Knox RB (1983) Callose and determination of pistil viability and incompatibility. Theor Appl Genet 67:1-10
Frankel R, Galun E (1977) Pollination mechanisms, reproduction and plant breeding. Springer-Verlag, Berlin Heidelberg New York
Heslop-Harrison J (1975) Incompatibility and the pollen-stigma interaction. Ann Rev Plant Physiol 26:403-425
Heslop-Harrison J, Heslop-Harrison Y, Knox RB (1973) The callose rejection reaction: A new bioassay for incompatibility in Cruciferae and Compositae. Incompat Newsletter 3:75-76
Hogenboom NG (1975) Incompatibility and incongruity: Two different mechanisms for the non-functioning of intimate partner relationships. Proc R Soc Lond 188B:361-375
Knox RB (1973) Pollen wall proteins: Pollen-stigma interactions in ragweed and Cosmos (Compositae) J Cell Sci 12:421-443.
Knox RB, Willing R, Ashford AE (1972) Role of pollen wall proteins as recognition substances in interspecific incompatibility in poplars. Nature (Lond) 237:381-383
Lewis D, Crowe LK (1958) Unilateral interspecific incompatibility in flowering plants. Heredity 12:233-256
Martin FW (1959) Staining and observing pollen tubes in the style by means of fluorescence. Stain Tech 34:125-128
Nettancourt D de (1977) Incompatibility in angiosperms. Springer-Verlag, Berlin Heidelberg New York
Olorode O (1970) The evolutionary implications of interspecific hybridization among four species of Zinnia sect. Mendezia (Compositae). Brittonia 22:207-216
Pullaiah T (1981) Studies in the embryology of Heliantheae (Compositae). Pl Sys Evol 137:203-214
Ronald WG, Ascher PD (1975) Self compatibility in garden Chrysanthemum: Occurrence, inheritance and breeding potential. Theor Appl Genet 46:45-54
Sampson DR (1962) Intergeneric pollen-stigma incompatibility in the Cruciferae. Can J Genet Cytol 4:38-49
Stephens LC, Ascher PD, Widmer RE (1982) Genetics of self incompatiblity in diploid Ageratum houstonianum Mill. Theor Appl Genet 63:387-394
Terry-Lewandowski VM, Bauchan GR, Stimart DP (1984) Cytology and breeding behavior of interspecific hybrids and induced amphiploids of Zinnia elegans and Zinnia angustifolia. Can J Genet Cytol 26:40-45
Torres DM (1963) Taxonomy of Zinnia. Brittonia 15:1-25
Vithanage HIMV, Knox RB (1977) Development and cytochemistry of stigma surface and response to self and foreign pollination in Helianthus annuus. Phytomorphology 27:168-179

# Pollen Ultrastructure And Development

# Pollen Sterility in Hybrids and Species of Oenothera

NOHER DE HALAC[1]

## 1. Introduction

In Oenothera hybrids and species there are different kinds of pollen sterilities. The aim of this chapter is to give an overview of the morphological characters in different pollen sterility types that are useful to establish a classification that can be used on a general way in the study of pollen sterilities in this genus.

Most of the research about pollen sterility in Oenothera done from a morphological point of view gives data on microsporogenesis, microgametogenesis and tapetum (Oehlkers 1927, Harte and Bissinger 1952, Laser and Lersten 1972, Noher de Halac 1975a and 1975b).

From the morphological differences observed during the development of sterile anthers several aspects can be inferred: 1. If there are taxonomical differences in a basically similar phenomenon (homologies). 2. If there are physiological correlations with other variables (nutritional state, environment, soil, etc.). 3. If there are genetical differences expressed.

There are three main problems related to pollen sterility studies:

1. The difficulty in the comparison between species due to the large number of elements that are analyzed in each one of the investigated examples.

2. The difficult discerning if a certain combination of events is due to specific differences or not.

3. The independent study of morphological and genetical aspects in pollen sterilities.

Since the genetic base of Oenothera pollen sterilities is known in the hybrids used in this research (Harte and Bissinger 1952), they are adecuate systems to get more insight about selection strategies towards pollen sterility. The study of pollen sterility in Oenothera is an attempt to overcome the difficulties originated in the separation of the morphological and genetical aspects.

The accurate knowledge of normal development is a starting point in every characterization of pollen sterility. Up from there the characters that show variation in each pollen sterility type are recognized. Those characters have to be identified, selected and listed. Each time that a new type of pollen sterility has to be characterized (in case that it would have been studied with the same methods) it can be compared to the other known species through the differences in their characters. In case that new variable characters should appear they can be added to the above mentioned list.

---

1. Centro de Microscopía Electrónica. Universidad Católica de Córdoba, Trejo 323, 5000 Córdoba, Argentina.

## 2. Species and hybrids used

The characters taken as starting points are those of the normal development of anthers in O. elata HBK ♀ x glazioviana MICHELI ♂ (complex combination: hookeri. velans). Pollen sterility characters are studied in the same hybrid and in O. biennis L ♀ x villosa THUNB ♂ (complex combination: flavens. stringens), O. odorata JACQ. and in O. aff. recurva DIETRICH.

The actualized names of the species growing in Argentina are taken from the work of Dietrich (1979). The names of the species used to obtain the hybrids are taken from the work of Raven, et al.(1979). The names of the chromosome complexes (Renner complexes) are those generally employed by Oenothera geneticians (Cleland 1972, Harte 1984).

The hybrid Oenothera elata x glazioviana is obtained by the cross of a pollen sterile plant (homozigous for the gene ster) with normal style (heterozigous for the gene br) and a plant with normal pollen (heterozigous for the gene ster) and short style (homozigous for the gene br). Four different genotypes come out form wich the genotypes 2 and 4 are used in this study.

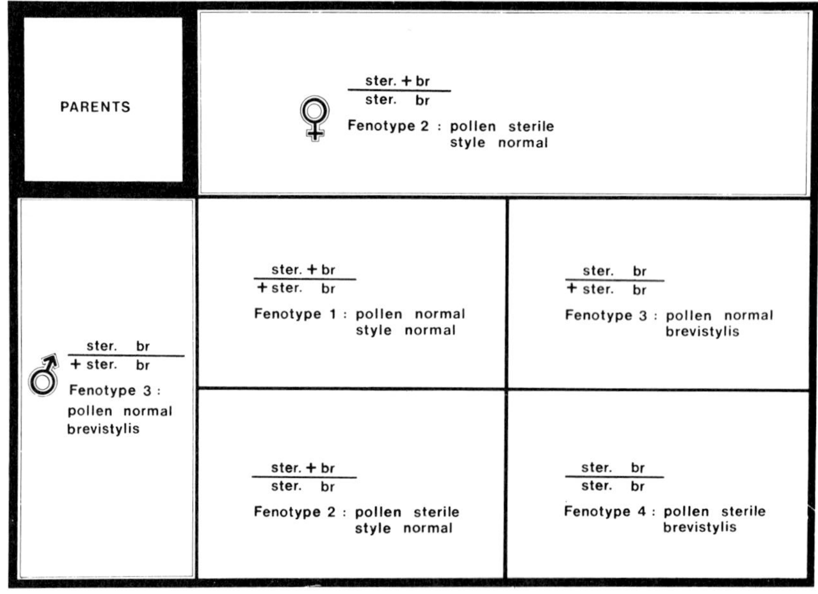


Fig. 1. Genotypes and phenotypes of the hybrid Oenothera elata HBK ♀ x glazioviana MICHELI ♂. Gene ster, when homozigous, pollen sterility. Gene br when homozigous, brevistylis.

Sterile plants of the hybrid Oenothera biennis x villosa contain the gene fr/fr in homozigous condition. This gen is linked with the gen S (Sulfurea) located in the chromosomic ring consisting of the ends 1-4 (Harte, personal communication). Only sterile anthers were analyzed in this work.

Both hybrids were obtained at the University of Cologne (W. Germany) by Prof. Dr. Cornelia Harte who, having started the study of this kind of pollen sterility in the fifties (Harte and Bissinger 1952), kindly supplied the material for this investigation.

Oenothera odorata JACQ is a species cultivated in the Province of Córdoba, the plants used in this study were growing in the town of Arroyito. Oenothera aff. recurva DIETRICH is a provisional name given to plants growing at 4000 m above sea level near the town of Paiquiquí in the Province of Catamarca. In the first species pollen dimorphism that corresponds to an

| CHARACTERS / HYBRIDS OR SPECIES | O. elata x glazioviana | O. biennis x villosa | O. odorata | O. aff. recurva |
|---|---|---|---|---|
| **GENERAL ASPECTS** | | | | |
| 1. 50 % normal and 50 % sterile plants in the population | yes | yes | no | no |
| 2. Normal and sterile pollen in each pollen sac | no | no | yes | ? |
| 3. Abnormalities during microsporogenesis | no | yes | no | no |
| 4. Abnormalities at the beginning of microgametogenesis | yes | yes | no | no |
| 5. Abnormalities only in mature pollen | no | no | yes | yes |
| **TAPETUM** | | | | |
| 6. Partially crushed down during microsporogenesis | no(yes) | yes | no | no |
| 7. Abnormal transfer phase | yes | yes | ? | ? |
| 8. SER hypertrophy | yes | no(yes) | ? | ? |
| 9. Lipids in plastids | no | yes | ? | ? |
| 10. Autolysis of cytoplasma | yes | yes | no | no |
| 11. Dark and lamellar inclusions | yes | no | no | no |
| **MICROSPOROGENESIS AND MICROGAMETOGENESIS** | | | | |
| 12. Callose during microsporogenesis | yes | no | yes | yes |
| 13. Too early deposition of the intine | no | yes | no | no |
| 14. Deposition of dark droplets on the primexine | yes | no | no | no |
| 15. Paracrystalline ektexine | no | no | yes | yes |
| 16. Endexine absent | no | yes | no | no |
| 17. Too thick endexine | no | no | no | yes |
| 18. Endexine dissolution | yes | no | no | no |
| 19. Dissolution of dark bodies | yes | no | no | no |
| 20. Lamellations on the surface of microspores | yes | no | no | no |
| 21. Pollen mitosis | no | no | ? | ? |
| 22. Pollen starch | no | no | ? | no |
| 23. Large and small pollen grains. | no | no | yes | no |
| 24. Empty pollen | no | no | no | yes |

Fig. 2. Coincidences and differences in pollen sterilities in hybrids and species of Oenothera (parenthesis indicate less frequent states, question marks indicate aspects not studied yet).

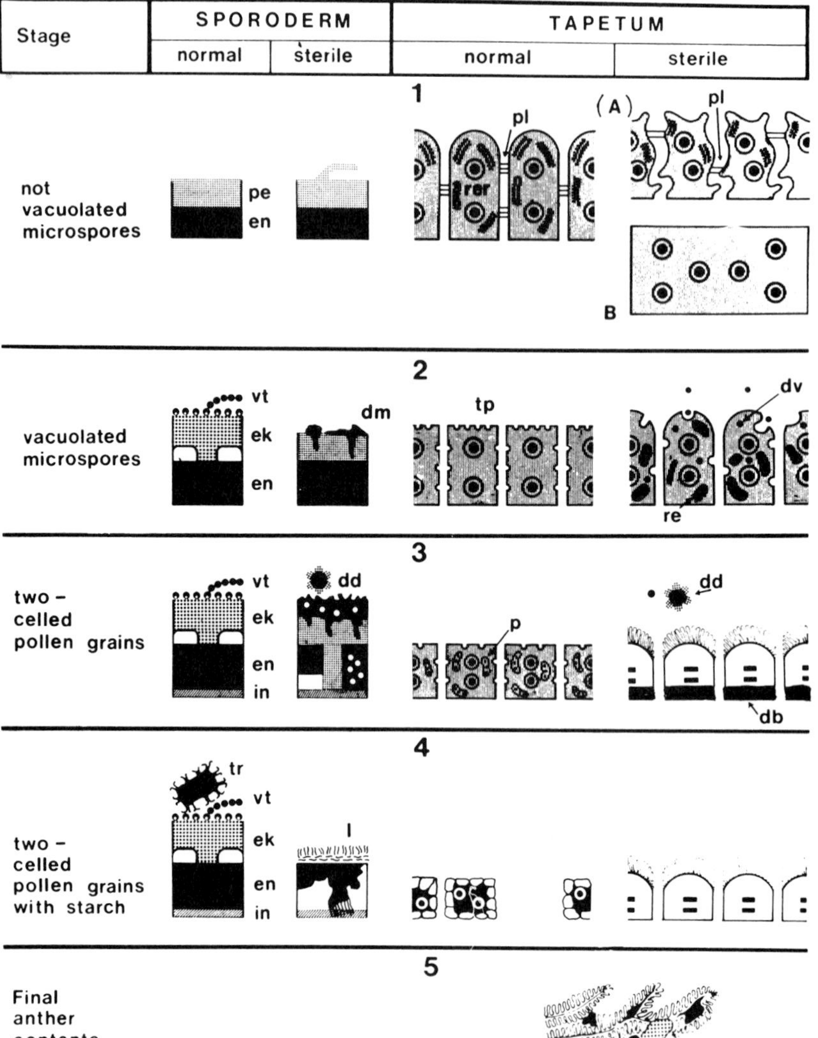


Fig. 3. Sporoderm and tapetal morphology during the stages of normal and sterile development in Oenothera elata x glazioviana. Each stage of microspores and pollen grains corresponds to a phase of tapetal cells: 1, pre-transfer phase which is modified in the sterile anthers remaining "in situ" (A) or crushing down (B). 2, transfer phase. 3, post-transfer phase. 4, vacuolation of cells. 5, reabsorption of cells. Abbreviations: db, dark bodies; dd, dark droplets; dm, dark material; dv, dark vesicles; ek, ektexine; en, endexine; in, intine; l, lamellations; pl, plasmodesmata; p, plastids with lipids; pe, primexine; re, endoplasmic reticulum; rer, rough endoplasmic reticulum; tr, tapetal remains; vt, viscin threads.

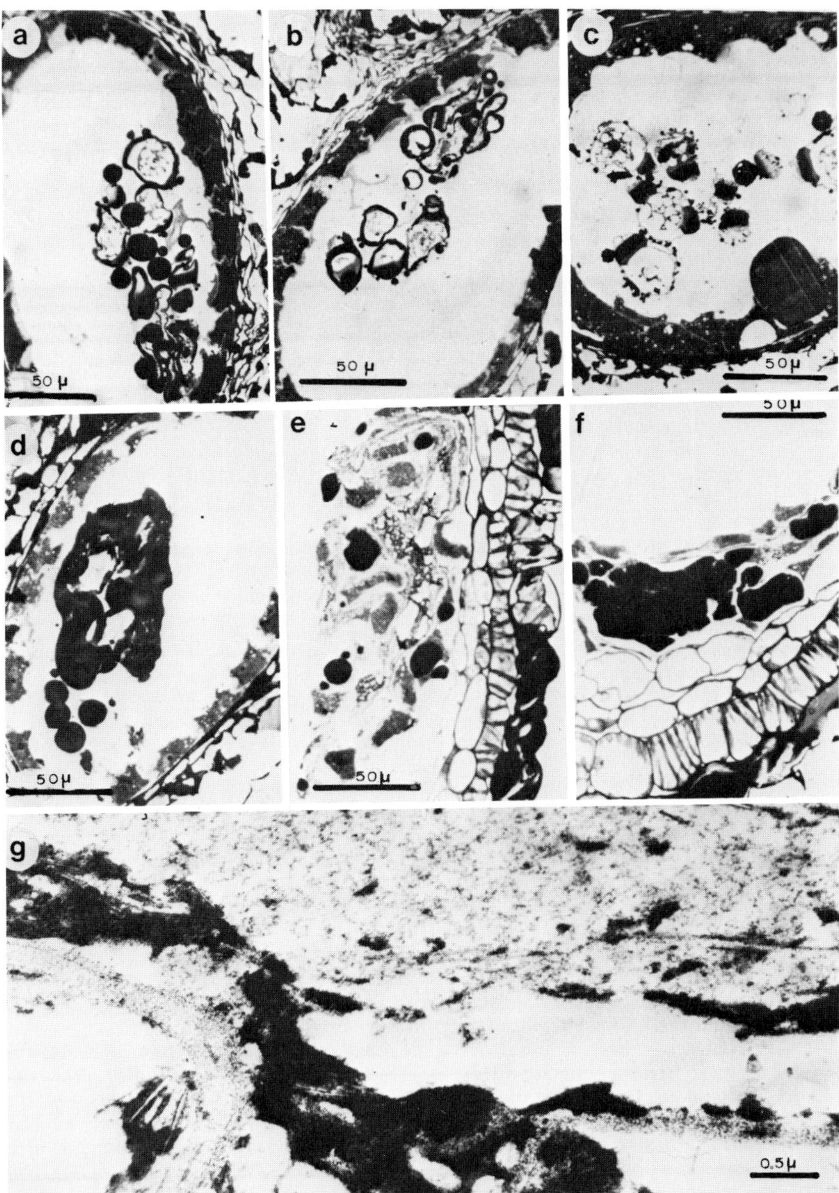


Fig. 4. Pollen sterility in *O. elata x glazioviana*. a, dark droplets in the pollen sac. b, microspores gradually sticking together. c, dark inclusions in the tapetum and dissolution of the endexine. d, dark material sticking the microspores together. e-g, final anther contents.

"active" and an "inactive" chromosomic complex is seen in all the pollen sacs. In the second one "empty" pollen grains are present in some anthers. Both kinds of pollen sterilities are correspondent to types found by Jean (1972 and 1984) in other Oenothera species. Detailed description of both cases will be published separately (Cismondi and Noher de Halac, in preparation).

Electron and light microscopy by methods published elsewere (Noher de Halac 1980) were used to collect the morphological data.

## 3. Character identification (Fig. 2)

A list of coincidences and differences between pollen sterilities in the four analyzed Oenothera entities is given in Fig. 2. Three groups of characters are considered: 1, General aspects. 2, Tapetum morphology. 3, Microsporogenesis and microgametogenesis.

Under the issue "general aspects" the percentage of sterile plants in the population and the presence of normal and sterile pollen in the same pollen sac are considered first and then the moment in wich abnormalities become evident is recorded.

In the item "tapetum" all important morphological evidences of variation are considered. Fig. 3 summarizes the normal and sterile development in O. elata x glazioviana showing the modifications occurring in the tapetal cells and the concomitant changes in the sporoderm.

Finally the item "microsporogenesis and microgametogenesis" includes all modifications affecting the microspores and pollen grains themselves during development. Characters 12-20 refer to sporoderm development while characters 21-24 correspond to modifications of the cytoplasm. Figs. 4 and 5 show the main aspects of development of pollen sterility in O. elata x glazioviana and Fig. 6 gives an idea of how pollen sterility in O. biennis x villosa develops.

## 4. Concluding remarks

On the basis of character variability, some concluding remarks on pollen sterility systems in Oenothera can be stated by unifying morphological and genetical criteria.

The gene fr/fr in homozigous condition codes for no callose deposition during microsporogenesis followed by the early intine formation in the microspores and the absence of endexine. The main modification shown by the tapetum is an alteration of the transfer phase that is concomitant with the lack of an ektexine with paracristallyne structure (Fig. 6). The tapetum is often partially crushed and the cell remains surround the microspores. It's remarkable that the capability of forming sporopollenin is blocked or absent in the microspores themselves and also in the tapetal cells.

The gene ster/ster codes for a tapetal modification that affects the transfer phase by replacing the eccrine secretion of lipidic precursor molecules assumed for sporopollenin (review: Buchen and Sievers 1981) for the granuloccrine secretion of lipid aggregates that form deposits on the microspores surface (Figs. 4 and 5). Concomitant to this alteration there is no presence of a paracrystalline structure of the ektexine. Then dissolution of the endexine and lipid aggregates follow in parallel to autolysis of tapetal cells instead of the usual post transfer phase (Fig. 4). The post transfer phase is characterized by the plastids containing lipids (Noher de Halac 1985b, Lombardo and Carraro 1976, Pacini and Casadoro 1981) wich seem to be generalized among the Angiosperms. The plastids are associated with the triphine formation which is absent in this hybrid. All evidences indicate that a deep modification of lipid metablism occurs in the anther affecting basically all steps of lipid formation: 1, eccrine secretion of tapetal cells. 2, paracrystalline structure of the ektexine. 3, triphine formation. 4, permanence of the endexine. In this context dark droplets and lamellations are interpreted

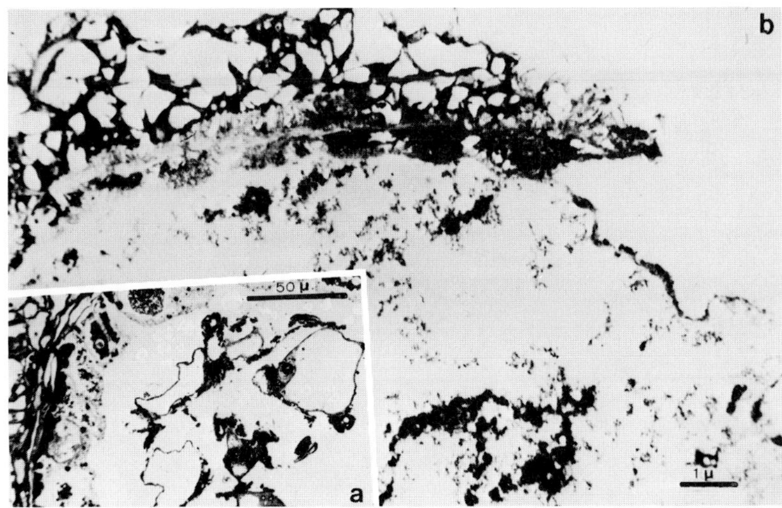


Fig. 5. Late stage of pollen degeneration in O. elata x glazioviana. a, Pollen sac showing the tapetum with inclussions and autolytic cytoplasm. Degenerating microspores are vacuolated and show the sporoderm reduced to a single layer of intine. b, Microspore surface at the same stage with a discrete portion of lamellar and amorphous material.

as being aggregates of sporopollenin monomers.

Taking the variable characters on account (Fig. 2) six main differences between pollen sterility in O. elata x glazioviana and O. biennis x villosa can be stated:

1. Pollen sterility becomes evident earlier in O. biennis x villosa (during microsporogenesis) than in O. elata x glazioviana (after release from the tetrad).
2. The lipids contained in the plastids of the tapetal cells are normally developed in O. biennis x villosa but are absent in O. elata x glazioviana.
3. Abnormal inclusions in the tapetal cells, as well as dark droplets and bodies and lamellations in the pollen sac are seen in O. elata x glazioviana but not in O. biennis x villosa.
4. Callose special wall is absent during microsporogenesis in O. biennis x villosa but normal in O. elata x glazioviana.
5. The timing of deposition of the intine is normal in O. elata x glazioviana but earlier than expected in O. biennis x villosa.
6. The behavior of the endexine is abnormal in both hybrids. In O. biennis x villosa no endexine is formed while in O. elata x glazioviana the deposition of the endexine is normal but it is later digested.

Both hybrids share a set of other abnormalities (Fig. 2) the most important of which are:

1. The modified transfer phase of the tapetum (eccrine secretion).
2. No paracristalline structure of the ektexine. This may be a consequence of the latter.
3. No pollen mitosis. That is the reason for assuming microspores sterility rather than pollen sterility.
4. No starch in mature grains.
5. Partially crushed tapetum during microsporogenesis. This fact is generalized in O. biennis x villosa, but exceptionaly seen in O. elata x glazioviana. It is a phenomenon commonly occurring after the post transfer phase that is moved

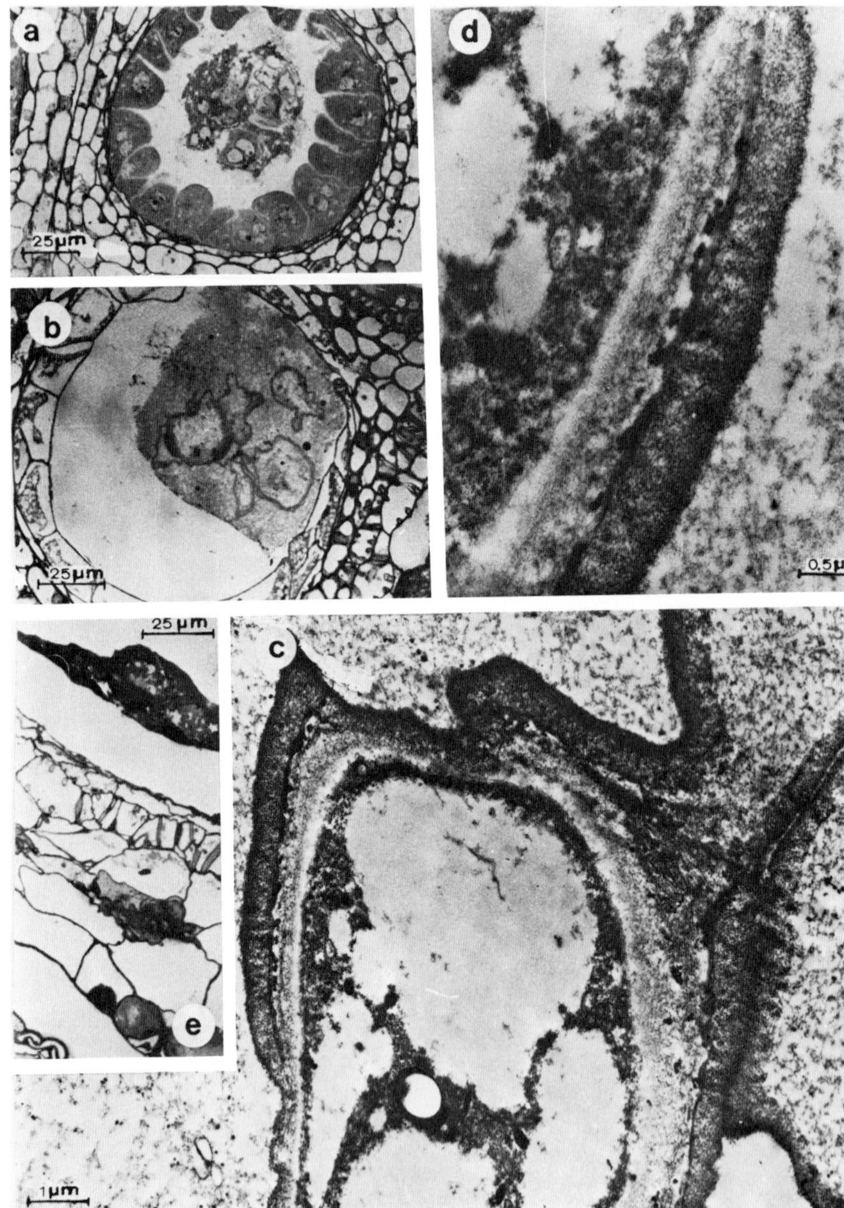


Fig. 6. Pollen sterility in *O. biennis x villosa*. a, absence of callose at the tetrad stage. b, tapetal cells are partially crushed down while the microspores have no ektexine at the moment of release from the tetrad. c-d, details of the sporoderm which is integrated by the intine and the primexine lacking the ectexine. e, final anther contents.

to an earlier stage of development in these hybrids.
6. The hipertrophy of the smooth endoplasmic reticulum is impressive in O. elata x glazioviana after the abnormal transfer phase of the tapetum. In O. biennis x villosa only some cells of the tapetum show the SER hypertrophy.
7. The autolysis of the tapetal cells that appears instead of the typical vacuolation which precedes the reabsorption of these cells during normal development.

In the other two species analyzed in this investigation pollen sterility becomes evident later during development affecting only the mature pollen. Another large difference with the hybrids shared by both species is the fact that the same plant has normal and abnormal pollen being both types in the same pollen sac in the case of O. odorata. No abnormalities in the tapetum nor sporoderm modifications in early stages of development have been found up to now. So the evaluated data (Fig. 2) show that there are more coincidences between O. odorata and O. aff. recurva than between any other pair of pollen sterilities.

In O. aff. recurva the sterile pollen is empty. The main differences with normal pollen is the death of the cytoplasm along with an overproduction of endexine while the tapetum presumably functions normally.

The mature pollen grains in O. odorata containing the "inactive" chromosomic complex show a size difference with the ones containing the "active" chromosome complex.-

Acknowledgements: I am gratefull to Prof. Dr. Cornelia Harte for supplying me the hybrids obtained in her laboratory. Also to Graciela Famá, Lucía Artino and Arturo Moya for their technical support and to Ariel Halac for help with the English version of the text.

References:

Buchen B and Sievers A (1981) Sporogenesis in plants In: Kiermayer O Cytomorphogenesis in Plants: 401-421. Springer Verlag. Berlin

Cleland RE (1972) Oenothera Cytogenetics and Evolution. Academic Press. London and New York

Dietrich W (1978) The South American Species of Oenothera sect. Oenothera (Raimannia, Renneria: Onagraceae). Ann Miss Bot Gard 64:425-626

Harte C (1984) Genetic control of the development of the haploid generation in Oenothera. Acta Soc Bot Pol 53:279-295

Harte C and Bissinger B (1952) Entwicklungsgeschichtliche Untersuchung der durch die Faktoren fr und ster bedingten Pollensterilität bei Oenothera. Z Vererbungslehre 84:251-269

Jean R (1971) La Paroi du Pollen d' Oenothera lamarckiana Bulletin de la Société de Botanique du Nord de la France 24:93-102

Jean R (1984) The genetics of pollen lethality in the complex heterozigote Oenothera-nuda. Biol Zentralbl 103:515-528

Laser KD and Lersten NR (1972) Anatomy and cytology of microsporogenesis in cytoplasmic male sterile Angiosperms. Botan Rev 38:425-454

Lombardo G and Carraro I (1976) Tapetal ultrastructural changes during pollen development I. Caryologia 29:114-125

Noher de Halac I (1980) Fine Structure of Nucellar Cells during Development of the Embryo Sac in Oenothera biennis L. Ann Bot 45:515-521

Noher de Halac I (1985a) Sterility of microspores in Oenothera hookeri. velans. In: Willemse MTM and van Went JL Sexual Reproduction in seed plants, ferns and mosses: 47-49. Pudoc. Wageningen

Noher de Halac I (1985b) Stages of tapetal cell development in sterile anthers of Oenothera hookeri. velans. In: Willemse MTM and van Went JL Sexual Reproduction in seed plants, ferns and mosses: 50. Pudoc.

Wageningen

Oehlkers F (1927) Entwicklungsgeschichte der Pollensterilität einiger Oenotheren. Z Vererbungslehre 43:265-348

Pacini E and Casadoro G (1981) Tapetum plastids of Olea europaea L. Protoplasma 106:289-296

# Fibrillar Structures in *Nicotiana* Pollen: Changes in Ultrastructure During Pollen Activation and Tube Emission

M. Cresti, P.K. Hepler*, A. Tiezzi, F. Ciampolini

## INTRODUCTION

In the course of analyzing the ultrastructure of *Nicotiana* pollen we have observed numerous crystalline-fibrillar bodies in the vegetative cytoplasm of mature and activated grains. Further studies during development revealed that these bodies change dramatically and eventually disappear, suggesting that they might be precursors for cytoskeletal components, e.g., microfilaments (MF) or microtubules (MT), that are elaborated during germination and participate in the growth process. Although fibrillar crystalloids have been reported in previous studies on the pollen (Franke et al., 1972; Hoefert, 1969), their developmental transformations have received little attention. The enclosed report, thus, provides a brief, preliminary description of these structures and their changes during pollen tube germination. We also provide preliminary results about the macromolecular composition of pollen grains as obtained by gel electrophoresis.

## MATERIALS AND METHODS

Pollen grains, obtained from dehiscing anthers of *Nicotiana tabacum*, were sown in culture medium (Brewbaker and Kwack, 1964) supplemented with 15% sucrose. Pollen grains or germinated pollen tube were harvested at sequential intervals following sowing and fixed in 3% glutaraldehyde in cacodylate buffer and postfixed in $OsO_4$ or in a mixture of $OsO_4$ and $K_3Fe(CN)_6$ (OsFeCN)(Hepler, 1981). Some material that had been postfixed in OsFeCN was also treated with tannic acid (McDonald, 1984). The material was dehydrated in acetone or ethanol and embedded in Spurr's low viscosity resin. Sections were analyzed with a JEOL Jem 100B operated at 80Kv.

----------

Dipartimento di Biologia Ambientale, Università di Siena (Italy)
* Department of Botany, University of Massachussetts, Amherst (USA)

## RESULTS

Fibrillar bodies commonly occur in the vegetative cytoplasm of the pollen grain of Nicotiana (Figs. 1 and 2). They are randomly scattered without any apparent association with other cytoplasmic organelles or inclusions. At low magnification they appear simply as electron dense bodies (Fig.1). However when viewed at higher magnification it becomes evident that they are composed of closely packed fine microfibrils in which the individual units measure approximately 4-7 nm in width (Fig. 2). Usually the bundles are viewed at an oblique angle relative to the fibrils and thus the superimposition of the individual elements tends to obscure the detailed substructure. Nevertheless, a sense of the inherent crystalline nature of these bodies can be gained from images revealing the presence of diagonal cross-striations (Fig. 3).

During germination there is a marked decline in the number of bodies as evidenced by their relative infrequency in young, growing pollen tubes, when compared to their numbers in mature grains. Accompanying their decline in number the individual fibrillar bundles appear to become frayed at their edge and loose aggregates of fine filaments seem to be in the process of sloughing free from the central crystalloid (Fig. 4). In addition it has become evident, especially following OsFeCN postfixation, that cisternae and vesicles of the Golgi apparatus become closely appressed to the sides of the fibrillar bundles (Fig. 3).

Working on the supposition that the bundles might be composed of actin MFs that become dispersed throughout the cytoplasm and participate in streaming we have carefully analyzed the cytoplasmic ultrastructure. Occasionally we observe one or a few filaments in the pollen tube cytoplasm that are similar in dimension to those associated with the bundles. Figure 5 for example, shows a few segments of single fibrils near the plasma membrane, and in a second example (Fig. 6) some fibrillar elements are observed deep within the pollen tube cytoplasm.

We have also considered the possible relationship of the fibrillar bundles to MTs. Cytoplasmic MTs have been observed in the generative cell (Fig. 7 and insert) where they occur in tightly clustered groups. Individual MTs may be separated by only few nm (about 5), and faint, but discernible cross-bridges are frequently observed. MTs have never been observed in the vegetative cytoplam of Nicotiana pollen grain or growing pollen tube.

Studies have also begun on the biochemical characterization of the macromolecular composition of the mature pollen grain. Preliminary results from gel electrophoresis reveal the presence of a host of proteins including ones that migrate with the same mobility as actin and tubulin. However, thus far no chemical identification can be given to the fibrillar bundles.

## DISCUSSION

The results show that crystalline-fibrillar bundles occur commonly in the vegetative cytoplasm of Nicotiana pollen. During activation and germination the bundles fray apart and eventually disappear. The central unaswered questions are as follows: what is the composition of the fibrillar material? and what role do these fibrils play during development? One attractive idea is that the bundles represent a storage form of F-actin and that during activation individual or small clusters of MFs are spun out from the bundles and dispersed thoughout the cytoplasm where they control cytoplasmic streaming. In support of this suggestion we direct attention to the fact that the dimensions of the unit fibrils within the bundles are similar to that for F-actin MFs. We also note that the dispersal of the fibrillar bundles corresponds temporally to the initiation of cytoplasmic streaming during pollen tube emission.

MFs composed of F-actin have been unequivocally demostrated in pollen tubes. Using protoplasts of pollen of Amaryllis, Condeelis (1974) has shown that filaments from burst cell preparations can be decorated with muscle heavy meromyosin. MFs, assumed to be composed of actin, have also been observed in Lilium (Franke et al., 1972). Based on the aforementioned study by Condeelis (1974) and on physiological experiments revealing that pollen tube cytoplasmic streaming is inhibited by the cytochalasins (Franke et al., 1972; Picton and Steer, 1981) there is ample reason to believe that streaming is generated by an acto-myosin system. Nevertheless it remains a question why so little F-actin is observed in the ultrastructural preparations. For example, in the studies of Picton and Steer (1981), which focus on the problem of the control of vesicle movement in pollen tubes of Tradescantia, no MFs are shown, despite the fact that the fixation and preservation of cytoplasmic detail appears to be very good. The prime reason may be that MFs are destroyed by the chemical fixation process, expecially by OsO4 (Maupin-Szamier and Pollard, 1978; Small and Langanger, 1981). While we have used a procedure, namely the OsFeCN method as modified by McDonald (1984), that appears to be among the most propitious for preserving MFs, still these structures may have been destroyed. Additional studies using alternate methods, such as antibody staining, are needed to resolve the problems concerning the presence and localization of action MFs.

A second hypothesis is that the fibrillar bundles consist of precursor material for MTs. This suggestion is not considered likely since MTs are only found in the generative cell while the fibrillar bundles occur only in the vegetative cytoplasm. Thus the fibrils, assuming that they were protofilaments of tubulin, would presumably have to be extensively broken down in order to be transported across the two plasma membranes that isolate the generative cell from the vegetative cytoplasm. Furthermore our observations indicate that the MTs of the generative cell are present by the time of pollen activation and thus their formation largely precedes the dispersal of the fibrillar bundles.

Finally the fibrillar bundles may be unrelated to either MTs or MFs. They could represent a storage form of protein that happens to crystallize into fibrous bundles. Work in progress is aimed at resolving this problem. Again we make reference to studies that employ fluorescent or gold labelled antibodies to the cytoskeletal proteins in order to see if they stain the fibrillar bundles. The observations presented in this current study thus represent our preliminary results about structural transformation of the fibrillar bundles. We hope that future investigations will allow us to determine the chemical composition of the bundles and to decipher their role in germination and growth of the pollen tube.

ACKNOWLEDGMENTS

The work was supported by CNR-Italy (bilateral project Italy-USA) and by "Cytomorphology Group" of CNR. Support for PKH has been provided in part by grants from the NSF (PCM 84-02414) and the NIH (GM 25120).

REFERENCES

Brewbaker J.L., Kwack B.H. (1964)- The calcium ion and substances influencing pollen growth. In: pollen physiology and fertilization. Linskens H.F., ed. Amsterdam, N.Y.: North Holland American Elsevier.

Condeelis J.S. (1974) - The identification of F actin in the pollen tube and protoplast of Amaryllis belladonna. Exptl. Cell Res. 88:436-439.

Franke W.W., W. Herth, W.J. Van Der Woude, D.J. Morré (1972)- Tubular and filamentous structures in pollen tubes: possible involvement as guide elements in protoplasmic streaming and vectorial migration of secretory vesicles. Planta 105;317-341.

Hepler P.K.(1981) - The structure of the endoplasmic reticulum revealed by osmium tetroxide-potassium ferricyanide staining. Europ. J. Cell Biol., 26:102-110.

Hoefert L.L. (1969) - Fine structure of sperm cells in pollen grains of Beta pollen. Protoplasma 68: 237-240.

Maupin-Szamier P., Pollard T.D.( 1978) - Actin filament destruction by osmium tetroxide. J. Cell Biol.77:831-852.

McDonald K. (1984) - Osmium ferricyanide fixation improves microfilament preservation and membrane visualization in a variety of animal cell types. J. Ultrastruct. Res. 86:107-118.

Picton J.M., Steer M.W. (1981) - Determination of secretory vesicle production rates by dictyosomes in pollen tubes of tradescantia using cytochalasin D. J. Cell Sci. 49:261-272.

Small J.V., Langanger G. (1981) - Organization of actin in the leading edge of cultured cells: influence of osmium tetroxide and dehydration on the ultrastructure of actin meshworks. J. Cell Biol. 91:695-105.

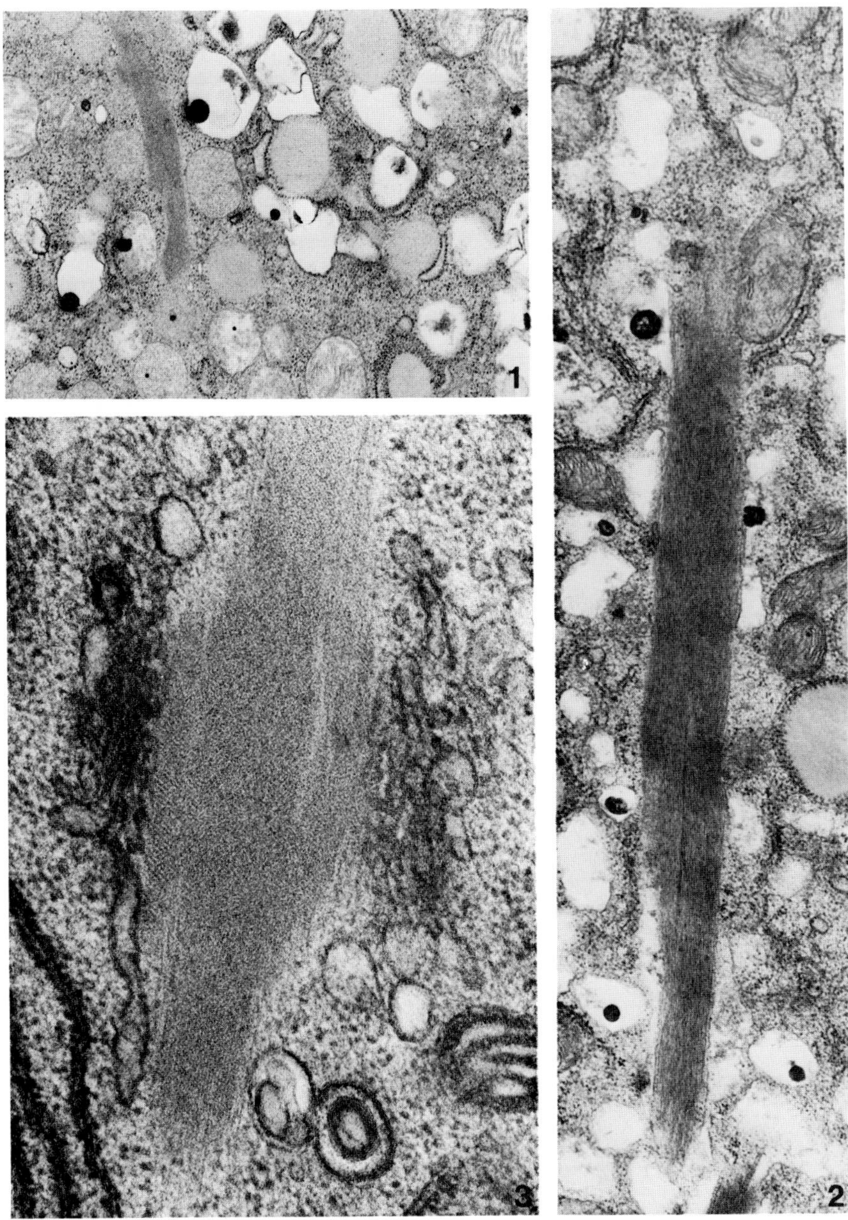

Figs.1 2. Fibrillar bodies in the vegetative cytoplasm of the mature inactivated pollen. Fig. 1 (X 13.300);Fig.2 (X 30.400)

Fig.3. Crystalline arrangements of the bodies. Golgi become closely appressed to the side of the fibrillar bundles. X 86.000

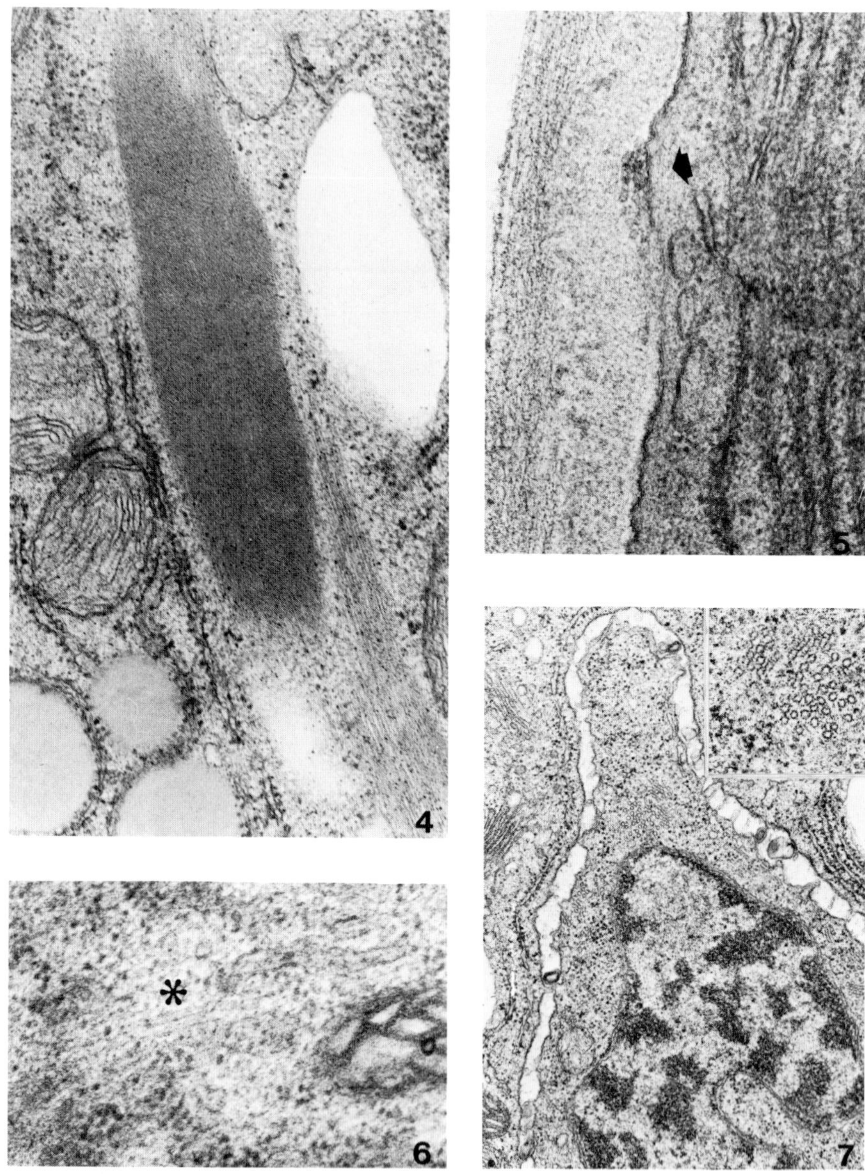


Fig.4. Activated pollen: individual fibrillar bundles appear to become frayed at their edge. X 50.000

Figs.5--6. Pollen tube: single fibril near the plasma membrane (arrow) and in the central part of the cytoplasm (asterisk). Fig. 5 (X 66.700); Fig. 6 (X 86.000)

Fig.7. Pollen tube: cytoplasmic microtubules in the generative cell. X 25.000; insert X 55.000

# The Male Germ Unit and Prospects for Biotechnology

C.A. McConchie and R.B. Knox[1]

If science moves in cycles then pollen biology is no exception. The widespread access to the electron microscope in the 1960s led this field to be dominated by ultrastructural studies of pollen wall development, and in the 1970s by an era when a functional approach developed. This later phase has been concerned with pollen-stigma interactions and the role of the male gametophyte in fertilization. The ready accessibility of pollen for studies of hydration, germination and penetration of the stigma, has meant that the role of the actual male gametes which perform fertilization has been almost totally neglected. Highlighting this attitude in the mid 1970s are the proceedings of the symposium held on the "Biology of the Male Gamete": (Duckett and Racey 1975) in which studies of sperm cells in fungi, algae, mosses, ferns through to animals are presented. However, the section on angiosperms deals entirely with pollen-wall proteins and self-incompatibility responses of the pistil. It is now in the 1980's that there has been a rediscovery of the male gametes of flowering plants, again ushered in by ultrastructural studies. This has opened the new field of sperm cell biology in relation to plant biotechnology.

---

[1] Plant Cell Biology Research Centre, School of Botany, University of Melbourne, Parkville, Victoria 3052, Australia.

## Concept of the male germ unit

The concept of the male germ unit in which the sperm cells and vegetative nucleus function as a single transmitting unit during reproduction has recently been developed (Dumas et al., 1984a,b). In flowering plants, all the DNA of male heredity, both nuclear and cytoplasmic, is linked together. This work was based on two-dimensional projections from thin sections of mature pollen of *Brassica oleracea*. Prior to this, it had been found that the two sperm cells are linked in *Plumbago* pollen tubes (Russell and Cass 1981), and in *Spinacia* pollen (Fig. 1, and Wilms and Van Aelst, 1983). In both these cases the sperm cells are closely associated in a distinctive manner with the vegetative nucleus. In *Spinacia* the projections from each of the sperms enter the enclaves of the vegetative nucleus. In *Plumbago*, like *Brassica*, only one of the sperms is associated with the vegetative nucleus; in the pollen grain, the sperm enters the enclaves (Fig. 1 and Russell 1984) and in the pollen tube it is wrapped around the vegetative nucleus (Russell and Cass 1981). The sperms are dimorphic in size, organelle content, and function in fertilization (Russell, 1984).

## The Male Germ Unit of Brassica campestris

The association of the sperm cells in *B. campestris* has been demonstrated using both manual and computer-assisted three-dimensional reconstruction of serial thin sections. The sperm cells are seen to be linked together with one of them having an elongate projection or tail that enters embayments of the vegetative nucleus (Fig. 1 and McConchie et al., 1984, 1985).

Recent quantitation of the sperm cell volumes using computerized image analysis techniques has shown that the sperm cell associated with the vegetative nucleus is larger than its partner and contains significantly greater numbers of mitochondria (Table 1). There are no plastids in either of the sperm cells.

pollen are closely associated, although not linked together (McConchie and Knox 1985). The sperm cells are dimorphic in terms of shape and size, and the longest sperm lies within pointed evaginations of the vegetative nucleus. This work confirms the prediction by Heslop-Harrison and Heslop-Harrison (1984) based on fluorescence microscopy that the sperm cells of pollen of Gramineae are not linked together as found in the dicot systems. Secondly, in the bicellular pollen system of *Rhododendron*, the elongate generative cell has terminal evaginations or tails which coil around the main body of the cell (Theunis et al., 1985). One of the these remains linked to a cell wall ingrowth of the intine. The vegetative nucleus is adjacent to the generative cell, but the tails have not been found within its enclaves in mature pollen.

We conclude that in a bicellular pollen system and in a tricellular monocot pollen system, there is no apparent linkage between the reproductive cells and the vegetative nucleus, negating a basic premise of the male germ unit concept (Dumas et al., 1984a,b). Nevertheless, the concept is based on the unit as target cells for non-random double fertilization. Provided the sperm cells are dimorphic in terms of their cell surface receptors, then the basic concept of a fertilization unit can be widened to encompass male-female recognition factors in the gamete cells. These could direct membrane fusion at fertilization.

## Sperm cell biotechnology

Figure 2. shows some of the possibilities offered by the male germ unit concept in the control and manipulation of fertilization. In order to characterize the pair of sperm cells, we can adopt many of the concepts that have been successfully applied to animal sperms, e.g. membrane zonation and capacitation. To achieve a fuller understanding of sperm cell biology, we have to be able to obtain access to the sperm cells, freeing them from their containing vegetative cell cytoplasm. This may be achieved by the general protocol given in Fig. 2. First, living pollen must be collected in quantity and used immediately, or frozen for storage in liquid nitrogen, which can pose significant problems, e.g. maize

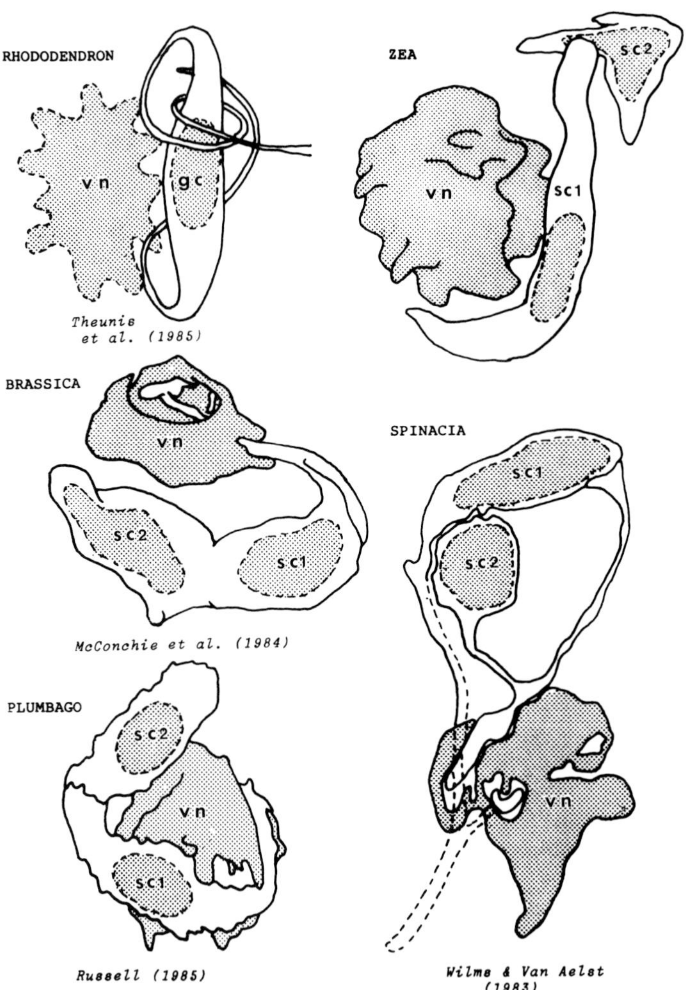


Fig. 1. Schematic diagrams of the male germ units in the pollen systems indicated, and adapted from the references sited. Abbreviations: vn, vegetative nucleus; sc, sperm cell; gc, generative cell.

## Sperm cell association in Zea and Rhododendron

In our laboratory, two other systems have been investigated by the technique of three dimensional reconstruction. Firstly in maize, Zea mays, the pair of sperm cells in mature

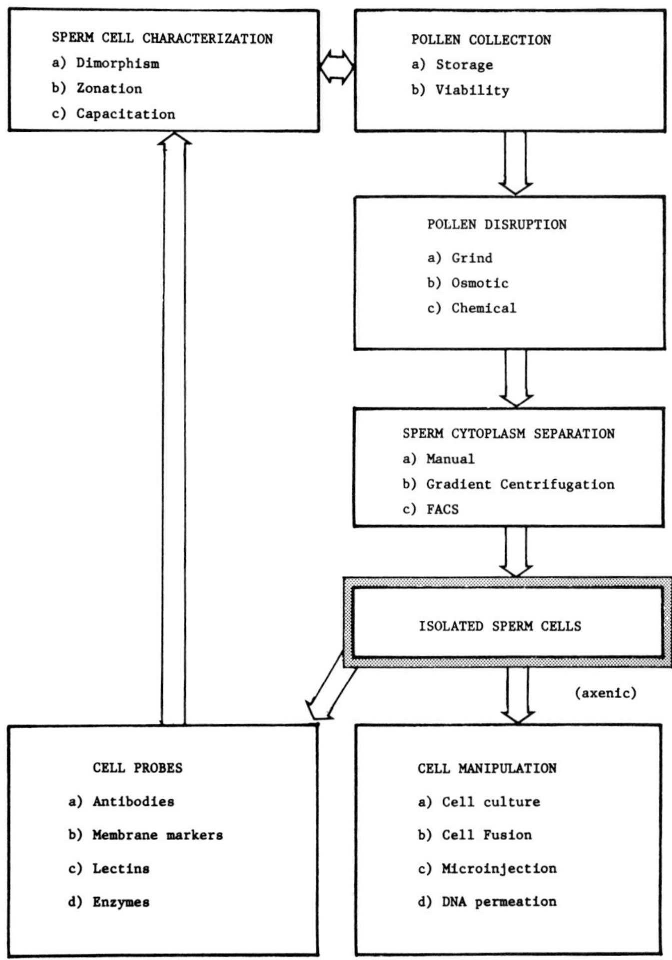


Fig. 2. Scheme showing selected applications of sperm cells in biotechnology.

pollen cannot at present be stored (see Knox 1984). Viability of the pollen before or after storage should be assessed and the FCR test is one method (see review by Heslop-Harrison et al., 1983), although we have found this method inapplicable to maize pollen. The mature pollen must then be disrupted, by grinding or partial homogenizing, by osmotic shock, or by chemical means, e.g. enzymic digestion. The sperm cells must then be separated from the cytoplasm of the vegetative cell by manual or fluorescent-activated cell-sorting, or by standard methods of gradient centrifugation.

Lafountain and Mascarenhas (1972) developed techniques for the isolation of the vegetative and generative nuclei of *Tradescantia* pollen. The way is now open to achieve the isolation of the male gametes. This will lead to the characterization by the probes of cell biology (Fig. 2) of the sperm cell surface, and result in identification and characterization of unique determinants that may be important in recognition of the female gametes. If the isolation procedure is axenic, then the way is open to culture sperm cells as a new source of haploid plantlets or tissue for genetic engineering.

## Conclusions

The sperm cell presents a new frontier in plant cell biology, and the male germ unit concept provides a useful basis to ask questions concerning the mechanism of double fertilization in flowering plants.

Table 1. Quantitative estimation of heritable organelle content in terms of cell volumes and numbers of mitochondria in pollen of *Brassica campestris*. Using the W test for two independent samples, P = 0.001 that the two sperms are from the same population for both parameters.

| Male Germ Units | Mitochondria | | Volume $\mu m^3$ | |
|---|---|---|---|---|
| | Sperm 1 | Sperm 2 | Sperm 1 | Sperm 2 |
| A | 22 | 4 | 10.98 | 6.09 |
| B | 30 | 9 | 11.97 | 6.82 |
| C | 31 | 11 | 12.12 | 6.20 |
| D | 15 | 4 | 11.20 | 8.78 |
| E | 23 | 10 | 10.56 | 8.59 |

References

Duckett J, Racey P (1975) The biology of the male gamete. Academic Press, London.

Dumas C, Knox RB, Gaude T (1984a) Pollen-Pistil Recognition: New concepts electron microscopy and cytology. Int Rev Cytol 90: 239-271.

Dumas C, Knox RB, McConchie CA, Russell SD (1984b) Emerging physiological concepts in fertilization. What's new In plant physiol 15: 17-20.

Heslop-Harrison J, Heslop-Harrison Y (1984) The disposition of gamete and vegetative-cell nuclei in the extending pollen tubes of a grass species, Alopecurus pratensis L. Acta Bot Neerl 33: 131-134.

Heslop-Harrison J, Heslop-Harrison Y, Shivanna KR (1983) The evaluation of pollen quality and a further appraisal of the fluorochromatic reaction (FCR) procedure. Theoret Appl Genet 67: 367-375.

Knox RB, (1984) The pollen grain. In B.M.Johri (ed) Embryology of Angiosperms. pp 197-271.

LaFountain KL, Mascarenhas JP (1972) Isolation of vegetative and cell generative nuclei from pollen tubes. Exptl Cell Res 73: 233-236.

McConchie CA, Jobson S, Knox RB (1984) Analysis of the ultrastructure of sperm cells of Brassica campestris by computer-assisted three-dimensional reconstruction. In EG Williams, RB Knox (eds) Pollination '84. pp 26-29, University of Melbourne.

McConchie CA, Jobson S, Knox RB (1985) Computer-assisted reconstruction of the male germ unit in pollen of Brassica campestris. Protoplasma (in press).

McConchie CA, Knox RB (1985) Three-dimensional reconstruction of the sperm cells of mature pollen of maize, Zea mays: unique chondriome and golgi apparatus. Planta (submitted).

Russell SD (1984) Ultrastructure of the sperm of Plumbago zeylanica. II. Quantitative cytology and three-dimensional organization. Planta 162: 385-391.

Russell SD, Cass DD (1981) Ultrastructure of the sperms of Plumbago zeylanica. I. Cytology and association with the vegetative nucleus. Protoplasma 107: 85-107.

Russell SD, Cass DD (1983) Unequal distribution of plastids and mitochondria during sperm cell formation in Plumbago zeylanica. In: Pollen: biology and implications for plant breeding, pp. 135-140. Mulcahy DL, Ottaviano E, (ed). Elsevier Biomedical Press, New York.

Theunis CH, McConchie CA, Knox RB (1985). Three-dimensional reconstruction of the generative cell and its wall connection in mature bicellular pollen of *Rhododendron* Planta (submitted).

Wilms HJ, Van Aelst AC (1983) Ultrastructure of spinach sperm cells in mature pollen. In: Erdelska O (ed.) Fertilization and Embryogenesis in Ovulated Plants. 105-112. Veda, Bratislava.

# On the Male Germ Unit in an Angiosperm with Bicellular Pollen, Hippeastrum vitatum

H. LLOYD MOGENSEN[1]

## 1 Introduction

Recent studies on plants with tricellular pollen have resulted in the emerging concept that the two male gametes of common origin, along with the vegetative nucleus, form a functional unit termed the "male germ unit" or "male fertilization unit" (Dumas et al., 1984; Heslop-Harrison and Heslop-Harrison, 1984). The basis for this terminology stems from observations which show that sister sperms are joined by a common wall and that at least one of the sperms forms a close association with the vegetative nucleus (Russell and Cass, 1981; Wilms and Van Aelst, 1983; Russell, 1984; Dumas et al., 1984, 1985; McConchie et al., 1985). It is speculated that such a unit would be effective in the tandem movement of the sperms from the pollen grain and in the pollen tube. Contacts between one of the sperms and the vegetative nucleus may be related to the orderly arrival of the sperms to their female target cells, thus setting the stage for directed fertilization where a given sperm may preferentially fertilize either the egg or the central cell (Dumas et al., 1984).

While physical connections between sperm cells may function in ensuring the synchronized movement of the sperms of plants with tricellular pollen, such a unit would seem

[1] Department of Biology, Box 5640, Northern Arizona Universtiy, Flagstaff, Arizona, 86011, USA

unnecessary in plants with bicellular pollen since the two sperms are formed in the pollen tube well after the time of pollination. If it is assumed that the forces resulting in sperm transport act equally on both sperms it would follow that they would remain in close proximity after their formation from division of the generative cell and during their passage in the pollen tube.

The present study was undertaken in order to test the validity of the concept of the male germ unit in a plant with bicellular pollen. Amaryllis was selected because the pollen tubes of this plant can be grown on a simple nutrient medium and because the generative cell, as well as the sperms, can be easily visualized in the living condition with the light microscope, without the use of stains.

## 2 Materials and Methods

Plants of *Hippeastrum vitatum* Herb., near the flowering stage, were obtained from M & G Flower and Bulb Corporation, Milwaukee, Wisconsin and placed in the greenhouse. *In vitro* observations were made from pollen tubes growing at room temperature on 10% lactose agar or in 10% lactose solution on a microscope slide under a coverslip supported at one end with a strip of filter paper. Material was observed either unstained in the living condition, or after staining with acetocarmine or aniline blue. *In vivo* material was obtained from hand pollinations. After the pollen tubes had grown for certain periods in the intact style, the styles were severed with a razor blade at measured lengths from the stigma and placed on 10% lactose agar in a covered petri dish. When the first pollen tubes emerged from the cut end of the style, they were fixed, intact, in 4% glutaraldehyde in phosphate buffer (pH 7.4), then a 2 mm segment of style containing the pollen tube tips was removed and processed further for electron microscopy. After 6 hrs. in glutaraldehyde at room temperature, the material was rinsed for two hrs. in phosphate buffer (3 changes), then post-fixed in 2% $OsO_4$ (phosphate buffer, pH 7.4) for two hrs. at room temperature, rinsed in buffer, dehydrated in an ethanol-acetone series and embedded in Spurr's resin.

Serial thick sections (3 µm) were cut with a diamond knife, observed with phase contrast microscopy, then re-embedded at the tip of epoxy blocks and serially ultrathin sectioned (Mogensen, 1971). Ultrathin sections were stained in an automatic LKB Ultrostainer with uranyl acetate and lead citrate, and observed with a JEM 7A transmission electron microscope.

## 3 Results

The generative cell of amaryllis pollen is approximately 30 µm long and 12 µm wide and can be readily visualized in the living condition with bright-field light microscopy. Pollen grains _in vitro_ begin germination within two hours. By about 4-6 hours, when the pollen tube is 400-600 µm long, the generative cell exits the pollen grain and enters the pollen tube. At about 15 to 20 hours after the pollen grains are placed in lactose solution, or on the stigma, the generative cell divides and soon produces the two male gametes. The vegetative nucleus was usually not visible in the living or the stained condition at the light microscope level.

Once generative cell division is complete, the sperm nuclei remain about 15 µm apart as they travel within the pollen tube approximately 100 µm from its tip. No connections between the cells were visible at the light microscope level in unstained or stained material of this study. However, at the electron microscope level, pollen tubes grown _in vivo_ for 60 hrs. (approximately 60% of the style length) show that the two sperms are connected along extensive areas of cytoplasmic projections between the cells (Figs. 1 & 4). Such a connection is maintained between the male gametes at least up to the time the pollen tubes have grown the full length of the style and are approaching the level of the uppermost ovules (about three days after pollination and 12 cm of style length). Each sperm nucleus is surrounded by a considerable amount of cytoplasm that is highly convoluted around most of its surface (Fig. 2).

The vegetative nucleus can be seen to have numerous embayments and to be in close association with cytoplasmic extensions of the leading sperm cell even after the pollen tube has grown the entire length of the style (Fig. 3).

**4 Discussion**

The present study extends the concept of the male germ unit to a plant with bicellular pollen and demonstrates that even though the sperms are not formed until the pollen tube has grown approximately 20% of the length of the style, there is still an apparent adaptive advantage for a physical connection between the male gametes as they continue their passage in the pollen tube. Dumas et al. (1984) suggest that the association of the sperm cells with each other and with the vegetative nucleus "may facilitate the transmission and order the arrival of the male gametes into the receptive cells of the female embryo sac." Clearly, only a limited amount of pollen tube cytoplasm can be accommodated at the site of discharge in the embryo sac, thus the sperms need not only be close together at this time, but must also be quite close to the pollen tube tip. Whether the order of arrival of the sperms to their female target cells is of importance is questionable since it is known that both sperms are ultimately in intimate contact with the plasma membranes of the egg and central cell before syngamy takes place (Mogensen, 1982; Russell, 1983), and in many plants it is very common for the sperms or the generative cell to exit the pollen grain ahead of the vegetative nucleus (Wylie, 1923; Chandra and Bhatnagar, 1974; Heslop-Harrison and Heslop-Harrison, 1984). If the sperm cells are pre-programmed to fuse with a given female gamete, as has been demonstrated in corn (Roman, 1948) and as also appears to be the case in *Plumbago* (Russell, 1984; Dumas et al., 1984), and if a cell surface recognition system is present at the gamete level, as proposed by Dumas et al. (1984), it would appear that the order of sperm arrival would be inconsequential as far as sperm specificity is concerned.

Results from studies on plants having morphologically distinct male germ units appear to be in conflict with the condition in barley (Mogensen and Rusche, 1985), and apparently other grasses (Karas and Cass, 1976; Hu et al., 1981; Schroder, 1983), where there is no physical contact either between the sperm cells or between sperm and vegetative nucleus. In barley the mature sperms (at anthesis) were not observed to be closer than 0.6 μm from each other, and the nearest a sperm and vegetative nucleus were observed was also 0.6 μm. Yet, the two sperms remain quite close to each other during pollen tube growth (Pope, 1937) and are both positioned within the intercellular space between the egg and central cell prior to gametic fusion (Mogensen, 1982). In the grass, Alopecurus pratensis, Heslop-Harrison and Heslop-Harrison (1984) observed that the sperm pairs remain close together during pollen tube extension, but they are separated from the vegetative nucleus by 19 to 64 μm.

Thus, some mechanism other than physical contact results in the near simultaneous arrival of the sperms to the embryo sac in plants where no connection exists between male gametes. In barley the majority of the volume within the pollen grain is occupied by numerous, large starch grains. The sperms and vegetative nucleus are consistently located in a smaller area of cytoplasm, containing little or no starch grains. It may be that this starch-free zone of cytoplasm acts as a confining space that keeps the sperms relatively close together as they exit the pollen grain and enter the pollen tube. Since the sperms of barley, unlike those of plants with physically linked sperms such as Plumbago, Brassica and spinach (Wilms and Van Aelst, 1983; Russell, 1984; Dumas et al., 1984, 1985; McConchie et al., 1985), are essentially identical in size and shape (Mogensen and Rusche, 1985), perhaps each gamete responds equally to the sperm-transporting forces, thereby eliminating the necessity of actual connections. Such a hypothesis would predict that all physically linked sperms are dimorphic and that all unconnected sperms are isomorphic.

An additional consideration regarding linked and unconnected sperms has to do with the phenomenon of hetero-fertilization. In Zea mays (Sprague, 1932) and barley (R. T. Ramage, unpublished) it has been shown genetically that in some cases the sperm fusing with the egg comes from a different pollen tube than the sperm that fuses with the polar nuclei of the same embryo sac. Presumably, hetero-fertilization would be less likely to occur in plants with physically linked sperms.

**Figure Descriptions**

Fig. 1-3. Electron micrographs of the male germ unit within the pollen tube. 1. Longisection of a pollen tube that had grown 60 hrs. and 7.5 cm in the style, showing the two sperm cells with numerous, overlapping extensions of cytoplasm between the cells. X2200. 2. Transverse section of a pollen tube after 70 hrs. and 12 cm of style growth, showing extensive, convoluted cytoplasm of the leading sperm cell. X5900. 3. Transverse section of the same pollen tube as in Fig. 2, showing the close association of the vegetative nucleus (VN) with cytoplasmic extensions (SC) of the leading sperm cell. X5900. SC= sperm cytoplasm; $SN_1$= nucleus of leading sperm cell; $SN_2$= nucleus of trailing sperm cell.

Fig. 4. Higher magnification of the area between sperm cells of Fig. 1, showing overlapping of sperm cytoplasmic extensions. Note that cytoplasmic continuity of the leading sperm cell can be traced from arrowhead #1 toward the trailing sperm cell to at least the point of arrowhead #2. $SN_1$= nucleus of leading sperm cell; $SN_2$= nucleus of trailing sperm cell; SC= sperm cytoplasm. X6200.

Acknowledgment. This work was supported by grant no. 83-CRCR-1-1270 from the United States Department of Agriculture, and by the Organized Research Fund, Northern Arizona University. I would like to thank Drs. J. Heslop-Harrison and Scott D. Russell for helpful suggestions on the manuscript.

**References**

Chandra, S. and S. P. Bhatnagar. 1974. Reproductive biology of Triticum. II. Pollen germination, pollen tube growth, and its entry into the ovule. Phytomorphology 24: 211-217.

Dumas, C., Knox, R. B. and T. Gaude. 1985. The spatial association of the sperm cells and vegetative nucleus in the pollen grain of Brassica. Protoplasma 124: 168-174.

Dumas, C., Knox R. B., McConchie, C. A. and S. D. Russell. 1984. Emerging physiological concepts in fertilization. What's New in Plant Physiology 15: 17-20.

Heslop-Harrison, J. and Y. Heslop-Harrison. 1984. The disposition of gamete and vegetative-cell nuclei in the extending pollen tubes of a grass species, Alopecurus pratensis L. Acta Bot. Neerl. 33: 131-134.

Hu, S. Y., Zhu, C. and L. Y. Xu. 1981. Ultrastructure of male gametophyte in wheat: 2. Formation and development of sperm cell. Acta Bot. Sin. 23: 85-91.

Karas, I. and D. D. Cass. 1976. Ultrastructural aspects of sperm cell formation in rye: Evidence for the cell plate involvement in generative cell division. Phytomorphology 26: 36-45.

McConchie, C. A., Jobson, S. and R. B. Knox. 1985. Computer-assisted reconstruction of the male germ unit in pollen of Brassica campestris. Protoplasma (in press).

Mogensen, H. L. 1971. A modified method for re-embedding thick epoxy sections for ultramicrotomy. J. Ariz. Acad. Sci. 6: 249-250.

Mogensen, H. L. 1982. Double fertilization in barley and the cytological explanation for haploid embryo formation, embryoless caryopses, and ovule abortion. Carlsberg Res. Comm. 47: 313-354.

Mogensen, H. L. and M. L. Rusche. 1985. Quantitative ultrastructural analysis of barley sperm, I. Occurrence and mechanism of cytoplasm and organelle reduction and the question of sperm dimorphism. Protoplasma (in press).

Pope, M. 1937. The time factor in pollen tube growth and fertilization in barley. J. Agr. Res. 54: 525-529.

Roman, H. 1948. Directed fertilization in maize. Proc. Nat. Acad. Sci. 34: 46-52.

Russell, S. D. 1983. Fertilization in Plumbago zeylanica: Gametic fusion and fate of the male cytoplasm. Amer. J. Bot. 70: 416-434.

Russell, S. D. 1984. Ultrastructure of the sperm of Plumbago zeylanica II. Quantitative cytology and three-dimensional organization. Planta 162: 385-391.

Russell, S. D. and D. D. Cass. 1981. Ultrastructure of the sperms of Plumbago zeylanica. I. Cytology and association with the vegetative nucleus. Protoplasma 107: 85-107.

Schroder, M. B. 1983. The ultrastructure of sperm cells in Triticale. In: Fertilization and embryogensis in ovulated plants. Ed. : O. Erdelska. pp. 101-104. Bratislava, Czech: Center of Biol. and Ecol. Sci. Slovak Acad. Sci.

Sprague, G. F. 1932. The nature and extent of hetero-fertilization in maize. Genetics 17: 358-368.

Wilms, H. J. and A. C. Van Aelst. 1983. Ultrastructure of spinach sperm cells in mature pollen. In: Fertilization and embryogenesis in ovulated plants. Ed. O. Erdelska. pp. 105-112. Bratislava, Czech.: Center of Biol. and Ecol. Sci. Slovak Acad. Sci.

Wylie, R. B. 1923. Sperms of Vallisneria spiralis. Bot. Gaz. 75: 191-202.

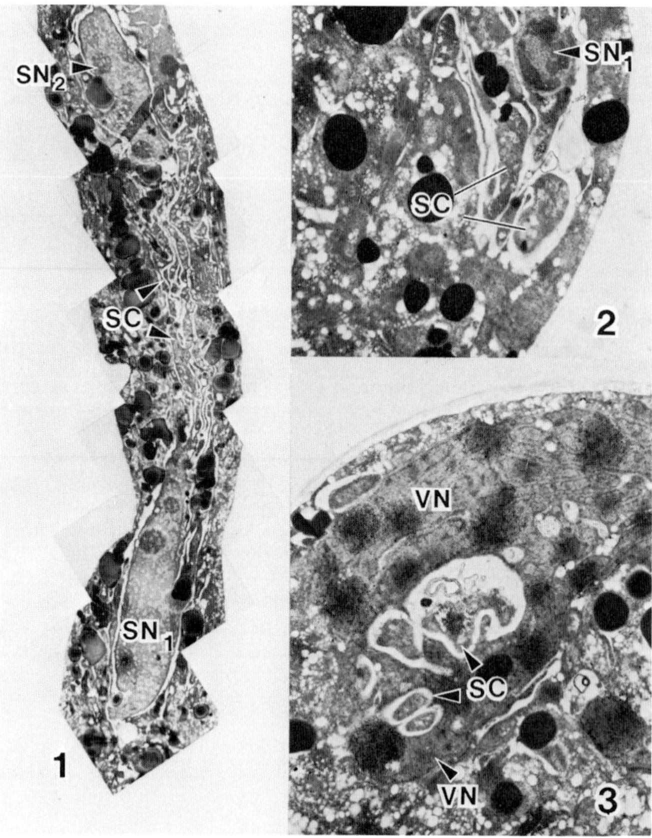


Fig. 1-3. Electron micrographs of the male germ unit within the pollen tube.

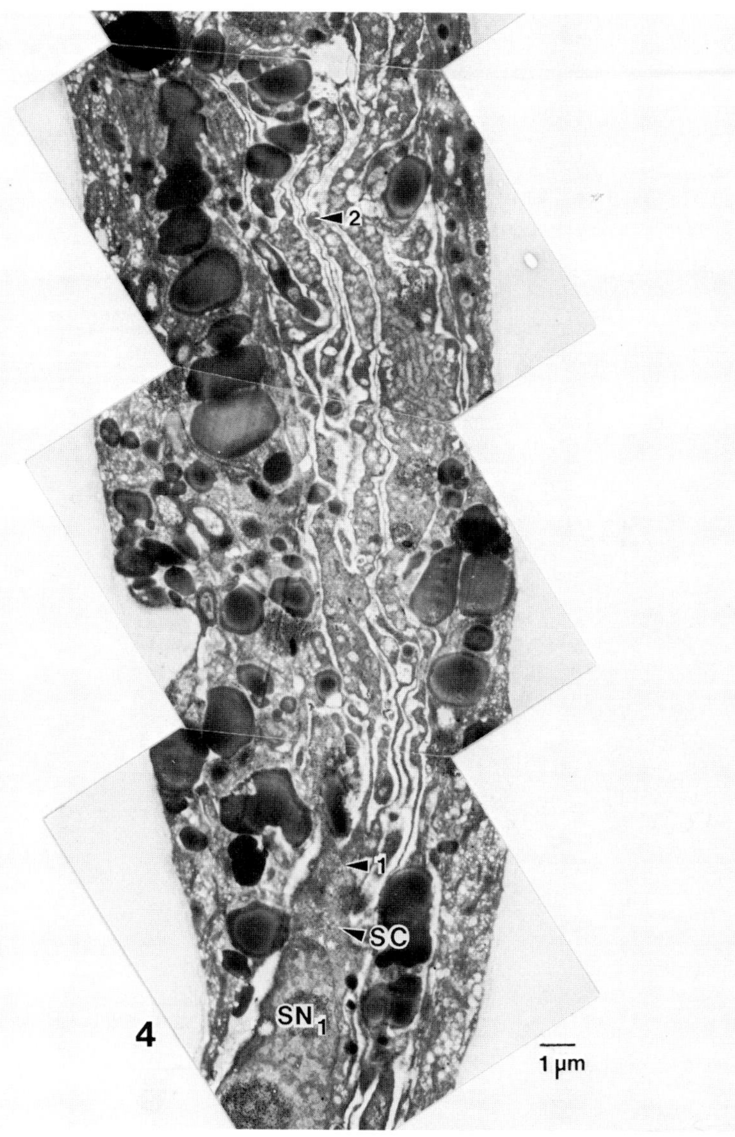


Fig. 4. Higher magnification of the area between sperm cells of Fig. 1, showing overlapping of sperm cytoplasmic extensions.

# Isolation of Spinach Sperm Cells: 1: Ultrastructure and Three-dimensional Construction in the Mature Pollen Grain

H. J. WILMS, H. B. LEFERINK-TEN KLOOSTER AND A. C. VAN AELST[1]

1 Introduction. Before isolating the sperm cells from pollen or pollen tubes of spinach, *Spinacia oleracea* L., we must know their ultrastructure, i.e. the organelle content and the membrane compilation and also the techniques with which the sperm cells can be grown *in vitro*, after which haploid plants can be regenerated from these cells. This chapter deals with the first part of this research, namely the structure, its organelle contents and the membrane compounds.

Sperm cells in mature pollen grains appear to have various structures, but all are specific for each plant species. In cotton (Jensen and Fisher, 1968), wheat (Zhu et al., 1980), *Plumbago zeylanica* (Russell, 1980, 1984, 1985; Russell and Cass 1981a, 1983), and *Brassica* (Dumas et al., 1984, 1985; McConchie et al., 1985) sperm cell-vegetative nucleus associations exist. The two sperm cells probably are dimorphic and have rather different patterns of male cytoplasmic transmission (Hagemann, 1976, 1981, 1983; Russell and Cass 1981a, b; 1983).

In *Spinacia* (Wilms, 1981, 1985; Wilms and Van Aelst, 1983) two, tailed sperm cells, both connected with the vegetative nucleus, are present in the mature pollen grain. The sperm cells have cytoplasmic connections with each other. In this chapter the continuing research on the 3D reconstruction will be presented in combination with a new freezing and etching scanning electron microscopy technique. With this method the outer structure of the different compounds of the sperm cells-vegetative nucleus association can be detected.

2 Sperm cell organization. In spinach sperm cells at the mature pollen grain stage mitrochondria are present, whereas plastids and dictyosomes lack. Both sperm cells have long slender projections. Earlier observations (Wilms and Van

---

[1] Dept Plant Cytol & Morphol, AU, Wageningen, The Netherlands

Aelst, 1983) state that these projections are encircled by the convoluted nucleus of the vegetative cell. Sometimes the cytoplasmic runners of the sperm cells seem to continue almost till the intine. The sperm cells are linked by a common cross "wall" (Fig. 5: right upper most cross section). Fig. 1 shows a section through a mature pollen grain at low magnification. The nucleus of the vegetative cell can be observed in the centre of the grain. The sperm cells are present in different parts and located in the region from the vegetative nucleus almost to the intine. The two sperm cells cannot be identified as different cells from individual sections. In this section, as in 98 % of the other sections, the connections between the two sperm cells cannot be seen. Only in a few sections this connection will be present. This is one of the reasons that sperm cell connections are not observed in general in other species. They will be identified more frequently by studying serial sections, as for example in _Plumbago_ (Russell, 1984).

In spinach these connections remain when the sperm cells-vegetative nucleus complex pass through the pollen tube (Wilms and Leferink, 1983; Wilms, 1985). In one case two sperm nuclei with much less cytoplasm within one common membrane are observed in the chalazal part of the penetrated degenerated synergid (Wilms, 1981). This phenomenon is still not yet understood. If this phenomenon occurs frequently the "two sperm nuclei within one membrane system" can be used for gaining individual cells with few organelles in _in vitro_ culture.

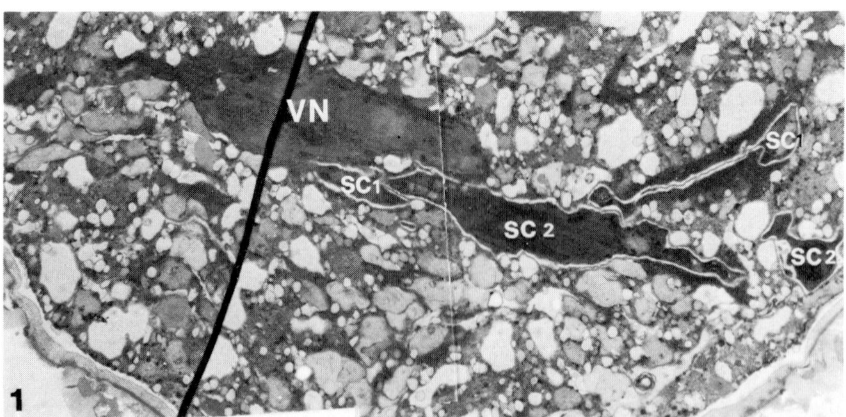


Fig. 1. Electron micrograph of mature pollen grain of _Spinacia_ showing the two sperm cells (SC1, SC2) and the vegetative nucleus (VN) in the vegetative cytoplasm.

3 Three dimensional reconstruction. Serial ultrathin sections of about 100 nm are cut and observed in an electron microscope. From these sections the sperm cells and the vegetative nucleus are photographed and printed at one magnification. The upper part of the sperm cells is not included. The prints are redrawn, i.e., the outlines of the sperm cells, their nuclei and organelles, and the vegetative nuclei are converted to digital X,Y locations.

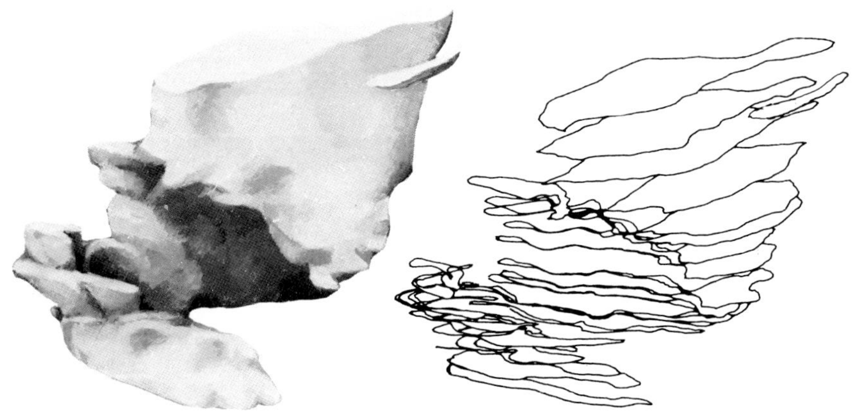

Fig. 2. Reconstruction of sperm cell 1 of Spinacia.

Fig. 3. Reconstruction of sperm cell 2 of Spinacia.

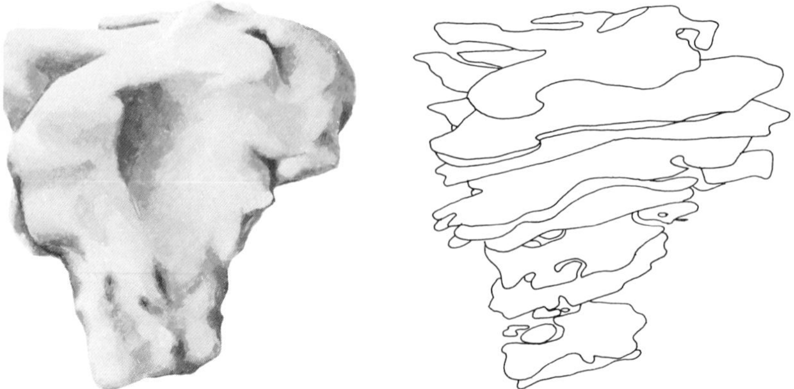

Fig. 4. Reconstruction of the vegetative nucleus of Spinacia.

The Z location is assigned on the basis of section number and section thickness. The drawings are then examined with the aid of the computer program for the Videoplan of Kontron. This allows observations of objects from an angle ($60^{o}$) and relative distance (see Figs. 2-4, right sides). Additional physical models of one sperm cell assemblage (Figs. 2-3) and the nucleus of the vegetative cell assemblage (Fig. 4) are constructed on the base of these rotated observations. The reconstruction of the two sperm cells and the nucleus of the vegetative cell can be put together. The cellular protrusion of, in this case, the associated sperm cell is found within clasping regions of the lobed vegetative nucleus and cannot be seen from this side. If the view is turned $180^{o}$ the long protrusion of sperm cell 2 can be partly seen in an evagination of the vegetative nucleus. The protrusion of sperm cell 1 is not present in this reconstruction because not all sections of the upper part have been prepared well. This protrusion has a direct relationship with the vegetative cytoplasm and not with its nucleus. Each sperm cell contains only a few mitochondria.

In Plumbago sperm dimorphism probably results from the polarized cytoplasmic conditions observable within the immature generative cell (Russell and Cass, 1983). Even the nuclei appear to be statistically different; the size of the nucleus in the sperm cell that is associated with the vegetative nucleus, is greater than that of the other sperm nucleus (Russell, 1984). Unequal distribution of plastids is also reported as early as during the generative cell formation in Impatiens pollen (Van Went, 1983).

These results indicate that each sperm cell is more differentiated than is often thought. Can it be that not only the cytoplasm and the nuclei of the sperm cells differ, but also their membranes and walls if present? To answer this question another method of preparing and observing is necessary.

4 Intracellular structures by SEM. Through the use of a recently developed preparative technique it is now possible to extend the application of scanning electron microscopy (SEM) to the structural aspects of sperm cell and vegetative nucleus development. The intracellular structures can only be observed by SEM when the cytoplasm of the vegetative cell is partly removed from the fractured surface of the cells.

The method used is an adaptation of the one devised by Tanaka and Mitsushima (1983) and used by them to study internal details of cells from a variety of animal tissues. With suitable modification the technique is equally successful with plant material (Blackmore et al., 1984; Blackmore and Barnes, 1985). Our pollen specimens are fixed, mounted on stubs with superglue ("Loctite", cyano acrylate) quick frozen, fractured whilst frozen in the Balzers Baf 400 freeze-etch unit and treated with a diluted solution of 0.01 % osmiumtetroxide in a hypotonic buffer. This removes the cytoplasmic matrix from the exposed cells and, after critical point drying, permits cell membranes and organelles to be observed in the scanning electron microscope.

The three dimensional images of this method enable a more accurate assessment of the sperm cells in relation to the vegetative nucleus. Fig. 5 shows a freeze-fractured pollen grain, which contains many spherical bodies, mainly starch-containing plastids and other organelles. The nucleus of the vegetative cell and the two sperm cells cannot be seen in the fractured pollen grain, but are present on the cut surface. The membranes of the sperm cells appear more smooth than the membrane of the vegetative nucleus. The sperm cells-

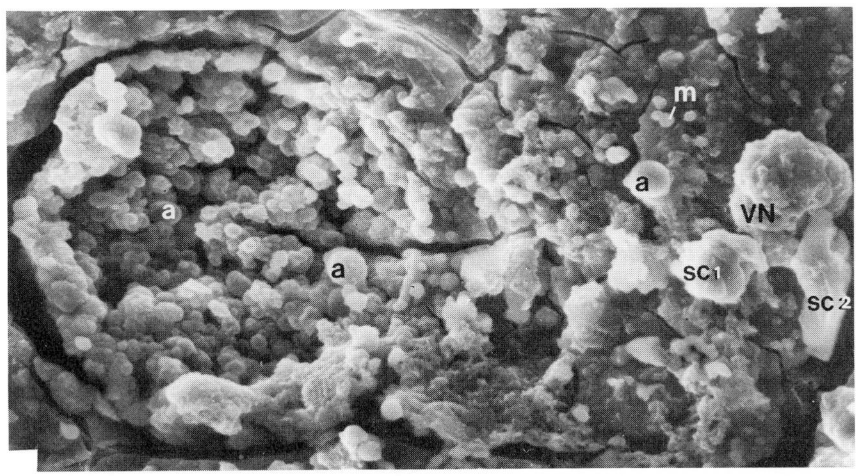


Fig. 5. Freeze-fractured and cytoplasmic macerated pollen grain embedded in superglue. At the right the vegetative nucleus (VN) and the sperm cells (SC1, SC2) can be observed outside the pollen grain on the cut surface. A - amyloplast; M - mitochodrium. Photograph by courtesy of Ir. C. J. Keijzer.

vegetative nucleus complex appears to be not connected with other membrane parts of the vegetative cell. The shape of the freeze-fractured complex has changed, the sperm cells are no longer in a close vicinity to each other over a long distance. Does this indicate that the shape of the complex is not stable? Is the connection between the sperm cells only important in relation to the transport mechanism in the pollen tube or is this to stabilize their position in the tube? More attention will have to be paid to the structures of the sperm cells. This can be done by improving the SEM-method, but also by using histochemical membrane staining methods for TEM.

Acknowledgement. The preparations of the 3D-reconstruction experiments were made with the participation of Dr. Rob Poehlmann, Dept. of Anatomy of the Medical Faculty, Leiden. The 3D-reconstructed drawings were performed by Allex Haasdijk. Prof. Dr. M.T. M. Willemse and Ir. C. J. Keijzer gave valuable comments in the course of the experiment and during the preparation of the manuscript. J. S. de Block corrected the English text and Mrs. G.G. van de Hoef-van Espelo typed the manuscript.

## References

Blackmore S, Barnes SH (1985) Protoplasma 126: 91-99

Blackmore S, Barnes SH, Claugher D (1984) J Ultrastructure Res 88: 215-219

Dumas C, Knox RB, Gaude T (1984) Int Rev Cytol 90: 239-272

Dumas C, Knox RB, Gaude T (1985) Protoplasma 124: 168-174

Hagemann R (1976) In: Genetics and Biogenesis of Chloroplasts and Mitochondria. Bucher T, Neupert W, Sebald W, Werner S (eds). Amsterdam: North Holland Publ., pp. 331-338

Hagemann R (1981) Acta Bot Soc Pol 50: 321-327

Jensen WA, Fisher DB (1968) Protoplasma 65: 277-286

McConchie CA, Jobson S, Knox RB (1985) In: Sexual Reproduction in Seed Plants, Ferns and Mosses. Willemse MTM, Van Went JL (eds.). Wageningen: PUDOC, pp. 147-148

Russell SD (1980) Science 210: 200-201

Russell SD (1984) Planta 162: 385-391

Russell SD (1985) In: Sexual Reproduction in Seed Plants, Ferns and Mosses. Williemse MTM, Van Went JL (eds). Wageningen: PUDOC, pp. 145-146

Russell SD, Cass DD (1981a) Protoplasma 107: 85-107

Russell SD, Cass DD (1981b) Acta Bot Soc Pol 50: 185-189

Russell SD, Cass DD (1983) In: Pollen: Biology and Implicatons for Plant Breeding. Mulcahy DL, Ottaviano E (eds). Elsevier, pp. 135-140

Tanaka K, Mitsushima A (1983) J Micr 133: 213-222

Van Went JL (1984) Theor Appl Genet 68: 305-309

Wilms HJ (1981) Acta Bot Neerl 30: 101-122

Wilms HJ (1985) In: Sexual Reproduction in Seed Plants, Ferns and Mosses. Williemse MT , Van Went JL (eds). Wageningen: PUDOC, pp. 143-144

Wilms HJ, Van Aelst AC (1983) In: Fertilization and Embryogenesis in Ovulated Plants. Erdelska O, et al. (eds). Bratislava: VEDA, pp 105-122

Wilms HJ, Leferink-ten Klooster HB (1983) In: Fertilization and Embryogenesis in Ovulated Plants. Erdelska O, et al. (eds). Bratislava: VEDA, p 239

Zhu C, Hu S, Xu L, Li X, Shen J (1980) Scientia Sinica 23: 371-376

# New Aspects of Sporopollenin Biosynthesis

A. K. PRAHL, M. RITTSCHER AND R. WIERMANN[1]

## Introduction

The chemical composition of sporopollenin is largely unknown as yet. According to a working hypothesis, sporopollenin is considered to be a polymer derived from carotenoids and/or carotenoid esters (Brooks and Shaw, 1968, 1977). In fact, there is some circumstantial evidence for sporopollenin being composed of the above mentioned compounds. On the other hand it should be stated that tracer experiments which were carried out upon higher plants and which were instrumental in the establishment of the carotenoid hypothesis do not support the carotenoid hypothesis (Green, 1973).

The following experiments were performed to elucidate the chemical composition of sporopollenin: 1. Starting from the working hypothesis that carotenoids and/or carotenoid esters are involved in the sporopollenin biosynthesis, we applied an inhibitor of carotenoid biosynthesis in order to examine whether a complete carotenoid biosynthesis is a crucial prerequisite for an intact sporopollenin accumulation. 2. Tracer experiments, which have been reported to date, have been restricted to uptake studies of a limited number of precursors applied over extremely long periods (e. g. more than 14 days; Green, 1973). Our labelling experiments were performed with the aim to improve the application technique and to vary the kind of precursors to a higher extent.

---

[1] Botanisches Institut der Westfälischen Wilhelms-Universität, Schloßgarten 3, 4400 Münster, Federal Republic of Germany.

## Material and Methods

For the experiments the following plants or plant material were used: **Curcubita pepo** "Gelber Zentner" (for inhibitor experiments) and **Tulipa** cv. Apeldoorn (for tracer experiments).

1. The application of an inhibitor of carotenoid biosynthesis
The herbicide Sandoz 9789 (a kind gift from Sandoz A. G. [Switzerland]), partly dissolved in acetone, was applied in appropiate concentration within a nutrient solution. The application by a wick was the most efficient method (for detail see Prahl et al., in press).
The inhibitor was applied during developmental stages in which the most intensive sporopollenin accumulation takes place (stage immediately after the degradation of the tetrades and the stage when binucleate microspores have developed). All flower buds with anthers in postmeiotic developmental stages were removed at the beginning of the experiments. The first new flower buds developed one week after application of the experimental treatment.
The sporopollenin content was determined gravimetrically.

2. Labelling experiments
The application of radioactive precursors, the isolation of an exine fraction, and the determination of the radioactivity was performed by methods followed Rittscher and Wiermann (in preparation). The radioactive precursors were applied directly at the anthers (see table 1).

## Results

1. The effect of Sandoz on sporopollenin accumulation
In order to estimate the effect of the inhibitor correctly it was necessary to prove whether the inhibitor was taken up by the plants and transferred to the site of sporopollenin biosynthesis, i. e. the loculus of the anthers and especially the pollen. We have used inhibitor-stimulated changes in carotenoid biosynthesis of the pollen as an attractive

indicator for the uptake and transport of the inhibitor into the anther loculus and to the pollen. The inhibitor Sandoz caused clear effects when applied.
Consequently, the effect of this inhibitor on sporopollenin accumulation was studied. Concerning the whole anthers, no significant differences in the amount of sporopollenin can be detected comparing control plants and those treated with the inhibitor (fig. 1). These investigations were extended to an analysis of the pollen itself. As shown in fig. 1 the amount of sporopollenin after treatment is similar to, or possibly even marginally higher, than the control, if dry weight is chosen as reference. If the amount of sporopollenin is expressed as µg/pollen grain a slightly lower content of sporopollenin is obtained for pollen from plants treated with the inhibitor (fig. 1).

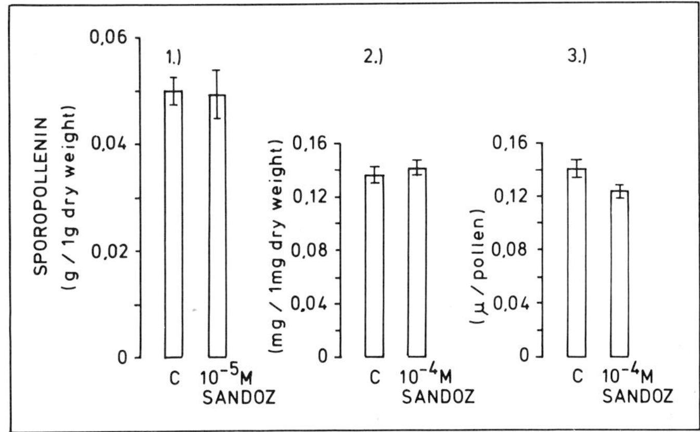


Fig. 1: The accumulation of sporopollenin in anthers 1.) and pollen 2.), 3.) of **Curcubita pepo** after application of Sandoz (C = control).

2. The incorporation of radioactive precursors into the exine fraction

In these experiments, the exine fraction was isolated by the method of Green (1973) and by new procedure using hydrolytic enzymes such as lipase, pronase, pektinase, cellulase, amylase

**a**

| Precusor(1) | Applied activity total cpm x $10^7$ | Applied activity specific $\frac{cpm \times 10^7}{mg\ precusor}$ | Activity remaining in the exine fraction total cpm x $10^3$ | Activity remaining in the exine fraction specific $\frac{cpm \times 10^3}{mg}$ | Incorporation rate(2) $\frac{1 \times 10^{-5}}{24\ h}$ |
|---|---|---|---|---|---|
| U-$^{14}$C-Tyr | 1.77 | 139.64 | 0.35 | 0.05 | 0.003 |
| 2-$^{14}$C-MVA | 2.09 | 2.66 | 0.07 | 0.01 | 0.045 |
| U-$^{14}$C-AA | 1.69 | 5.43 | 0.91 | 0.10 | 0.175 |
| 2-$^{14}$C-MA | 3.66 | 3.89 | 2.96 | 0.33 | 0.847 |
| 2-$^{14}$C-p-CA | 0.49 | 0.32 | 0.19 | 0.03 | 0.987 |
| 1-$^{14}$C-Glucose | 1.80 | 1.93 | 1.78 | 0.33 | 1.709 |
| U-$^{14}$C-Phe | 1.73 | 2.09 | 6.62 | 1.25 | 5.965 |

(1) MVA = Mevalonic acid, AA = Acetic acid, MA = Malonic acid, p-CA = p-Coumaric acid (Na-salts)

(2) Incorporation rate = $\frac{cpm/mg\ exine\ fraction}{cpm/mg\ appl.\ precursor \times 24\ h}$

**b**

| Precusor(1) | Applied activity total cpm x $10^7$ | Applied activity specific $\frac{cpm \times 10^7}{mg\ precusor}$ | Activity remaining in the exine fraction total cpm x $10^3$ | Activity remaining in the exine fraction specific $\frac{cpm \times 10^3}{mg}$ | Incorporation rate(2) $\frac{1 \times 10^{-5}}{24\ h}$ |
|---|---|---|---|---|---|
| U-$^{14}$C-Tyr | 1.83 | 142.76 | 0.50 | 0.04 | 0.003 |
| 2-$^{14}$C-MVA | 2.18 | 2.77 | 0.11 | 0.01 | 0.029 |
| U-$^{14}$C-AA | 1.42 | 4.58 | 1.02 | 0.09 | 0.205 |
| 2-$^{14}$C-MA | 3.74 | 6.66 | 4.63 | 0.31 | 0.463 |
| 2-$^{14}$C-p-CA | 0.29 | 0.32 | 0.19 | 0.02 | 0.463 |
| 1-$^{14}$C-Glucose | 1.94 | 2.08 | 1.80 | 0.18 | 0.848 |
| U-$^{14}$C-Phe | 1.76 | 2.12 | 11.45 | 0.86 | 4.064 |

(1) MVA = Mevalonic acid, AA = Acetic acid, MA = Malonic acid, p-CA = p-Coumaric acid (Na-salts)

(2) Incorporation rate = $\frac{cpm/mg\ exine\ fraction}{cpm/mg\ appl.\ precursor \times 24\ h}$

Table 1: Incorporation of labelled precursors into the exine fraction of **Tulipa** cv. Apeldoorn. The sporopollenin was isolated by a method of Green (a) and a procedure using hydrolytic enzymes (b).

and amyloglucosidase. Among the precursors applied the highest incorporation is achieved with phenylalanine, whereas mevalonic acid is only weakly incorporated. A substantial incorporation is obtained with glucose, malonic acid and p-coumaric acid, too (table 1.a).
The same tendancy in the degree of incorporation is observed if the sporopollenin fraction is prepared by an enzymatic method (table 1.b).

**Discussion**

Recently it could be demonstrated that the use of inhibitors is an appropriate method to elucidate metabolic pathways in plants (cf. L-α-aminooxy-ß-phenylpropionic acid [Amrhein and Gerhardt, 1979]). Therefore, as an obvious approach, it was studied by using an inhibitor of the carotenoid biosynthesis whether an intact carotenoid biosynthesis system is a prerequisite for an undisturbed sporopollenin accumulation.
The application of Sandoz resulted in extensive changes in the UV-spectra of the carotenoid extracts of pollen. In contrast, the effects on sporopollenin accumulation caused by the inhibitor were very minor. Therefore, we suggest that severe interference in carotenoid biosynthesis does not result in a subsequent drastic inhibition of sporopollenin biosynthesis.
As shown above, incorporation rate into the so-called sporopollenin fraction was highest when the aromatic amino acid phenylalanine was supplied as labelled precursor. Although the degradation of the labelled polymer and the subsequent analysis of the degradation products has not yet been carried out we propose that phenylalanine, as an immediate precursor of the phenolic metabolism, is involved in sporopollenin biosynthesis. This conclusion is in good agreement with results of experiments in which "native" sporopollenin was degraded by nitrobenzene oxidation (Pew, 1955). By this procedure, we obtained considerable amounts of aromatic compounds (unpublished results).
Summarising the results we conclude that sporopollenin is a highly complex biopolymer. Based on the information available the possibility cannot be excluded that the terpenoid metabo-

lism is somehow involved in sporopollenin biosynthesis. However in contrast to earlier assumptions (Brooks and Shaw, 1968) the metabolism of phenolics is definitely an essential part of sporopollenin biosynthesis.

**Acknowledgements**

This studies were financially supported by the Deutsche Forschungsgemeinschaft (Wi 386/7-1). We thank Dr. Mukherjee for revising the English.

**References**

Amrhein N, Gerhardt J (1979) Superinduction of phenylalanine ammonia lyase in gherkin hypocotyls caused by the inhibitor L-α-aminooxy-ß-phenylpropionic acid. Biochem Biophys Acta 583: 434-442

Brooks J, Shaw G (1968) Chemical structure of the exine of pollen walls and and a new function for carotenoids in nature. Nature 219: 532-533

Brooks J, Shaw G (1977) Recent advances in the chemistry and geochemistry of pollen and spore walls. Trans Bose Res Inst 40: 19-38

Green D (1973) A radiochemical study of spores and sporopollenins. Ph D Thesis, University of Bradford

Pew JC (1955) Nitrobenzene oxidation of lignin model compounds, spruce wood and spruce "native lignin". J An Chem Soc 77: 2831-2833

Prahl AK, Springstubbe H, Grumbach K, Wiermann R (1985) Studies on sporopollenin biosynthesis: The effect of inhibitors of carotenoid biosynthesis on sporopollenin accumulation. In press

# The Use of Gaspé Variety for the Study of Pollen and Anther Development in Maize

Dennis E. Hourcade, Michael Bugg and Dale F. Loussaert

## 1 Introduction

The male reproductive tissue in *Z. mays*, from anther and pollen development to pollen germination, has been well characterized at the plant and cell level (Bonnett, 1948; Weatherwax, 1955; Cheng, Greyson and Waldon, 1979). In addition, biochemical studies describing the enzymes contained in mature pollen grains are abundant (see Miller, 1982). Recently, the developmental programs that underlie anther differentiation (Abbott, Ainsworth and Flavell, 1984) and pollen germination (Mascarenhas *et al.*, 1984) have begun to yield to molecular investigation.

We have initiated a program for the biochemical study of pollen and anther development in *Z. mays*. We have used the variety Gaspé for this work, taking advantage of its short life cycle and stature to establish in a single growth chamber a continuous supply of plant tissue at every stage of development. A brief description of pollen and anther development in Gaspé follows including an analysis of alpha-amylase activity and starch content of these tissues.

## 2 Materials and Methods

*Plant Material.* The Gaspé strain used for this experiment was derived from Gaspé stock (Pioneer Seed Company) by successive inbreeding with selection for rapid pollen development and ease of growth in an environmental chamber. It was grown in a highly controlled environment (16 hr light/ 8 hr dark; 20 degrees C day/ 17 degrees C night; 800 microEinsteins light intensity; 40% humidity). Six inch standard pots were filled with one inch of Promix and then to within one inch of the top with fine vermiculite supplemented with 1/4 tsp Micromax, 1 tsp Osmocote 14-14-14 and 1 tsp Peters Slow Release 14-7-7 (Hummert Seed Company, St. Louis). This mixture was soaked, the seeds were planted, and the pots were not watered again until the plants had emerged. One plant was grown per pot. Under these conditions pollen shed began to occur about 28 days after emergence and lasted several days. Pollinations were done by hand and seeds were mature about 4 weeks later. In order to have a continuous supply of pollen and anther tissue at every stage, 12 pots were planted three times per week. This resulted in 100-300 mg of fresh pollen every day. Anthers were dissected from plants of various age groups. They were either analyzed

Monsanto Company, 700 Chesterfield Parkway, Chesterfield, Missouri 63198

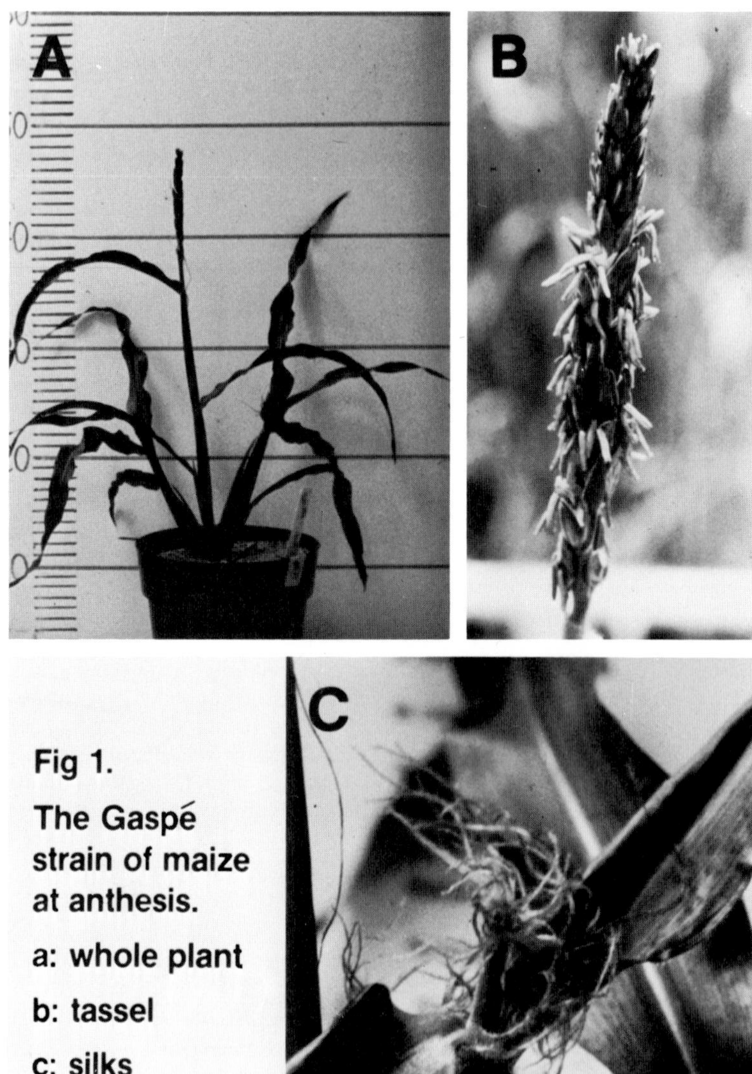


**Fig 1.**
**The Gaspé strain of maize at anthesis.**
**a: whole plant**
**b: tassel**
**c: silks**

immediately or were frozen in liquid nitrogen and lyophilized at -35 degrees C for several days and stored with desiccation at -20 degrees C. Most of the analysis was done with material from the lower half of the tassel where pollen development was more advanced.

*Biochemistry.* Pollen and anther tissue were homogenized using a 5 ml conical glass homogenizer (Bellco). Alpha-amylase was assayed by a modification of the procedure of Doehiert and Duke (1983), starch was measured by the method of Swank et al. (1982) or of Outlaw and Manchester (1979), and protein was assayed by the method of Bradford (1976).

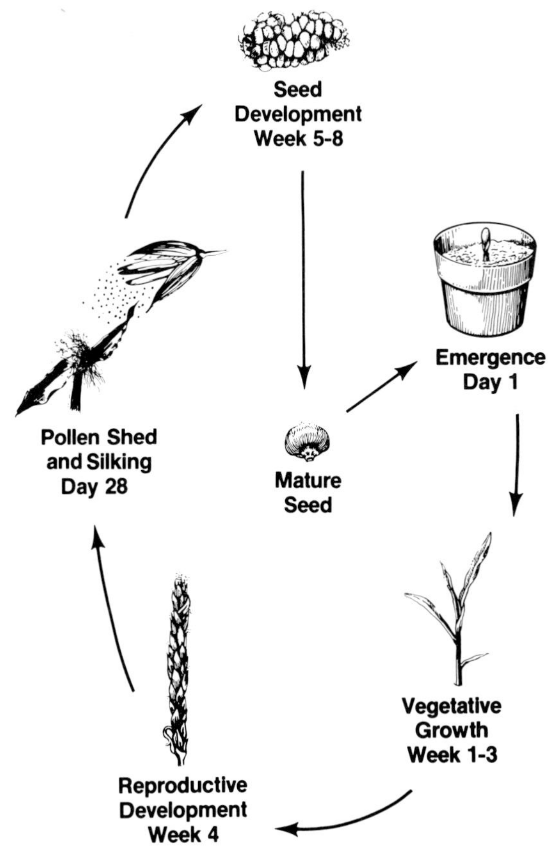


Fig 2. The life cycle of Gaspé.

## 3 Results

The life cycle of Gaspé is shorter than that of standard varieties and the plant size is greatly reduced (Figures 1,2). The plant undergoes primarily vegetative growth for three weeks after emergence and then reproductive growth predominates. Pollen shed occurs at 28-30 days, and silking usually takes place at the same time. Seed development occurs during the next 3-4 weeks. The mature plant is about 40 cm tall and produces one or two ears with 50-100 kernels per ear.

Each anther produces about 2000 pollen grains. Meiosis is found in the male reproductive tissue approximately 12-16 days after emergence, when the anthers are 1.5-3.0 mm in length. The microspores remain uninucleate from 17-22 days and the anthers elongate to 5-6 mm. About day 23 the anthers have reached 8 mm in length and the pollen begins to fill with starch. This is also when mitosis 1 takes place. During the last two or three days before anthesis the anthers turn yellow and then brown.

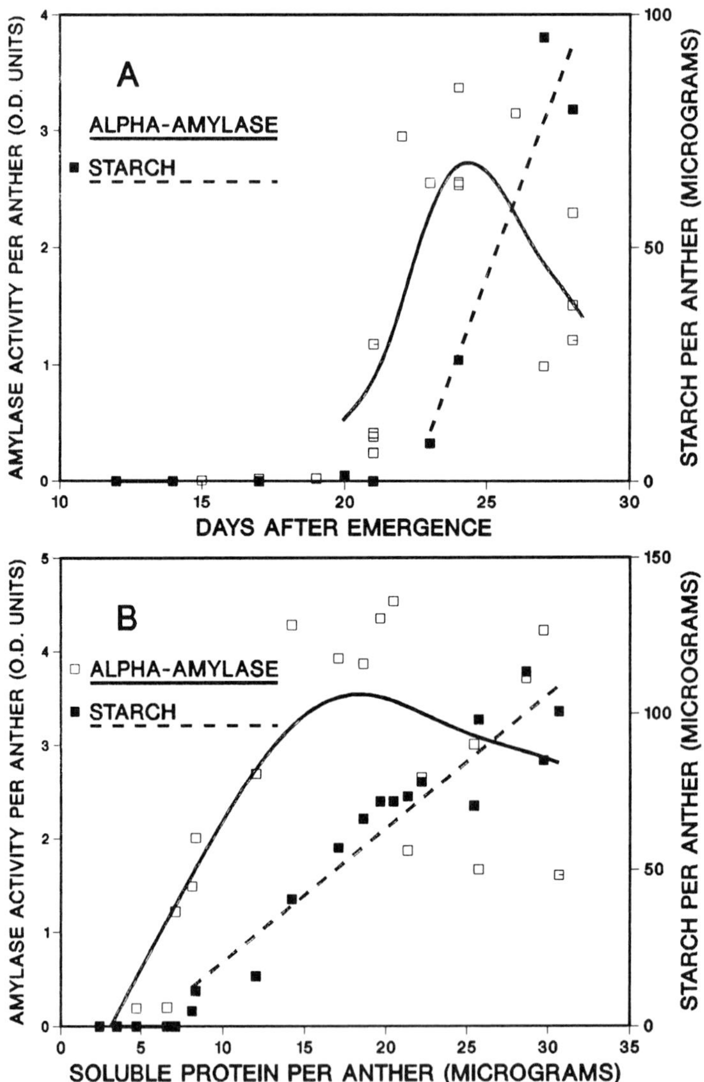


Fig 3 a-b. Alpha-amylase activity and starch content of developing Gaspé anthers. a: Measured with respect to days from emergence; b: Measured with respect to soluble protein content.

Alpha-amylase and starch were assayed in anther tissues during pollen development (Figure 3a). Amylase first appears around day 21, increases sharply within two days and then drops to about a third of its maximum. Starch appears at day 23, consistent with cytological determinations, and increases until pollen is shed.

Total anther protein was also used as a parameter for studying anther developmental sequences (Fig 3b). The results were similar to those in Figure 3a, where the age of the plant was used.

There are a number of cell types in the developing anther: The pollen is surrounded by four layers of somatic cells and the anther lobes are connected by additional tissue (Cheng, Greyson and Walden, 1979). While mature pollen can be isolated as it is shed, the analysis of developing pollen is complicated by these somatic cells. Our approach to this problem was to freeze-dry anthers first and then manually separate the somatic tissue from the pollen. With this technique we succeeded in isolating most of the somatic anther tissue without substantial pollen contamination and this allowed us to localize the alpha-amylase and starch during development (Fig 4). Most of the alpha-amylase is located in the somatic tissue while most of the starch is found in the pollen. Similar results were obtained when mature pollen was analyzed: Pollen contains substantial starch but little alpha-amylase activity.

| | TOTAL ANTHER | SOMATIC TISSUE ( % of ANTHER) | MATURE POLLEN ( PER ANTHER) |
|---|---|---|---|
| ALPHA AMYLASE (ACTIVITY) | 1.94 +/- 1.20 | 1.32 +/- 0.81 (72% +/- 23%) | 0.158 +/- 0.017 |
| STARCH (MICROGRAMS) | 70.0 +/- 17.9 | 20.3 +/- 3.8 (30% +/- 5.8%) | 33.8 +/- 7.3 |

Fig 4. Alpha-amylase activity and starch content in anther tissues. Individual anthers were cut in half. One half was used to determine total alpha-amylase or starch while the remaining half was dissected and used to determine somatic levels. All values were normalized to activity or micrograms per anther. Mature pollen was normalized from estimates of pollen number per anther and pollen number per sample.

## 4 Discussion

The Gaspé variety of *Z. mays* is a convenient tool for the study of the male reproductive system of maize: It grows quickly and easily in an environmental growth chamber, takes up less space than tall varieties and produces sufficient pollen for biochemical investigations. In principle, this strain should be useful for biochemical analysis of any phase of the maize life cycle.

A brief description of pollen and anther development has been given and the anther tissue has been analyzed for alpha-amylase activity and starch content during this period. The majority of the amylase activity was found in the somatic tissue in addition to the small amount found in the pollen. The function of this activity in anther development is of potential interest since it could indicate a biochemical relationship between the somatic tissue and the developing pollen grains. The somatic tissue could be a source of energy-rich metabolites for the synthetic activities in the pollen.

Two methods were used to monitor developmental time: plant age and anther protein content. While both approaches gave similar results, the second should, in principle, be more precise since it is not affected by variation among anthers on the same tassel.

Abbott, Ainsworth and Flavell (1984) analyzed esterases in maize anthers during development. They reported that at mitosis 1 esterases disappear from somatic tissue and new ones appear in the pollen. This is close to the time in Gaspé when amylase activity begins to appear in the somatic tissue and starch is syn-thesizedin the pollen. Thus, a substantial change in genetic activity may occur in maize anthers at this time, both in pollen and somatic tissues. The variety Gaspé, with several advantages for biochemical study, is an ideal system for further investigation of this developmental period.

*Acknowledgment.* The authors thank Drs. Tom Soong, Tom Skokut, Peter Mascia, David Ho and Ms. Jill Manchester and Toni Armstrong for their interest in this work.

## References

Abbott AG, Ainsworth, CC, Flavell, RB (1984) Characterization of anther differentiation in cytoplasmic male sterile maize using a specific isozyme system (esterase). Theor Appl Genet 67:469-473.

Bonnett, OT (1948) Ear and tassel development in maize. Annals Missouri Bot Garden 35:269-287.

Bradford, M (1976) A rapid and sensitive method for the quantitation of microgram quantities of protein utilizing the principle of protein-dye binding. Anal Biochem 72:248-254.

Cheng, PC, Greyson, RI, Walden, B (1979) Comparison of anther development in genic male-sterile (*ms-10*) and in male-fertile corn (*Zea mays*) from light microscopy and scanning electron microscopy. Can J Bot 57:578-596.

Doehlert, DC, Duke, SH (1983) Specific determination of alpha-amylase activity in crude plant extracts containing beta-amylase. Plant Physiol. 71:229-234.

Mascarenhas, NT, Bashe, D, Eisenberg, A, Willing, RP, Xiao, C-M, Mascarenhas, JP (1984) Messenger RNAs in corn pollen and protein synthesis during germination and pollen tube growth. Theor Appl Genet 68:323-326.

Miller, PD (1982) Maize pollen: collection and enzymology. In: Sheridan, WF (ed) Maize for Biological Research. Plant Molecular Biology Association, 279-293.

Outlaw, WH, and Manchester, J (1979) Guard cell starch concentration quantitatively related to stomatal aperature. Plant Physiol 64:79-82.

Swank, JC, Below, FE, Lambert, RJ, Hageman, RH (1982) Interaction of carbon and nitrogen metabolism in the productivity of maize. Plant Physiol. 70:1185-1190.

Weatherwax, P (1955) Structure and development of reproductive organs. In: Sprague, GF (ed) Corn and Corn Improvement. Academic Press, 89-121.

Pollen Physiology and Metabolism

# Energy Metabolism in Petunia hybrida Anthers: A Comparison Between Fertile and Cytoplasmic Male Sterile Development

R.J. BINO[1], S.J. DE HOOP[2], G.A.M. VAN MARREWIJK[2] AND J.L. VAN WENT[1]

## 1 Introduction

Cytoplasmic male sterility (cms) in petunia associates with aberrations in mitochondrial DNA (Kool et al. 1985). The cms-specific alterations in DNA are revealed in mitochondria purified from vegetative cells (Kool et al. 1985). However, the structural effects of cms-plasmatype only become manifest during anther development (Bino 1985a). The mitochondrial genome encodes for a number of proteins, which are involved in energy-generating processes (Dillon 1981). Hence, defects in mitochondrial DNA may affect the energy supply of cells containing cms-plasmatype.

Many metabolic reactions are depended on the energy status of a cell. An index of the energy status is the adenylate energy charge: AEC= [(ATP)+(ATP+ADP)]/2(ATP+ADP+AMP). According to Atkinson (1968), AEC ratio modulates activity of various metabolic sequences related to energy utilization and regeneration. The AEC can have values ranging from 0 (all AMP) to 1 (all ATP), but in normally metabolizing cells and tissues the AEC value is usually higher than 0.8 (Pradet 1982). In the present study, the energetic balance of anthers of fertile and cms petunia is determined at different stages of flower bud development.

## 2 Materials and Methods

**Plant Materials.** Two idiotypes of Petunia hybrida were used in this study, i.e., the male fertile cv. "Blue Bedder" (BBF), and the cms "Blue Bedder" (BBS) described by Van Marrewijk (1969). BBF and BBS are highly isogenic. The plants were cultivated in a growth chamber under a regime of 16 h light, 8 h dark at 17 $^{o}$C.

[1] Department of Plant Cytology and Morphology, Agricultural University, Arboretumlaan 4, 6703 BD Wageningen, The Netherlands

[2] Institute of Plant Breeding, (IvP), Agricultural University, P.O. Box 386, 6700 AJ Wageningen, The Netherlands

**Extraction Methods.** In order to determine quantitatively the amounts of adenine nucleotide mono-, di- and triphosphates, we used perchloric acid to inactivate hydrolytic enzyme activity. A problem with the PCA extraction method is, that the activity of phosphatases (which hydrolyze ATP and ADP into AMP) may partly be restored when the extract is neutralized (Pradet 1982). As a consequence, adenylate ratios (ATP/ADP, ATP/AMP or AEC) may be low compared to results obtained with other extraction methods, such as formic acid dissolved in ethanol or trichloroacetic acid dissolved in diethyleter (Pradet 1982). However, when these methods were applied, quantitative recoveries of ATP, ADP and AMP were not as satisfactory as those obtained with the perchloric acid extraction (Ikuma and Tetley 1976). Hence, for studies whith small amounts of plant material, perchloric acid turned out to be more suitable. According to Ikuma and Tetley (1976), ATP hydrolyzing activity in solanaceous plants was optimal at pH 5, whereas no activity was detected below pH 3 and above pH 9. To circumvent the hydrolysis of ATP and ADP, we followed the extraction procedure as proposed by Ikuma and Tetley (1976), and maintained the pH of the tissue extract below 3 throughout the extraction procedure to adjust the pH to 8.5 prior to the quantitative assay.

Plant material for extraction was prepared in the following manner. The length of a dissected flower bud was measured. One anther was used to determine the stage of development. The other four anthers were directly frozen in liquid $N_2$, whereafter they were pulverized in 65 μl ice-cold 1 N $HClO_4$ (final pH below 3). After 10 min at 4 $^{o}$C, the mixture was centrifuged at 13,000 g for 3 min; 40 μl of the supernatant was pipetted into a test tube containing 240 μl buffer (0.06 M Tricine/$MgSO_4$, with 1 % (w/v) $KHCO_3$, pH 7.6). The pH of the extract was directly adjusted to 8.5 with 1 N KOH, and the $KClO_4$ was pelleted in the cold (2,500 g, 1 min).

**Determination of Adenine Nucleotide Levels.** For adenine nucleotide phosphates determination, we used the methods described by Hoekstra (1979) and adapted them for small amounts of plant material. (ATP + ADP), and (ATP + ADP + AMP) were determined after enzymatic conversion of ADP and AMP into ATP. For (ATP + ADP) determination, 40 μl of the extract was diluted in buffer (40 l 0.06 M Tricine/$MgSO_4$, pH 7.6), containing 40 μl 0.125 % (w/v) hydrated phosphoenolpyruvate (Sigma) and 2.75 % (v/v) pyruvate kinase (EC 2.7.1.40) (Sigma). For (ATP + ADP + AMP) determination, 1.6 % (v/v) myokinase (EC 2.7.4.3) (Boehringer) was added to the above mentioned reaction mixture. The extracts were incubated for 30 min at 35 $^{o}$C. The resulting ATP was determined by the luciferin - luciferase assay, using a luminometer 1250 (LKB - Wallac). Twenty μl of a concentrated ice-cold firefly lantern extract (Boehringer), was injected into a small vial containing 20 μl of the ATP extract and 0.3 ml 0.02 M Tricine/$MgSO_4$ buffer (pH 7.6). Exactly 10 s after injection, bioluminoscence was measured for 6 s at 18 $^{o}$C. Samples with and without an internal standard were alternately counted. In each extract, the ATP, (ATP + ADP), and (ATP + ADP + AMP) were assayed 3 times. The amounts of ADP and AMP were determined by difference.

**Respiration Measurements.** $O_2$ consumption of anthers was measured polarographically with a Clark-type $O_2$ electrode. Eight to 15 anthers were inserted in a reaction chamber containing 1 ml 0.1 M mannitol and 0.5 mM $CaSO_4$ at 24 $^{o}$C. $O_2$ consumption was measured for 15 min and respiration rates were calculated according to Hoekstra (1979).

## 3 Results and Discussion

**Effect of Plasmatype on Anther Fresh Weight.** First structural aspects of abnormal anther development in BBS plants became apparent at leptotene stage of prophase I (Bino 1985b). Initial aberration was represented by the presence of large vacuoles in the cytoplasm of tapetal cells (Bino 1985b). At meiosis I, sporogenesis arrested and meiocytes and tapetal cells degenerated. During premeiosis, however, BBF and BBS development was indistinguishable (Bino 1985a). Table 1 indicates that at similar flower bud length anther fresh weights of fertile and sterile plants were similar until meiosis. As degeneration progressed, anther fresh weights of sterile plants decreased in comparison to fertile-type anthers. The relation between stage of development and flower bud length is in accordance with the data obtained by Van Marrewijk and Suurs (1985).

**Table 1.** Effect of plasmatype on anther fresh weight at different stages of development (values are means ± SE, of 500 to 1,000 anthers). The data are representative of a number of experiments which gave similar results.

| | Developmental stage | Bud length (mm) | Anther fresh weight (mg) |
|---|---|---|---|
| BBF | premeiosis | 1.0 - 1.7 | 0.15 ± 0.02 |
| | meiosis | 1.8 - 2.8 | 0.56 ± 0.04 |
| | postmeiosis: tetrads and microspores | 2.9 - 4.5 | 1.08 ± 0.07 |
| BBS | premeiosis | 1.0 - 1.7 | 0.15 ± 0.03 |
| | meiosis: first aspects of degeneration | 1.8 - 2.8 | 0.42 ± 0.05 |
| | postmeiosis: sporogenous and tapetal tissues degenerated | 2.9 - 4.5 | 0.54 ± 0.06 |

**Adenine Nucleotide Levels in BBF and BBS Anthers.** ATP and ADP contents in BBF anthers increased from premeiosis to postmeiosis (Fig. 1 A and B). At premeiosis and meiosis, ATP and ADP levels in BBF and BBS anthers were similar. During postmeiosis, ATP and ADP contents in BBS anthers were significantly lower than the amounts found in fertile-type anthers. In contrast, the AMP contents in BBS anthers exhibited a significant increase during meiosis (Fig. 1 C). This increase was coincident with degeneration of sporogenous and tapetal tissues. When the process of abortion was completed, the AMP contents in BBS anthers declined sharply. Total amounts of adenine nucleotides in BBF and BBS anthers were similar at premeiosis and meiosis (Fig. 1 D). At postmeiosis, (ATP + ADP + AMP) levels in BBS anthers decreased significantly in comparison to fertile-type anthers. The change in total adenine contents was comparable with the observed differences between fresh weights of BBF and BBS anthers.

Dissimilarities in adenylate contents between BBF and BBS anthers were associated with degeneration of sporogenous and tapetal tissues. At premeiosis, ATP, ADP and AMP levels were similar.

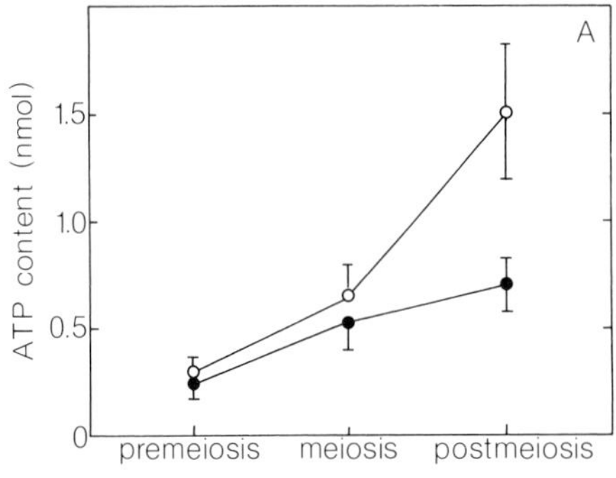


**Fig. 1 A - D.** Adenine nucleotide phosphates contents in BBF (-o-) and BBS (-●-) anthers at different stages of development. Each point represents the mean (± SE) of 7 to 11 separate extractions of four anthers of one flower bud. The fifth anther was used to determine the stage of development. **A**, ATP contents. **B**, ADP contents. **C**, AMP contents. **D**, (ATP + ADP + AMP) contents.

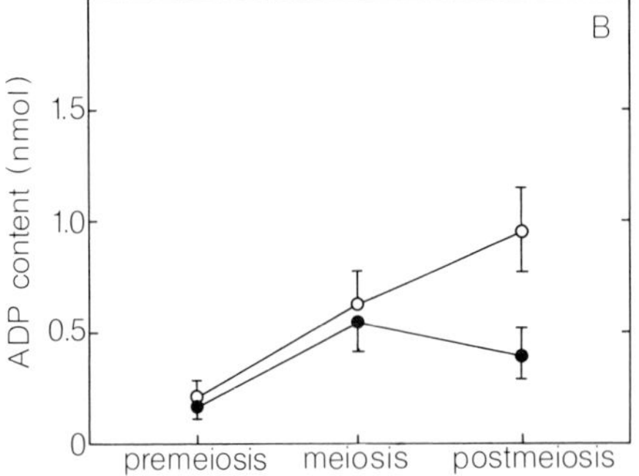


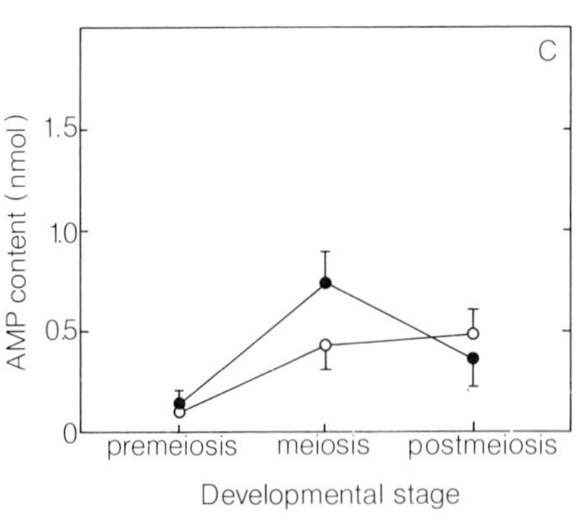


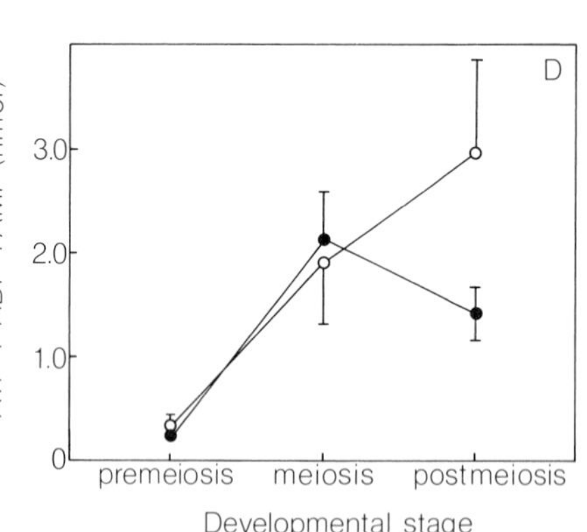

**Table 2.** AEC values of anthers of fertile and sterile plants at different stages of development. Values represent the mean $\pm$ SE of n extractions.

| Stage | BBF | n | BBS | n |
|---|---|---|---|---|
| premeiosis | 0.64 $\pm$ 0.04 | 11 | 0.61 $\pm$ 0.03 | 11 |
| meiosis | 0.56 $\pm$ 0.04 | 7 | 0.46 $\pm$ 0.02 | 9 |
| postmeiosis | 0.69 $\pm$ 0.04 | 10 | 0.67 $\pm$ 0.04 | 8 |

**AEC ratios of BBF and BBS Anthers.** To ascertain the effectiveness of the extraction method, we determined adenylate levels in leaf, and in combined leaf and anther extracts. AEC ratio obtained in petunia leaf tissue was high (AEC = 0.83 $\pm$ 0.05; n = 2). Combined leaf and anther extracts did not show any evidence for an anther-specific increase in phosphatase activity. Table 2 indicates that AEC ratios of fertile- and sterile-type anthers were similar during premeiosis and postmeiosis. At the meiotic interval, AEC values of BBF and BBS anthers decreased significantly. The drop in AEC ratio of BBS anthers exceeded the decrease in BBF anthers ($P < 0.025$). The difference in AEC ratios of fertile- and sterile-type anthers was correlated with the increase in AMP contents of BBS anthers at meiosis (Fig. 1 C).

At all stages of development, AEC values of BBF and BBS anthers were low in comparison to the ratios obtained in petunia leaves and other normally metabolizing cells or tissues (Pradet 1982). According to Atkinson (1968), low AEC ratio may exhibit a disparity between energy-generating and energy-utilizing systems. Metabolic activity of anther tissues was reported by Porter et al. (1983), who found that the amounts of mRNA in _Lilium_ meiocytes varied considerably as meiosis progressed. Williams and Heslop-Harrison (1979) observed a similar variation in the synthetic activity of _Lilium_ and _Rhoeo_ tapetal cells during meiosis. Possibly, the low AEC ratios of petunia anthers reflected the special metabolic state of tapetal and sporogenous tissues.

**Respiratority Rate of BBF and BBS Anthers.** $O_2$ consumption of BBF anthers increased during sporogenesis (Table 3). At premeiosis and meiosis, respiration rates of BBS and BBF anthers were similar. However, during postmeiosis, $O_2$ consumption of BBS anthers decreased significantly.

**Table 3.** Effect of plasmatype on $O_2$ consumption (pmol/min/anther) at different stages of development. Values represent the mean $\pm$ SE of n separate measurements.

| Stage | BBF | n | BBS | n |
|---|---|---|---|---|
| premeiosis | 50.3 $\pm$ 9.7 | 3 | 51.8 $\pm$ 4.7 | 2 |
| meiosis | 81.4 $\pm$ 4.7 | 4 | 74.6 $\pm$ 8.7 | 3 |
| postmeiosis | 91.7 $\pm$ 10.8 | 5 | 62.3 $\pm$ 1.6 | 2 |

**Conclusions.** Before the structural aspects of degeneration in BBS anthers became apparent, we could not establish a difference in energy metabolism between fertile- and sterile-type anthers. At premeiosis, adenylate contents, AEC ratios, respiration rates and fresh weights were similar. At meiosis, however, AMP levels of BBS anthers increased and AEC values declined. Low AEC ratios may inhibit ATP-utilizing pathways and stimulate ATP-generating pathways (Atkinson 1968). However, the direct effects on microsporogenesis are unknown. At postmeiosis, sporogenous and tapetal tissues were aborted and metabolic activity of sterile anthers was reduced.

A disadvantage of the used methods was, that tissue specific differences were difficult to determine. Nevertheless, a tissue specific determination of cytochrome c oxidase activity could not establish cms-associated changes between BBF and BBS anthers either (Bino et al. 1985).

*Acknowledgment.* We thank Mr. P.A. Van Snippenburg for drawing the figures. We are grateful to Prof. M.T.M. Willemse and Dr. F.A. Hoekstra for many helpful discussions. RJB thanks the "Fonds Landbouw Export Bureau 1916/1918" for grant No. 95(B).

## References

Atkinson DE (1968) The energy charge of the adenylate pool as a regulatory parameter. Interaction with feedback modifiers. Biochemistry 7: 4030-4034

Bino RJ (1985a) Histological aspects of microsporogenesis in fertile, cytoplasmic male sterile and restored fertile Petunia hybrida. Theor Appl Genet 69: 423-428

Bino RJ (1985b) Ultrastructural aspects of cytoplasmic male sterility in Petunia hybrida. Accepted by Protoplasma (in press)

Bino RJ, De Hoop SJ, Van Der Neut A (1985) Cytochemical localization of cytochrome oxidase in anthers of cytoplasmic male sterile Petunia hybrida. In: Willemse MTM, Van Went JL (eds) Sexual reproduction in seed plants ferns and mosses. Pudoc, Wageningen, pp 44-46

Dillon LS (1981) Energy-oriented organelles and activities. In: Ultrastructure, macromolecules, and evolution, Plenum Press, New York, pp 375-440

Hoekstra FA (1979) Mitochondrial development and activity of binucleate and trinucleate pollen during germination in vitro. Planta 145: 25-36

Ikuma H, Tetley RM (1976) Possible interference by an acid-stable enzyme during the extraction of nucleoside di- and triphosphates from higher plant tissues. Plant Physiol 58: 320-323

Kool AJ, De Haas JM, Mol JNM, Van Marrewijk GAM (1985) Isolation and physicochemical characterization of mitochondrial DNA from cultured cells of Petunia hybrida. Theor Appl Genet 69: 223-233

Porter EK, Parry D, Dickinson HG (1983) Changes in poly(A) RNA during male meiosis in Lilium. J Cell Sci 62: 177-186

Pradet A (1982) Oxidative phosphorylation in seeds during the initial phases of germination. In: Khan AA (ed) The physiology and biochemistry of seed development, dormancy and germination. Elsevier Biomedical, Amsterdam, New York, Oxford, pp 347-369

Van Marrewijk GAM (1969) Cytoplasmic male sterility in petunia. I. Restoration of fertility with special reference to the influence of environment. Euphytica 18: 1-20

Van Marrewijk GAM, Suurs LCJM (1985) Characterization of cytoplasmic male sterility in Petunia X hybrida (Hook) Vilm. In: Willemse MTM, Van Went JL (eds) Sexual reproduction in seed plants, ferns and mosses. Pudoc, Wagenigen, pp 39-43

Williams EG, Heslop-Harrison J (1979) A comparison of RNA synthetic activity in the plasmodial and secretory types of tapetum during the meiotic interval. Phytomorphology 29: 370-381

# Water Content, Membrane State and Pollen Physiology

C. DUMAS, C. KERHOAS, G. GAY AND T. GAUDE[1]

## 1 Introduction

The living pollen grain is generally a dehydrated organism except where pollination is through the medium of water. The pollen grains are largely insulated from their surroundings by a complex pollen wall and their water content varie considerably between low water content (W.C.) species e.g. 6% in poplar pollen, 8% in Brassica oleracea and 60% in Zea mays. In the literature, little information is available on pollen longevity, generally estimated directly by the ability to get the seed (see Knox 1984; Kerhoas and Dumas in press). We demonstrated previously there is no water content expenditure in viable pollen of Brassica under isothermic conditions, suggesting that the water content is regulated within certain limiting temperatures and relative humidities (R.H.) (Dumas et al. 1983). This characteristic is perhaps mediated by the plasma membrane or the wall of the pollen grain surrounded by its pollen coat. This may be an important adaptation for pollen survival during dispersal.

Some experiments have been carried out to correlate water content and pollen viability. And the correlation between F.C.R. (Heslop-Harrison and Heslop-Harrison 1970) and germinability was found to be highly significant (Shivanna and Heslop-Harrison 1981). In addition a new method based on Nuclear Magnetic Resonance spectrometry was used to test the water content evolution which occur during the loss of viability in Brassica (Dumas et al. 1983). This non destructive method provided a good tool to demonstrate that the loss of viability was correlated with the loss of water content and with the F.C.R. data.

Three key problems remain to solve:

1. What is the water state into the pollen grain and its evolution

1 Université Cl. Bernard-LYON I, R.C.A.P., UM CNRS 380024, 69622 Villeurbanne Cedex 02. FRANCE

during the loss of viability?

2. Does any correlation exist between water content and plasma membrane structure in living pollen grain?

3. How the pollen hydration is controlled onto the stigma surface? In our laboratory these different aspects of pollen physiology are currently in progress in order to built a good tool for further genetic manipulations using the pollen as a vector including pollen quality definition, pollen storage, pollen germination *in vitro* and *in vivo*, sperm cells modification occuring during fertilization mechanisms,...

## 2 Water state evolution during pollen ageing

In previous papers we mentionned the efficiency of $^{1}$H-N.M.R. studies to analyze water content evolution (Dumas et al. 1983) and water state with the help of the spin-spin relaxation time ($T_2$)(see Dumas and Gaude 1983; Duplan and Dumas 1984).

A model has been employed which includes two types of cellular water, bound water (low $T_2$ values) and free water (high $T_2$ values). Thus, in a pollen grain population, we may represent this phenomenon by the following formula:

$$\frac{1}{T_2} = \frac{P_f}{T_2(b)} + \frac{Pb}{T_2(f)}$$

Where b: bound water molecules

f: free water molecules

$T_2$: proton spin-spin relaxation time

A joined program allows to envisage 3 speeds of water loss that many mean 3 hydric compartments in pollen organisms (Kerhoas et al., unpublished data and Kerhoas et al., in this volume). This idea is consistent with a previous data obtained in *Brassica* and NMR-curve composed by 3 parts (see Dumas et al. 1983), in two models: *Brassica* (WC 8%) and pumpkin (WC 45-50%).

In addition F.C.R. data are closely correlated to relaxation time $T_2$, especially when relaxation time is greatly decreased. These different hydric compartments could confirm that the loss of viability is correlated to the appearance of more and more free water. In addition, this water state behaviour is closely related to decreasing pollen viability, rather than decreasing total water content (Kerhoas et al., in this book). In fact, the loss of bulk water determines the beginning of plasmalemma destructuration and the decrease of pollen viability.

These observations confirm:

- the occurence of "bulk water" in plasmalemma integrity (Kerhoas and Dumas, in press)
- the occurence of membrane integrity in a viable pollen (see Kerhoas et al., in this book).

Then the first hydric compartment would be mainly the "bulk water compartment", the second compartment begins to lose its water content after the plasmalemma goes on destructuration. And, all water would be freed from the dead pollen in the third compartment (Kerhoas et al., unpublished data).

## 3 Membrane state in living pollen grain

When internal membranes possess little water for integrity maintenance, thay go on destructuration and pollen die. The general importance of membrane alterations during ageing and senescence have been pointed out by Mazliak (1983). These views are consistent with the validity of FCR test for the plasma membrane integrity (Heslop-Harrison et al. 1984; Dumas et al. 1984a,b).

Nevertheless some recent data obtained in the laboratory on different types of pollen with various W.C. seem to infirm a general concept of membrane state in dry biological system proposed by Simon (1978)(Kerhoas et al., unpublished data and Kerhoas et al., in this book).

## 4 Pollen hydration and pollen-stigma recognition

Recently, pollen hydration on the stigma has been considered as an active phenomenon rather than a simple osmotic phenomenon. Several authors demonstrated the absence of pollen hydration when the distance species was greater (see Knox 1984). In this way, we demonstrated the presence of the structured and organized layer called Exinic Outer Layer (EOL)(Gaude and Dumas 1984) and pointed out its possible key role in this critical step of the pollen-stigma recognition. Recently, we established an ATPase activity strongly increasing after a specific pollen-stigma contact (Gaude and Dumas, in preparation). This enzyme involvement could control the water flow from the stigma cells to the dehydrated pollen grain.

Prospects

Because N.M.R. results are statistic results for a population of pollen grains some data are quite difficult to interprete. In order to overcome this difficulty we are performing a new microprobe allowing to get a specific signal from a single pollen grain (Kerhoas et al., unpublished data) and a new technique to analyze proteins of a pollen grain (Gay et al., in this book) during ageing.
Water content remains a very exciting area for pollen physiologists according to new possible techniques to use the pollen grain as a vector for plant transformation.

References

Dumas C, Gaude T (1983) Stigma-pollen recognition and pollen hydration. Phytomorphology 30: 191-201.
Dumas C, Duplan JC, Said C, Soulier J.P (1983) $^{1}$H Nuclear Magnetic Resonance to correlate water content and pollen viability. In D.L Mulcahy and E Ottaviano (Eds) Pollen: Biology and Implications for Plant Breeding. Elsvier Science Publ. Co., pp 15-20.
Dumas C, Duplan J.C, Gaude T, Said C (1984a) Cytologie et physico-chimie, deux approches complémentaires pour tester la viabilité pollinique. In Vème Sympos. Internat. sur la Pollinisation Versailles, 27-30 sept. 1983. INRA Publ. (Ed) n°21, pp 415-421.
Dumas C, Knox R.B, Gaude T (1984b) Pollen-pistil recognition: new cone concepts from electron microscopy and cytochemistry. Internat. Rev. Cytol 90: 239-272.
Duplan J.C, Dumas C (1984) Viabilité pollinique et conservation du pollen In Hervé Y and Dumas C (Eds) Incompatibilité pollinique et Amélioration des Plantes. ENSA Rennes Publ., pp 40-50.
Gaude T, Dumas C (1984) A membrane-like structure on the pollen wall surface in Brassica. Ann Bot 54; 821-825.
Gay G, Kerhoas C, Dumas C Micro isolectric focusing of pollen grain proteins in Cucurbita pepo. In Mulcahy D.L and Ottaviano E (Eds) Biotechnology and Biology of Pollen. Springer Verlag (in press).
Heslop-Harrison J, Heslop-Harrison Y (1970 Evaluation of pollen viability by enzymatically induced fluorescence; intracellular hydrolysis of fluorescein diacetate. Stain Technol 45: 115-120.
Heslop-Harrison J, Heslop-Harrison Y, Shivanna KR (1984) The evaluation of pollen quality, and a further appraisal of the fluorochromatic (FCR) test procedure. Theor Appl Genet 67: 367-375.
Kerhoas C, Gaude T, Gay G, Dumas C Pollen cryofracture. In Mulcahy DL and Ottaviano E (Eds) Biotechnology and Biology of Pollen. Springer Verlag (in press).
Kerhoas C, Gay G, Duplan JC, Dumas C. Water content evolution in Cucurbita pepo during aging: a NMR study. In Mulcahy DL and Ottaviano E (Eds) Biotechnology and Biology of Pollen. Springer Verlag (in press).
Kerhoas C, Dumas C. Nuclear Magnetic Resonance and pollen quality. In Jackson JF and Linskens HF (Eds) Modern Methods in Plant Analysis. Academic Press (in press).

Knox RB (1984) Pollen-pistil interactions. In Linskens HF and Heslop-Harrison (Eds) Cellular Interactions. Encyclop. Plant Physiol., new Ser. 17. Springer Verlag, pp 508-608.
Mazliak P (1983) Plant membrane lipids. Changes and alterations during aging and senescence. In Lieberman M (Ed) Post-Harvest Physiology and Crop Preservation. Plenum Publ. Co, pp 123-140.
Shivanna KR, Heslop-Harrison J (1981) Membrane state and pollen viability Ann Bot 47: 759-770.
Simon EW (1978) Membranes in dry and imbiding seeds. In Crowe JH and Clegg JS (Eds) Dry Biological Systems. Academic Press, pp 205-224.

# Do Anti-Oxidants and Local Anaesthetics Extend Pollen Longevity during Dry Storage?

FOLKERT A. HOEKSTRA AND JAN H.M. BARTEN[1]

## 1 Introduction

When not deposited at the right place, on the stigma, mature pollen may lose vitality by inappropriate metabolic processes at elevated moisture contents (Hoekstra and Bruinsma 1975a,1980). In general, however, pollen have the ability of surviving short periods of dehydration at the end of their maturation in the anther and during dispersal. This requires a special condition of pollen membranes as they apparently do not disintegrate upon dehydration, in contrast to membranes from shoots and leaves.

Upon incubation in an artificial germination medium loss of vitality occasionally occurs. That a membrane-related problem might be involved was deduced from the fact that fluorescein is not retained inside the plasmalemma (Shivanna and Heslop-Harrison 1981; Heslop-Harrison et al. 1984). Fluorescein is the fluorescent product split by esterases inside the grain from the vital stain fluorescein diacetate (Heslop-Harrison and Heslop-Harrison 1970). In many species equilibration of the dry pollen in humid air prior to germination *in vitro* prevents this vitality loss (Lichte 1957; Hoekstra and Bruinsma 1975b; Visser et al. 1977; Shivanna and Heslop-Harrison 1981), particularly when imbibition occurs in the cold (Hoekstra 1984). Under condition of cold imbibition metabolites were shown to leak from dry *Typha* pollen (Hoekstra 1984) and $K^+$ from a number of other dry pollen species (unpublished results). Here, too, humid air treatment prior to imbibition considerably improves germinability and prevents leakage.

In contrast to the large body of papers on molecular mechanisms of membrane deterioration during storage of seeds, little has been published in this field with pollen. Recently Hoekstra and van Roekel (1985) noted the close correlation between loss of pollen vitality during dry storage at 22°C and increase in potassium leakage, irrespective of the widely divergent longevities of the pollen species concerned. After a 150 days storage period, the composition of fatty acids in the phospholipids of 9 pollen species does not differ significantly from that in fresh pollen. However, some decrease in phospholipid content and threefold increase in free fatty acids (FFA's) were observed. Longevity of the various species

[1]Dept of Plant Physiol, AU, Arboretumlaan 4, 6703 BD Wageningen, Holland

differs considerably and is negatively correlated with the linolenic acid content (C18:3) of total lipid (Hoekstra and van Roekel 1985).

As the marked increases in FFA's were analysed after most of the pollen species had died (Hoekstra and van Roekel 1985), in this chapter we have pursued simultaneous analysis during loss of vitality. In microsomal membranes from dehydration-sensitive germinating soybean seeds, free radical treatment in vitro brings about significant changes in the lipid phase transition temperature, a decrease in phospholipid content, and an increase in FFA's, but no change in degree of fatty acid saturation of total lipid (Senaratna and McKersie 1985). Because of the similarity of phenomena during dehydration in vivo, Senaratna and McKersie (1984,1985) claim that free radical-induced de-esterification of phospholipids is underlying the deterioration of membranes. In this paper we report on the inability of the natural and synthetic antioxidants, α-tocopherol and butylated hydroxytoluene (BHT), and of local anaesthetics that are known to inhibit phospholipase activity (Vojnikov et al. 1983), to prevent accumulation of FFA's and to extend pollen longevity.

## 2 Materials and Methods

Male flowers of Typha latifolia L. and flower buds of Papaver rhoeas L. were collected from field populations in the neighborhood of Wageningen, The Netherlands. Closed anthers of Narcissus poeticus L. and Impatiens glandulifera Royle were from the garden of the laboratory of Plant Physiology. After drying for one day on the laboratory bench, pollen was sieved from male flowers or anthers, and subsequently air-dried to moisture contents of 5-6% (on a FW basis), and then stored at -25$^{o}$C.

Prior to germination tests in vitro, pollen was pretreated in humid air for at least 2 h at 20$^{o}$C, except for Impatiens pollen, for which 15 min was sufficient. Germination in vitro was performed by mixing pollen with a drop of liquid medium, then spreading the drop on small agar plates ( 34 mm diameter), followed by blotting off the surplus of liquid. The liquid medium was according to Hoekstra and van Roekel (1983). The solid medium contained 0.6% agar and 0.2M sucrose in half the concentration of the other liquid medium components. Tube emergence and growth were examined by light microscopy, by viewing from the bottom side through the closed petri dishes, 1, 1.5, and 4 h after imbibition.

One application of α-tocopherol (Serva cat # 36570), butylated hydroxytoluene (BHT) and lidocain via hexane was performed by washing the dry pollen twice in hexane, followed by mixing with the desired concentration of the chemical for 3 min (6 mg pollen/ml hexane), removal of the excess solvent, and evaporation. The other application was via liquid germination medium. For this purpose pollen was equilibrated in a water-saturated atmosphere for 2 h. After mixing for 3 min, pollen was filtered and spread out on paper for rapid drying to 5% moisture content. Suspensions of the antioxidants in liquid media were made by mixing with acetonic solutions, avoiding the concentration of the acetone in the medium to exceed 0.1%. Storage was at 24$^{o}$ in a thermostated incubator.

Lipid extracts were made by pressing 25-30 mg samples of dry pollen in 20 ml $CHCl_3$-MeOH (2:1), containing 0.01% BHT, through a french pressure cell at 20,000 psi until complete disruption. After 5 min of ultrasonic treatment, the extracts were clarified by filtration , and the solvents removed by evaporation under vacuum. After readdition of 1 ml $CHCl_3$, aliquots were mixed with a freshly prepared ethereal solution of diazomethane. After evaporation of the surplus solvent, the methylated free fatty acids were determined by GC-analysis. Prior to disruption, an internal standard of heptadecanoic acid (C17:0) was added.

GC-analysis of fatty acid methylesters was performed on a Varian Series 1800 Gas Chromatograph equipped with a flame ionization detector, and coupled to an electronic integrator. The stainless steel column used ( 4 mm diameter x 180 cm) was packed with 10% Chrompack-Sil 88 on 100-120 mesh Chromosorb WHP. The oven temperature was 200$^{o}$C, total run time 15 min. The carrier gas was $N_2$ at 20 ml.min$^{-1}$.

## 3 Results and Discussion

The four pollen species in Fig 1 were selected on account of their considerably different longevities under conditions of dry storage, i.e. 5-6% moisture content at 24$^{o}$C. In an attempt to understand the origin of this variation in longevity we analysed the content of FFA's in the different pollen species during the storage period (dashed curves in Figure 1). Apart from *Impatiens*, the decrease of vitality coincided or was even preceded by an increase in FFA's. The use of antioxidants or inhibitors of phospholipase activity was meant to elucidate the mechanism by which these FFA's are generated. Furthermore, by applying antioxidants a double purpose was served: on the one hand, we would be able to establish whether a causal relationship exists between the accumulation of FFA and the decrease of vitality and, on the other hand, positive effects on storability would be of practical value, as it is for some seeds (Kaloyereas et al. 1961; Barnes and Berjak 1978; Basu and Dasgupta 1978; Woodstock et al. 1983). By preliminary experimentation, the maximum dose of a compound was established which could safely be applied before harmful effects were noticed. In Fig. 1 the results are shown on longevity and FFA accumulation of hexane-applied tocopherol, BHT, and the phospholipase inhibitor, lidocain. Generally the pattern of the curves does not differ much from that of the untreated control for both viability and FFA content. Only in *Papaver*, hexane washing apparently removed part of the FAA, which presumably was sticking at the outside of the grain. It can be concluded that hexane washing does not impair storability of pollen and, furthermore, that application of the above-mentioned chemicals via hexane does neither effectuate an extension of longevity, nor prevent the accumulation of FFA's. Such negative results limit a possible use of tocopherol, BHT, and lidocain applied via hexane, in pollen biotechnology.

It is, however, still uncertain whether application via hexane allowed the compounds to penetrate into the pollen interior. By using supra-optimal concentrations, it was established that upon incubation of the pretreated pollen in a germination medium the compounds exerted effects, possibly by penetration afterwards during imbibition. To avoid such difficulties, the chemicals had preferably to be administered via liquid

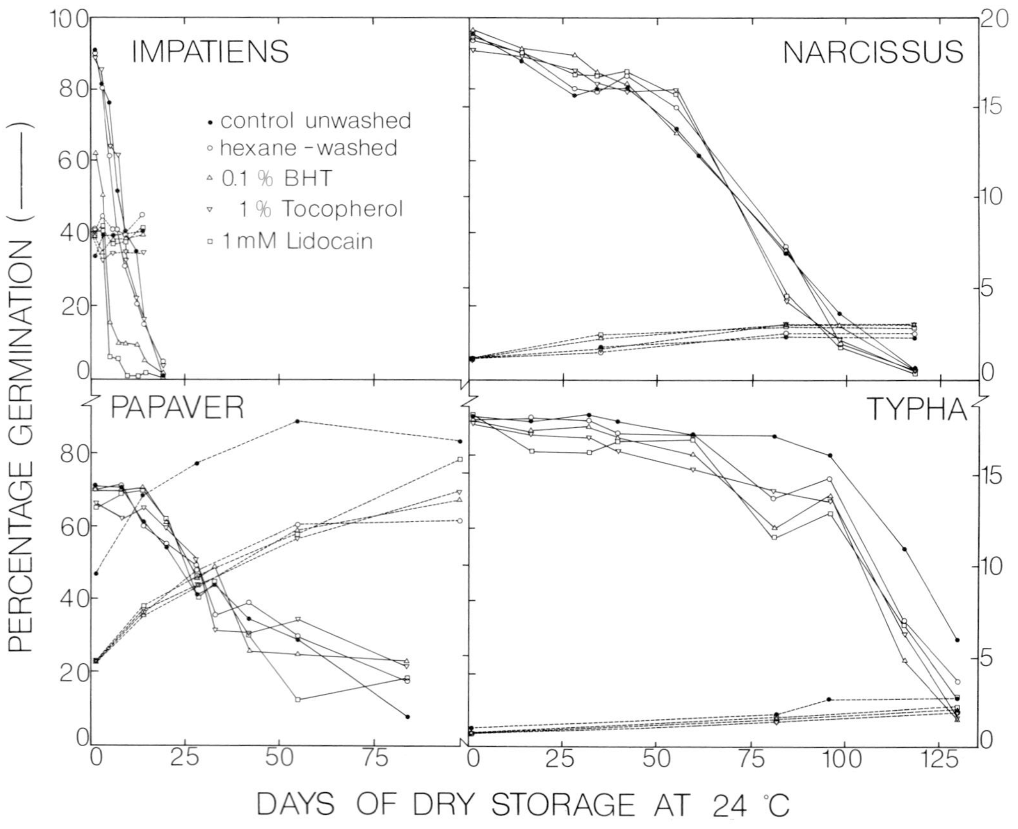


**Fig. 1.** Effects of pretreatment with tocopherol (1%), butylated hydroxytoluene (BHT) (0.1%) and lidocain (1 mM), applied via hexane solutions on pollen viability and free fatty acid (FFA) content in the course of dry storage at 24°C. FFA contents are averages of duplicate methylations and GC-analyses.

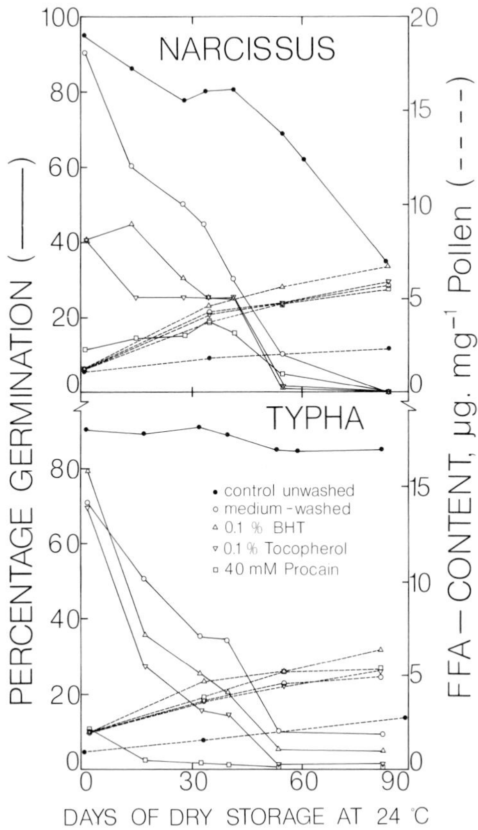


**Fig. 2.** Effects of pretreatment with tocopherol (0.1%), BHT (0.1%) and procain-HCl (40 mM), applied via liquid germination medium (3 min) and subsequent dehydration, on pollen viability and FFA content in the course of dry storage at 24°C. FFA contents are averages of duplicate methylations and GC-analyses.

medium. Viability is generally preserved following short term incubation in germination medium and dehydration, particularly in pollen species having relatively long lag periods, i.e. pollen with delayed tube emergence (Hoekstra 1983). As Impatiens and Papaver rapidly start tube emergence, 2 and 15 minutes after imbibition in the medium, respectively, we selected Narcissus (40 min) and Typha (70 min) for this purpose. Fig 2 shows that, in contrast to hexane-washed pollen, longevity of medium-washed pollen was considerably reduced, with a concomitant more rapid accumulation of FFA's. Despite their effectiveness as radical scavengers (Roubal 1970) suspensions of tocopherol and BHT failed to extend longevity and to curtail FFA accumulation. The same holds for the water-soluble phospholipase inhibitor, procain-HCl. The substances even turned out to harm viability. This renders the compounds unsuitable as adjuvants in, for instance, the processing of bee-collected pollen baskets into a dry powder (see Verhoef and Hoekstra, this volume). That injury to membranes was involved in the reduced longevity after the medium washing and redrying, is suggested by the excessive leakage of potassium upon re-incubation in the germination medium (Table I).

During dry storage the FFA's apparently are generated by a de-esterification process that cannot be blocked by free-radical scavengers or by inhibitors of phospholipase activity. As yet a causal relationship of rise in FFA's and fall in vitality remains obscure, although the accelerated and simultaneous manifestation of both processes after medium washing is suggestive in that direction. FFA's in the membrane are claimed to increase the phase transition temperature and may contribute to the formation of gel phase domains, and ultimately to the loss of viability (cf. Senaratna et al. 1984). What remains peculiar are the different levels of FFA's in the different species, and the varying rates

**Table I** Effect of pretreatment (3 min) with liquid germination medium on the ability of *Narcissus* and *Typha* pollen in the course of dry storage at 24°C, to retain potassium upon re-incubation in the medium. Potassium was determined by flame photometry.

| Treatment | Days of Dry Storage | | | |
|---|---|---|---|---|
| | 0 | 33 | 55 | 125 |
| | $K^+$-Efflux, in % of Original Content | | | |
| *Narcissus* | | | | |
| Medium-washed | 26 | - | 84 | - |
| 0.1% Tocopherol (via medium) | - | - | 82 | - |
| 0.1% BHT (via medium) | - | - | 88 | - |
| Control, unwashed | 22 | - | - | 92 |
| *Typha* | | | | |
| Medium-washed | 50 | 119 | 100 | - |
| 0.1% Tocopherol (via medium) | - | 90 | 112 | - |
| 0.1% BHT (via medium) | - | 99 | 93 | - |
| Control, unwashed | 49 | - | 53 | 69 |

of accumulation during dry storage. In *Impatiens* pollen only a slight, non-significant rise was noticed (Fig 1 A). If FFA's are underlying the loss of membrane integrity, then we may assume that they act differently in different fosfolipid backgrounds. Alternatively, the rise in the FFA contents might result from a harmful process in membrane lipids, which ultimately leads to loss of integrity.

It is concluded that in the course of dry storage at 24°C FFA's accumulate, generally prior to loss of vitality, and further that antioxidants and local anaesthetics cannot prevent these processes.

## References

Barnes G, Berjak P (1978). Proc Elec Microsc Soc South Afr 8:95-96
Basu RN, Dasgupta M (1978). Indian J Expt Biol 16: 1070-1073
Heslop-Harrison J, Heslop-Harrison Y (1970). Stain Technol 45:115-120
Heslop-Harrison J, Heslop-Harrison Y, Shivanna KR (1984). Theor Appl Genet 67:367-375
Hoekstra FA (1983). In: Pollen: Biology and Implications for Plant Breeding pp 35-41. Mulcahy DL, Ottaviano E, eds, Elsevier
Hoekstra FA (1984). Plant Physiol 74:815-821
Hoekstra FA, Bruinsma J (1975a). Physiol Plant 34:221-225
Hoekstra FA, Bruinsma J (1975b). Z Pflanzenphysiol 76:36-43
Hoekstra FA, Bruinsma J (1980). Physiol Plant 48:71-77
Hoekstra FA, van Roekel T (1983). Plant Physiol 73:995-1001
Hoekstra FA, van Roekel T (1985). Suppl Plant Physiol 77:#669
Kaloyereas SA, Mann W, Miller JC (1961). Econ Bot 51:213-217
Lichte HF (1957). Angew Bot 31:1-28
Roubal WT (1970). J Am Oil Chem Soc 47:141-144
Senaratna T, McKersie BD, Stinson RH (1984). Plant Physiol 76:759-762
Senaratna T, McKersie BD, Stinson RH (1985). Plant Physiol 77:472-474
Shivanna KR, Heslop-Harrison J (1981). Ann Bot 47:759-770
Visser T, de Vries DP, Welles GWH, Scheurink JAM (1977). Euphytica 26:721-728
Vojnikov VK, Luzova GB, Korzun AM (1983). Planta 158:194-198
Woodstock LW, Maxon S, Faul K, Bass L (1983). J Amer Soc Hort Sci 108: 692-696

# Protein Pattern of Apple Pollen in Culture: Effect of Actinomycin D

A. Speranza, G.L. Calzoni and N. Bagni*

## 1 Introduction

Many pollen species have been studied as influenced by the translation inhibitor actinomycin D (Dexheimer,1968; Mascarenhas, 1975). General evaluations are hard to be drawn because requirement of new RNA and protein synthesis for either tube emergence or/and growth is strictly related to and varies with different levels of physiological evolution of pollen. However, some recent data on tobacco pollen (Čapková et al.,1983; Tupý,1983) suggested that the growth inhibitory effect of actinomycin D should not only be mediated via blocking of transcription. Indeed, a probable interference with the development and function of structures involved in pollen tube wall synthesis was hypothesized.

In this chapter we describe the evolution of protein pattern in the binucleate apple pollen under the influence of actinomycin D. We considered the proteins of cytoplasm, those released in culture and a fraction which so far has received little attention (Dashek and Harwood,1974; Li Yi-qin et al.,1983; Li Yi-qin and Linskens,1983), that is the wall-bound proteins of grains and tubes. We thought that mainly from studying the latter some light could be shed on the interesting problem of actinomycin D action.

## 2 Protein extraction and separation

Pollen of Malus domestica Borkh. cv.Starkrimson germinated

*Institute of Botany, University of Bologna, Italy

in liquid culture in the dark according to Calzoni et al. (1979); in some cases 40 µM actinomycin D was added to the medium as solution in 0.01 M phosphate buffer, pH7.

Pollen ungerminated or germinated was suspended in 0.05 M phosphate buffer pH 6, and ruptured by passing twice through a French Press at 100 Kg/cm$^2$. Cell walls were pelleted and washed at 0°C according to Li Yi-qin et al. (1983). After each washing the protein content in the supernatant was determined (Lowry et al., 1951) until no protein was detectable. Then the walls were suspended twice in 1 M NaCl for 3 h at 0°C, the two saline extracts were pooled,dialyzed, liophylized and resuspended in electrophoresis buffer.

Cytoplasmic proteins were precipitated from supernatant of ground pollen with cold ethanol below 0°C; then they were dialyzed, liophylized and resuspended in electrophoresis buffer. Proteins released in the culture medium were also precipitated with the same procedure.

SDS-polyacrylamide gel (15%) electrophoresis was performed according to Laemmli (1970); proteins were visualized with Coomassie blue, glycoproteins with PA-silver staining (Dubray and Bezard, 1982).

## 3 Cell wall proteins

Isolation of grain and tube cell walls was carefully made to eliminate residual cytoplasm and plasmalemma by repeated washing until the preparation was free of contaminant materials. The fraction of wall bound proteins we considered was operationally defined as those present in salt extract of

Table 1.Proteins in cell wall and cytoplasm of ungerminated or germinated apple pollen, and proteins released in culture (a: control; b: plus actinomycin D ).

| TIME (min) | PROTEINS (µg/mg pollen) | | | | | |
|---|---|---|---|---|---|---|
| | CELL WALL | | CYTOPLASM | | CULTURE MEDIUM | |
| | a | b | a | b | a | b |
| 0 | 5 | - | 58 | - | - | - |
| 60 | 15 | 7 | 65 | 50 | 16 | 7 |
| 120 | 22 | 12 | 12 | 26 | 19 | 4 |

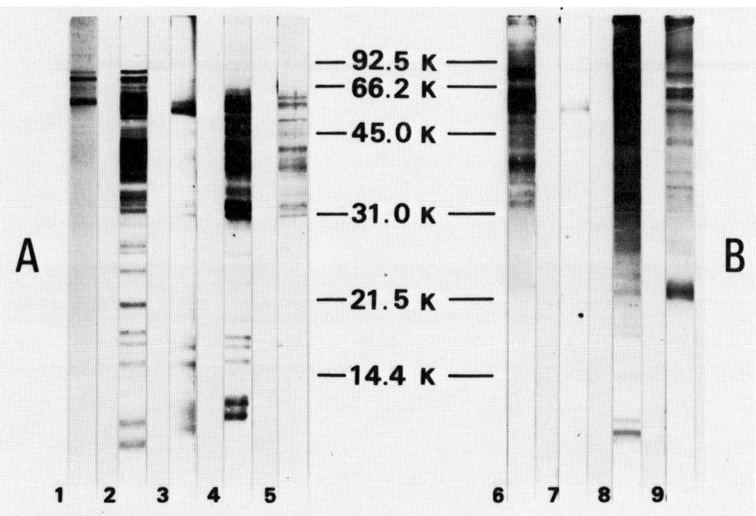


Fig.1. SDS-polyacrylamide gel electrophoresis of cell wall-bound proteins in apple pollen. A: Coomassie blue stain, B: silver stain. Lane 1: ungerminated pollen. Control pollen: lane 2 and 6, 60 min germination; lane 4 and 8, 120 min. Pollen treated with 40 µM actinomycin D: lane 3 and 7, 60 min; lane 5 and 9, 120 min germination.

the walls (not covalently bound). Proteins rapidly increased in the germinating control, whereas very few proteins (about 50% less) were bound to walls of tubes growing in the presence of actinomycin D (Table 1).

Gel electrophoresis separation (Fig.1) showed that in the ungerminated grains very few bands were present, thereafter tube emergence led to the appearance of many new bands whose molecular weight was spread over a wide range from 92.5 K to 12 K. Particularly one protein band (58 K) greatly increased in density at 60 min. At 120 min some bands disappeared, others increased in density; in the zone of low weights at least two new were present.

The actinomycin D strongly affected wall protein pattern, which remained dramatically lacking of bands in spite of tube outgrowth.At 120 min some increase of the number of bands was evident, but fewer and lower in density than in the control.

Most of bands detected with Coomassie blue reacted positively after specific staining of glycoproteins (Fig.1). Owing to the high sensitivity of the stain, some bands became more evident after this procedure.

## 4 Cytoplasmic and culture medium proteins

Transition from ungerminated state to 60 min growth in normal medium caused some increase of cytoplasmic protein content (Table 1); some new bands appeared mainly in the lower weight zone (Fig.2). Thereafter, a rapid and strong decrease took place at 120 min as well in the protein quantity as in the number of electrophoretic bands, which appeared grouped in a very narrow weight range. The presence of actinomycin D seemed to cause only quantitative differences in the pattern of cytoplasmic proteins at 60 min (Fig.2), after that it promoted a higher protein level and more numerous bands than in the control at 120 min.

The release of proteins starts as soon as grains enter in contact with the medium, becoming more conspicuous as tubes outgrow and elongate through exocytotic apical secretion.

Protein release by apple pollen increased during germination in the normal medium, whereas it seemed to be reduced by actinomycin D (Table 1); gel pattern shows only a small group of bands at 60 min, and a further reduction in number and intensity of bands at 120 min germination (Fig.3).

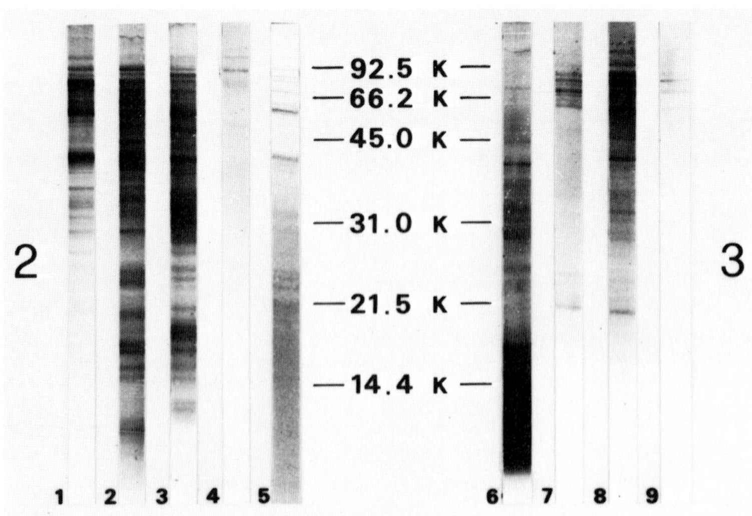


Fig.2 and Fig.3. SDS-polyacrylamide gel electrophoresis of cytoplasmic proteins of ungerminated or germinated apple pollen (2), and proteins released in culture (3). For legends of lanes 1-9 see Fig.1.

## 5 A model for actinomycin D action

Protein dynamics in the communicating compartments of cytoplasm, tube wall and outer medium was heavily affected by the actinomycin D. In particular, the analysis of cell wall-bound proteins revealed that tubes grown in the presence of the antibiotic were poorly arranged (Fig.1).

Actinomycin D interferred with organization and secretory function of rough endoplasmic reticulum vesicles in barley aleurone cells (Vigil and Ruddat,1973); an inhibitory effect on cell wall regeneration was observed in Chlamidomonas (Robinson and Schloesser,1979); changes in cytoplasmic and membrane-bound structures were described in cultured cells of African green monkey kidneys (Benedetto et al.,1979). The Čapková's hypothesis of an actinomycin interference with structures involved in pollen tube wall synthesis was also supported by data indicating a great reduction of protein release from tobacco pollen tubes simultaneous with the growth inhibition (Čapková et al.,1983). Also data of Tupý (1983) evidenced a transcription unrelated inhibition on the growth of tobacco pollen by means of actinomycin D.

The present observations on the apple pollen seem to agree with such a model of action. If the antibiotic somehow interferes with functioning of the secretory vesicles responsible for surface growth of tube walls, it could well result a disorder in protein insertion into these latter. Thereafter, a certain protein accumulation in the cytoplasm and a reduction of protein release in the culture (Table 1, Fig.2 and 3) might be coherent with such a model.

Finally, it was hypothesized that mainly the glycoproteins newly appearing after germination could play a role in the maintenance and elongation of tube wall (Li Yi-qin et al.,1983). Moreover, some band differences in wall-bound proteins between self- and cross-pollen tubes were observed in Lilium longiflorum (Li Yi-qin and Linskens,1983). Thus, these glycoproteins could also be involved in cellular recognition as key element in the pollen-stigma interactions.

Acknowledgement. Work supported by funds of Ministero della Pubblica Istruzione, Italy . Many thanks to Mr. N. Mele for photographic assistance.

References

Benedetto A, Cassone A, Delfini C (1979) Resistance of African green monkey kidney cell lines to actinomycin D: drug uptake in 37 RC cells after persistent inhibition of transcription. Antimicrob Agents Chemother 15: 300-312

Calzoni GL, Speranza A, Bagni N (1979) In vitro germination of apple pollens. Scientia Hortic 10: 49-55

Čapková V, Hrabětová E, Tupý J (1968) Reduction of leucine efflux and protein release from tobacco pollen tubes in culture by actinomycin D in the absence of calcium. Biochem Physiol Pflanzen 178: 521-527

Dashek WV, Harwood HI (1974) Proline, hydroxyproline and lily pollen tube elongation. Ann Bot (London) 38: 422-427

Dexheimer J (1968) Sur la synthèse d'acid ribonucléique par les tubes polliniques en croissance. CR Acad Sci D 267: 2126-2128

Dubray G, Bezard G (1982) A high sensitive periodic acid-silver stain for 1,2- diol groups of glycoproteins and polysaccarides in polyacrylamide gels. Anal Biochem 119: 325-329

Laemmli UK (1970) Cleavage of structural proteins during the assembly of the head of bacteriophage T4. Nature (London) 227: 680-685

Li Yi-qin, Croes AF, Linskens HF (1983) Cell-wall proteins in pollen and roots of Lilium longiflorum: extraction and partial characterization. Planta 158: 422-427

Li Yi-qin, Linskens HF (1983) Wall-bound proteins of pollen tubes after self- and cross-pollination in Lilium longiflorum. Theor Appl Genet 67: 11-16

Lowry OH, Rosebrough NJ, Farr AL, Randall RJ (1951) Protein measurement with the Folin phenol reagent. J Biol Chem 193: 266-275

Mascarenhas JP (1975) The biochemistry of angiosperm pollen development. Bot Rev 41: 259-314

Robinson DG, Schloesser UG (1978) Cell wall regeneration by protoplasts of Chlamidomonas. Planta 141: 83-92

Tupý J (1983) Transcription activity and the effects of transcription inhibitors in tobacco pollen culture. In: Fertilization and Embryogenesis in Ovulated Plants, Veda, Bratislava, pp. 133-136

Vigil EL, Ruddat M (1973) Effect of gibberellic acid and actinomycin D on the formation and distribution of rough endoplasmic reticulum in barley aleurone cells. Plant Physiol 51: 549-559

# Ion Localisation in the Stigma and Pollen Tube of Cereals

J. S. Heslop-Harrison[1]

Cytochemical techniques and energy dispersive X-ray analysis have been used to follow the distribution of calcium, an element known to be essential for normal pollen germination and tube growth, in the tubes of *Zea*, *Pennisetum* and *Secale*, during germination and early tube extension. No tip gradient is present comparable with that seen in *Lilium*. A parallel study on the stigmas of the same grasses has shown little calcium is present but the stigmas are very rich in potassium.

Calcium ions are essential for the *in vitro* germination of pollen and growth of pollen tubes of many species (Johri & Vasil, 1961; review Rosen, 1968). Many aspects of calcium function in the pollen-tube require clarification, but the evidence suggests that it is involved in some rather direct way in regulating the tip-growth system which could account for the chemotropic response the calcium ion elicits in some pollen tubes (Mascarenhas & Machlis, 1962). The present chapter provides a first report of the results of the distribution of calcium and other ions in the stigma and pollen of three cereal species, *Zea mays* L., *Pennisetum americanum* (L.) Leeke and *Secale cereale* L.

The distribution of calcium in actively growing pollen tubes of *Zea* was investigated cytochemically with chlorotetracycline (CTC), which produces a fluorescent complex with $Ca^{2+}$ ions (e.g., Caswell, 1979). The pollen was germinated in a minimal medium containing 15 or 20 %sucrose, 1 mM $H_3BO_3$ and 1 mM $Ca(NO_3)_2$, and at the required lengths the tubes were transferred to a slide

[1] Plant Breeding Institute, Trumpington, Cambridge, U.K.

and the medium replaced by a solution of 1 mM CTC in germination medium. After 10-20 min, the distribution of fluorescence in the tube was examined by fluorescence microscopy. In all tubes showing normal tip morphology, the distribution of fluorescence was uniform throughout the terminal non-vacuolate part of the tube. The plasmalemma, particulate inclusions and membrane aggregates showed strong fluorescence, while the cytoplasmic matrix showed generalised, weaker, activity. The particulate inclusions were likely to be mainly mitochondria.

The distribution of fluorescence observed over the terminal 200 µm of an actively extending tube of Pennisetum is shown in Fig. 1. There is no indication of a basipetally-declining gradient of fluorescence immediately behind the tip. If CTC-fluorescence is indeed providing a true indication of calcium distribution, this implies that in the Zea pollen tube there is no special concentration in the tip region during this phase of growth. Similar distributions were observed in the tubes of Pennisetum and Zea.

Energy-dispersive X-ray analysis (EDX) was used to examine the distribution of calcium in the grass pollen tube. Pre-hydrated pollen of Pennisetum was germinated in liquid medium. Samples of the suspension were withdrawn when the required tube-length had been attained, transferred to microcentrifuge tubes, rinsed, suspended in a drop of water and transferred directly to carbon SEM stubs before rapid drying. The whole procedure took under a minute. The rinsing burst some of the pollen tubes, but many remained intact and appeared morphologically normal. The procedure effectively removed the sucrose and presumably most of the inorganic constitutents of the medium from the surfaces, leaving the free and bound ions inside the intact tubes.

The samples dried onto the stub surface were examined without coating. EDX spectra were acquired before surface charging was visible in the SEM. The scanned fields (c. 70 $\mu m^2$) were adjusted to lie within the width of the tube. Because the pollen tube is uniformly cylindrical, the X-ray emmision spectra were unlikely to be affected by variation in surface topology or varying depth of penetration of the electron beam from the SEM. A background spectrum was taken from a field near the specimen, within the area where the residual rinsing water had dried down. This was subtracted from the smoothed spectra from the specimens.

Figures 2-3 show the ion distribution over two parts of a tube of Pennisetum c. 120 µm in length. They confirm the CTC localisations, showing that there is no appreciable gradient in calcium from the tube tip all the way back to the grain.

There is an obvious contrast between these results and those obtained by

Reiss & Herth (1978) from the pollen tubes of Lillium longiflorum. These authors observed a strong gradient of CTC-induced fluorescence in the immediate sub-apical region, with much more calcium, as indicated by CTC-staining, present in the terminal 50 μm than in the rest of the tube. In a later investigation, Reiss et al. (1983) examined calcium distribution in L. longiflorum using proton, rather than electron, induced X-ray emission. Their spectra show a great accumulation of calcium in the tube tip, an observation confirming their cytochemical finding and also verifying the earlier observations of Jaffe et al. (1975).

The difference in calcium distribution between the grass pollen tube and that of Lilium is related to a difference in growth physiology. The tubes share a common method of extension-growth, new wall being continually laid down at the tip through the insertion of polysaccharide-containing wall precursor particles. In Lilium these are abundantly present in the apical 5-10 μm of the tube, and the stock is constantly supplemented during growth by dictyosome activity in the sub-apical region. Mitochondria are also concentrated in the sub-apical region, so that the tube tip shows a marked zonation in organelle content. In the grasses, the pollen tubes extend at rates 5-15 times greater than that attained by tubes of Lilium . This high rate of growth is achieved initially by the utilisation of reserves of polysaccharide particles (P-particles) deposited in the grain while it is still in the anther. Since the P-particles are distributed throughout the length of the tube during the early stages of growth, and since there is little dictyosome activity in this period, the cytoplasm in the terminal part does not show any marked organelle zonation comparable with that seen in the lily tube. Evidently the distribution of calcium reflects these differences in apical zonation.

Three different techniques were used to examine the distribution of ions in the stigma of Pennisetum: cytochemical localisation, EDX analysis and spectrometry. In a male sterile genotype, BE23, the stigmas remain exerted for 2-3 days, and since they can be harvested readily in an uncontaminated state, analysis by inductively coupled argon-plasma flame spectrometry (ICAP) as well as by the micro methods was possible. For ICAP, 0.5 g of stigmas were ground in distilled water, and then the particulate material was precipitated by centrifugation. Only the ions present in the cytosol were preserved; the analysis would take no account of organelle or membrane-bound calcium or boron. In two extraction series, c. 3500 μg/g-fresh-weight of potassium was found in the stigma soluble fractions, 620 μg/g phosphorus, and 140 μg/g magnesium. In other runs with higher concentrations of extract, 1.7 +/- 0.3 μg/g of calcium and 1.3 +/- 0.3 μg/g boron were measured, both some 2000 times less than the potassium level.

For EDX analysis, ovaries and attached, receptive stigmas were removed from the plants. The ovary was then pressed firmly against a carbon SEM stub. The liquid expressed held the stigma in place in the microscope and provided a conductive pathway which prevented charging of the uncoated specimen during observation. Spectra were acquired from areas of the ovary top, stigma base and trichome region (Figs. 4-5). High levels of potassium ions were found, notably in the trichome region. Chloride ions (not detectable by the ICAP spectrometer) were also found to be present in high concentrations, but not in amounts sufficient to suggest that potassium was present principally as KCl (Fig. 6). The implication is that one or more additional, organic, anions are present in the stigma tissue as well as the chloride. In the trichomes of the receptive region of the stigma, potassium was by far the most frequent atom detected, although towards the top of the ovary the amounts of calcium, phosphorus and sulphur (presumably in proteins and nucleic acids) increased considerably.

Figure 7 shows the cytochemical localisation of potassium ions using the potassium cobaltinitrite technique (Pearse, 1972). Potassium is evidently abundant in the receptive trichomes.

It is strikingly evident that notwithstanding the well-proven requirement of the grass pollen tube for boron and calcium when cultured _in vitro_, the amounts actually available in the stigma in soluble form are remarkably low. If calcium were to be held tenaciously in walls, intercellular material and organelles in the receptive papillae, it would not be detected in the ICAP analysis, although, given the tenuous nature of the tissue, it should contribute to the EDX calcium peak of these regions.

The results of all three techniques indicate high levels of potassium in the receptive cells, and since an almost immediate consequence of pollination is to increase the permeability of cells in the vicinity of the pollen contact, it might be expected that this element would be present in some quantity in the medium bathing the germinating grain and the emerging tube. Preliminary observations showed that $K^+$ concentrations above 5 mM virtually stopped rye pollen germination, while even 0.1 mM K had a slightly inhibitory effect on germination and subsequent tube growth.

The most likely function of potassium in the stigma is as a major componenet of the osmoticum which maintains turgidity of the stigma over the period of flower opening--several days in the case of _Pennisetum_. It is often exerted into an atmosphere which is often highly desiccating. It is obvious that, since the grasses are wind-pollinated, the stigma must not restrict local airflow; and indeed its feather-like structure is adapted to ensure the maximum air-sampling capacity. Yet the stigmas of grasses like _Pennisetum_ can retain

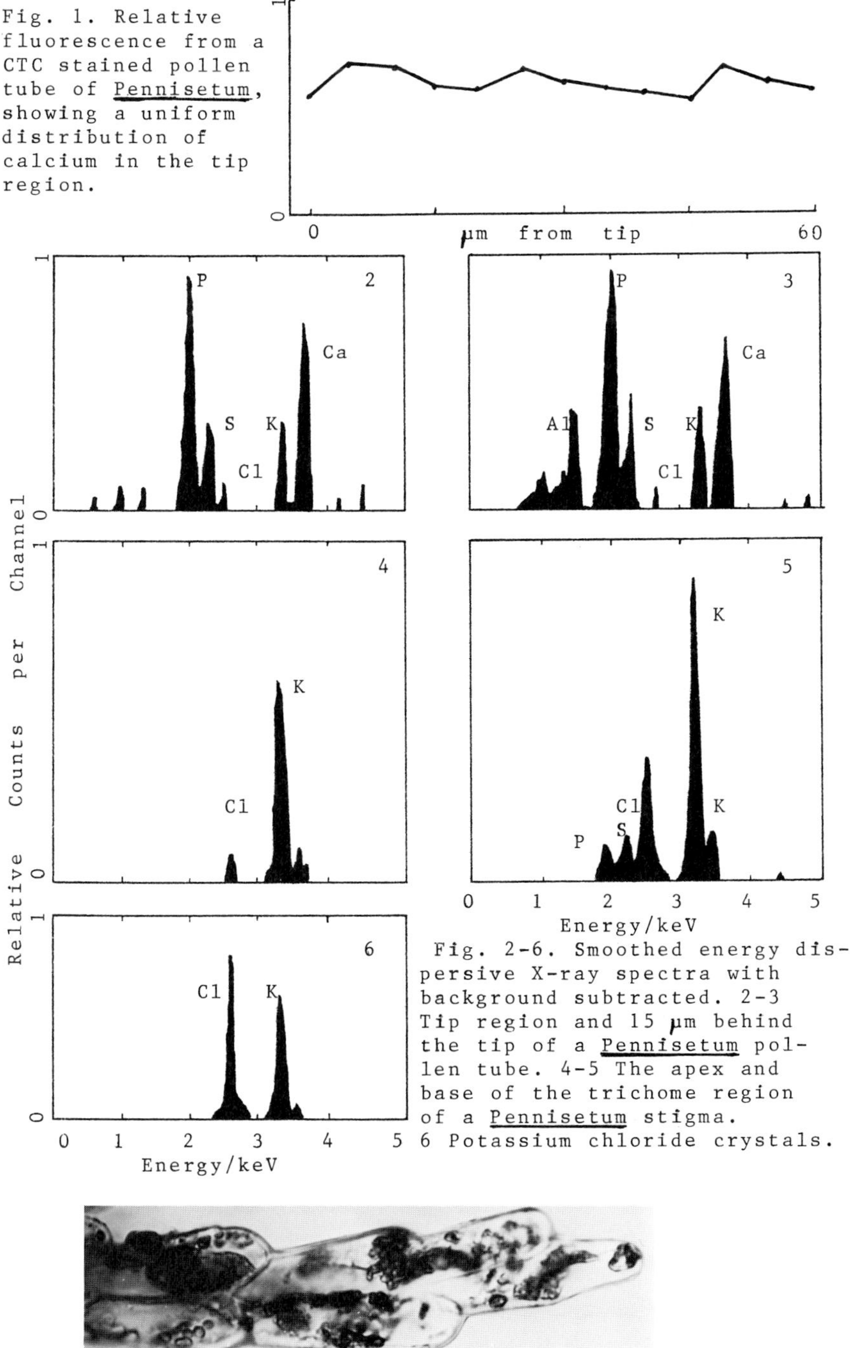


Fig. 1. Relative fluorescence from a CTC stained pollen tube of Pennisetum, showing a uniform distribution of calcium in the tip region.

Fig. 2-6. Smoothed energy dispersive X-ray spectra with background subtracted. 2-3 Tip region and 15 µm behind the tip of a Pennisetum pollen tube. 4-5 The apex and base of the trichome region of a Pennisetum stigma. 6 Potassium chloride crystals.

Fig. 7. Precipitated crystals of potassium cobaltinitrite showing potassium presence in the Pennisetum stigma.

their turgidity and maintain suitable surfaces to trap pollen and nurture emerging pollen tubes. Turgor is maintained even although the papillate tips of the receptive trichomes are clad with a thin and discontinuous cuticle. At maturity, the tonoplast is also discontinuous, implying that turgor depends exclusively on the retentivity of the plasmalemma. As noted above, this is rapidly lost after the capture of pollen, at which time the receptive parts of the stigmas become flaccid. The movement of potassium ions into the tissues may be a factor contributing to the early rapid unfolding and extension of the stigma cells.

Ackknowledgments. I thank Peterhouse, Cambridge, U.K. for the award of a William Stone Research Fellowship, and Prof. J., Dr. Y., Heslop-Harrison, Aberystwyth, U.K. and Dr. B. Reger, Russell Research Center, Athens, Georgia, U.S.A. for collaboration.

## References

Bruyn JA de (1966) Physiol Plantarum 19: 322-327, 365-376.

Caswell AH (1979) Int Rev Cytol 56: 145-181.

Heslop-Harrison JS, Heslop-Harrison J, Heslop-Harrison Y, Reger BJ (1985) Proc Roy Soc London, B: in proof.

Jaffe LA, Weisensell MH, Jaffe LF (1975) J Cell Biol 67: 488-492.

Johri BM, Vasil IK (1961) Bot Rev 27: 325-381.

Mascarenhas JP, Machlis L (1962) Am J Bot 49: 482-489.

Pearse AGE (1972) Histochemistry--Theoretical and Applied. Third Edition. Churchill, London.

Reiss H-D, Herth W (1978) Protoplasma 97: 373-377.

Reiss H-D, Herth W, Schnepf E (1983) Protoplasma 115: 153-159.

Rosen WG (1968) Rev Plant Physiol 19: 435-462.

# Phytases of Germinating Lily Pollen

D. B. Dickinson and J. J. Lin[1]

## 1 Introduction

Phytic acid is a well known constituent of seeds, where it acts as a phosphorus and myo-inositol reserve (Loewus 1983). Phytate is stored in the protein body as the mixed salt of potassium and magnesium and lesser amounts of calcium, iron, and manganese. Upon germination phytate is degraded, supplying the seedling with these constituents. Phytase is a phosphomonoesterase that hydrolyses phytate to give $P_i$ and lower phosphoric esters of myo-inositol.

Recently phytic acid was identified in pollen of many plant species (Jackson et al. 1982), and its metabolism in *Petunia* pollen was studied (Jackson and Linskens 1982; Helsper *et al*. 1984). Significant quantities (0.05 to 2.1%) of phytic acid were observed in pollen from plants with style lengths greater than 5 mm, while little or none was found in pollen from composites and grasses with very short styles. There is a high demand for cell wall and membrane precursors during pollen tube growth, and phytic acid is a logical source of precursors (Helsper *et al*. 1984).

The literature on phytic acid metabolism in pollen is very limited, so we have studied this process in germinating lily pollen. Phytases were isolated from mature ungerminated and germinated lily pollen and were partially purified and characterized.

## 2 Phytase extraction and assay

Pollen of *Lilium longiflorum* was germinated in a 0.29 M glucose culture medium (Dickinson, 1967) for various times and was harvested by filtering through nylon cloth (20-40 um apertures). Pollen samples were ground 5 min in a mortar with 0.1 M Na acetate buffer, pH 5.2, at 0°C. The clarified supernatant from two centrifugations (each 10,000*g*, 10 min) was dialyzed in 2X1 l of 0.1 M Na acetate buffer,

---

[1]Dept. of Horticulture, University of Illinois, Urbana, Illinois, USA. Present address of J. Lin is Dept. of Biology, Washington University, St. Louis, MO, USA

pH 5.2, for 18 to 24 h. The dialysate was centrifuged and assayed for phytase. The standard reaction mixture (0.5 ml) contained 2.5 mM $MgCl_2$ and 0.5 mM Na phytate in 0.1 M PIPES (pH 6.5) or 2.5 mM Na phyate in 0.1 M Na acetate (pH 5.2). The reaction was initiated with enzyme, and incubation was 0.5 to 2 h at 30°C. The reaction was stopped with 50 ul of 50% cold trichloroacetic acid. $P_i$ was then determined colorimetrically (Chen et al. 1956).

## 3 Phytase characterization

The optimal pH of phytase from ungerminated pollen was about 5.0 (Fig. 1a), while the phytase from pollen germinated 2 h had optimal activity at pH 6.5 (Fig. 1b). Activity was largely independent of [$Mg^{2+}$]. Assays were carried out at various phytic acid concentrations to insure that optimal or near optimal substrate levels were used in the standard assays. For phytase isolated from pollen germinated 2 h, activity increased as the phytic acid concentration was raised to 0.5 mM, with progressively increasing inhibition at higher concentrations (Fig. 2a). The estimated $K_m$ for phytic acid was approximately 0.2 mM. Phytase of ungerminated pollen showed a more complex pattern, with activity approaching a plateau from 1 to 7.5 mm phytic acid and then increasing sharply to a peak at 10 mM, with considerably less activity at 15 mM (Fig. 2b).

Fig. 1a (above left). Effect of pH on the activity of crude phytase from ungerminated lily pollen. Enzyme (3.2 ml) was prepared from 1 g pollen. Reaction mixtures contained 2.5 mM Na phytate. The actual pH of each assay tube was plotted in the graph. Each point is the average of duplicate assays that contained 100 ul of enzyme and were incubated for 2 h. Differences between duplicates ranged from 3 to 8%. Na acetate (-◆-), PIPES (-◇-), and BICINE (-■-).
Fig. 1b (above right). Effect of pH on the activity of crude phytase isolated from pollen germinated for 2 h. Enzyme (3.1 ml) was prepared from 620 mg pollen. Reaction mixtures contained 0.5 mM Na phytate and were incubated for 1 h. Each point is the average of duplicate assays, each with 80 ul of enzyme. The range between duplicates was 1 to 8%.

Fig. 2a (below left). Determination of optimal phytic acid concentration for phytase from ungerminated lily pollen. The pH of the assay mixture was 5.2. Each point is the mean of duplicates, each with 50 ul of enzyme. The range between duplicates was 2.6 to 8.4%.
Fig. 2b (below right). Determination of optimal phytic acid concentration for phytase from pollen germinated 2 h. Enzyme (3.6 ml) was prepared from 800 mg pollen. The standard assay conditions were used with various levels of Na phytate in 0.1 M PIPES buffer, pH 6.5. Each point is a mean of duplicate assays, each with 50 ul of enzyme. The range between duplicates was 0 to 6%.

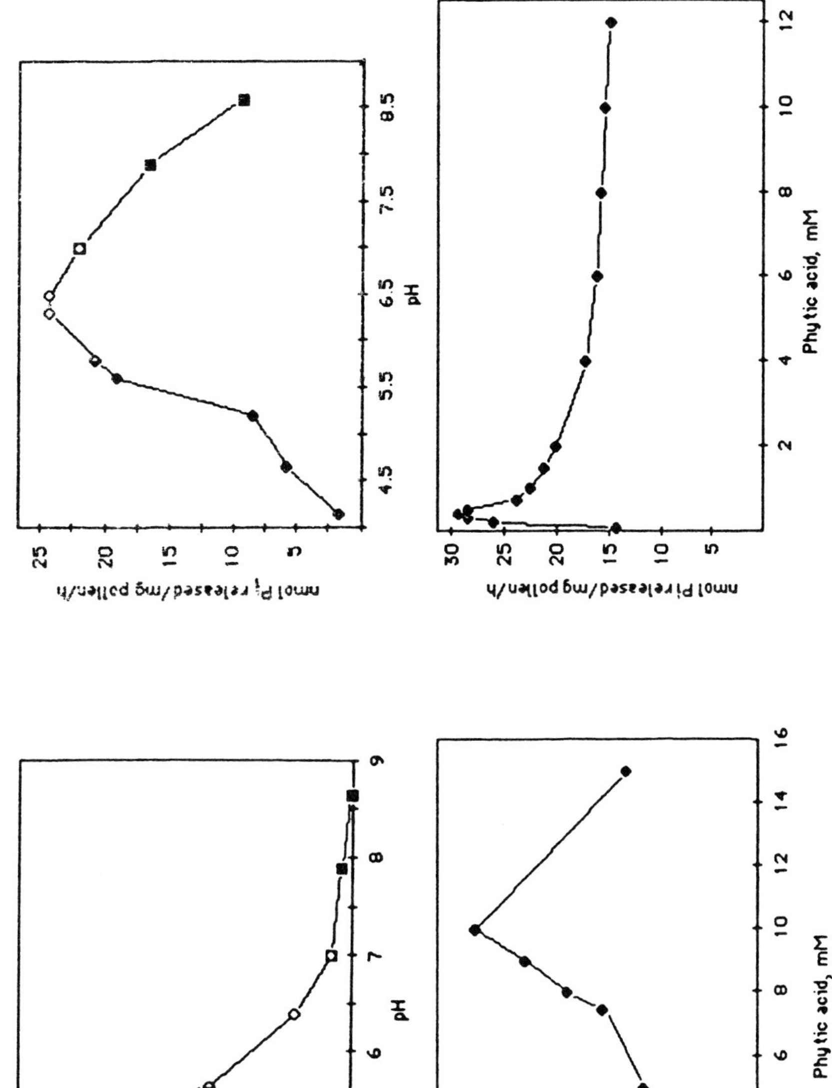
nmol Pi released/mg pollen/h
pH
nmol Pi released/mg pollen/h
pH
nmol Pi released/mg pollen/h
Phytic acid, mM
nmol Pi released/mg pollen/h
Phytic acid, mM

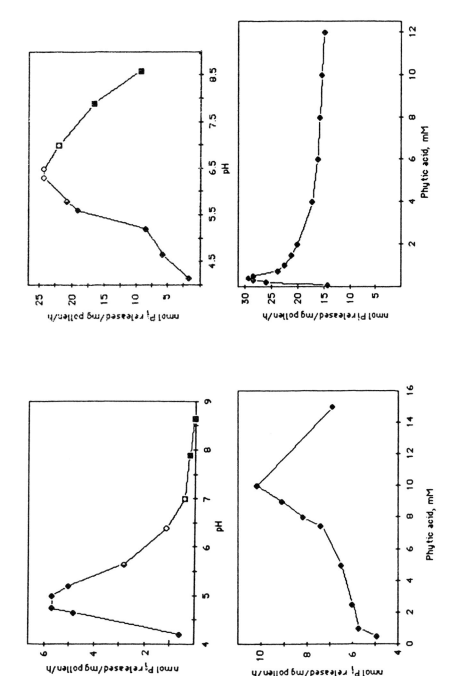

4 Phytase activity during germination

Phytase was determined on extracts of pollen germinated for various times. Some pH 5.0 activity was present in ungerminated pollen, and activity at that pH increased several-fold during germination (Fig. 3a). Mature ungerminated pollen as well as pollen harvested before anthesis contained no activity at pH 6.5, but activity appeared by 1 h of germination and increased 6-fold by 2 h (Fig. 3b). Pollen germinated 90 min had about the same activity at pH 6.5 and 5.0, but activity was higher at pH 6.5 after 90 min.

Fig. 3a (above). Time-course of phytase activity at pH 5.0 during lily pollen germination. Each point is a mean of duplicates. Other details are given in the text.
Fig. 3b (below). Time-course of phytase activity at pH 6.5 during lily pollen germination.

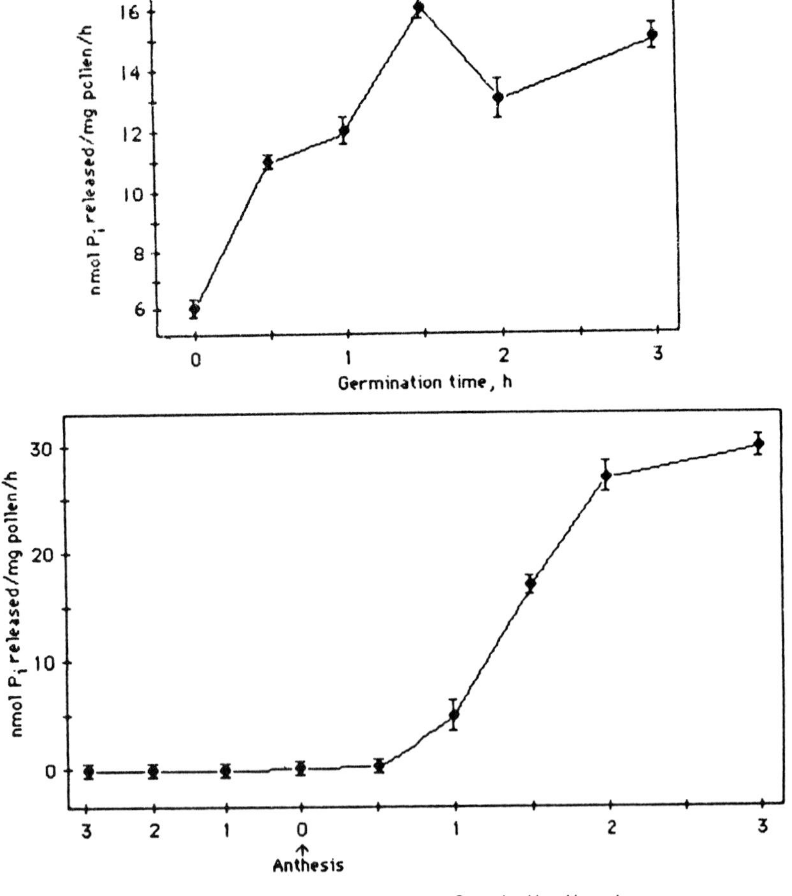

Phytase preparations from ungerminated and from pollen germinated 2 h were partially purified by heat (2 min, 59°C) and ammonium sulfate (45-65%) and were then subjected to HPLC using a DEAE column and a tris acetate gradient. The phytase with optimum activity at ph 5.0 was eluted at 0.3 M tris acetate, and a second peak containing the pH 6.5 activity was eluted at 0.5 M. The latter peak was absent from ungerminated pollen and appeared during germination (Lin 1985). The appearance of this second peak was prevented by incubating pollen with the protein synthesis inhibitor cycloheximide, which also caused a drastic reduction in the rate of phytate disappearance during germination.

## 5 Discussion

Most pollen enzymes studied to date are plentiful in the ungerminated pollen and exhibit little or no increase in activity during germination, so the rise in pH 6.5 phytase activity from a negligible level in ungerminated pollen to a high level during germination is quite unusual and is similar to that reported for phytase of *Petunia* pollen (Jackson and Linskens 1982). Additional work is needed to learn whether this increase is due to *de novo* synthesis of lily pollen phytase or an activation process that depends on protein synthesis. The pH 6.5 phytase is probably needed for normal growth of the lily pollen tube, so appearance of this enzyme might be affected in instances where tube growth does not proceed normally. A seed posphatase that is specific for phytate is associated with protein bodies (Dalling and Bhalla 1984). The pH 6.5 pollen phytase is more specific for phytate than is the pH 5.0 enzyme (Lin 1985), so the former may be localized with its substrate. Both phytate and phytase may be present in the single-membrane organelles reported earlier (Southworth and Dickinson 1981). These organelles contain densely staining bodies which may be phytate and which disappear during germination of the lily pollen.

*Acknowledgement* Dr. David Ho is thanked for advice and help.

## References

Chen PS, Toribara TY, Warner H (1956) Microdetermination of phosphorus. Anal Chem 28:1756-1758.

Dalling MJ, Bhalla PL (1984) Mobilization of nitrogen and phosphorus from endosperm. In: Murray DR (ed) Seed Physiology 2. Germination and Reserve Mobilization. Academic Press, pp 163-199.

Dickinson DB (1967) Permeability and respiratory properties of germinating pollen. Physiol Plant 20:118-127.

Helsper JPFG, Linskens HF, Jackson JF (1984) Phytate metabolism in petunia pollen. Phytochemistry 23:1841-1845.

Jackson JF, Linskens HF (1982) Phytic acid in *Petunia hybrida* pollen is hydrolyzed during germination by a phytase. Acta Bot Neerl 31:441-447.

Jackson JF, Jones G, Linskens HF (1982) Phytic acid in pollen. Phytochemistry 21:1255-1258.

Jackson JF, Kamboj RK, Linskens HF (1983) Localization of phytic acid in the floral structure of *Petunia hybrida* and relationship to incompatibility genes. Theor. Appl. Genet. 64:259-262.

Lin JJ (1985) Metabolism of Phytic Acid in Pollen of *Lilium longiflorum* Thunb. and Mathematical Modeling of Wall Polysaccharide Formation during Pollen Tube Growth. Ph.D. thesis, University of Illinois, 183 p

Loewus FA (1983) Phytate metabolism with special reference to its *myo*-inositol component. Recent Adv Phytochem 17:173-192.

Southworth D, Dickinson DB (1981) Ultrastructural changes in germinating lily pollen. Grana 20: 29-35.

# Polyamine Biosynthesis in Germinating Apple Pollen

N. BAGNI*, D. SERAFINI-FRACASSINI*, P. TORRIGIANI* AND V.R. VILLANUEVA**

## 1 Introduction

Previous work on apple pollen showed that the genes of rRNA tRNA and probably mRNA were active during germination and that aliphatic polyamines, plant growth substances of higher plants (Bagni et al. 1982; Galston et al. 1982), displayed profiles similar to those of RNA and proteins (Bagni et al. 1981). It has been shown, in fact, that biosynthesis of RNA and polyamines precedes tube emergence (Bagni et al. 1981) and that a polyamine-mediated control of RNAse activity occurs (Speranza et al. 1984).

In order to further investigate the hypothesis of a possible role of polyamines, which are known to be released with proteins into the medium during the progression of germination, in the progamic phase of the fertilization processes (Speranza and Calzoni 1980), in the present study we examine their biosynthesis and the activity of their biosynthetic enzymes during apple pollen germination, subject until now little investigated (Bagni et al. 1981).

## 2 Pollen Culture and Germination

Mature pollen of Malus domestica Borkh. cv. Starkrimson was collected as previously described (Calzoni et al. 1979) from

* Istituto di Botanica, Università di Bologna, Bologna (Italy)
**Institut de Chimie des Substances Naturelles,CNRS, Gif-sur-Yvette (France)

plants grown in experimental plots of Defendi Farm, Altedo (Bologna), and stored in glass tubes at -20°C in the presence of NaOH pellets. Under these conditions viability (Hoekstra and Bruinsma 1975) and germination (Calzoni et al. 1979) were higher than 90% (even two years after the harvest). Germination was carried out as described (Calzoni et al. 1979) using mass culture in 5 cm-wide Petri dishes. For the determination of polyamines and amino acids and of enzyme activity 40 mg samples of pollen were used.

## 3 Polyamine, Amino Acid and Enzyme Determination

Polyamine and amino acid separation and determination were carried out by automatic ion-exchange column chromatography (Adlakha and Villanueva 1980). Ornithine was detected and determined using a DC 6 ($Li^+$ form) Durum column. 185 kBq of L-[U-$^{14}$C]arginine (12.43 GBq/mmol), obtained from Radiochemical Centre, Amersham, were added to the medium of each sample at the beginning of the germination procedure. The incorporated radioactivity was measured in a liquid scintillation counter LS 1800 (Beckman).

Arginine decarboxylase (ADC), ornithine decarboxylase (ODC) and S-adenosylmethionine decarboxylase (SAMDC) activities were measured as described elsewhere with minor modifications (Bagni et al. 1983). For ADC and ODC determination samples of pollen were suspended in a 100 mM Tris-HCl buffer containing 25 µM pyridoxal phosphate, 2.5 mM 1,4-dithiothreitol, 50 µM ethylene diamine tetracetic acid, at different pH values ranging from 7.1 to 8.9, and then disrupted in a Potter-Elvehjem homogenizer. 100 mM Tris buffer was not able to completely buffer the pH in the pollen homogenate, which increased the pH values by about 0.2-0.3 units; higher molarities of Tris buffer were not used to avoid secondary effects. 7.4 kBq of the labelled compounds, respectively L-[U-$^{14}$C]arginine (12.43 GBq/mmol) and L-[U-$^{14}$C] ornithine (10.54 GBq/mmol), were added to each assay tube.

For SAMDC activity determination, samples of pollen were homogenized as described above in a 200 mM Tris-HCl buffer at pH ranging from 7.1 to 8.9; in this case the pollen homogenate decreased the pH values by 0.2-0.3 units. 7.4 kBq of S-adenosyl-L-[carboxyl-$^{14}$C]methionine (2.22 GBq/mmol) were added to each assay tube. In Fig. 3 pH values measured after pollen homogenization are reported. Each sample was repeated five times for each experiment.

## 4 Polyamine Biosynthesis and Enzyme Activity

Since arginine is the main amino acid stored in mature pollen (Bagni et al. 1978) as well as in other plant organs such as tubers (Bagni et al. 1980) and a precursor of putrescine synthesis, an experiment was carried out in which $^{14}$C arginine was utilized to investigate polyamine biosynthesis. In Figs 1 and 2 arginine uptake and incorporation into other compounds related to polyamine biosynthesis and the specific activity of the different polyamines during pollen germination are shown. Arginine uptake incresed till 60 min after the beginning of germination and thereafter decreased during the last period of tube elongation . If we consider the endogenous content of free arginine in mature pollen, about 4.2 mM (Bagni et al. 1978), the uptake of labelled arginine, whose concentration was 0.4 µM in the germination medium, occurred against a concentration gradient. The maximum arginine uptake occurred at 60 min of germination; at this time it is known from the literature that while the free arginine level rapidly drops, bound arginine increases (Bagni et al. 1981); moreover the incorporation of arginine into putrescine showed a sharp increase between 60 and 120 min of germination. The pattern of incorporation of arginine into putrescine and spermine confirms previous data (Bagni et al. 1981); the presence of labelled agmatine and N-carbamylputrescine indicates the presence of the ADC pathway, one of the two ways leading to putrescine synthesis (Slocum et al. 1984). In particular agmatine, the product of arginine decarboxylation, showed a sharp peak at 30 min in correspondence to the rapid synthesis of putrescine. An exogenous supply of agmatine (1 and 10 nmol/mg pollen) in the medium inhibited arginine uptake and enhanced its endogenous accumulation, and the production of N-carbamylputrescine,

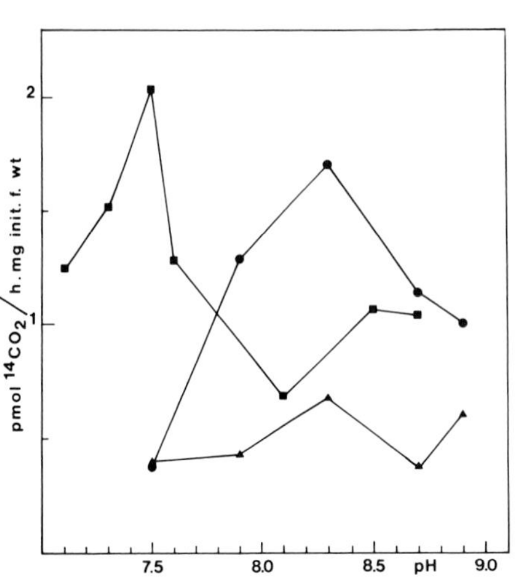


Fig. 1. $^{14}$C arginine uptake (■) and incorporation in ornithine (●), agmatine (▲) and N-carbamylputrescine (◆) during the progression of germination in apple pollen

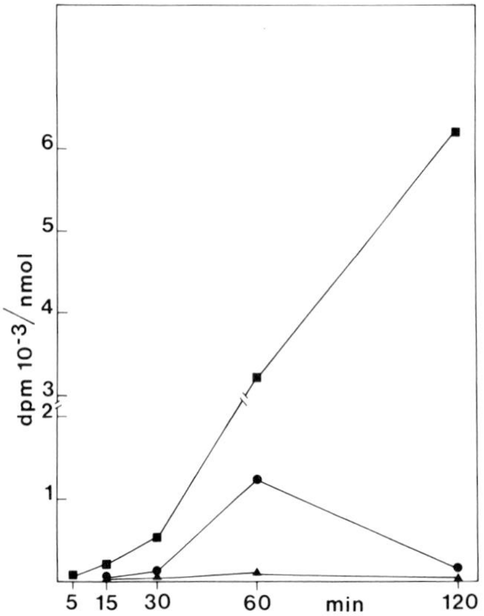

Fig. 2. $^{14}$C arginine incorporation in polyamines during the progression of germination in apple pollen; putrescine ■, spermidine ▲, spermine ●

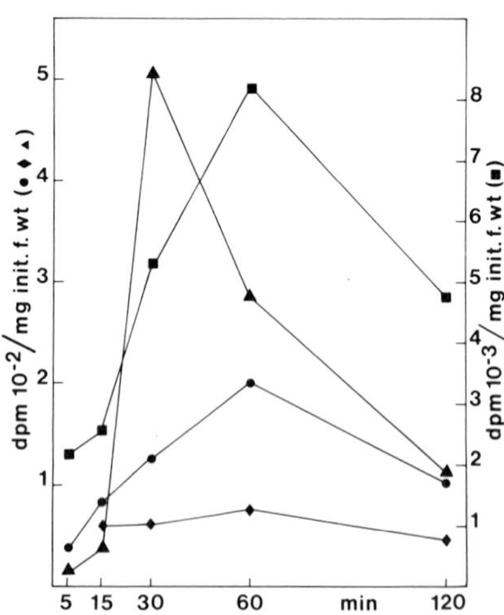


Fig. 3. Arginine decarboxylase (●), ornithine decarboxylase (▲) and S-adenosylmethionine decarboxylase (■) pattern in relation to different pH in mature apple pollen

putrescine and spermidine. However we cannot exclude the possibility of the occurrence of a citrulline decarboxylase activity leading to the formation of N-carbamylputrescine and then putrescine, as reported in few plant systems (Speranza and Bagni 1978). In addition, the presence of labelled ornithine, derived from arginine, indicates an arginase activity that also furnishes the substrate for ODC activity.

In ungerminated apple pollen both ADC and ODC activities were present and exibited an optimum pH at 8.3 (Fig. 3), as also found for other plants such as Helianthus tuberosus (Bagni et al. 1983); also SAMDC activity was detectable with an optimum pH at 7.5 (Fig. 3). Assuming that in the enzyme assays the substrate conditions were as close as possible to the endogenous levels in the cells, the activity of the three enzymes resulted of the same magnitude; at the optimum pH, ADC activity was two-fold higher than ODC activity. Preliminary experiments showed that in germinating apple pollen (30 min) the optimum pH of ADC and ODC was different from that in mature pollen, being shifted to more basic values. The different behaviour with respect to the pH might be due to the involvement of different enzymes in the two physiological stages. Also an enzyme compartmentation in the cell (Torrigiani et al. 1985) can be hypothesized.

The presence of both ADC and ODC, demonstrated in many other plant systems (Slocum et al. 1984), is reported for the first time in pollen. Evidence exists that the two enzymes play a role in different physiological stages of the plant organ or tissue (Bagni et al. 1983; Slocum et al. 1984) suggesting a high degree of regulation of putrescine synthesis.

In conclusion polyamine biosynthesis is very active at the beginning of germination before tube emergence, as well as during the period of tube elongation. This fact might be also related to the release of polyamines into the medium during tube elongation suggesting that polyamines play an important role in fertilization processes probably through their action on macromolecule synthesis.

Acknowledgment. This research was supported by contract from the Consiglio Nazionale delle Ricerche (Italy) in collaboration with Centre Nationale de la Recherche Scientifique (France).

## References

Adlakha RC, Villanueva VR (1980) Automated ion-exchange chromatographic analysis of usual and unusual polyamines J Chromatogr 187: 442-446

Bagni N, Serafini-Fracassini D, Villanueva VR, Adlakha RC (1978) Contenuto in poliammine, etanolammina e arginina nel polline di melo. Ortoflorofrutticoltura Ital 5: 470--476

Bagni N, Malucelli B, Torrigiani P (1980) Polyamines, storage substances and abscissic acid-like inhibitors during dormancy and very early activation of Helianthus tuberosus tuber tissue. Physiol Plant 49: 341-345

Bagni N, Adamo P, Serafini-Fracassini D, Villanueva VR (1981) RNA, proteins and polyamines during tube growth in germinating apple pollen. Plant Physiol 68: 727-730

Bagni N, Serafini-Fracassini D, Torrigiani P (1982) Polyamines in cellular growth processes in higher plants In: Wareing PF (ed) Plant Growth Substances 1982. Academic Press London, pp 473-482

Bagni N, Barbieri P, Torrigiani P (1983) Polyamine titer and biosynthetic enzymes during tuber formation of Helianthus tuberosus. J Plant Growth Regul 2: 177-184

Calzoni GL, Speranza A, Bagni N (1979) In vitro germination of apple pollen. Sci Hortic 10: 49-55

Galston AW, Kaur-Sawhney R (1982) Polyamines: are they a new class of plant growth regulators? In: Wareing PF (ed) Plant Growth Substances 1982. Academic Press London, pp 451-462

Hoekstra FA, Bruinsma J (1975) Respiration and vitality of binucleate and trinucleate pollen. Physiol Plant 34: 221-225

Slocum RD, Kaur-Sawhney R, Galston AW (1984) The physiology and biochemistry of polyamines in plants. Arch Biochem Biophys 235: 283-303

Speranza A, Bagni N (1978) Products of L-$^{14}$C-carbamoylcitrulline metabolism in Helianthus tuberosus activated tissue Z Pflanzenphysiol 88: 163-168

Speranza A, Calzoni GL (1980) Compounds released from incompatible apple pollen during in vitro germination Z Pflanzenphysiol 97: 95-102

Speranza A, Calzoni GL, Bagni N (1984) Evidence for a polyamine-mediated control of ribonuclease activity in germinating apple pollen. Physiol Veg 22: 323-331

Torrigiani P, Serafini-Fracassini D, Biondi S, Bagni N (1985) The compartmentation of polyamines and their biosynthetic enzymes in plant cells. In: Selmeci L, Brosnan ME, Seiler N (eds) Recent Progress in Polyamine Research. Akademiai Kiado Budapest and VNU The Nederlands. In press

# Control of Protein Release from Germinating Pollen

J.F. JACKSON AND R.K. KAMBOJ[1]

The pollen of many angiosperms release proteins during germination and pollen tube extension, which could play a role in the complex interaction between pollen and stigma (Heslop-Harrison, 1975; Pacini, Franchi and Sarfatti, 1983). Some of these proteins at least have been classified as sporophytic or gametophytic, and appear to be released from the pollen grain wall (Heslop-Harrison *et al.*, 1975). It has long been thought that the pollen tubes digest their way through pistil tissue by means of enzymes (Robbers & Floss, 1968; Hsu & Anderson, 1971), an hypothesis which gains support from electron microscopic studies (Anderson & Saini, 1974; Ogunlana *et al.*, 1970).
It is no surprise then that enzyme activity has been shown to be associated with the released proteins (Poddubnaya-Arnoldi *et al.*, 1959; Mäkinen & Brewbaker, 1967; Matousek & Tupy, 1983), while the pollen of at least one species contains enzymes that can break down pectins (Kroh & Loewus, 1978). Several of the proteins have antigenic properties in some species (Heslop-Harrison, Knox & Heslop-Harrison, 1974).

The controls operating on protein release during pollen tube extension must then have important implications for interaction between pollen and style and the rate of growth of pollen tube. It was observed by Stanley & Linskens (1964, 1965) that protein was steadily given up to the medium during *Petunia hybrida* pollen germination, while Kirby and Vasil (1979) investigated washings of Petunia pollen which yielded proteins. It was accepted in all these cases that protein diffuses out passively from the pollen. We describe here further investigations with *Petunia hybrida* pollen which set out to describe the controls operating on protein release. It is suggested that

---

[1] Agricultural Biochemistry Department, Waite Agricultural Research Institute, University of Adelaide, Glen Osmond, South Australia 5064, Australia

protein export is an energy-driven process, and that boron and calcium influence greatly the rate of release, except for newly synthesized protein which is only slightly affected by these inorganic ions.

Methods.

Pollen from *Petunia hybrida* L clone W166H was collected and stored as described previously (Jackson & Linskens, 1979). After 1 h in a chamber at 100 % humidity, pollen was germinated for up to 5 h in an Erlenmeyer flask on a shaking water bath as described previously (Kamboj, Linskens and Jackson, 1984). Total proteins were estimated by the method of Bradford (1976). Labelling of proteins was carried out by adding $^{35}$S-methionine (1390 Ci/mmol) to the germination medium while protein samples for two dimensional gel electrophoresis (O'Farrell, 1975) were precipitated from the spent medium with two volumes of ethanol. Autoradiographs were prepared by exposing the dried gels to x ray film at -15 $^{o}$C.

Results and Discussion.

When *Petunia hybrida* pollen from clone W166H is cultured in 10% sucrose, no germination or tube growth is observed as long as precautions are taken to exclude boron (e.g. use of plastic Erlenmeyer flasks and pipettes). Under these conditions massive amounts of proteins are released to the medium (Table 1), the amounts rising steadily with time. At 3 h, 5 mg protein has been given up by 35 mg pollen. When small amounts of boric acid are added, pollen germination and tube length increases dramatically, while protein release is reduced , but much more slowly (Table 1). Calcium salts also reduce the amount of protein released, however quite high concentrations are required for significant effects, while calcium has no effect on germination and tube growth for this clone of *Petunia hybrida* (Table 1). Similar effects of boron and calcium on protein release from pollen of *Nicotiana tabacum* have been described by Capkova-Balatkova Hrabetova and Tupy (1980).

We have already shown that protein release is energy-dependent in Petunia pollen in the presence of boron (Kamboj, Linskens and Jackson, 1984). As shown in Table 1, "energy poisons" like DNP prevent protein release in the absence of boron also, and similar experiments in the presence of calcium (and absence of boron) show that DNP is equally effective in this case. We conclude that protein release from *Petunia*

Table 1. Effect of various compounds on protein release, germination and tube length after 3 h of culture of *Petunia hybrida* W166H pollen. The basic culture solution contained 35 mg pollen in 5 ml of 10% sucrose and was shaken at 25 °C.

| Compound | Concentration | Protein released (Bradford) | Germination % | Tube length µm ± standard deviation |
|---|---|---|---|---|
| none | - | 0.347 | 0 | 0 |
| $H_3BO_3$ | 0.0001% | 0.274 | 13 | 148 ± 23 |
| $H_3BO_3$ | 0.01% | 0.090 | 35 | 215 ± 70 |
| $Ca(NO_3)_2$ | 1 mM | 0.125 | 0 | 0 |
| $H_3BO_3$+$Ca(NO_3)_2$ | 0.01%+1 mM | 0.053 | 35 | 210 ± 80 |
| $H_3BO_3$+DNP | 0.01%+0.05 mM | 0.015 | 0 | 0 |
| $H_3BO_3$+CHI | 0.01%+100µg/µl | 0.02 | 34 | 75 ± 55 |
| $H_3BO_3$+NEM | 0.01%+0.1 mM | 0.01 | 0 | 0 |
| $H_3BO_3$+IAA | 0.01%+0.5 mM | 0.025 | 15 | 90 ± 51 |
| DNP | 0.05 mM | 0.02 | 0 | 0 |

DNP, dinitrophenol; CHI, cycloheximide; NEM, N-ethylmaleimide; IAA, indoleacetic acid.

*hydrida* pollen is an energy-driven process, which can be reduced, but not eliminated, by boron and calcium. Only the so-called "energy-poisons" can virtually eliminate the release, although very low amounts of protein release have been seen with pollen stored for several years and which was no longer viable.

Several other compounds inhibit protein export from *Petunia hybrida* pollen. Auxins exert a considerable effect, although only at relatively high concentrations. At 0.5 mM, IAA and 2,4-D both show significant inhibition. Another type of compound found to inhibit protein release is the sulphydryl-binding reagent, such as N-ethylmaleimide (NEM, Table 1), or phenylarsine oxide, which is equally effective. In the germinating barley embryos, active peptide transport is also severely inhibited by these thiol reagents (Smith and Payne, 1983). Taken together with our own observation that thiol reagents inhibit energy-linked transport of uridine and some other nucleosides into Petunia pollen (Kamboj and Jackson, 1984), this fact tends to support the notion that protein release is energy-driven.

To explain the above facts, we suggest that protein release from *Petunia hybrida* pollen is not merely a passive diffusion, but rather goes by way of Golgi-derived secretory vesicles which make their way to the plasma membrane and there give up proteins to the outside after vesicle fusion and exocytos s. To complete the cycle of membrane flow, fragments of membrane released by an endocytosis process, migrate back to the Golgi area. These processes are known to be involved in pollen tube extension (Picton and Steer, 1983). As known from work with animal cells, several steps in this process are energy-dependent, including packaging of proteins into secretory granules and exocytosis itself. Many of these proteins exported may be destined for the cell wall that is built up during pollen tube extension; we would argue that boron is needed for this process, so that in the absence of boron, proteins normally inserted into the wall are released to the culture medium. However, protein export cannot be completely stopped, and so we suggest that much of the protein released during exocytosis as the pollen tube is extended, escapes outside the tube and interacts with the tissue of the pistil, helping to make way for further pollen tube elongation.

The above hypothesis may well apply to the bulk of the protein stored in the pollen grain at anthesis. However, during pollen germination, a small amount of protein is newly synthesized. We find that this too is exported; however it cannot be controlled to the same extent as stored protein, so that the amount secreted is independent of boron and calcium added to the culture medium. It is however stopped, as expected, by the energy poisons, and by cycloheximide. Cycloheximide also inhibits bulk (stored) protein release (Table 1), suggesting a role for some protein synthesis in the secretion process involving stored proteins as well.

References

Anderson JA, Saini MS (1974) Tetrahedron Letter, 2107.
Bradford MM (1976) Anal Biochem 72,248.
Capkova-Balatkova V, Hrabetova E, Tupy J (1980) Biol Plant 22,294.
Heslop-Harrison J (1975) Proc Roy Soc B 190,275-299.
Heslop-Harrison J, Knox RB, Heslop-Harrison Y (1974) Theor Appl Gen 44, 133.
Heslop-Harrison J, Knox RB, Heslop-Harrison Y, Mattson, O (1975) In: Duckett SG, Racey PA (eds) The Biology of the Male Gamete, Academic Press, London, pp 188-202.
Hsu JC, Anderson JA (1971) Biochim Biophys Acta 230,518.
Jackson JF, Linskens HF (1979) Mol Gen Genetics 176,11.
Kamboj RK, Linskens HF, Jackson JF (1984) Ann Bot 54,647.
Kamboj RK, Jackson JF (1984) Plant Phys 75,499.
Kirby EG, Vasil IK (1979) Ann Bot 44,361.
Mäkinen Y, Brewbaker JL (1967) Phys Plant 20,477.
Matousek J, Tupy J (1983) Plant Sci Lett 30,83.
O'Farrell PH (1975) J Biol Chem 250,4007.
Ogunlana EO, Wilson BJ, Tyler VE, Ramstad E (1970) Chem Comm 775.
Pacini E, Franchi GG, Sarfati G (1983) Ann Bot 47,405.
Picton JM, Steer MW (1983) J Cell Sci 63,303.
Poddubnaya-Arnoldi VA, Tsinger NV, Petrovskaya IP, Polunina NN (1959) Recent Adv Bot 1,682.
Robbers JE, Floss HG (1968) Arch Biochem Biophys 126,967.
Smith DJW, Payne JW (1983) Febs Lett 160,25
Stanley RG, Linskens HF (1964) Nature 203,542.
Stanley RG, Linskens HF (1965) Phys Plant 18,47.

# Metabolic Role of Boron in Germinating Pollen and Growing Pollen Tubes

RUPINDER J.K. SIDHU AND C. P. MALIK[1]

An essential role of boron has been persistently confirmed for angiosperms, especially in plants which possess well-developed xylem. Based on the available studies, the following putative major roles for this microelement may be assigned: I, at the whole plant level (control of growth and differentiation; II. at the physiological level(regulation of membrane permeability absorption and translocation of sugar), and III. at the biochemical level(control of enzymes concerning metabolism of carbohydrates, polyphenols and lignin, auxin and nucleic acids biosynthesis)(Lewis, 1980). The primary and secondary roles of boron are not clearly delineated and an integrated picture in this regard is still illusive.

Present studies utilize pollen and pollen tubes as a convenient system for boron investigations since it is non-photosynthetic, lacks lignin etc. A direct relation between boric acid concentration(up to 10 ug/ml) and enhancement of tube length was noticed. This is attributed to enhancement in tube growth rate. Boron had a continuous effect on growth rate. It caused early appearance of pollen tubes and stimulated growth in the initial phases while the later phases were not affected. Addition of boron affected the tube growth in the first 60 min, whereafter its requirement diminished. Data from pre- and post-treatment indicated that initial supply of boron for 60 min induced sufficient stimulation compared with 90 or 105 min post-treatment. Boron also affected fresh weight, dry matter and moisture percentage.

Ethrel could replace the boron effect, but together did not produce a synergestic or additive effect. Compared with uracil, addition of thymine to the culture medium, stimulated tube length suggesting its role in nucleotide and nucleic acid metabolism by increased precursor availability.

The uptake of (U-$^{14}$C) sucrose as affected by boron was

[1]Department of Botany, Punjab Agricultural University, Ludhiana India.

studied at three stages of tube growth. First, $^{14}C$-sucrose uptake with increasing boron concentrations was not linear and gave no correlation as a function of tube growth. Second, boron reduced $^{14}C$-sucrose uptake in the later phases of tube growth. Growing pollen tubes metabolize actively and require considerable amount of sugars for respiration and tube growth. Maximum $^{14}C$-sucrose uptake was noticed during the period of activation rather than tube growth. In fact, boron addition at 10-20; 50-60 and 110-120 min decreased the uptake implying that boron may be involved in processes other than absorption.

Boron greatly reduced the leakage of carbohydrates, amino acids and proteins from the soaked and growing pollen. Boron possibly regulates the permeability and integrity of membranes.

The effect of boron on the activity patterns of glycosidases, acid phosphatase and invertase(cytoplasmic and wall fractions) was studied. Boron did not affect the cytoplasmic glycosidases. The present studies do not implicate the role of boron through glycosidases. We have studied the activity of pentose phosphate pathway by two methods: first, by the estimation of glucose-6-P-DH activity; second, by calculating the ratio of $CO_2$ evolved from 1-$^{14}C$ or 6-$^{14}C$-glucose; and third, through cytochemical method by seeking a distinction between the levels of utilization of NADPH for electron transport involved with hydroxylation and biosynthetic activity(Altman, 1972). In the first 15 min, PMS addition did not significantly alter the activity in the absence of boron. Addition of boron, enhanced G-6-PDH activity enormously at all the stages when PMS was added. This points towards enhanced biosynthetic processes in the presence of boron. Boron decreased the $C_6/C_1$ ratio and the decrease was more in the initial 0-15 min of incubation (activation) than later phases. It appears that the carbon flow through PPP is stimulated at the expense of EMP pathway.

<u>Amaryllis</u> pollen has high amount of stored malate. The level of G-6-PDH is controlled <u>in vivo</u> by NADP:NADPH ratio. Conceivably, enhancement of PPP reduces the level of malate decarboxylation or <u>vice versa</u> by controlling this ratio. The amounts of NADP-malic enzyme, malate and $C_6/C_1$ ratios indicate that the addition of boron stimulates the production of NADPH via PPP while in the absence of boron, NADPH is supplemented

by the decarboxylation of malate by NADP-malic enzyme.

Boron was also noticed to alter the activities of irreversible enzymes of glycolysis and oxidative pentose phosphate pathway(i.e. hexokinase, phosphofructokinase, pyruvic kinase, G-6-P dehydrogenase). Apparently, boron interacts in same way to control the activities of these enzymes.

Boron adversely affected the non-photosynthetic $CO_2$ fixation as revealed by reduced levels of PEP-carboxylase, malate dehydrogenase as well as in vivo $^{14}CO_2$ fixation. Amaryllis pollen has 14% of stored lipids and enhancement of glyoxylate pathway by boron(as evidenced by increased isocitrate lyase activity) may explain the high level of malate.

Boron in the medium reduced the capacity for starch synthesis which is attributed to the rapid rate of glucose utilization in other pathways. Boron increased ATP content at all the stages while ADP and AMP were not significantly altered. In the activation phase, ADP and AMP are utilized for phosphorylation to ATP.

Boron increased the IAA content of Amaryllis pollen possibly by regulating IAA oxidase activity. It is also probable that enhanced PPP by boron provides more erythrose-4-P which reacts with PEP leading to the synthesis of IAA precursor tryptophan.

In higher plants, the role of boron in the production of phenols and lignification is well documented. In pollen tubes, which constitute a non-lignified system, boron appears to act differently. Addition of boron decreased the total phenols in Amaryllis pollen and stimulated the activity of catechal oxidase markedly(Fig. 1). The role of this enzyme in the formation of superoxide radical, required for the formation of ethylene from methanol, has been suggested. In Amaryllis pollen this role appears to be enacted.

Based on large number of experiments, we have attempted to assess the precise metabolic role(s) of boron. Some of the inferences appear to converge towards some of conclusions(see Fig. 1). Thus, lack of boron caused high accumulation of phenolics, low catechol oxidase, partitioning of carbon plant via glycolytic and PPP, etc. Pollen tube walls contain B-1-3-linked glycan or callose and during tube growth through the style, there is production of ethylene (Lee et al., 1978). It may not be

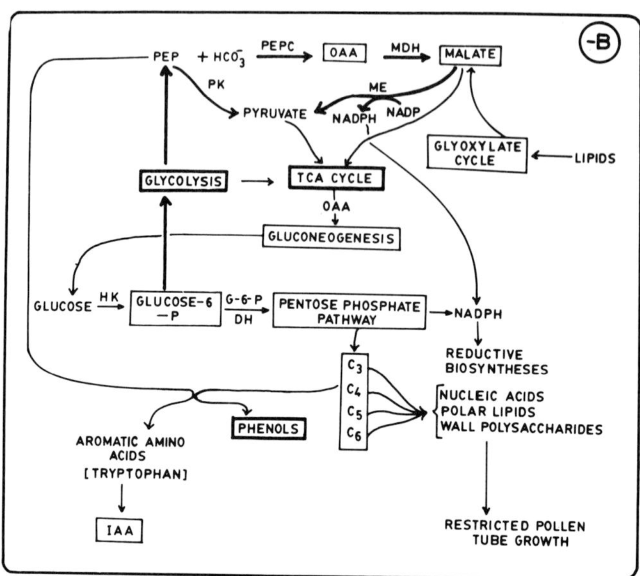


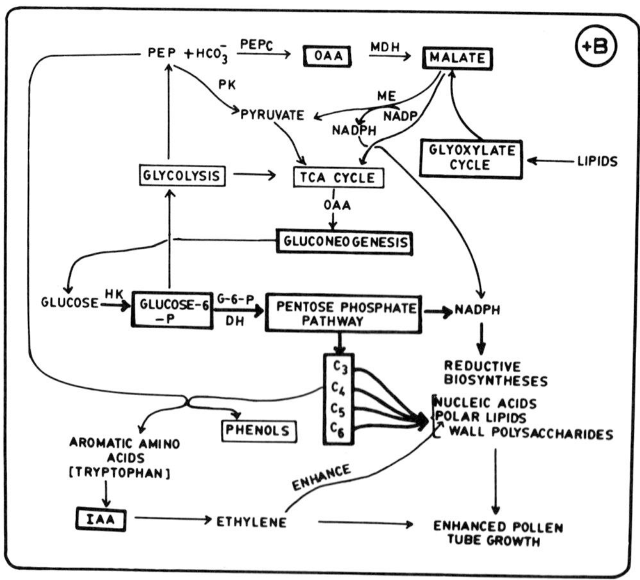


Fig.1. Schematic representation of possible roles of Boron in the germinating *Amaryllis* pollen at 60 min. Thick lines indicate higher operation of the particular pathway relative to the thin lines.

possible to involve boron as an essential element for lignin biosynthesis since pollen tube lack it. Our experimental data pertaining to pollen do not support the thesis of Lewis(1980) that "boron is required by only those plants in which there is well developed lignified super apaplast - the xylem".

Our studies indicate that boron has a primary role in metabolic events concerning shifts in carbohydrate oxidation (Fig.1).

Since G-6-P is oxidized by both EMP and PPP pathways addition of boron somewhat regulates the extent of carbon flux between the two pathways. EMP-TCA-oxidative phosphorylation sequence is the principal route for ATP production whereas PPP produced the reducing power(NADPH) in conjunction with the NADP malic enzyme in the cytocol. Boron increases the carbon flux through the PPP at the expense of EMP, decreases carboxylation of PEP and pyruvate kinase whereas it increases isocitrate lyase in the glyoxylate cycle concerned with the production of malate which can be fed to the mitochondria for sustaining respiration. Boron also decreased the total phenols but markedly stimulated the catechol oxidase activity resulting in the production of o-diphenols which are known to inhibit IAA oxidase and thus building up the level of auxins. Additionally, the enhanced synthesis of tryptophan via activation of PPP could lead to the increased endogenous synthesis of auxins. It is probable that increased production of IAA leads to the production of ethylene which reportedly enhances pollen tube growth (Stanley and Linskens, 1974). The inhibition of nonphotosynthetic $CO_2$ fixation by $C_2H_4$ is plausible in view of the fact that the osmoticum is readily available from the medium and whatever amount of malate that is formed is metabolized either in the mitochondria or in the cytoplasm by NADP-malic enzyme producing NADPH. The regulation of tube extension by ethylene gets further support from the observation that the effect of boron in eliciting increased pollen tube length is replaced by ethylene. Ethrel has been shown to increase polar lipid synthesis in germinating pollen(Bhandal and Malik, 1981) to achieve active tube extension. The enhancement of PPP by boron besides producing NADPH also produces the carbon phosphorylated sugars which are precursors to the wall pentoses and the nucleic acids.

References

Altman FP (1972) Quantitative dehydrogenase histochemistry with special reference to the pentose shunt dehydrogenases. In "Progress in Histochemistry and Cytochemistry", Vol 4. Gustav Fischer Verlag Stuttart, Portland. Oreg. U.S.A.

Bhandal IS, Malik CP (1981) Total and polar lipid biosynthesis during growth of Crotalaria juncea pollen tubes. Phytochem 20: 429-432

Lee CW, Erickson HT, Janaick J (1978) Chasmogamous and cleistogamous pollinations in Salpiglossis sinuata. Physiol Plant 43: 225-230.

Lewis DH (1980) Boron lignification and the origin of vascular plants. A unified hypothesis. New Phytol 84: 209-229

Stanley RG, Linskens HF (1974) Pollen: Biology, biochemistry management" Springer-Verlag, Berlin, Beidelberg, New York

Table 1 Boron effect on relative participation of glycolysis and PP pathway. Data based on the $^{14}CO_2$ release by pollen (10 mg) germinated in G-1-$^{14}$C and G-6-$^{14}$C.

| Medium | Incubation period (min) | G-1-$^{14}$C (PPP) | G-6-$^{14}$C (EMP) (dpm) | $C_6/C_1$ ratio |
|---|---|---|---|---|
| PE | 15 | 1163 | 3788 | 3.26 |
| PE + B | " | 2946 | 705 | 0.24 |
| PE | 60 | 4261 | 6299 | 1.48 |
| PE + B | " | 6182 | 5594 | 0.91 |
| PE | 120 | 1728 | 552 | 0.32 |
| PE + B | " | 6417 | 2182 | 0.34 |

Table 2 Boron affecting the activity of glucose-6-phosphate dehydrogenase(µg Neotetrazolium/mg pollen/h) with and without PMS.

| Incubation time(min) | PMS | PE | PE + B |
|---|---|---|---|
| 15 | - | 64.5 | 42.6 |
| " | + | 69.0 | 148.5 |
| 60 | - | 52.5 | 75.0 |
| " | + | 132.0 | 204.0 |
| 180 | - | 42.0 | 48.0 |
| " | + | 159.0 | 159.0 |

# Complex Carbohydrates at the Interacting Surfaces during Pollen-Pistil Interactions in Nicotiana alata

P.J. Harris, A.L. Rae, A.M. Gane, A. Gell, A. Bacic, G-J. van Holst, M.A. Anderson and A.E. Clarke[1]

## 1 Introduction

As part of a broad programme to study the molecular basis of fertilization, we have investigated the complex carbohydrates at the interacting surfaces of the pollen tube and pistil in Nicotiana alata (Solanaceae), an ornamental tobacco. This plant exhibits gametophytic incompatibility and self-incompatibility genes (S-genes) operate to prevent in-breeding. Pollen behaviour is controlled by its own genotype and pollen tubes are arrested if the allele carried by the pollen matches one of the two alleles in the pistil tissue (for review see de Nettancourt 1977). In a compatible mating, pollen grains germinate on the surface of the stigma and produce pollen tubes which penetrate the surface of the stigma and grow through the central solid transmitting tissue of the style to the ovary. As the pollen tube grows, callose plugs are laid down at regular intervals cutting off the growing tip from the spent pollen grains; this gives the pollen tube a ladder-like appearance. At maturity, the transmitting tissue cells are elongated and connected by plasmodesmata at their transverse walls into vertical files which are separated by secreted intercellular matrix material through which the pollen tubes grow. In contrast, in an incompatible or self mating, growth of the pollen tube is arrested just below the stigma, in the top 2-3 mm of the style. The tips of the inhibited pollen tubes swell and sometimes burst; there is also a characteristic deposit of callose immediately behind the tip.

---

[1] Plant Cell Biology Research Centre, School of Botany, University of Melbourne, Parkville, Victoria 3052, Australia

## 2 Complex Carbohydrates of the Pollen Tube Cell Wall

In spite of the important role played by the pollen tube, chemical studies on the cell-wall polysaccharides of pollen tubes are restricted to only a few species (see Harris *et al.* 1984 for review). In each case, glucose was the most abundant monosaccharide, with the second most abundant being either galactose or arabinose. We have investigated the composition of the cell-wall polysaccharides of *N. alata* pollen tubes grown *in vitro* (Rae *et al.* 1985). The predominant neutral monosaccharides in acid hydrolysates of the cell walls were glucose and arabinose; other monosaccharides were present in small amounts. Uronic acids, as determined by the method of Blumenkrantz and Asboe-Hansen (1973), comprised 2.8% w/w of the cell wall. The positions of the linkages between the component neutral monosaccharide residues were determined by methylation analysis. The results of this analysis and of cytochemical studies are consistent with, but not proof of, the presence of two major polysaccharides in the cell wall: a (1→3)-β-D-glucan with branch points through C(O)6 and C(O)2 (callose) and a (1→5)-α-L-arabinan with branch points through C(O)2. The results also indicate the presence of some cellulose.

Callose is usually detected as material which fluoresces when treated with aniline blue. The fluorochrome is an impurity in aniline blue and has been synthesized chemically. Treatment of germinated pollen of *N. alata* with the synthetic aniline blue fluorochrome caused the pollen tube cell walls and the plugs present in the pollen tubes to fluoresce intensely. However, the cell wall at the tips of the pollen tubes did not fluoresce. The synthetic fluorochrome is primarily specific for (1→3)-β-D-glucans, although some (1→3),(1→4)-β-D-glucans also bind the fluorochrome (Evans *et al.* 1984; Stone *et al.* 1984). Thus the fluorescence induced by this fluorochrome in the cell wall and plugs of the pollen tubes, taken together with the linkage analysis data, is consistent with the presence of a (1→3)-β-D-glucan. The absence of fluorescence at the pollen tube tip indicates the absence of (1→3)-β-D-glucan at the tip.

Arabinans with a backbone of (1→5)-α-L-linked arabinofuranosyl residues and with branching through C(O)2 and/or C(O)3 have been isolated from the pectic fraction of vegetative cell walls of a number of dicotyledons (Darvill *et al.* 1980). Insight into the location of the arabinan within the pollen tube wall of *N. alata* is given by immunoelectron microscopy of sections taken from pollen tubes grown both *in vivo* and *in vitro*. Sections were treated with mouse monoclonal antibody directed to terminal α-L-arabinofuranosyl residues (Anderson *et al.* 1984) and then with colloidal gold-labelled goat anti-mouse IgG. Gold particles were found on the outer fibrillar wall layer, but not the inner electron-lucent wall layer which resembles callose in other situations (Fig. 1a and b). Although the two-layered wall is seen in both compatible and incompatible pollen tubes, it is more distinct in the thicker walls of incompatible tubes.

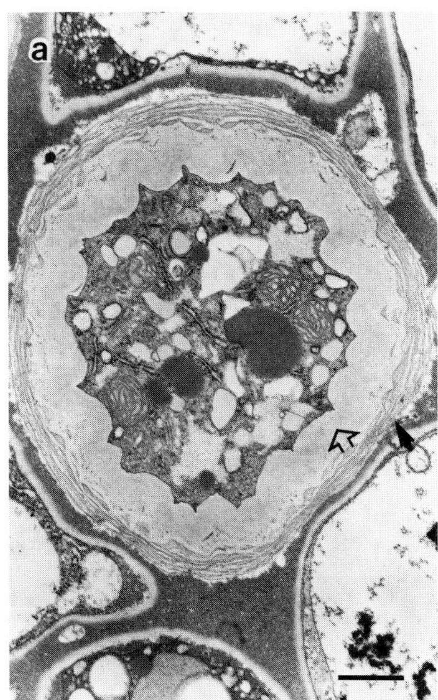


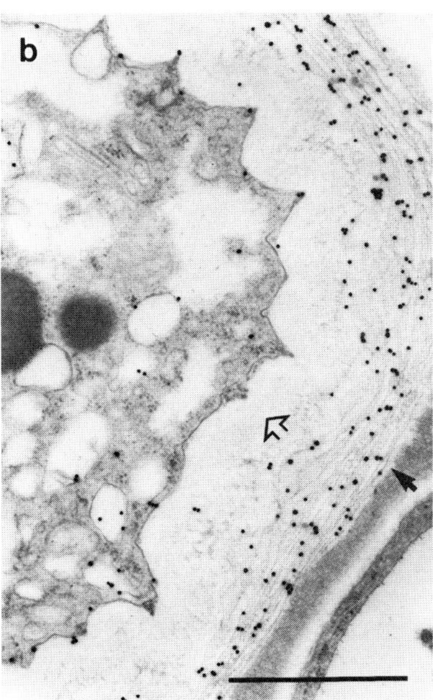


Fig. 1 a,b. Electron micrographs of transverse sections of an incompatible pollen tube in a style, 4-5 mm below the stigma surface, 12 h after pollination. a: Section stained with uranyl acetate and then lead citrate. b: Section treated with mouse monoclonal antibody directed to α-L-arabinofuranosyl residues, then with colloidal gold-labelled goat anti-mouse IgG, followed by uranyl acetate.

The open arrows show the inner electron-lucent wall layer and the solid arrows show the outer fibrillar wall layer. Bars = 1 μm. Figure prepared by Ms Ingrid Bönig.

## 3 Complex Carbohydrates of the Pistil

Arabinogalactan-proteins (AGPs) are the dominant secreted high molecular weight material in the intercellular matrix of the transmitting tissue of the style through which the pollen tubes grow. These proteoglycans account for at least 60% of the total soluble high molecular weight carbohydrate in the style of _N. alata_ (Gell _et al._ 1985). A number of proteins and glycoproteins are also present in the matrix which include S-genotype-associated glycoproteins. The nature of these is discussed in Clarke et al. (1985). AGPs have been found in the styles and stigmas of flowers of many plant families (Hoggart and Clarke 1984). However, they are not restricted to these tissues, but are of widespread occurrence in many plant tissues and secretions (Clarke et al. 1979; Fincher et al. 1983). AGPs contain a high proportion of carbohydrate which is covalently linked to a protein backbone rich in hydroxyproline, serine and alanine. The carbohydrate moiety consists mainly of galactosyl and arabinosyl residues. It has a (1→3)-linked β-D-galactan backbone, branched through C(O)6 to (1→6)-β-D-galactan side chains which are terminated mainly by α-L-arabinofuranosyl and β-D-galactopyranosyl residues. Sections stained with the β-glucosyl Yariv reagent (Yariv et al. 1962), which binds to and precipitates many AGPs (Jermyn and Yeow 1975), indicate that the AGPs are localized in the intercellular matrix material of the style transmitting tissue of N. alata. This localization has been confirmed by immunoelectron microscopy using monoclonal antibodies directed to terminal α-L-arabinofuranosyl and β-D-galactopyranosyl residues (Sedgley et al. 1985).

We have isolated AGPs from extracts of N. alata stigmas (genotypes $S_2S_2$ and $S_3S_3$) by affinity chromatography using the galactosyl-specific IgA myeloma antibody J539 (Glaudemans 1975) and subsequent fractionation by gel filtration. The AGPs from both genotypes contain arabinose and galactose as the major monosaccharides in a ratio of approximately 1:2 respectively, low levels of glucuronic acid (0.5-1.0% w/w) and are associated with a small amount of protein which has a high content of hydroxyproline (Gell et al. 1985).

The amount and concentration of AGP in the style and stigma of N. alata at different stages of development has been measured using a new, sensitive method (van Holst and Clarke 1985a) which is a modification of the single radial immunodiffusion technique described by Mancini et al. (1965) and relies on the specific interactions of AGPs with the β-glucosyl Yariv reagent. In the stigma, the concentration of AGP increases between petal formation and flower maturity, whereas in the style, the concentration of AGP remains constant throughout flower development. The AGPs in the style and stigma of N. alata at different stages of development have also been separated and compared using a new crossed-electrophoretic method. This method is related to crossed immunoelectrophoresis and AGPs separated electrophoretically in the first direction are then run electrophoretically into a gel containing β-glucosyl Yariv reagent in the second direction (van Holst and Clarke 1985b). Using this method, it has been shown that the stigma and style contain different families of AGPs, the proportions of which change during development (Gell *et al.* 1985). However, the physiological role of AGPs in the style and stigma is unknown. It is possible that they provide carbohydrate precursors for the growing pollen tube cell wall (Loewus and Labarca 1973), or they may be implicated in the expression of identity of individual tissues or cell types (Clarke *et al.* 1979). The relationship, if any, between the style and stigma AGPs and the *S*-allele-associated glycoproteins is unknown.

The intercellular matrix, through which the pollen tubes grow, originates from the cells of the transmitting tissue and is secreted through the walls of these cells. We have, therefore, begun a study of the cell walls of the transmitting tissue cells. A preliminary analysis has been carried out on cell walls isolated from whole styles (Gane *et al.* 1985). This indicated that cellulose and xylans are the dominant polysaccharides. However, these are probably mainly located in the thick, lignified walls of the epidermal cells. A method is currently being developed for the isolation of walls of transmitting tissue cells so that we can establish the wall structure.

The way in which products of identical *S*-alleles carried in the pollen and the style interact to cause arrest of pollen tube growth within the style is not understood. Possible mechanisms have been discussed recently (Harris *et al.* 1984; Clarke *et al.* 1985).

## References

Anderson MA, Sandrin MS, Clarke AE (1984) A high proportion of hybridomas raised to a plant extract secrete antibody to arabinose or galactose. Plant Physiol 75: 1013-1016.

Blumenkrantz N, Asboe-Hansen G (1973) New methods for quantitative determination of uronic acids. Anal Biochem 54: 484-489.

Clarke AE, Anderson RL, Stone BA (1979) Form and function of arabinogalactans and arabinogalactan-proteins. Phytochemistry 18: 521-540.

Clarke AE, Anderson MA, Bacic A, Harris PJ, Mau S-L (1985) Molecular basis of cell recognition during fertilization in higher plants. J Cell Sci (in press).

Darvill AG, McNeil M, Albersheim P, Delmer DP (1980) The primary cell walls of flowering plants. In: Tolbert NE (Ed) The Biochemistry of Plants, Vol. 1: ("The Plant Cell"). Academic Press, pp 92-162.

Evans NA, Hoyne PA, Stone BA (1984) Characteristics and specificity of the interaction of a fluorochrome from aniline blue (Sirofluor) with polysaccharides. Carbohydr Polymers 4: 215-230.

Fincher GB, Stone BA, Clarke AE (1983) Arabinogalactan-proteins: structure, biosynthesis and function. Ann Rev Plant Physiol 34: 47-70

Gane AM, Harris PJ, Clarke AE (1985) Composition of the cell walls of styles of Nicotiana alata (Otto and Link). In preparation.

Gell A, Bacic A, Clarke AE (1985) Arabinogalactan-proteins from Nicotiana alata styles. In preparation.

Glaudemans CPJ (1975). The interaction of homogeneous, murine myeloma immunoglobulins with polysaccharide antigens. Adv Carbohydr Chem Biochem 31: 313-346.

Harris PJ, Anderson MA, Bacic A, Clarke AE (1984) Cell-cell recognition in plants with special reference to the pollen-stigma interaction. Oxford Surveys of Plant Molecular and Cell BIology 1: 161-203.

Hoggart RM, Clarke AE (1984) Arabinogalactans are common components of angiosperm styles. Phytochemistry 23: 1571-1573.

Jermyn MA, Yeow YM (1975) A class of lectins present in the tissues of seed plants. Aust J Plant Physiol 2: 501-531.

Leowus F, Labarca C (1973) Pistil secretion product and pollen tube wall formation. In: Loewus F (ed) Biogenesis of Plant Cell Wall Polysaccharides. Academic Press, pp 175-193.

Mancini G, Carbonara AO, Heremans JF (1965) Immunochemical quantitation of antigens by single radial immunodiffusion. Immunochemistry 2: 235-254.

de Nettancourt D (1977) Incompatibility in Angiosperms. Springer-Verlag.

Rae AL, Harris PJ, Bacic A, Clarke AE (1985) Composition of the cell walls of Nicotiana alata (Link and Otto ) pollen tubes. Planta (in press).

Sedgley M, Blesing MA, Bonig I, Anderson MA, Clarke AE (1985) Localization of antigenic glycoconjugates in styles of Nicotiana alata, an ornamental tobacco. Protoplasma (in press).

Stone BA, Evans NA, Bonig I, Clarke AE (1984) The application of Sirofluor, a chemically defined fluorochrome from aniline blue, for the histochemical detection of callose. Protoplasma 122: 191-195.

van Holst G-J, Clarke AE (1985a) Quantification of arabinogalactan-protein in plant extracts by single radial gel diffusion. Anal Biochem (in press).

van Holst G-J, Clarke AE (1985b) Organ-specific arabinogalactan-proteins of Lycopersicon peruvianum (Mill) demonstrated by crossed electrophoresis. Plant Physiol (submitted).

Yariv J, Rapport MM, Graf L (1962) The interaction of glycosides and saccharides with antibody to the corresponding phenylazo dyes. Biochem J 85: 383-388.

# Biphasic pollen tube growth in Plumbago zeylanica

SCOTT D. RUSSELL[1]

## 1 Introduction

Tricellular pollen is typified by complex recognition requirements, rapid pollen tube growth, and high respiratory rates, whereas bicellular pollen has less complicated recognition processes, slower pollen tube growth and relatively lower respiratory rates (Hoekstra, 1983). In a comparison of bicellular and tricellular pollen tube growth rates, Mulcahy and Mulcahy (1983) reported that in vivo pollen tube growth rates in bicellular pollen were clearly biphasic, but in tricellular pollen growth was monophasic. In biphasic pollen, an initial slow period of pollen tube growth occurred that was not accompanied by the formation of callose plugs. However, coincident with the formation of the sperm cells, the physiology of the pollen tube apparently changed dramatically, resulting in increased growth rates, the creation of callose plugs along the length of the pollen tube, and much higher metabolic rates. Apparently, the metabolic requirements of bicellular pollen also change at this time, as pollen tubes grown in vitro cease growth entirely at this time and do not form sperm cells. Mulcahy and Mulcahy (1983) suggested that the physiology of tricellular pollen was directly comparable to the second phase of bicellular pollen growth and inferred an evolutionary significance for this comparison.

The present study reports biphasic pollen tube growth in the tricellular and evolutionarily advanced species Plumbago zeylanica, a plant that has been the subject of numerous previous studies and for which the organization of the sperm cells is particularly well known (Russell and Cass, 1981, 1983; Russell, 1984). This plant displays a strongly biphasic pattern of pollen tube growth despite its production of tricellular pollen. This form of biphasic pollen tube growth, however, appears not to relate to specific changes in pollen physiology, but to changes in the structure of the stylar transmitting tissue in which the pollen elongates.

## 2 Materials and Methods

Plants of Plumbago zeylanica L. were grown at 18°-23°C with 16 hr days in growth chambers. Flowers were emasculated prior to anthesis and pollinated generously with freshly opened anthers from other flowers.

[1] Dept. of Botany and Microbiology, Univ. of Oklahoma, Norman, OK 73019

This resulted in between 10 and 25 pollen grains adhering to each of 5 stigma lobes and reduced flower-to-flower variation in pollen tube growth rates. Stigmas, styles and ovaries were collected and fixed 2, 3, 4, 6, and 8 hr after pollination using chemical and physical fixation.

Chemically fixed materials were immersed at room temperature in 3% glutaraldehyde in 0.067M phosphate buffer (pH 6.8) for 6 to 8 hr. Tissue was rinsed in buffer and fixed in 2% osmium tetroxide in the same buffer, dehydrated in ethanol followed by propylene oxide, and embedded in low viscosity resin (Russell, 1983).

Physical fixation was conducted by immersing selected tissues in 12% methylcyclohexane in isopentane cooled to near liquid nitrogen temperatures (Russell and Cass, 1981). Dehydration was conducted in acetone at -55°C using acetone for 3 hr, followed by fixation in 2% osmium tetroxide in acetone for 24 hr and several changes of acetone for an additional 48 hr (Browning and Gunning, 1977). Tissue was then gradually warmed to room temperature. Acetone was replaced with low viscosity resin, and the tissue was embedded. Pollen tube lengths were calculated using sectioned material. Stylar squashes similar to those prepared by Mulcahy and Mulcahy (1983) were inadequate for visualizing pollen tubes in _P. zeylanica_.

Material was sectioned at 2-6 um on dry glass knives using a Reichert Ultracut, dried on gelatin-coated slides (Jensen, 1962), mounted with glycerol and observed using a Leitz Dialux 20 equipped with differential interference contrast microscopy (DIC). Selected sections were stained for insoluble carbohydrates using the periodic acid-Schiff's (PAS) reaction (Jensen, 1962) and photographed using brightfield microscopy.

Quantitative data was obtained by either direct measurement of cells using phase contrast microscopy or photomicrographs. Cell dimensions in the transmitting tissue were obtained using material stained with the PAS reaction for insoluble carbohydrates on freeze-substituted material (Russell and Cass, 1981) and photographed using brightfield microscopy. All statistics represent the means followed by ± the standard error of the mean.

## 3 Results

Germination of the pollen occurs at 20-30 min following pollination and results in the formation of an aggressively growing pollen tube (Fig. 1). During the first hours, the pollen tube is narrow and pollen tube growth is extremely rapid (Tables 1 and 2). It is not unusual during this phase for the vegetative nucleus to lead the two sperm cells by up to 60 μm (Russell and Cass, 1981). During the first 2-1/2 hr of growth, the pollen tube elongates to a length of up to 23,000 μm (Table 1) at average speeds as rapid as 212 μm/min (Table 2). The volume rate of growth may be up to 4000 $\mu m^3$/min (Table 2) and remains nearly consistently at this high rate until the pollen tube is within 800 μm of the uniovulate ovary.

At this time, the rapidity of previous growth decreases to less than 5 μm/min (Table 2) and the tip of the pollen tube swells to an average of over 3 times its previous diameter (Table 1). From its location at 3 hr after pollination to the site where the sperm cells are released from the successful pollen tube is about 1000 μm, a distance which requires typically another 5-1/2 hr for the pollen tube to travel (Table 1; Russell, 1983). The pollen tube reaches its widest tip diameter about 4 hr after pollination and its speed of growth slows to a mean of 214 μm/hr

(Table 2). Although the speed of pollen tube growth rises slightly to a mean of 241 µm/hr at 6 hr after pollination, this increase is apparently an effect of a decrease in the diameter of the tube. In terms of the volume growth rate, expressed as volume of tube added per minute, growth becomes progressively slower throughout development (Table 2). Volume growth reaches a low of less than 60 $\mu m^3$/min at 8 hr after pollination.

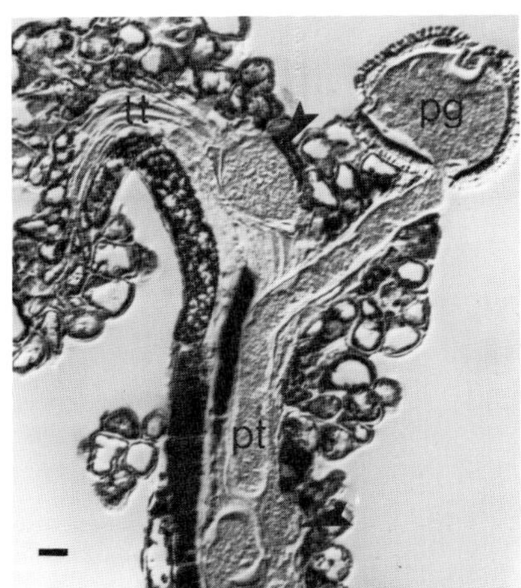


**Figure 1.** DIC of median longitudinal section of the pollinated stigma of P. zeylanica, showing a pollen grain (pg), tube (pt) and discharged pollen cytoplasm (arrowheads) of an abortive tube. In upper left, the reflexed tip of the stigma and thin layer of transmitting tissue (tt) are visible. Scale bar = 20 µm.

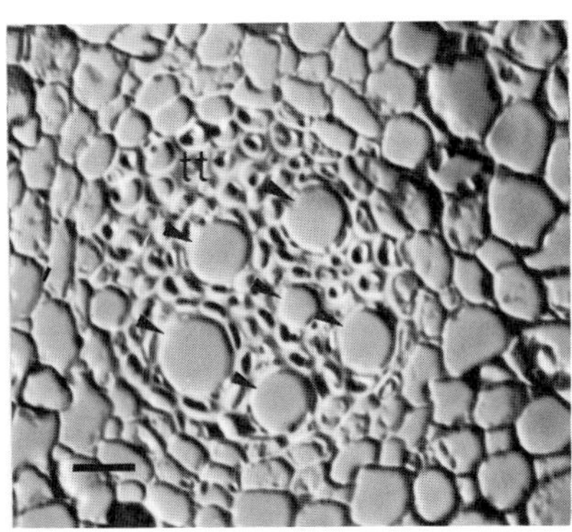


**Figure 2.** DIC of mid-stylar transmitting tissue (tt) in cross section. Six pollen tubes (arrowheads) are evident. Scale bar = 10 µm.

**Table 1.** Average length and diameter of the pollen tube of P. zeylanica at selected time intervals following pollination.

| Time[a] | Distance[b] | Diameter[c] |
|---|---|---|
| 0 | 23,550 ± 109[d] | |
| 2 | 4,438 ± 338 | 6.0 ±0.4 |
| 3 | 830 ± 74.4 | 11.0 ±1.7 |
| 4 | 616 ± 34.7 | 18.5 ±1.7 |
| 6 | 134 ± 31.4 | 14.8 ±1.1 |
| 8 | 0 ± 0.0 | 11.4 ±1.0 |

[a] In hours after pollination. The two leading pollen tubes were measured for 2-6 hr; at 8 hr, only the leading tube was measured.
[b] In micrometers from the ovule.
[c] In micrometers.
[d] From the middle of the stigma. Exact location of pollen varies by ±250 µm depending on where pollen lands on the stigma.

**Table 2.** Average in vivo pollen tube growth rates of P. zeylanica at selected time intervals following pollination.

| Time[a] hr | Distance[b] µm | µm/hr | µm/min | Volume[c] $\mu m^3$/min |
|---|---|---|---|---|
| 2 | 19,112 | 12,741 | 212.4 | 4002.9 |
| 3 | 3,608 | 3,608 | 60.1 | 3809.6 |
| 4 | 214 | 214 | 3.6 | 639.7 |
| 6 | 482 | 241 | 4.0 | 459.9 |
| 8 | 134 | 67 | 1.1 | 74.4 |

[a] In hours after pollination.
[b] In micrometers, as measured from the previous mean.
[c] Rate of pollen tube growth in terms of volume increase.

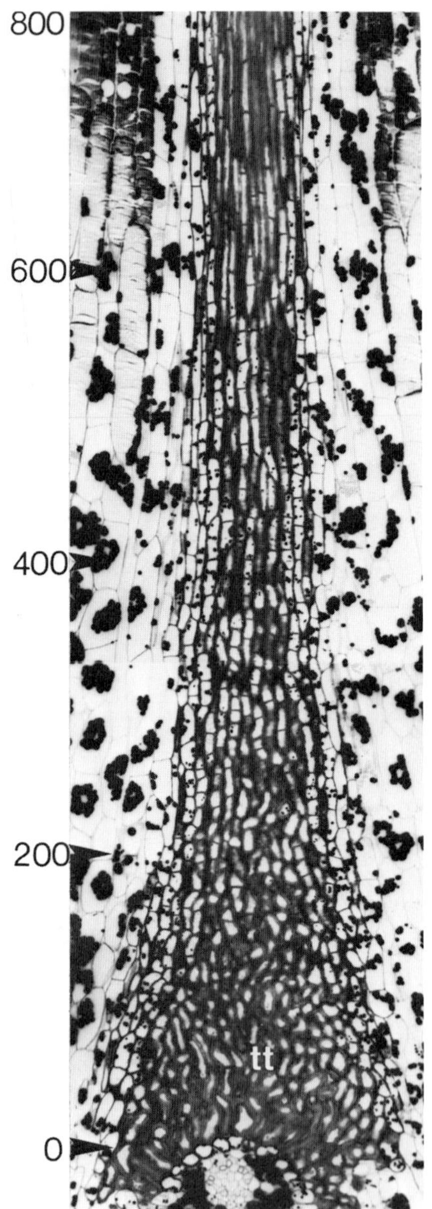


**Figure 3.** Median longitudinal section illustrating the heterogeneity of transmitting tissue (tt) at 800 um to the ovule. Stained with PAS reaction for insoluble carbohydrates. Scale in micrometers.

**Table 3.** Average cell sizes in the transmitting tissue of P. zeylanica as a function of distance from the ovule.

| Location[a] | Length[b] | Width[b] |
|---|---|---|
| mid-style[c] | 157.0±17.2 | 3.46±0.11 |
| 800 um | 58.8± 7.2 | 5.17±0.25 |
| 600 | 36.3± 4.4 | 6.83±0.29 |
| 400 | 21.9± 2.4 | 8.77±0.42 |
| 200 | 14.0± 1.3 | 9.57±0.46 |
| 0 | 18.5± 2.1 | 10.45±1.04 |

[a] Distance from ovule in micrometers
[b] In micrometers.
[c] Approx. 4,000 μm from the ovule.

Concomitant with the change in pollen tube growth rates 3 hr after pollination is an abrupt change in the organization of the stylar transmitting tissue. From the five-lobed stigma to the confluent style, the transmitting tissue consists of axially aligned cells, forming a classical closed transmitting tissue. At the stigma, the tissue may be composed of as few as 1 or 2 cells in cross section (Fig. 1), but at the mid-style, the transmitting tissue may be as many as 12 cells thick beneath 9 layers of external tissue, with pollen tubes travelling exclusively in the transmitting tissue (Fig. 2). Transmitting tissue cells are long (mean: >150 μm in length) and narrow (mean: 3.46 μm in diameter, Table 3), cylindrical cells with inclined end walls. This 45:1 ratio of length to width in these aligned cells presumably promotes linear growth and provides an optimal medium for the passage of pollen tubes as judged from the growth rates in Tables 1 and 2.

From 1000 μm to the ovary to the ovule itself, transmitting tissue cells become consistently shorter and increase in diameter (Fig. 3, Table 3). The length to width ratio varies from 11.4:1 at 800 μm to a minimum of 1.5:1 at 200 μm from the ovule. The orientation of these cells is axial near 800 μm, but becomes skewed near the ovule (Fig. 3). The lower style (near and entering the ovary) is much larger in diameter (Fig. 3), containing transmitting tissue up to 20 cells thick next to the funiculus (Fig. 3) and surrounded by increasingly numerous layers of protective tissue near the summit of the ovary. The transmitting tissue directly appressed to the funiculus and micropyle of the ovule forms an obturator of stylar origin (Fig. 3). The twisted orientation of these transmitting tissue cells may necessitate the presence of chemotropic substances as the pollen tubes in this region often travel at high angles relative to the stylar axis (Russell and Cass, 1981). The lower region of the style is presumably a far less optimal environment for the continuation of rapid tube elongation as evidenced from pollen tube growth rates (Table 2).

**4 Discussion**

The tricellular pollen of P. zeylanica clearly exhibits two distinct phases of pollen tube growth which apparently relate to gradual changes in the structure of the transmitting tissue through which it grows. Narrow, axially-aligned cells with thick, highly pectic cell walls and highly inclined end walls may be expected to provide minimal resistance to the growth of pollen tubes. As the cells are axially aligned, the direction of growth does not serve to shear transmitting tissue cells. The abundant pectic materials also tend to promote the separation of transmitting tissue cells from one another with the passage of a pollen tube. Near the ovule, the transmitting tissue has relatively thinner cell walls and shorter, often axially-twisted cells that would be expected to provide greater resistance to the passage of pollen tubes.

A wave of degeneration of stylar transmitting tissue cells often precedes or accompanies the passage of the pollen tube in Plumbago, as has been previously reported in cotton (Jensen and Fisher, 1970). If degenerative changes in transmitting tissue are triggered on a cellular basis (as electronmicrographic evidence has suggested) and are required either mechanically or nutritionally to promote pollen tube growth, the rate of response would be strongly correlated to the relative length of transmitting tissue cells. Longer cells would begin to degenerate farther in advance of the pollen tube, thereby promoting more rapid elongation. The present data, illustrating an abrupt change in pollen tube structure, elongation rate, and diameter in relation to changes in stylar tissue, tends to support the possibility of sporophytic control of pollen tube growth rates by transmitting tissue structure.

Heterogeneous transmitting tissue in P. zeylanica appears to arise in part through the differential division of cells during elongation of the style. The more centrally-located transmitting tissue cells elongate with less frequent mitotic divisions during development than the more isodiametric cells of the surrounding gynoecial tissues. Near the base of the style, a lesser degree of elongation would result in transmitting tissue cells closer to their immature size and shape. At the ovary, the longer transmitting tissue cells near the ovule result from the later expansion of papillate transmitting tissue that occurs once the circintropous ovule reaches its final position. The remaining space between the transmitting tissue and the ovule is occupied by the expansion of these cells.

If there is an evolutionary advantage in a biphasic pattern of growth as occurs in P. zeylanica, it may relate to several important features of the plant's biology. Since P. zeylanica is a uniovulate, self-fertile plant, the pollen tubes are consequently competing for a single goal. Speed of the pollen tubes, presumably of utmost importance for pollen tube success in a uniovulate plant, is controlled in such a manner that there is a high rate of pollen tube abortion in the stigma causing a relatively limited number of highly fit pollen tubes to compete in the lower style for entry into the ovule. As the anthers tend to open within 30 min after floral anthesis, the plant is almost exclusively self-pollinating. In the described system, such competition in the lower style could represent a veritable gauntlet of physiological challenges to the pollen tubes that may be beneficial in selecting a superior genotype in a typically self-pollinating plant.

Acknowledgements. I wish to thank Rahmona A. Thompson and Susan M. Heinrichs for technical assistance. This research was supported in part by NSF grants PCM-8208466 and PCM-8409151. Use of the facilities of the Samuel Roberts Noble Electron Microscopy Laboratory is gratefully acknowledged.

**References**

Browning AJ, Gunning BES (1977) An ultrastructural and cytochemical study of the wall-membrane apparatus of transfer cells using freeze-substitution. Protoplasma 93:7-26.

Hoekstra FA (1983) Physiological evolution in angiosperm pollen: possible role of pollen vigor. In: Mulcahy DL, Ottaviano E (eds) Pollen: biology and implications for plant breeding. Elsevier Biomedical, pp 35-41.

Jensen WA (1962) Botanical histochemistry. WH Freeman & Co.

Jensen WA, Fisher DB (1970) Cotton embryogenesis: the pollen tube in the stigma and style. Protoplasma 69:215-235.

Mulcahy GB, Mulcahy DL (1983) A comparison of pollen tube growth in bi- and trinucleate pollen. In: Mulcahy DL, Ottaviano E (eds) Pollen: biology and implications for plant breeding. Elsevier Biomedical, pp 29-33.

Russell SD (1982) Fertilization in Plumbago zeylanica: entry and discharge of the pollen tube in the embryo sac. Can J Bot 60:2219-2230.

Russell SD (1983) Fertilization in Plumbago zeylanica: gametic fusion and fate of the male cytoplasm. Amer J Bot 70:416-434.

Russell SD (1984) Ultrastructure of the sperm of Plumbago zeylanica. II. Quantitative cytology and three-dimensional organization. Planta 162:385-391.

Russell SD, Cass DD (1981) Ultrastructure of the sperm of Plumbago zeylanica. I. Cytology and association with the vegetative nucleus. Protoplasma 107:85-107.

Russell SD, Cass DD (1983) Unequal distribution of plastids and mitochondria during sperm cell formation in Plumbago zeylanica. In: Mulcahy DL, Ottaviano E (eds) Pollen: biology and implications for plant breeding. Elsevier Biomedical, pp 135-140.

# Absence of 10-Hydroxy-2-Decenoic Acid (10-HDA) in Bee-Collected Pollen

HANS C.M. VERHOEF AND FOLKERT A. HOEKSTRA[1]

## 1 Introduction

Much attention has been paid to the survival of bee-collected pollen in pellets as well as those sticking to body hairs. This interest stems from plant breeders concern to prevent contamination when using bees for pollination purposes in isolation cages on the one hand and, on the other hand, one would like to use bee-collected pollen to improve seed set and fruit growth in orchards.

Excellent germinability *in vitro* of freshly bee-collected pollen was noted by several investigators, depending on weather conditions and flight behaviour of the foraging bees (cf. Griggs et al. 1953; Klungness et al. 1983). In contrast to hand-collected apple pollen, pellet pollen rapidly loses viability, both under refrigeration (7°C) (Kremer 1949) and at ambient temperature, particularly when tested for its ability to bring about fruit set (Singh and Boynton 1949). Storability can be improved by direct cooling with the aid of dry ice (Griggs et al. 1951), or by drying the pellets prior to deep freezing (Johansen 1956). Bees can still carry viable pollen collected on a previous trip (Free and Durrant 1966). However, pollen generally fails to survive when sticking on bees which are confined to their hive for 12 h (Kraai 1962). From this it may be concluded that bees harm pollen, possibly by adding a substance during collection.

A fatty acid occurs in the mandibular glands of honeybees (Callow et al. 1959; Barker et al. 1959b), which has antibiotic properties against many bacteria and fungi (Blum et al. 1959). This fatty acid is identical to 10-hydroxy-2-decenoic acid (10-HDA) which was found earlier to be the main lipid component in royal jelly and worker food (Butenandt and Rembold 1957). 10-HDA does not occur in hand-collected pollen of *Zea*, *Alnus* and *Pinus*, or nectar (Barker et al. 1959a).

In contrast to fresh pollen, pellets contain inhibitory substances for pollen germination, as do ethereal extracts from bee heads (Lukoschus and Keularts 1968). Purification of the inhibitory lipid from pollen pellets showed the inhibitory activity to be confined to the free fatty acid fraction after elution from florisil columns. From this, Lukoschus and

[1] Dept of Plant Physiol, AU, Arboretumlaan 4, 6703 BD Wageningen, Holland

Keularts (1968) assumed the inhibitory lipid to be identical to 10-HDA.

Keularts and Linskens (1968) analysed the inhibitory effects of fatty acids with carbon chain lengths of 10-12 atoms on germination *in vitro* and respiration of pollen. Among these fatty acids, 10-HDA is particularly inhibitory at concentrations above 0.1 mM. Extracts from comb-stored pollen also reduce pollen respiration. Not aware of this paper, Iwanami et al. (1979a) reported inhibitory effects of 10-HDA on the germination of *Camellia* pollen *in vitro*. Apart from inhibition of tube elongation, Iwanami et al. (1979b) also found inhibition of pollen tube mitosis.

Because of the practical significance of a possible 10-HDA inhibition when re-using pollen pellets for pollination purposes we determined the occurrence of 10-HDA in pollen pellets with GC-MS. In contrast to our expectations we failed to detect 10-HDA in fresh pollen pellets and those packed in combs.

## 2 Materials and Methods

Pollen pellets were obtained by means of traps mounted onto the hive. Traps were emptied daily and the pellets were immediately frozen. Bees and comb-stored pellets (bee bread) were obtained through the Dept of Entomology. Collection and handling of *Typha latifolia* L. pollen, and the composition of the liquid germination medium were as described by Hoekstra and Barten (this volume).

Pollen pellets and bee bread (0.5 g) were ground in MeOH until completely dissolved, followed by the addition of $CHCl_3$ (final ratio 1:2 v/v). Bees were ground with sand in MeOH-$CHCl_3$ (1:2,v/v). The lipid extracts were clarified and dried by filtration over anhydrous $Na_2SO_4$. Heptadecanoic acid was used as an internal standard.

For GC-analysis, 10-HDA and free fatty acids (FFA's) have to be made more volatile by methylation. This was done by mixing aliquots of the extracts with a freshly prepared ethereal solution of diazomethane. GC-analysis of FFA-methyl esters was performed as described by Hoekstra and Barten (this volume). On this particular GC column, the methylester of 10-HDA was infinitely retained. However, excellent separation was obtained on a FID-equipped Perkin Elmer 8320 capillary gas chromatograph, with a 26 m fused silica column, inner diam. 0.32 mm, liquid phase CP Sil 5 CB, and having a film thickness of 1.3 $\mu$m. Carrier gas was helium at a pressure of 62 KPa. Oven temperature was 200$^{o}$C, after 15 min temperature-programmed to 225$^{o}$C at 5$^{o}$C/min, total run time 30 min.

Samples of 10-HDA, one purified from royal jelly and the other produced by organic synthesis, were the generous gifts of Professor RW Shuel, Dept of Environmental Biol, Univ of Guelph, Guelph, Ont. Canada.

## 3 Results and Discussion

Peak #1, in gas chromatogram D (Fig 1) represents the methyl ester of 10-HDA purified from royal jelly, well separated from a number of fatty acid

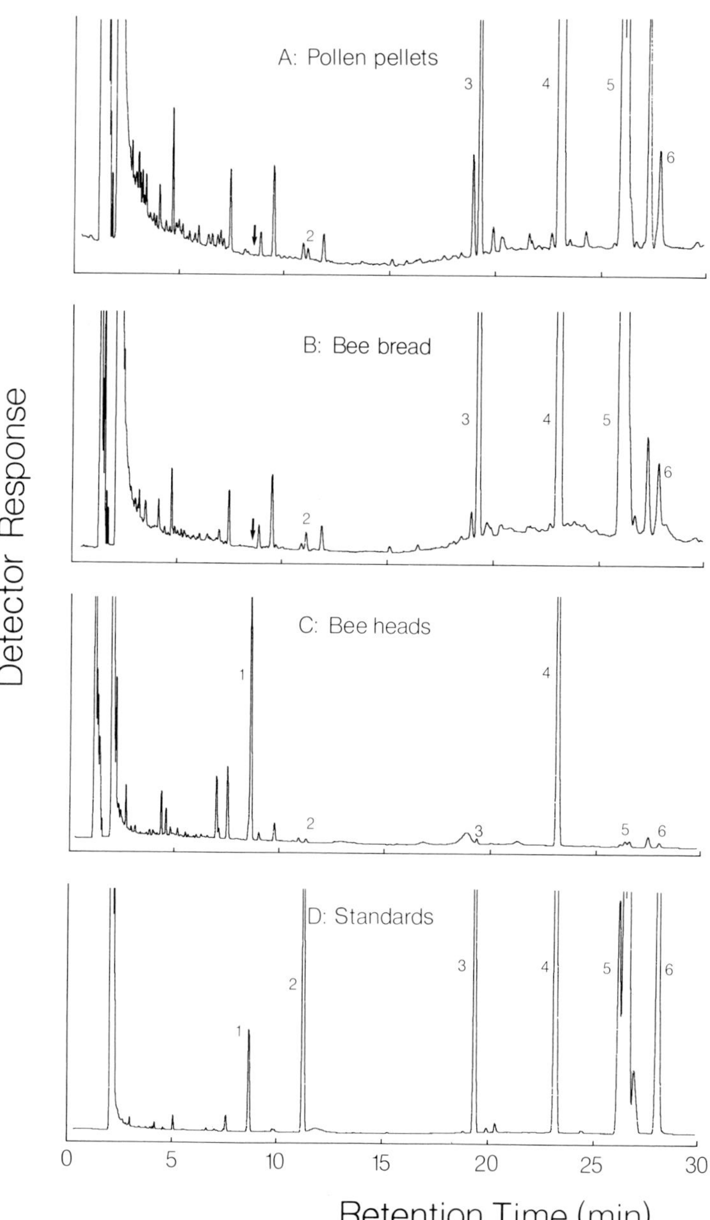


**Fig. 1A-D** Examples of gas chromatograms using a fused silica capillary column. A: methylated extract of pollen pellets (mainly Crataegus); B: methylated extracts of bee bread (mainly Crataegus); C: methylated extract of 50 bee heads; D: methylated standards. Numbers indicate methylesters of 1: 10-HDA; 2: myristic acid (14:0); 3: palmitic acid (16:0); 4: margeric acid (internal standard 17:0); 5: unresolved peaks of linolenic, linoleic, and oleic acids (18:3, 18:2, and 18:1); 6: stearic acid (18:0). Arrows indicate the position of 10-HDA, methyl ester.

methyl ester standards. The identity of peak #1 was verified by means of mass spectrometry. The royal jelly acid was further shown to be identical to a synthetic standard, and both direct probe MS and NMR identified this compound as 10-HDA.

The above mentioned GC-analysis enabled the search for a possible presence of 10-HDA in bee-collected pollen, which was earlier suggested by Lukoschus and Keularts (1968). Fig 1A shows that in MeOH-$CHCl_3$ extracts from freshly bee-collected pellets, there was no detectable peak at the position of 10-HDA methyl ester. This was independent of the season of collection and of the pollen species involved. The same holds for extracts from pollen pellets stored for some time in the comb (bee bread) (Fig. 1B). By adding margeric acid (C17:0) as an internal standard prior to extraction, the amounts of 10-HDA and FFA's per gram dry weight could be calculated. It turned out that there was less than 0.5 and 1.2 $\mu$g 10-HDA per gram (dry weight) of pollen pellets and bee bread, respectively. The chance that 10-HDA was missed in the MeOH-$CHCl_3$ extraction is minimal as other FFA's having a chain length of 16:0, 18:0, 18:1, 18:2 and 18:3 were abundantly present, particularly in the bee bread. We nevertheless performed an extraction in acetone, in which 10-HDA is very soluble, with similar results as the MeOH-$CHCl_3$ extraction (chromatogram not shown). That, indeed, 10-HDA was present in bee heads is shown in Fig. 1C. In contrast, the rest of the body practically lacked the compound.

We conclude that bees neither add substantial amounts of the antibiotic substance 10-HDA during collection of pollen, nor are they doing so while packing pellets in the combs. The efficiency of bees as pollinators would not have been favoured by a possible mixing of pollen with 10-HDA, although application of 10-HDA onto stigmas does not prevent fertilization (Keularts and Linskens 1968). In contrast, ant pollination systems are remarkably rare (Beattie et al. 1984), possibly resulting from the addition of substances as 2-hydroxydecanoic acid (Schildknecht and Koob 1971).

Considerable amounts of FFA's were present in pellets and bee bread. It might be possible that bees are the source of these FFA's that are inhibitory to pollen germination as well (unpublished results). However, the fresh pollen pellets contained amounts that are comparable with those occurring in fresh hand-collected pollen (approximately 0.1%, Hoekstra and Barten, this volume). In contrast to fresh pollen pellets, bee bread had much higher FFA contents. It is unlikely that this excessive amount originates from bees, as Hoekstra and Barten (this volume) report that dry storage of pollen (6% moisture content) at 24$^{o}$C led to a similar accumulation of FFA's, assumed to be related to the loss of viability. As the moisture content of the stored pellets was 16.4%, and the temperature in the hive can reach over 30$^{o}$C, it is to be expected that accumulation of FFA's due to degradation of endogenous lipids will proceed more rapidly. This might explain why pollen viability in the hive so rapidly deteriorates (Maurizio 1944). Elevated FFA concentrations in bee bread might also be the cause of the inhibitory effect of bee bread extracts on pollen respiration, as observed by Keularts and Linskens (1968).

To analyse whether very small amounts of 10-HDA have physiological significance, pollen of *Typha latifolia* L. was germinated in a series of 10-HDA concentrations. Table I shows that at and above 21.2 $\mu$g per ml germination medium $O_2$ uptake decreased, and from 42.5 also germination

**Table I.** Effect of 10-HDA injected into the medium as an acetonic solution on germination in vitro and respiration of Typha latifolia pollen. Humidified pollen (moisture content 32%; 100mg) was incubated in 20 ml liquid medium (pH=5.0). After 1 h, an aliquot was filtered, homogenized (french press) and extracted for determination of 10-HDA uptake from the medium (per g dry weight).

| 10-HDA applied to germination medium | 10-HDA recovered from pollen | Germination in vitro | $O_2$-Uptake |
|---|---|---|---|
| $\mu g.ml^{-1}$ medium | $\mu g.g^{-1}$ | % | $nmol.h^{-1}.mg^{-1}$ |
| 170.0 | 3583 | 0 | 222 |
| 127.5 | - | 0 | 221 |
| 85.0 | 2619 | 0 | 265 |
| 64.8 | - | 19 | 347 |
| 42.5 | 1668 | 51 | 341 |
| 21.2 | 525 | 84 | 515 |
| 8.5 | 240 | 86 | 623 |
| 4.2 | 101 | 87 | 597 |
| 0.8 | - | 85 | 629 |
| 0 | 0 | 86 | 626 |

(and tube growth) were inhibited. Extraction of the pollen in $MeOH-CHCl_3$ after 1 h of incubation indicated that at a concentration of 21.2 µg/ml germination medium, 525 µg 10-HDA was recovered per gram pollen (dry weight). This is 500 times more than the amount that was maximally present in pollen pellets. Apart from the germination experiments of Table I at pH 5.0, another experiment was done at pH 4.6, because 10-HDA has greater inhibitory effect at low pH (Iwanami et al. 1979b). However, experiments at pH 4.6 gave similar results. We conclude that the very low 10-HDA concentrations that might be present in pollen pellets cannot have physiological significance.

The absence of 10-HDA in pellets opens the possibility of succesfully re-using pellets for pollination purposes. After removing the sticky sugary substances in the pellets by washing them in germination medium for 3 min, bee-collected pollen of Crataegus monogyna Jacq. was filtered, dried, and stored as a dry powder in the deep freezer. Prior to germination in vitro the pollen was humidified in humid air. Similarly as shown earlier (Hoekstra 1983; Hoekstra and Barten this volume) this redried pollen had well retained germinability in vitro.

Acknowledgments: We are most grateful to Professor RW Shuel, Univ of Guelph, Canada for providing 10-HDA samples. We thank Dr J Beetsma, Dept of Entomol, for advise and providing equipment, and Dr MA Posthumus, Dept of Org Chem for the help with MS. We could not have done this research without the spontaneous help of the Mrs GP Lelyveld and W Ch Melger, Dept of Org Chem, and E Vermeer of our own Dept. We are indebted to Tineke van Roekel for excellent technical assistance.

**Summary**

**During pollen collection and storage of pollen pellets in the comb, bees do not add lipid substances such as 10-hydroxy decenoic acid and other free fatty acids.**

**References**


Barker SA, Foster AB, Lamb DC, Hodgson N (1959a) Identification of 10-hydroxy-2-decenoic acid in royal jelly. Nature 183: 996

Barker SA, Foster AB, Lamb DC, Jackman LM (1959b) Biological origin and configuration of 10-hydroxy-2-decenoic acid. Nature 184: 634

Beattie AJ, Turnbull C, Knox RB, Williams EG (1984) Ant inhibition of pollen function: A possible reason why ant pollination is rare. Amer J Bot 71: 421-426

Blum MS, Novak AF, Taber III S (1959) 10-Hydroxy-2-decenoic acid, an antibiotic found in royal jelly. Science 130: 452-453

Butenandt A, Rembold H (1957) Ueber den Weiselzellenfuttersaft der Honigbiene I. Isolierung, Konstitutionsermittlung und Vorkommen der 10-hydroxy-2-decensaure. Z Physiol Chemie 308: 284-289

Callow RK, Johnston NC, Simpson J (1959) 10-Hydroxy-2-decenoic acid in the honeybee (Apis mellifera). Experientia 15: 421-422

Free JB, Durrant AJ (1966) The transport of pollen by honey-bees from one foraging trip to the next. J Hort Sci 41: 87-89

Griggs WH, Vansell GH, Reinhardt JF (1951) The germinating ability of quick-frozen, bee-collected apple pollen stored in a dry ice container. Amer Bee J 91: 470

Griggs WH, Vansell GH, Iwakiri BT (1953) The storage of hand-collected and bee-collected pollen in a home freezer. Proc Amer Soc Hort Sci 62: 304-305

Hoekstra FA (1983) Physiological evolution in Angiosperm pollen: Possible role of pollen vigor. In: Pollen: Biology and Implications for Plant Breeding pp 35-41. Mulcahy DL, Ottaviano E eds, Elsevier Sci Publ Co.

Iwanami Y, Iwamatsu M, Okada I, Iwadare T (1979a) Comparison of inhibitory effects of royal jelly acid and myrmicacin on germination of Camellia sinensis pollens. Experientia 35: 1311-1312

Iwanami Y, Okada I, Iwamatsu M Iwadare T (1979b) Inhibitory effects of royal jelly acid, myrmicacin, and their analogous compounds on pollen germination, pollen tube elongation, and pollen tube mitosis. Cell Structure and Function 4: 135-143

Johansen C (1956) Artificial pollination of apples with bee-collected pollen. J Econ Entomol 49: 825-828

Keularts JLW, Linskens HF (1968) Influence of fatty acids on petunia pollen grains. Acta Bot Neerl 17: 267-272

Klungness M, Thorp R, Briggs D (1983) Field testing the germinability of almond pollen (Prunus dulcis). J Hort Sci 58: 229-235

Kraai A (1962) How long do honey-bees carry germinable pollen on them? Euphytica 11: 53-56

Kremer JC (1949) Germination tests on the viability of apple pollen gathered in pellets. Proc Amer Soc Hort Sci 53: 153-157

Lukoschus FS, Keularts JLW (1968) Eine weitere Function der Mandibeldruse der Arbeiterin von Apis mellifera L. Production eines die Pollenkeimung hemmenden Stoffes. Z Bienenforschung 9: 333-343

Maurizio A (1944) Wie lange bleibt der Pollen in den Bienenwaben keimfahig? Verhandlungen Schweizerischen Naturforschungsgeselschaft 124: 128-129

Schildknecht H, Koob K (1971) Myrmicacin, the first insect herbicide. Angew Chemie Int Ed 10: 124-125

Singh S, Boynton D (1949) Viability of apple pollen in pollen pellets of honeybees. Proc Amer Soc Hort Sci 53: 148-152

# Gametophytic Ecology

# Pollen and People

SOPHIE C. DUCKER[1], R. BRUCE KNOX[1]

The great Indian embryologist *Panchanan Maheshwari* (1950) pointed out that until the end of the sixteenth century the occurrence of sex in plants was totally denied and even the mention of it was regarded as inappropriate and obscene. We know that in antiquity agricultural practices involved certain rites using pollen. The wellknown Assyrian relief of the dusting of pollen on the date palms by mythological figures shows a practice employed as a fertilization process - fertilization in its nutritional sense to make the crop more fertile.

During the Middle Ages, botanical writers did not observe plants per se but saw them only in their relationship to man. They were useful in medicine, in agriculture, or as decorations in art or in gardens. Plants as scientific objects were not recognised until the Renaissance liberated man's thought from the conventions of the classics. During the sixteenth century, descriptions of plants were first made, including the different parts of the flower.

In this review, we trace the discovery of pollen. We have recently given a complete account of the discovery of pollen structure and function (Ducker and Knox, 1985). Here, we focus on aspects related to the biotechnology and ecology of pollen, and present in detail material relating to historical figures not previously reviewed.

---

1 Botany School, Melbourne University, Parkville 3052, Victoria, Australia.

## The discovery of pollen

The awareness of stamens began with the work of the German *Leonhart Fuchs* (1501-1566), who used stamens in the descriptions of plants in his herbal. The stamen and the pistil were first described by *Hieronymus Bock* (1498-1554) in 1552. He defined a stamen as "the little apex on the hair, where the summit is sustained" - and also used the Latin term for the pistil, "the thing in the middle of the flowerbell the *pistillum*".

The godfather of pollen is undoubtedly *Valerius Cordus* (1515-1544). This German medico is only perpetuated by the posthumous publications of his manuscripts and lectures by his students (Greene, 1983; Sprague and Sprague, 1939). At the age of 29 he died of fever in Rome when on a plant collecting tour with his students in Italy. He called the "stalks" of anthers *stamina*. He took particular notice of dust *rubiginosus pulviusculus* which lily anthers shed before collapsing. He saw the same "dust" in other flowers and assigned it a name, the name which it has since borne "*luteo polline conspersa*" (sprinkled with yellow pollen).

The development of the microscope by *Hooke* and *van Leeuwenhoek* enabled the Italian, *Marcello Malpighi*, and the Englishman, *Nehemiah Grew* to study the differing colours, shapes and sizes of pollen (Fig. 1). Later in the century, the sexuality of plants and the definition of parts and genders were experimentally investigated by *Rudolf Camerarius* (Fig.1). He defined the function of stamens and pistils and recognized male and female plants, saying that these came from different seeds.

Similar experimental evidence was produced by *Johann Gottlieb Gleditsch* (Fig. 1). The long-established date palm in the Berlin glasshouse did not bear fruit until Gleditsch pollinated the mature female flowers with pollen brought from an inflorescence of the same species of date palm growing in Leipzig, a journey that had taken nine days by specially hired couriers. It was *Konrad Sprengel* who discovered "the great secret of nature". He demonstrated insect pollination and investigated its floral adaptations. Poor Sprengel was so interested in his botanical studies, that he neglected

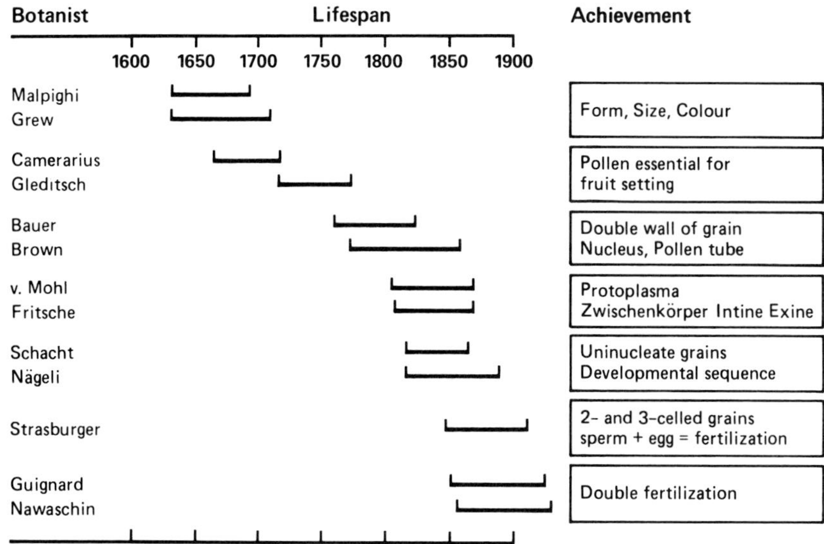


Fig. 1. Scheme showing the principal workers, their life spans and their contributions in the field of pollen biology from 1600 until 1900.

his job as a schoolmaster, was sacked and died in absolute poverty.

His nephew, *Kurt Sprengel*, is of equal importance. In 1812 he described pollen of all shapes and sizes and ornamentations. He showed for the first time compound pollen grains - the pollinia of orchids and the viscin threads between the pollen grains of *Fuchsia* and other Onagraceae.

## The forgotten contribution of Filippo Cavolini

In Naples, *Filippo Cavolini* (1756-1810), a rich lawyer, was interested in marine life. He published a large number of papers on marine organisms and is famous for elucidating the life history of jelly fish (Monticelli, 1910). He appears to have been the first scientist to develop an experimental transplant set up for growing plants of the seagrass *Posidonia oceanica* in submerged containers in Naples harbour. His observations show his ingenuity and remarkable powers of observation with the help of only a low power microscope. His writings and illustrations of seagrasses are all the more interesting because they are not only the first publications on the pollen of these plants, but also the recognition that they have so differently shaped pollen to other flowering plants.

The extraordinary Cavolini described and illustrated filiform pollen and observed its cottonwool-like appearance in the sea. The genus *Zostera* had been described by Linnaeus but not so the pollen, and other genera of seagrasses like *Amphibolis* on which we have done most of our work, were described as marine algae by the famous Carl Agardh in 1823. Cavolini showed that seagrasses were true flowering plants with a pistil, stamens and filiform pollen. Further he showed that their seeds had the ability to germinate. He thus acknowledged the fact that water could be the agent for pollination.

In 1788, at the age of 22, Cavolini gave an illustration of the stamens of *Ficus carica* and showed the emitted pollen. Incidentally Valerius Cordus (mentioned earlier)was the first to observe the stamens of the fig. Cavolini realised that there were male, female and gallflowers in the same inflorescence and that a wasp was the pollinator. He regarded this "bizarre manner of pollination by the small animal as a great help to man", realising that the wasp was laying its eggs in the florets.

Cavolini, like his very rich father, was trained as a lawyer. He also studied the classics, physics, mathematics and botany. His botanical professor was Domenico Cirillo (1739-1799) who had been appointed to the first chair of botany at Naples. When Cirillo was murdered in 1799, Cavolini became his successor to the Chair of Natural History,but died of typhoid in 1810. Cavolini's discoveries of flowering and pollination of seagrasses were acclaimed by botanists all over Europe.

## The structural and functional analysis of pollen

It was in the nineteenth century that some of the questions involving pollen structure and function leading to fertilization were to be answered (Ducker and Knox, 1985).

The first contribution concerns the dual nature of the pollen wall, and the emergence of pollen tubes, which we attribute to *Ferdinand Bauer* and *Robert Brown* (Fig. 1). The patterning of the outer wall in terms of pores and ornamentation, and the thin, transparent inner wall layer were described by the German morphologist *Hugo von Mohl*. He also named the proteinaceous ground substance of cells "*Protoplasma*" (Fig. 1). *Carl Fritsche* called the wall layers exine and intine. He carried out ingenious studies that demonstrated the different chemical components in these two wall layers, introducing the term *Zwischenkörper* for the hyaline lens-shaped wall structure underlying each germinal aperture (Fig. 1). He classified the various types of pollen (Ducker and Knox, 1985).

Although Robert Brown in 1833 had noted the appearance of what we now recognise as tetrads of microspores, the study of pollen development received its major impetus from the work of *Hermann Schacht* and *Carl Nägeli*. Both observed the callose special wall that surrounds the meiocytes and tetrads of microspores (Fig. 1), while Schacht is notable for the first use of polarised light to distinguish cellular features and inclusions of pollen. He also introduced sectioning to botanical microscopy, a technique that greatly aided his studies of pollen development.

In the mid-nineteenth century, there was also a change in the concept of pollen-stigma interactions. *Schleiden* in 1846, believed that seed-setting arose from the descending "female" pollen tubes in the ovary. This view was corrected by Schacht in 1859, who showed first the role of the stigma in receiving the pollen, and later the penetration of pollen tubes through the style and ovary tissues, to enter the embryo sac in *Gladiolus* and *Polygonum* (Ducker and Knox, 1985).

## The cellular basis of pollen

In 1879, *Frederick Elfving*, the Finnish phycologist, reported on his cytological studies of pollen, which he had carried out on a visit to Jena in the laboratory of *Eduard Strasburger*. Elfving illustrated and recorded the bicellular and tricellular nature of different pollen types, but it was not until the publications of Strasburger that these cells were given names: the generative cell as the progenitor of the pair of sperm cells, lying wholly within the vegetative cell (Fig. 1).

At the same time, detailed cytological studies,mainly of bicellular pollen types, were carried out by the Frenchman *Leon Guignard*. Like Elfving, he observed the division of the generative cell in the pollen tube. Both Guignard in 1879 and Strasburger in 1884 had extended their cytological studies to encompass fertilization. At first, they considered it a single event: the sperm nucleus fused with the egg nucleus to give the first cell of the embryo. The discovery of double fertilization was first reported by the Russian *Sergius Nawaschin* (Fig. 1) in 1898.

There is little doubt that the understanding of pollen structure and function in the nineteenth century proceeded logically - with the juxtaposition of great minds and the latest in light microscopic technology. Most of the cytological features of pollen seen by our predecessors have been confirmed today with the advent of electron microscopy. Some, such as the cellular nature of bicellular and tricellular types of pollen were ignored for nearly a century (Maheshwari, 1950; Ducker and Knox, 1985). Others, such as the centrosomes or polar bodies seen so remarkably by Guignard in 1879, have yet to be found, although they are a feature of the sperm cells of lower plants.

We are grateful to Prof. Ettore Pacini for his assistance in obtaining literature concerning Cavolini.

## References

Ducker SC, Knox RB (1985) Pollen and pollination: a historical review. Taxon 34:401-419.

Greene EL (1983) Landmarks of botanical history. Egerton FN (ed) Vol 1 and 2. Stanford University Press, Stanford.

Maheshwari P (1950) An introduction to the embryology of angiosperms. McGraw Hill, New York.

Monticelli T (ed) (1910) Opere di Filippo Cavolini, Ristampa a cura della Societa dei Naturalisti in Napoli. Detken e Rocholl, Napoli.

Sprague TA, Sprague MS (1939) The herbal of Valerius Cordus. J. Linn. Soc. London Bot. 52:1-113.

# Evidence For and Against Pollen Tube Competition in Natural Populations

ALLISON A. SNOW

Do pollen deposition rates lead to gametophytic competition in natural populations? Despite a recent surge of interest in the evolutionary implications of pollen tube competition (Stephenson and Bertin 1983, Wilson and Burley 1983, this volume), information on its occurrence is scant. Demonstrating that this process takes place in nature is difficult. Detailed information is needed on rates of pollen arrival, pollen tube growth, and fertilization. A further problem is that plant-pollinator interactions are often complex; spatial and temporal variation in pollinator service may preclude broad generalizations. Nevertheless, it is essential that we examine the potential for pollen-pollen competition if its significance is to be understood.

The purpose of this chapter is to compare pollination rates at natural populations of three species: *Passiflora vitifolia*, *Epilobium canum*, and *Cassia reticulata* (Table 1.). All of these species require animal vectors to effect pollination. I conclude that the opportunity for pollen tube competition ranged from infrequent to common, based on measurements of pollen deposition rates and seed set.

## METHODS

Experimental studies of *Passiflora*, *Epilobium*, and *Cassia* were conducted between 1980 and 1985. Only the salient features of each study are described below. Further details can be found in Snow (1982), Snow and Roubik (1985), and Snow (1985).

Table 1. CHARACTERISTICS OF THE STUDY SPECIES.

| | Passiflora | Epilobium | Cassia |
|---|---|---|---|
| Location | Costa Rica | Calif., USA | Panama |
| Frequency of Pollen Tube Competition | RARE | INTERMEDIATE | COMMON |
| Pollinator | Hummingbirds | Hummingbirds | Bees |
| Max. No. Seeds per Fruit | 275 | 21 | 36 |
| No. Pollen Grains for Full Seed Set | 450 | 80 | 50 |

TOTAL POLLEN DEPOSITION - Pollen tube competition is uncommon in populations where seed set per fruit is pollen-limited. Since the potential for pollen-limitation is relatively easy to assess, this information was obtained for all three species. First, the threshold amount of pollen required for full seed set was determined. I then collected naturally pollinated flowers, counted the number of pollen grains per stigma, and estimated seed set as a function of pollen load. For *Passiflora*, which is self-incompatible, all flowers on a plant were emasculated to prevent contamination with self pollen. *Epilobium* and *Cassia* are self-compatible.

RATES OF POLLEN DEPOSITION - When seed set per fruit is not pollen-limited, we need to know whether the timing of pollen arrival forces pollen tubes to compete for available ovules. If pollen grains arrive gradually, many ovules may be fertilized by a random population of pollen grains, rather than by those with rapid germination and growth. Rates of pollen deposition were not measured for *Passiflora* because most of these one-day flowers were pollen-limited.

Pollen arrival on *Epilobium* stigmas was monitored by counting the number of tetrads present at 2 hr intervals. Pollen tetrads are large enough to be counted *in situ* with a

10X hand lens. In 1983 and 1984, more than 400 stigmas were censused between 0700 and 1900 hrs for 2 - 3 days each.

Rates of pollen deposition could not be measured directly on the one-day flowers of *Cassia*. Instead, I used a microscope to count the number of pollen grains that were transferred during single visits by solitary bees. Previously bagged flowers were exposed to single visits, re-bagged, and collected the following day for counts of germinated pollen (Snow and Roubik 1985).

## RESULTS AND DISCUSSION

FREQUENCY OF POLLEN TUBE COMPETITION - Pollen-limitation of seed set per fruit varied among species (Table 2). Most *Passiflora* flowers that received enough pollen to set fruit (i.e., > 25 grains) did not obtain enough for full seed set (450 grains). Hand-pollination experiments performed in 1981 (Snow 1982) and again in 1985 (unpubl. data) confirmed that seed set per fruit was pollen-limited. Thus, the opportunity for pollen tube competition at this population was rare.

Table 2. FREQUENCY OF POLLEN LIMITATION.

| | | Proportion of flowers that were pollen-limited | N* |
|---|---|---|---|
| *Passiflora* | 1981 | .68 | 92 |
| *Epilobium* | 1982 | .30 | 159 |
| | 1983 | .39 | 367 |
| *Cassia* | 1982 | .04 | 121 |

* Number of flowers with at least 1 pollen grain per stigma; *Passiflora* flowers with insufficient pollen for fruit set (< 25 grains) were not included.

Most Epilobium flowers received enough pollen for full seed set in both 1982 and 1983, but 30 - 40 % did not (Table 1.). Flowers with < 20 tetrads per stigma almost always set fruit, even if only one tetrad was deposited. In 1983 I compared pollen loads of different individuals. Surplus pollen was deposited on 80-90 % of the stigmas of some plants, and on only 15-20 % of others (Snow 1985).

Epilobium tetrads were transferred in sticky clumps connected by viscin threads. However, only 43 % of the "visits" by hummingbirds transferred clumps of surplus pollen (an increase of >5 tetrads was considered to result from a hummingbird visit; Fig. 1A). About 70 % of the stigmas with >5 tetrads (N = 239) received their pollen in one load. At the other 30 %, a second load arrived, usually at least 4 hrs. later and often on the following day. Calculations described in Snow (1985) showed that approximately half of the seeds from all flowers in the study population were fathered by competing pollen tubes.

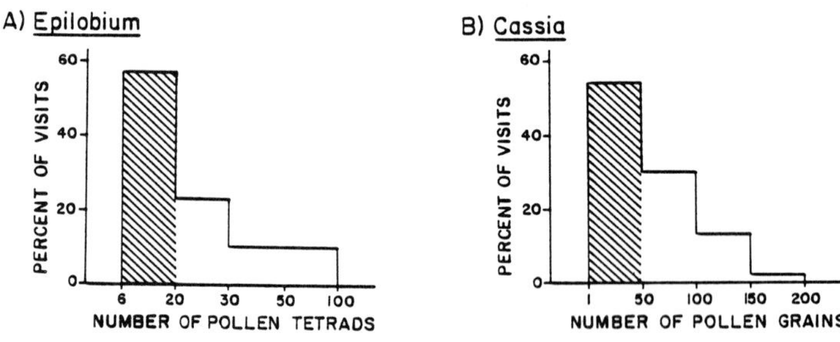


Figure 1. AMOUNT OF POLLEN DEPOSITED PER VISIT. Shaded area indicates insufficient pollen for full seed set. In Epilobium, an increase of > 5 tetrads was termed a "visit"; N = 357. Cassia data are from single visits to 304 flowers.

Pollen-limitation of seed set was rare at flowers of Cassia (Table 2.) About half of the single visits supplied more than enough pollen for full seed set, presumably leading to pollen tube competition (Fig. 1). Further opportunity for competition resulted from multiple visits. Pollen accumulated on stigmas (compare data in Table 2 and Fig. 1), and several visits to the same flower probably occurred during peak foraging activity. I conclude that pollen tube competition was likely at most of the flowers in this population. More precise estimates of the frequency of competition would require data on the timing of deposition and variation in pollen tube growth rates.

INTENSITY OF COMPETITION - The intensity of gametophytic competition can be influenced by pollination rate, style length, variation in pollen tube growth rate, and other factors (Mulcahy 1983). In both *Epilobium* and *Cassia*, single visits sometimes transferred 3 - 4 times the amount of pollen required for full seed set (Fig. 1). Although *Cassia* flowers were visited more frequently, the much longer style of *Epilobium* (5 cm) may intensify competition by allowing small differences in pollen tube growth rate to be expressed.

One method of quantifying the intensity of competition is to determine what proportion of the male gametes is excluded from access to ovules. At *Epilobium* flowers that received surplus pollen, about 20 - 60 % of the tetrads were outcompeted, and sometimes as many as 80 % (Snow 1985). Another approach involves quantifying pollen tube growth rates. Mulcahy et al. (1982) estimated that natural pollination rates selected for pollen tubes that were 34 % faster than the mean in *Geranium maculatum*.

Pollen tube competition can result in gametophytic selection if genetically based variation in pollen performance leads to nonrandom fertilization. Previous investigators have varied the intensity of pollen tube competition by manipulating the size of the pollen load (Fingerett 1979, Mulcahy 1979, Stephenson et al. 1985), or the distance required for pollen tube growth (Mulcahy and Mulcahy 1975, Mulcahy 1974). These treatments generally result in faster progeny growth rates due to gametophytic selection. However, treatments that mimic *natural* levels of pollination have not been employed. The data presented here provide a starting point for understanding the dynamics of pollen deposition in natural systems.

Pollen tube competition is but one method by which gametophytic selection can occur. Other modes of selection which do not require excessive pollination include incompatibility systems and other male-female interactions. The opportunity for pollen tube competition in natural populations is variable, as outlined in this chapter, and further studies are needed to assess its evolutionary significance.

## REFERENCES


Fingerett, E.R. 1979. Pollen competition in a species of evening primrose, _Oenothera organensis_ Munz. Master's Thesis, Washington St., Pullman, WA.

Mulcahy, D.L. 1974. Correlation between speed of pollen tube growth and seedling height in _Zea mays_ L. Nature 249:491-493.

Mulcahy, D.L. 1979. The rise of the angiosperms: a genecological factor. Science 206:20-23.

Mulcahy, D.L. 1983. Models of pollen tube competition in _Geranium maculatum_. _In_ L. Real (Ed.), Pollination biology. Academic Press, N.Y.

Mulcahy, D.L., and G.B. Mulcahy. 1975. The influence of gametophytic competition on sporophyte quality in _Dianthus chinensis_. Theor. Appl. Genet. 46:277-280.

Mulcahy, D.L., P.S. Curtis, and A.A. Snow. 1982. Pollen competition in a natural population. _In_ C.E. Jones and R.J. Little (Eds.), Handbook of experimental pollination biology. Van Nostrand-Reinhold, N.Y.

Snow, A.A. 1982. Pollination intensity and potential seed set in _Passiflora vitifolia_. Oecologia 55:231-237.

Snow, A.A. 1985. Pollination dynamics in _Epilobium canum_ (Onagraceae): consequences for gametophytic selection. Am. J. Bot. _in press_.

Snow, A.A., and D.W. Roubik. 1985. Pollen transfer by bees visiting two tree species in Panama. Biotropica _in press_.

Stephenson, A.G., and R.I. Bertin. 1983. Male competition, female choice, and sexual selection in plants. _In_ L. Real, (Ed.), Pollination biology. Academic Press, N.Y.

Stephenson, A.G., J. Winsor, and L. Davis. (this volume).

Willson, M.F., and N. Burley. 1983. Mate choice in plants. Princeton Univ. Press, Princeton, N.J.

# Pollen Competition in Aureolaria pedicularia

J. RAMSTETTER AND D. L. MULCAHY[1]

## 1 Introduction

Pollen quantity and quality are important determinants of reproductive success in natural populations of flowering plants. Recent discussions of these topics include studies of the occurrence of pollen competition (Mulcahy et al. 1983, Snow, this volume), the effect of pollen composition on offspring quality (Schemske and Pautler 1984), pollen competition as an important factor in fruit and seed abortion (Lee 1984), the significance of pollen competition in the evolution of the angiosperms (Mulcahy 1979), and the influence of pollen characteristics on its quality (Stanton, this volume). However, the extent to which pollen competition occurs and how details of the pollination process affect pollen competition are largely unknown. Our study addresses these questions in the annual *Aureolaria pedicularia*.

## 2 Materials and Methods

Several parameters potentially determining the degree of pollen competition were studied in two populations of *Aureolaria pedicularia* (L.) Raf. (Scrophulariaceae) near Amherst, MA during the late summer of 1984. These included 1) the number of pollen grains deposited per pollinator visit, 2) the number of ovules per flower, 3) the rate of pollination, and 4) style length in this highly branched, many-flowered plant of open oak woods.

---

[1]Botany Department, University of Massachusetts, Amherst, MA 01003 USA

The number of pollen grains deposited by bumblebees, the only effective pollinators observed, was estimated for visits from ten bumblebees to different flowers. Flowers of *Aureolaria pedicularia* generally open for one day only, and the flowers used had not yet opened. After opening the flowers by hand and finding that anthers were not touching the stigma, a single bumblebee visit was allowed. The entire style was removed from the flower and the pollen grains transfered from the stigma to a fuschin-stained glycerin jelly (Beattie 1971). Pollen grains of *A. pedicularia* were readily distinguishable when projected and were then counted. Four stigmas from unvisited flowers were checked in the same manner to ensure that pollen grains were not present on stigmas without pollination. The number of ovules per flower was estimated by counting all ovules in three flowers from different plants.

Pollination rate was determined by recording the amount of time between visits to individual flowers. Observation periods were approximately one-half hour long; total observation time was four hours and 24 minutes. From six to eight flowers were observed at any one time, and they were chosen from most plants in the population and from most positions on plants from September 2 to September 14. Air temperatures ranged from 11.5°C to 19° C. Bumblebees were active during weather ranging from full sun to overcast conditions with light rain; they were seen pollinating flowers as early as 6 a.m. and as late as 7:30 p.m.

Pollen tube growth was assessed by hand-pollinating 16 emasculated flowers on a single plant with pollen from a different individual. The flowers were bagged to prevent additional pollination. Four styles were collected at each of four different times: 7, 12, 22, and 31 hours after pollination. Air temperature ranged from 14° C at the time of pollination to 19° C during the collection period. The styles were prepared and observed under fluorescent microscopy according to the methods of Mulcahy et al. (1983). The percentage of the style traveled by the pollen tube front was recorded and averaged for all styles collected after each time period. Style length was measured in six additional plants also.

## 3 Results

The mean number of pollen grains deposited per pollination was 447, while the mean number of ovules per flower was 210 (Table 1).

| | Mean ± S.E. |
|---|---|
| Pollen grains/visit | 447 ± 130 |
| Ovules/flower | 210 ± 15 |

Table 1. The mean number (± standard error) of pollen grains deposited per bumblebee visit (n=10) and the number of ovules per flower (n=3) in *Aureolaria pedicularia*.

Bumblebees deposited from 94 grains in the smallest load to 1453 grains in the largest load. Of the four stigmas checked for pollen grains immediately after opening, but without being pollinated, three had no pollen grains and one had five pollen grains. Mean time between pollinator visits for 34 flowers receiving at least two visits during an observation period was 6.7 minutes (n=90) (Figure 1).

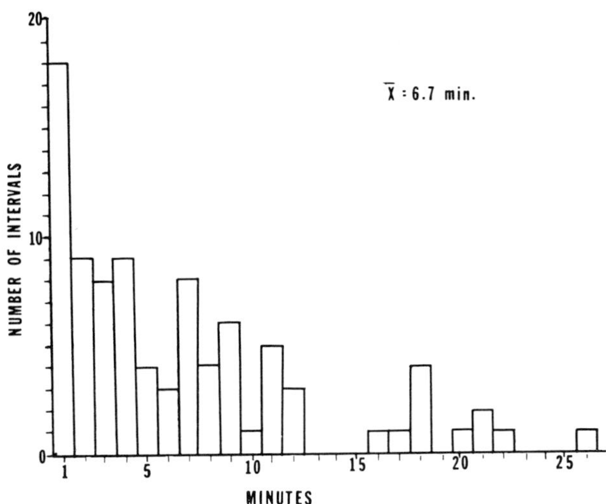


Figure 1. The amount of time between pollinator visits to *Aureolaria pedicularia* in 34 flowers (n=90). Mean time between visits was 6.7 minutes.

Of a total of 44 flowers observed, six flowers received a single visit and four flowers received no visits during the time observed. Estimated visitation rate in the 44 flowers ranged from 0 visits to 15 visits per hour ($\bar{X}$=4.7 visits per hour).

Mean style length was 2.53 cm, and pollen tube fronts traversed this distance to the ovary at some time between 12 and 22 hours (Figure 2).

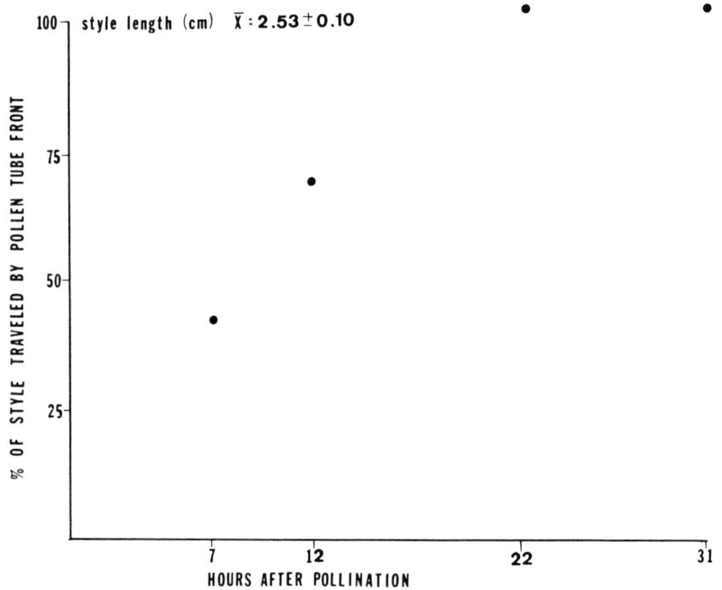


Figure 2. Rate of growth of pollen tubes in *Aureolaria* styles as represented by the percentage of the style traveled by the pollen tube front. Data are means for four styles collected at each of four times: 7, 12, 22, and 31 hours after pollination.

Seven hours after pollination, the pollen tube front had traveled approximately 40% of the style and after 12 hours approximately 60% of the style. Pollen tube fronts in the styles collected after 22 and 31 hours were at the base of the ovary; some pollen tubes were seen entering the ovules.

4 Discussion

Large numbers of pollen grains per pollination in relation to the number of ovules per flower, frequent bumblebee visits, and relatively long styles, in which pollen tube growth rate may be slow, create the potential for pollen competition in *Aureolaria pedicularia*. For larger pollen loads (those greater than the number of ovules), competition may occur within a load. For a load with fewer pollen grains than the number of ovules per flower, competition may result between loads if the slower pollen tubes from the initial pollination are overtaken by faster tubes from successive loads.

The importance of pollen competition in *Aureolaria pedicularia* remains speculative. Clearly there is variation in the size of pollen loads and the rate of visitation, and this will lead to variation in the degree of pollen competition. Additionally, variation in pollen tube growth rate was not measured, nor was the effect of multiple pollinations on pollen tube growth rate. These factors will affect the possibility of slow growing pollen tubes being overtaken by more rapidly growing tubes. Another unknown factor is the number of pollen grains required to result in full seed set. However, our preliminary data do permit a general discussion of pollen competition in a natural population. Using mean values for the number of pollen grains per visit and number of visits per hour and assuming 12 hours of pollination per day, an individual flower would receive 25,211 pollen grains. This is 120 times the mean number of ovules per flower and well in excess of estimates of the amount of pollen required for full seed set. Based on a small sample of hand-cross and self-pollinated flowers, *A. pedicularia* appears to be self-compatible, so even if self pollen grains represent many of the grains deposited by bumblebees, they may still contribute to the pool of grains competing for ovules. After 12 hours, the pollen tube front had traversed only 60% of the style; some degree of overlap will exist in the contribution to fertilization by pollen tubes from different bumblebee visits depending on the variance in pollen tube growth rate. In this hypothetical scenario, pollen competition both within and between pollen loads may be an important phenomenon. Even in the case where a flower always receives the minimum number of grains per visit and only a single visit per hour for twelve hours, the number of pollen grains will still be 5.4 times the mean number of ovules. Refinements of these measures will enable us to predict more accurately the extent to which pollen competition occurs; monitoring the vigor of offspring from different competition regimes will reveal any impact of genetically determined differences in pollen tube growth rate.

## References

Beattie AJ (1971) A technique for the study of insect-borne pollen. Pan Pacific Ent 47: 82.

Lee TD (1984) Patterns of fruit maturation: a gametophyte competition hypothesis. Am Nat 123: 427-432.

Mulcahy DL (1979) The rise of the angiosperms: a genecological factor. Science 206: 20-23.

Mulcahy DL, Curtis PS, Snow AA (1983) Pollen competition in a natural population. In Handbook of Experimental Pollination Biology, CE Jones and RJ Little eds. Van Nostrand Reinhold Co. Inc., NY, NY, USA.

Schemske DW, Pautler LP (1984) The effects of pollen composition on fitness components in a neotropical herb. Oecologia 62: 31-36.

# Pollination Intensity, Fruit Maturation Pattern, and Offspring Quality in Cassia fasciculata (Leguminosae)

T.D. LEE AND A.P. HARTGERINK[1]

## 1 Introduction

Individuals of many seed plants produce more pistillate flowers and initiate more fruits than they can mature with available resources; thus flower and fruit abortion are common (Stephenson 1981). Relatively little is known, however, about the factors that determine which flowers give rise to mature fruits and which do not. In some species, such as *Cassia fasciculata* (Leguminosae), flowers receiving light pollen loads (few grains/stigma) are less likely to produce mature fruits than those receiving heavy loads (Lee and Bazzaz 1982).

We were interested in how this fruit maturation pattern affected fitness of the parent plant. Specifically, we hypothesized that offspring in typically abortive, lightly pollinated ovaries were less vigorous than those from typically maturing, heavily pollinated ovaries. This assertion is based on the greater amount of competition among pollen tubes in the latter. When pollen compete for ovules, fast growing pollen tubes are relatively more successful than slow ones in fertilizing eggs and, if fast tubes confer their vigor on resulting embryos, then, on average, progeny produced under heavy pollination should be more vigorous (Mulcahy 1979, and refs. therein).

It is difficult to test this hypothesis if lightly

[1] Department of Botany and Plant Pathology, University of New Hampshire, Durham, NH, USA 03824

pollinated ovaries always abort. Preliminary observations on *C. fasciculata*, however, indicated that such ovaries would mature under certain environmental conditions. Here we (1) demonstrate that exposure to short days induces plants to mature lightly pollinated ovaries and (2) compare fitnesses of progeny from light and heavy pollinations.

## 2 Environmental Control of Fruit Maturation Pattern

Individuals of *C. fasciculata* were grown in clay pots (15 cm diam.) in a glasshouse at Durham, NH, USA, under natural light. Just prior to flowering, plants were randomly assigned to one of two treatments. In the long day treatment, 8 plants were exposed to natural light supplemented and extended to a 16 hr photoperiod with 5, 200 W incandescent lamps which yielded a flux density at plant height of ca. 120 uE $m^{-2}$ $s^{-1}$. In the short day treatment, 8 plants were exposed to natural and artificial light, as above, but for only 9 hrs; they were then enclosed in a "dark-box". Air temperatures in the two treatments remained within 2-3 C of each other during the course of the experiment.

On each plant, randomly selected flowers were pollinated with either pure pollen (PP) or pollen diluted 6:1 with talc (DP). The talc dilution resulted in less pollen on stigmas and fewer seeds per fruit (Lee and Bazzaz 1982). Fruit initiation, growth, and date of maturation of fruits were recorded. An 'initiated fruit' was defined as an ovary that remained on the plant for at least 5 d after pollination (unpollinated flowers abort in 2-3 d). The experiment was concluded after 120 days.

Under long days, pollination treatment affected both the proportion of flowers initiating fruits and the proportion of initiated fruits maturing. The proportion of flowers initiating fruits was much lower with DP than with PP (Table 1), and fruits initiated with DP had a significantly lower chance of reaching maturity within 90 d of initiation than did those initiated with PP ($X^2$ on raw data = 17.2, d.f. = 1, $p < 0.005$;

Table 2). Differences in proportions of fruits maturing were reflected by rates of fruit growth. Fruits from DP took 50% longer to reach maximum elongation (59 vs. 40 d).

Table 1. Proportions of flowers, receiving pure and dilute pollen, that initiated fruits under short and long photoperiods*.

| | PROPORTION OF FLOWERS INITIATING FRUITS | | |
|---|---|---|---|
| | PURE POLLEN | DILUTE POLLEN | n |
| SHORT DAYS | 0.93 (0.19) | 0.89 (0.09) | 7 plants |
| LONG DAYS | 0.89 (0.09) | 0.58 (0.16) | 8 plants |

*F(interaction) = 7.86, $p < 0.01$; regression on arcsin square-root transformed data. Standard deviations (non-transformed data) in parentheses.

Table 2. Proportions of fruits, initiated with pure and dilute pollen, that matured within 90 days of initiation under long and short photoperiods*.

| | PROPORTION OF INITIATED FRUITS MATURING | |
|---|---|---|
| | PURE POLLEN | DILUTE POLLEN |
| SHORT DAYS | 1.00 (51) | 1.00 (44) |
| LONG DAYS | 0.87 (79) | 0.55 (51) |

* Data from all plants lumped. Number of fruits per treatment in parentheses.

In contrast, under short days, pollen dilution affected neither the probability of fruit initiation nor the probability of initiated fruits maturing. Approximately 90% of all pollinated flowers initiated fruits, and 100% of these fruits matured, regardless of pollen treatment (Tables 1, 2). For fruits that matured, only 11 days were required to reach maximum fruit length, again, regardless of pollen treatment.

Our results suggest that fruit initiation and maturation patterns in _C. fasciculata_ are not fixed; lightly-pollinated flowers and fruits initiated from them are differentially aborted under long days but not under short days. Note, that our experiment does not resolve whether daylength *per se* or total radiation load is the cue that controls this pattern.

## 3 Fitness Effects of Pollen Dilution

To determine whether or not seeds resulting from lightly pollinated ovaries produced plants with lower vigor than those resulting from heavy pollination, we grew 4 *C. fasciculata* individuals in a glasshouse and established a 9 hr photoperiod as in the previous experiment, but without supplemental lighting. When flowering commenced, 2 plants ('recipients') were chosen to receive pollen and the remaining 2 were used as pollen 'donors'. Each recipient received pollen from only one donor. All flowers on each recipient were pollinated with either pure pollen (PP) or pollen diluted 6:1 with talc (DP). Fruits were collected and fresh weights of individual seeds were obtained on a Mettler AE 163 balance. Seeds were soaked in 70% sulfuric acid and then manually scarified to stimulate germination. For each recipient, half of the seeds from PP and DP were sown individually in clay pots (12.5 cm diam.). This was referred to as the non-competitive experiment. Remaining seeds were sown in vermiculite and, as seedlings emerged, groups of 4 were transplanted to clay pots (12.5 cm diameter). Each pot received 2 seedlings from PP and 2 from DP arranged in a square with seedlings from the same pollination regime diagonal to one another. This was referred to as the competitive experiment. Both experiments were placed on benches in a glasshouse during the summer of 1984. We recorded date of emergence of seedlings and, at the end of the experiment, we noted plant height and, on a per plant basis, number of seeds produced, total seed weight, pericarp weight, and stem weight.

Seeds from DP were 8% heavier than those from PP (11.13 vs. 10.27 mg, fresh weight; $F = 51.8$, $p < 0.001$, $n = 397$ seeds). This trend is not surprising as fruits resulting from DP contained fewer seeds. Emergence after scarification was approximately 80% for seeds from both pollination regimes in the non-competitive experiment ($X^2$ on raw data = 0.08, d.f. = 1, $p > 0.5$).

Plants arising under the two pollination treatments performed differently in the competitive and non-competitive experiments. Under non-competitive conditions, plants arising under DP tended to be taller, heavier, and had greater seed production than those arising under PP, though only the difference in height was significant (Table 3). In contrast, under competitive conditions, plants from PP were significantly more vigorous than those from DP (Table 3). Though significant, these differences were not great, ranging from 6-12%, and pollination treatment and recipient plant together typically accounted for only 15-20% of the variance in plant performance. When variance due to initial seed weight was removed using ANCOVA, plants arising under pure pollination outperformed their counterparts in seed no., seed weight, and total weight by 13-15%, suggesting that their advantage would have been greater had seeds from the two pollination treatments started with the same weight. Even with the inclusion of initial seed weight, ANCOVA accounted for no more than 30% of the variance in plant performance.

Table 3. Height, seed production, and total weight of plants from pure and dilute pollinations under competitive (164 plants) and non-competitive (183 plants) conditions.

| | HEIGHT (cm) | SEED NUMBER | TOTAL SEED WEIGHT | TOTAL PLANT WEIGHT |
|---|---|---|---|---|
| NON-COMPETITIVE | | | | |
| DILUTE POLLEN | 80 | 416 | 3.40 | 9.68 |
| PURE POLLEN | 76 | 391 | 3.16 | 9.00 |
| P* | <0.02 | NS | NS | NS |
| COMPETITIVE | | | | |
| DILUTE POLLEN | 67 | 192 | 1.38 | 3.73 |
| PURE POLLEN | 71 | 213 | 1.52 | 4.17 |
| P* | <0.01 | <0.05 | <0.05 | <0.02 |

* Analysis of covariance, covariate = recipient plants.

**4 Discussion**

Individuals of *C. fasciculata* differentially abort lightly pollinated ovaries. We have shown here that such ovaries can become mature fruits and that seeds from these fruits, despite an initial size advantage, may be outperformed by seeds from heavily pollinated ovaries, at least under competitive conditions. These differences in performance were probably due to the intensity of gametophyte competition at pollination. However, we have not excluded the possibility that talc, the diluting agent, may have had an effect. Our results are similar to those of McKenna and Mulcahy (1983), who found that the effects of pollen competition in *Dianthus chinensis* became evident only when seedlings were grown under interspecific competition. They are also relevant to the notion that fruit abortion in *C. fasciculata* is an adaptation allowing elimination of inferior offspring during their development (Stephenson 1981, Lee 1984).

*Acknowledgements*. We thank B. Hood, C. Gitschier, S. Reed, and S. Hon for assistance with the research. Funding was provided by the New Hampshire Agricultural Experiment Station (Hatch 273).

**References**

Lee TD (1984) Patterns of fruit maturation: a gametophyte competition hypothesis. Amer Natur 123:427-432.

Lee TD and Bazzaz FA (1982) Regulation of fruit maturation pattern in an annual legume, *Cassia fasciculata*. Ecology 63:1374-1388.

McKenna MA and Mulcahy DL (1983) Gametophytic competition in *Dianthus chinensis*: effect on sporophytic competitive ability. pp. 419-424 *in* DL Mulcahy (ed) Pollen: biology and applications in plant breeding. Elsevier, New York.

Mulcahy DL (1979) The rise of the angiosperms: a genecological factor. Science 206:20-23.

Stephenson AG (1981) Flower and fruit abortion: proximate causes and ultimate functions. Annu Rev Ecol Syst 12:253-279.

# Variation of Reproductive Success Rates of Ovule and Pollen Deposited upon Stigmas According to the Different Number of Pollen on a Stigma in Angiosperm

H. NAMAI AND R. OHSAWA[1]

## 1 Introduction

In this Chapter,we would like to bring up a new concept,namely the "Reproductive Success Rate(s)"(abbreviated as RSR) of ovule and pollen deposited upon stigmas. The RSRs of ovule and pollen on stigmas mean the usual seed set percentage and the percent number of seeds to the number of pollen on a stigma,respectively. The concept of "RSR" is surely an important and convenient part of the terminology used,to understand true pollen competition,especially postpollination competition.

In many species,far more pollen are frequently deposited up on stigmas than needed to fertilize all ovules contained in the ovary(Levin and Berube 1972;Mulcahy et al.1983). Huxley (1942) explained that competition among pollen in pistils is likely to occur,and suggested that the competition should result in the rapid growth of pollen tubes. The number of pollen on a stigma influences the pollen competition,which in turn exerts variuos effects on the pollen tube development(Brewbaker and Majunder 1961;Jennings and Topham 1971; Ottaviano et al.1975) and the fruit and seed set percentage(Mulcahy et al. 1975).

When less pollen grains are deposited on the stigma the variations among the resultant seedling growth is greater;the resultant seedling variation is not simply an artifact of the variation of the seed size,even though the mechanism of this phenomena is uncertain(Ter-Avanesian 1978). On the other hand, seeds resulting from intense competition owing to excessive

[1]Institute of Agriculture and Forestry,University of Tsukuba,Tsukuba Science City,Ibaraki,305 Japan

pollen deposition produce seedlings that are significantly more vigorous(Mulcahy and Mulcahy 1975;Mulcahy et al.1975). However,few studies supply corresponding,detailed information on the variances,in the RSRs,of ovules in a flower and the pollen deposited upon its stigma , according to the quantity of pollen deposited on the stigmatic surface. The RSRs should be strongly correlated to the number of self- and/or cross-pollen on a stigma,that is the pollen-ovule ratio,and fruit and seed set percentage. The terminology "pollen-ovule ratio" used here is consistent with Cruden(1977) who interpreted it as the ratio of whole released pollen to whole ovules in each flower. Here,we will use it as the ratio of pollen deposited upon a stigma to ovules in each flower.

Needless to say,male gametic competition in plants is occurring in the two following scope:

(1) Prepollination competition; competition among pollen donor plants for access to specific stigmas,

(2) Postpollination competition; competition among pollen on specific stigmas for access to ovule in ovaries.

The purpose of this chapter is to present a summary of the information on the RSRs of ovule and pollen deposited upon the stigma,with special reference to the correlation between gametic competitions and the quantity of pollen deposited upon stigmas in postpollination. We consider that studies of these phenomena are very important not only for improving plant evolusion,but also for the progress of plant breeding.

## 2 Current Research

In this section,we shall describe two scope of our current research on the RSRs of ovules and pollen deposited on a stigma in Brassica juncea(leaf mustard),a flower having about twenty ovules per pistil,and Fagopyrum esculentum(buckwheat), a flower having only one ovule per pistil. Usually,the former is self-compatible but the latter is heterostylistic self-incompatible.

The plants used for artificial pollination experiments were grown in pots,in an insect-proof glasshouse. In Brassica,various quantities of self-pollen were roughly applied to the stigmas with fine wire tools(0.1mm and 0.5mm in diameter). Two days after pollination,all stigmas were removed and the number of pollen deposited on each stigma was counted microscopically by aceto-carmine squash method. In Fagopyrum, various quantities of reciprocal legitimate pollen were accurately applied to the stigmas of plants with long and with short styles by a fine wire instrument(0.3mm in diameter) under an operation microscope.

The experiments with insect pollination in Brassica were carried out in isolation cages with 12 plants. Shimahanaabu (Eristalis serealis, an ally of the flower fly), artificially reared, were used as a pollinator.

(1) Brassica juncea

In the experiments of artificial self-pollination(Table 1), the RSR of ovules increased from 13.5 in the flowers with a few self-pollen to 81.4%, as compared with those having larger numbers. Contrary, the RSR of pollen deposited on stigmas decreased notably from 23.7 to 2.1%. As seen in Fig.1-A, the flowers with much the same number of self-pollen as the number of ovules per pistil showed about 30% on the average in both the RSRs. Moreover, the flowers with 5 to 20 self-pollen, set one or more seeds, were about 40% in the RSR of pollen deposited on stigmas(Fig.1-B). The pollen-ovule ratios of the flowers with about 20 self-pollen were almost 1:1. In addition, the results of insect pollinations, self and crossed, were slightly higher in both the RSRs than that of the artificial self pollination; but showed fairly similar to those of self pollination(Fig.2).

Under the natural conditions, the number of pollen deposited on a stigma seemed to be more than 500(Ohsawa and Namai 1984a, b). Table 2 shows the variations of expected number of seeds obtained from 500 self-pollen with different number of pollen

Table 1. Effect of the number of pollen on a stigma on the reproductive success rates(RSRs) of ovule and selfed pollen in Brassica juncea cv.Hakarashina

| Number of pollen deposited | Pollen ovule ratio | Number of flowers observed | % pod set | RSR of ovule | RSR of pollen |
|---|---|---|---|---|---|
| 1 - 20 | 0.6 | 55 | 78.2 | 13.5 | 23.7 |
| 21 - 40 | 1.5 | 61 | 83.6 | 24.9 | 17.0 |
| 41 - 60 | 2.7 | 34 | 97.1 | 33.1 | 12.4 |
| 61 - 100 | 4.1 | 40 | 90.o | 46.3 | 11.3 |
| 101 - 150 | 6.2 | 49 | 93.9 | 56.8 | 9.2 |
| 151 - 200 | 8.9 | 35 | 94.3 | 55.1 | 6.2 |
| 201 - 400 | 18.0 | 12 | 100.0 | 77.4 | 4.3 |
| 401 -1000 | 30.5 | 13 | 100.0 | 81.4 | 2.1 |

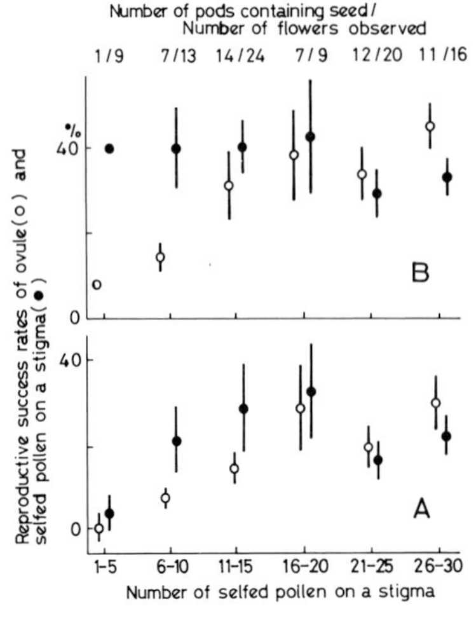


Fig.1. Correlation between the number of artificially selfed pollen on a stigma and RSRs of ovules and the pollen on the stigma of Brassica juncea cv.Hakarashina. A:all flowers pollinated, B:flowers only set seeds.

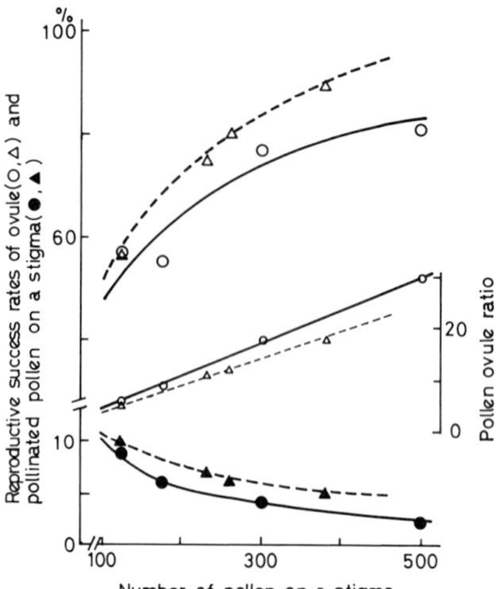


Fig.2. Correlation between the number of pollen deposited on a stigma and RSRs of ovules and artificially selfed and insect pollinated pollen on the stigma of B. juncea. Circle:self pollination in cv. Hakarashina. Triangle:insect pollination in cv.Kikarashina.

per stigma,from calculation of our data of the RSR of ovules in B.juncea cv.Hakarashina(Table 1 and Fig.1). Pollination of 500 pollen on a flower(pollen-ovule ratio=25:1) produces only 16 seeds in all,but about 25 flowers with about 20 pollen per stigma(pollen-ovule ratio = ca 1:1)produce nearly 100 seeds.

(2) Fagopyrum esculentum

Although the RSR of ovules increased clearly from 40% in the

Table 2. Variation of expected number of seeds obtained from 500 pollen grains by self pollination with different number of pollen grains per stigma

| Number of pollen per stigma | Possible no. of flowers pollinated | Expected no. of seeds obtained |
|---|---|---|
| 5 | 100 | 62.1 |
| 10 | 50 | 117.5 |
| 20 | 25 | 122.3 |
| 25 | 20 | 100.4 |
| 50 | 10 | 66.2 |
| 100 | 5 | 51.6 |
| 250 | 2 | 31.0 |
| 500 | 1 | 16.3 |

Calculated from the data of RSR of ovules in Brassica juncea.

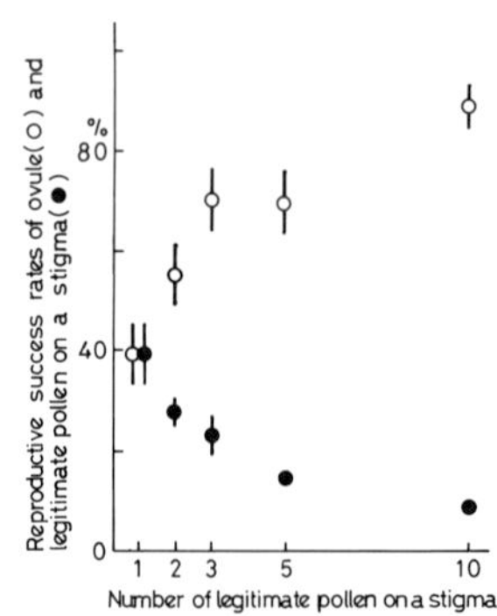


Fig.3. Correlation between the number of legitimate pollen on a stigma and RSRs of ovules and legitimate pollen on the stigma of Fagopyrum esculentum cv.Botan-soba.

flowers with only one legitimate pollen to more than 80% in those with 10 legitimate pollen,the RSR of legitimate pollen deposited on stigmas decreased notably from 40% to less than 10%(Fig.3). The pollen-ovule ratio of flowers with one legitimate pollen is just 1:1.

## 3 Discussion and Conclusion

Even if self- and cross-pollen are equally deposited from the donors at the same ratio by insect pollinators,pollen competition in postpollination should occur under either of the following cases; (1) many flowers receiving various kinds of pollen are unable to bear any seeds,and (2) more viable pollen is deposited on the stigmas than there are ovules in the ovary (Stephenson and Bertin 1983).

From our results and limited available data on the number of viable pollen deposited on the stigmas(Fig.4 and 5),the flowers showed more than 80% in RSR of ovules,when excessive pollen are deposited. Pollen on the stigmas in various plant species tend to have a law RSR percentage,but the RSRs of pollen are highest in all the species when pollen-ovule ratios, viz.the number of pollen deposited upon a stigma to the number of ovules in each flower,are just or about 1:1.

We conclude that the greater variation in growth behavior among the resultant plants,from limited pollination(Ter-Avanesian 1978),ought to be derived through a higher RSR of pol-

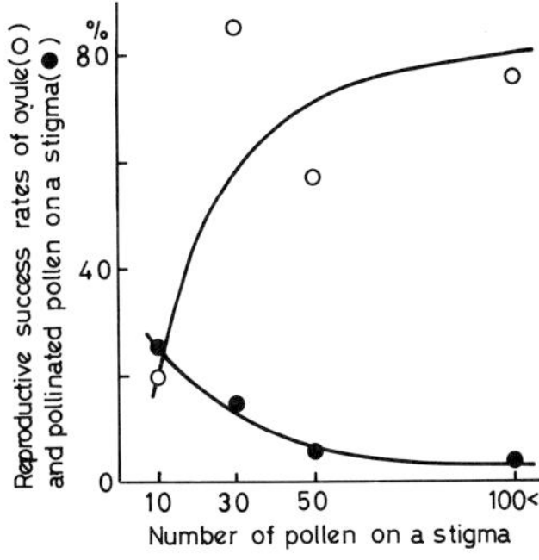


Fig.4. Correlation between the number of artificially crossed pollen on a stigma and RSRs of ovules and the pollen on the stigma of *Geranium maculatum*(adapted from Mulcahy et al.1983). Ten ovules but a maximum of five seeds per flower.

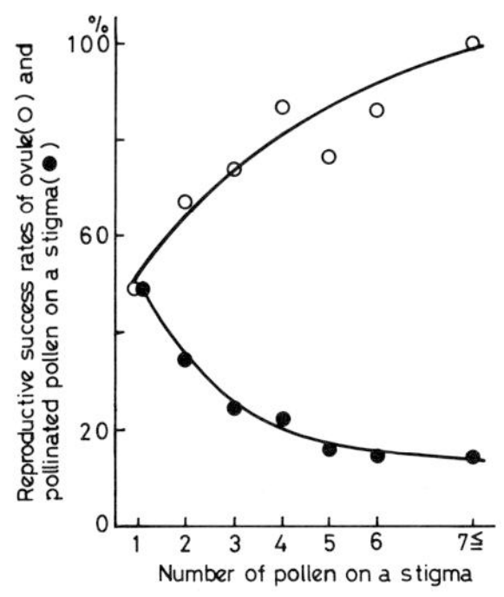


Fig.5. Correlation between the number of artificially selfed pollen on a stigma and RSRs of ovules and the pollen on the stigma of *Mirabilis jalapa*(adapted from Cruden 1977). Only one ovule per flower.

len deposited on stigmas,under limited pollination coditions wherein about 1:1 in pollen-ovule ratio. Furthermore,it appeared that those with limited pollination also give rise to the even intensity of the RSRs of ovules and pollen deposited on stigmas,and to the lowest intensity of gametic competition in postpollination. The gametic competition,we presume,have an important potential for an evolutional change and stability in the plant population. Therefore,we should emphasize the necessity for closely considering the quantity of pollen deposited on stigmas,viz.the number of pollen on a stigma to the number of ovules per flower,when we study the RSRs,and the gametic competition in plant species.

References

Brewbaker J L and Majumder S K (1961) Cultural studies on the pollen population effect and the self-incompatibility inhibition.Am J Bot 48:457-464.

Cruden R W (1977) Pollen-ovule ratio: A conservative indicator of breeding systems in flowering plants. Evolution 31:32-46.

Huxley J (1942) Evolution,the Modern Synthesis. Harper,New York.

Jennings D L and Topham P B (1971) Some consequences of raspberry pollen dilution for its germination and or fruit development.New Phytol 70:371-380

Levin D A and Berube D E (1972) Phlox and Colius: The efficiency of a pollination system. Evolution 26:242-250.

Mulcahy D L,Mulcahy G B(1975) The influence of gametophytic competition on on sporophytic quality in Dianthus chinensis. Teor Appl Genet 46:277-280.

Mulcahy D L,Mulcahy G B and Ottaviano E (1975) Sporophytic expression of gametophytic competition in Petunia hybrida. In "Gamete Competition in plants and Animals" (Mulcahy D L ed) pp227-232. North-Holland PublAmsterdam

Mulcahy D L,Curtis P S and Snow A A (1983) Pollen competition in a natural population.In "Handbook of Experimental Pollination Biology" (Jones C E and Little R J ed) pp.330-337.Scientific and Academic Editions.New York.

Ohsawa R and Namai H (1984a) The relations between number of pollinators and number of pollen on stigma in Brassica crops under isolation cages. Japan J Breed 34(Suppl 2):74-75.

Ohsawa R and Namai H (1984b) Using insect pollinators improves seed yield in Brassica juncea under isolation cages.Cruciferae Newsletter 9:43-44.

Ottaviano E,Golra M S and Mulcahy D L (1975) Gametic and intergametophytic influences on pollen tube growth. In "Gamete Competition in Plants and Animals"(Mulcahy D L ed) pp.125-130. North-Holland Publ,Amsterdam.

Stephenson A G and Bertin R I (1983) Male competition,female choice,and sexual selection in plants. In "Pollination Biology"(Real L ed) pp.109-149. Academic Press,New York.

Ter-Avanesian D V (1978) The effect of varying the number of pollen grains used in fertilization. Teor Appl Genet 52:77-79.

# Effects of Pollen Load Size on Fruit Maturation and Sporophyte Quality in Zucchini

A.G. STEPHENSON[1], J.A. WINSOR[2], AND L.E. DAVIS[1]

## 1 Introduction

Many plant species, including many economically important species, regularly produce far more flowers than mature fruits (Lloyd 1980; Stephenson 1981; Sutherland and Delph 1984). Since these species are physiologically incapable of providing the necessary resources to develop mature fruits from all of their flowers, they commonly abort a sizeable portion of their immature fruit crops in order to match their fruit production to their available resources (see Stephenson 1981). Moreover, fruit abortion in many species is selective (i.e., non-random). For example, fruits with a below average number of developing seeds are commonly the most likely to abort (see review by Stephenson and Bertin 1983). Because differences in seed number among the developing fruits on a plant are often due to differences in the number of pollen grains deposited by the pollinators, Lee (1984) hypothesized that, by aborting fruits with low seed numbers, plants may actually improve the average quality of their offspring by eliminating those fruits in which there was little pollen competition for access to the ovules. In this chapter, we report that common zucchini (*Cucurbita pepo*) selectively aborts fruits on the basis of seed number. Moreover, the progeny from fruits produced by high pollen loads (containing a full complement of seeds) are more vigorous than the progeny from fruits produced by low and medium pollen loads (containing less than a full complement of seeds). This study

[1] Department of Biology, Pennsylvania State University, University Park, PA 16802

[2] Department of Biology, Pennsylvania State University, Altoona, PA 16603

is a portion of a larger project investigating the regulation of offspring quality in zucchini by non-random fertilization and selective fruit abortion.

## 2 Plant Description, Study Site and Methods

One hundred and thirty Black Beauty Bush Variety Zucchini plants (Agway Inc., Lot R-8, V-893) were grown two meters apart (5 rows of 26 plants) in a field at The Pennsylvania State University Agricultural Experimental Station at Rock Springs in Centre Co., PA. This variety is a monoecious short-internode vine with indeterminate growth and reproduction. The flowers are born individually, and each plant produces about 8-12 pistillate flowers (4 to 6 mature fruits) and 20-30 staminate flowers over the course of the growing season.

In order to determine if zucchini selectively aborts fruits on the basis of seed number, we varied the number of pollen grains deposited onto the stigmas of 15 plants. On 5 of the 15 plants, the first pistillate flower produced by each plant received a low pollen load (240 ± 36 pollen grains); the second pistillate flower on each plant received a medium pollen load (2 times the low load); and the third flower on each plant received a high pollen load (stigma saturated with pollen). We then repeated this pollination cycle on the successive flowers produced by each plant. On 5 more of the 15 experimental plants, we rotated the pollen loads in the following cycle: medium then high then low. On the final 5 experimental plants, the order of the pollination cycle was high then low then medium pollen loads. Each pistillate flower was covered with a cheesecloth bag from one day before anthesis to two days after anthesis to exclude pollinators. Each day, we collected one staminate flower from a minimum of 10 plants at the field site. The pollen from these flowers was removed with a small brush, placed into a plastic container, thoroughly mixed, and carefully applied to the stigmas using a stainless steel rod with a diameter of 1.5 mm (for details see Winsor et al., in prep.). The developing fruits were tracked throughout the growing season and, in October, the mature fruits were harvested. The seeds of each fruit were counted, air dried, and stored at room temperature in a paper envelope.

To assess the vigor of the seedlings produced by different pollen loads, we performed the following study in March 1984. First, we identified all of the seeds weighing .13 to .15 g from one high pollen load fruit and one low or medium pollen load fruit from each of the three plants. We specifically selected pairs of fruits from each plant that were produced during the same week of the growing season. Consequently, variance due to seed size and seasonal effects, such as the age of the fruit at the time of the first killing frost, are minimized. We then planted each seed in a 4 liter pot containing equal parts peat, perlite, and vermiculite; we randomly assigned each pot to a bench site in the greenhouse; we added fertilizer weekly and watered when necessary. Finally, we recorded the number of days to seedling emergence, the number of leaves at 20 and 30 days after emergence, and the dry weight of the above-ground parts at 30 days after emergence.

## 3 Results and Discussion

On the 15 experimental plants, only 18% of the low pollen load flowers and only 28% of the medium pollen load flowers produced a mature fruit, while 71% of the high pollen load flowers produced mature fruits. A chi-square test reveals that the probability that a flower will produce a mature fruit is not independent of the size of the pollen load ($\chi^2 = 40.6$; df = 2; $p < 0.0001$). Clearly, zucchini selectively matures the fruits from the high pollen load flowers. Moreover, an analysis of variance (GLM; SAS 1982) reveals that there are significant ($p < 0.0001$) differences in the number of seeds per fruit produced by the three pollen load sizes. The low, medium and high pollen loads produced 13.0 ± 7.8 ($\bar{X}$ ± S.E., N=11), 77.4 ± 30.7 (N=17), and 289.6 ± 26.5 (N=44) seeds per fruit, respectively. Because the fruits from low and medium pollen loads contained fewer seeds than the fruits from high pollen loads, it is reasonable to assume that the fruits from low and medium pollen loads were produced under conditions of little or no pollen tube competition.

In the greenhouse study of seedling vigor, we knew the maternal parent (one of three plants) and the size of the pollen load that produced each seedling. Consequently, we were able to examine, using an analysis of variance, the effects of parent and pollen load size on the number of days required for seedling emergence, the number of leaves at 20 and 30 days after emergence, and dry weight of each plant 30 days after emergence (Table 1). The analysis reveals that seedling vigor is strongly influenced by both the maternal parent and the size of the polen load that produced the seed (Table 1). That is,

Table 1. Zucchini Seddling Vigor. F values and probabilities from an analysis of variance (GLM: SAS 1982) for vigor of seedlings produced from three pollen load sizes on three maternal parents.

| | Degrees of Freedom | Days to Emergence | Number of Leaves | | Dry Weight |
|---|---|---|---|---|---|
| | | | 20 Days | 30 Days | |
| Maternal Parent | 2,155 | 8.42 $p<0.001$ | 4.01 $p<0.05$ | 6.43 $p<0.01$ | 1.75 N.S. |
| Size of Pollen Load | 2,155 | 6.35 $p<0.01$ | 11.17 $p<0.001$ | 6.59 $p<0.01$ | 3.57 $p<0.05$ |
| Parent x Pollen Load Interaction | 1,144 | .32 N.S. | .03 N.S. | 2.10 N.S. | 1.12 N.S. |

seeds produced by high pollen loads (high pollen competition) give rise to more vigorous seedlings than do seeds produced by either low or medium pollen loads (without pollen competition). For example, the seedlings produced by the high pollen fruit on Plant 6 (one of the three plants used in this study) emerged more rapidly, had more leaves at 20 and 30 days after emergence, and had a greater dry weight than those seedlings produced by low pollen loads (Table 2).

Table 2. Vigor of Zucchini Seedlings Produced by Maternal Parent 6. $\bar{X} \pm$ S.D.

| Size of Pollen Load | N | Days to Emergence | Number of Leaves | | Dry Weight (Grams) |
|---|---|---|---|---|---|
| | | | 20 Days | 30 Days | |
| Low | 27 | 9.1±1.5 | 7.5±1.2 | 8.1±1.1 | 6.66±2.37 |
| | | * | *** | ** | * |
| High | 16 | 7.8±1.8 | 9.4±1.5 | 9.3±1.4 | 8.04±2.46 |

*$p<0.05$; **$p<0.01$; ***$p<0.001$

Mulcahy (1979) proposed that vigorous microgametophytes (fast growing pollen tubes) produce vigorous progeny. Support for Mulcahy's proposal comes from recent studies showing that a large portion of the microgametophytic genome is transcribed, that the genotype of the pollen tube influences its growth rate, and that many (60% or more) of the genes expressed by the microgametophyte are also expressed by the sporophyte (see review by Mulcahy 1979; also Tanksley et al. 1981; Willing and Mascarenhas 1984; and several chapters on "Gene Expression in Pollen" in this book). Our data, which show a relationship between the degree of pollen competition and seedling vigor, are also consistent with Mucahy's proposal. That is, when few pollen grains are deposited onto a stigma (low and medium pollen loads) both the fast and the slow pollen tubes affect fertilization while only the most vigorous pollen tubes achieve fertilization when many grains are deposited onto a stigma (high pollen loads).

This study is also consistent with Lee's (1984) hypothesis that plants can improve the average quality of their offspring by "over-producing" flowers and then selectively aborting those fruits with the fewest seeds (fruits in which there was little or no pollen tube competition for access to their ovules). Our finding has profound evolutionary implications for natural populations where plants have little control over the number or the genotype of the pollen grains deposited onto their stigmas. Selective abortion can be viewed as a means of regulating offspring quality by influencing the paternal parentage of the progeny. In this regard, selective abortion is similar to self-incompatibility systems and other pollen-style interactions that lead to non-random fertilization. Finally, agricultural researchers, who are attempting to improve yield by decreasing the levels of fruit abortion, should be cautioned that reductions in fruit abortion may be accompanied by reduced offspring vigor.

Acknowledgments. We thank George Conrad, his staff and the Department of Horticulture for use of the Agricultural Experiment Station, V.A. Borowicz and S. T. Stephenson for field assistance, R. Voytko for assistance in the laboratory and B. Devlin and C. Schlichting for comments on a previous draft. This project was supported by Hatch and General Funds from The Pennsylvania State University Agricultural Experiment Station, Project 2683.

## References

Lee TD (1984) Patterns of fruit maturation: A gametophyte competition hypothesis. Amer Nat 123: 427-432

Lloyd DG (1980) Sexual strategies in plants. I. An hypothesis of serial adjustment of maternal investment during one reproductive session. New Phytol 86: 69-80

Mulcahy DL (1979) The rise of the angiosperms: a genological factor. Science 206: 20-23

SAS Institute Inc (1982) SAS User's Guide: Statistics. Cary, NC

Stephenson AG (1981) Flower and fruit abortion: Proximate causes and ultimate functions. Annu Rev Ecol Syst 12: 253-279

Stephenson AG, Bertin RI (1983) Male competition, female choice and sexual selection in plants. In: Real L (ed) Pollination Biology. Academic Press, pp 109-149

Sutherland S, Delph LF (1984) On the importance of male fitness in plants: Patterns and fruit set. Ecology 65: 1093-1104

Tanksley SD, Zamir D, Rick CM (1981) Evidence for extensive overlap of sporophytic and gametophytic gene expression in *Lycopersicon esculentum*. Science 213: 453-455

Willing RP, Mascarenhas JP (1984) Analysis of the complexity and diversing of mRNAs from pollen and shoots of *Tradescantia*. Plant Physiol 75: 865-868

# Pollen Heteromorphism as a Tool in Studies of the Pollination Process in Pontederia cordata L.

Spencer C.H. Barrett and Lorne M. Wolfe[1]

## 1 Introduction

Heterostylous breeding systems provide useful experimental material for studies of mating system evolution and the influence of natural selection on floral form. Recent work has focused on the functional significance of the syndrome of traits that comprise the heterostylous syndrome (Ganders 1979). The conspicuous size differences among pollen types in heterostylous species enables a distinction to be made between compatible and incompatible pollen present on stigmas. Species studies of stigmatic pollen loads have been undertaken to evaluate whether the stamen-style polymorphism promotes phenotypic disassortative pollination among the floral morphs as Darwin (1877) originally proposed. Studies of deposition patterns have largely ignored other components of the pollination process and as a result little quantitative information is available on the pollination biology of most heterostylous species.

As part of a comprehensive study of tristylous breeding systems in the monocotyledonous family Pontederiaceae, we have examined in detail the pollination biology of several Pontederia species. In this chapter we review some of this work and then describe several field experiments concerned with the dynamics of pollen transport in P. cordata. We organise our discussion by addressing three questions: 1) How effective are insect visitors at removing pollen from the

---

1 Department of Botany, University of Toronto, Toronto, Ontario, Canada M5S 1A1

three stamen levels of tristylous flowers? 2) Does the complementary placement of sexual organs in tristylous flowers result in significant pollen partitioning on the bodies of pollinators? 3) How much compatible and incompatible pollen is deposited on individual stigmas by pollinators?

## 2 Floral Biology of Pontederia

Pontederia is composed of five species of long-lived, clonal, emergent aquatics native to the New World. Details of the tristylous breeding systems and reproductive biology of populations can be found in Barrett (1977), Price and Barrett (1982), and Glover and Barrett (1983). Most work has been conducted on N. American populations of *P. cordata*. Figure 1 illustrates salient features of the tristylous syndrome.

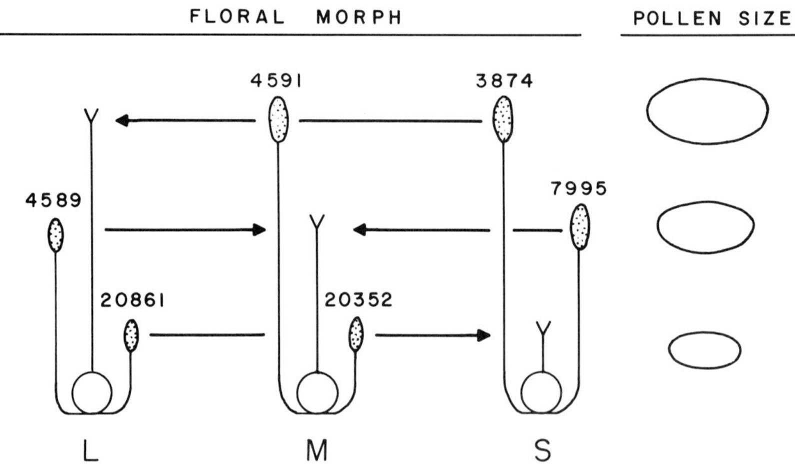


**Fig. 1** Schematic diagram of the relationship between pollen size, pollen production, and stamen-style polymorphism in *Pontederia cordata*. After Price and Barrett (1984).

Three points are worth noting: 1) There is virtually no overlap in the size of pollen produced by the three stamen levels. 2) Large pollen production differences occur among anthers and morphs with approximately 2/3 of all pollen produced in populations originating from short-level anthers and mid-level anthers of the S morph producing twice as much pollen as the corresponding anthers of the L morph (Price and Barrett 1982; Barrett et al. 1983). 3) There is variation in the expression of trimorphic incompatibility among morphs with the M morph considerably more self-compatible than the L and S morphs (Barrett and Anderson, 1985). These three peculiarities seem to be general features of the tristylous syndrome of Pontederia species.

Pontederia flowers are pollinated by a range of bees especially long-tongued solitary bees in tropical regions (Ancyloscelis and Florilegus spp.) and Bombus spp. and Melissodes apicata in North America. The blue, mauve, or white, uniovulate flowers are produced in showy spicate inflorescences, and anthesis usually lasts for 6-8 hours. A study of the dynamics of pollination in Ontario populations of P. cordata (Wolfe 1985) revealed that the daily events associated with pollination occur very rapidly. At peak flowering virtually all flowers in populations are pollinated between 2-3 hours from the beginning of anthesis and pollen tubes can be detected at the base of the style soon after. Experimental study of the number of compatible pollen grains required to ensure seed set with regularity indicate that 3-5 per flower are sufficient. The low pollen requirement and high bumblebee densities that service populations in many N. American populations result in near maximal seed set. Attempts to elevate seed set above open-pollinated controls by hand cross-pollinations have been unsuccessful, suggesting that ovules are rarely pollen limited at least where bumblebees are the primary pollinators.

## 3 Pollen Removal

Notwithstanding the growing awareness of the importance of male function to plant fitness, few studies have quantified pollen removal from stamens. Such considerations may be particularly significant in P. cordata since early bee visits to flowers are likely to be important in increasing male mating success. We examined how stamen position influences the quantity of pollen removed from previously unvisited flowers following a single bumblebee (B. griseocollis) visit. Pollen grain counts to visited and unvisited flowers were made using a hemacytometer. The results (Table 1) indicate that 1/3-1/2 of the total number of pollen grains produced by flowers are removed by the first bee visit. Relative to the total number of pollen grains in unvisited flowers, significantly different fractions of pollen are removed from the three stamen levels (l 68.5%, m 50.0%, s 37.5%). Measurements of the rate of pollen removal during the anthesis period of flowers indicate that within 60 minutes of anther dehiscence more than 1/2 of all pollen within a population is removed from stamens by pollinators.

**Table 1** Pollen removal and pollen deposition in previously unvisited flowers of Pontederia cordata following a single visit by Bombus. Removal data based on 20 flowers per morph, values are the mean percentage of pollen grains removed from each stamen level. Deposition data from 60, 85, and 58 stigmas of the L, M, S morph, respectively (L.M. Wolfe and S.C.H. Barrett unpublished data).

| | Floral Morphs | | |
|---|---|---|---|
| | L | M | S |
| A) Pollen removal | | | |
| l anthers | - | 64 | 73 |
| m anthers | 61 | - | 39 |
| s anthers | 32 | 43 | - |
| Total | 36 | 45 | 52 |
| B) Pollen deposition | | | |
| % visits with no deposition | 26 | 21 | 39 |
| Mean number of pollen grains deposited/stigma | 38 | 68 | 18 |
| % compatible pollen grains | 59 | 22 | 25 |

## 4 Pollen Partitioning on Bees

The three stamen levels in Pontederia flowers may be expected to contact different parts of a bee's body during nectar feeding. The idea of pollen partitioning is in fact central to Darwin's hypothesis on the function of the stamen-style polymorphism. A critical issue in evaluating the hypothesis is whether significant partitioning is maintained during foraging activity.

To verify that pollen is initially located on different parts of an insect's body simple observations on the distribution of pollen on dead bees inserted into flowers can be undertaken. When this was done using bumblebees and *P. cordata* flowers a high degree of segregation of the pollen types was observed with most *l*, *m*, and *s* pollen on the abdomen, head and proboscis, respectively. Under field conditions, grooming activities as well as variable body orientations during entry and exit from flowers are likely to disturb these patterns. The amount of mixing of pollen types was investigated in three taxa of bees that commonly visit flowers of *P. cordata* (*Bombus* spp., *Apis mellifera*, *Melissodes apicata*). Pollen counts were made by sampling portions of the body using a uniform size cube of fuchsin-glycerine jelly mounted on a pin. The cube was pressed onto different parts of a bee's body and then melted on a microscope slide and the pollen types counted. Fifteen free-foraging individuals of each taxon were sampled in this manner; Figure 2 illustrates the results. As can be seen the

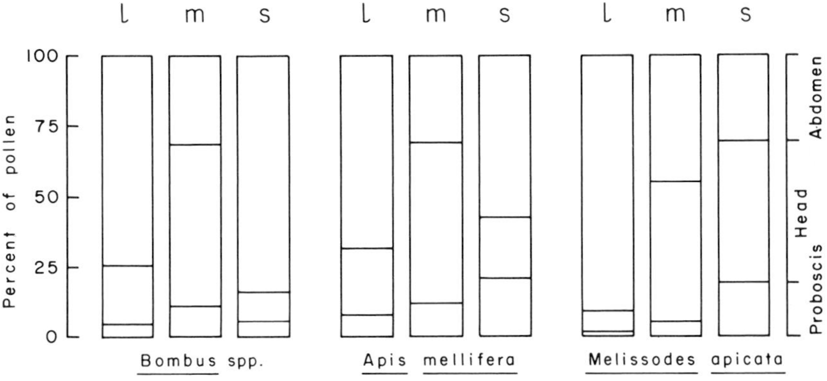


**Fig. 2** Percentage of *Pontederia cordata* pollen types on body parts of three pollinator groups. N = 15 bees per taxon. (L.M. Wolfe and S.C.H. Barrett, unpublished data.)

pollen types are not deposited similarly on each body part. Pollen from long-level anthers (l) was most abundant on the abdomen, and m pollen on the head; s pollen differed in abundance with bee group. In bumblebees and honeybees s pollen was most common on the abdomen, whereas in *Melissodes apicata* it was most frequent on the head. The data indicate that l and m pollen tends to remain in greatest concentration on the body parts where it is initially deposited, whereas s pollen tends to be displaced backwards from the proboscis to more posterior parts of the bee. The total number of pollen grains deposited on the proboscis was much lower than on the other regions sampled. This is probably due to the smaller surface area of the proboscis, for adherence of pollen, as well as the tendency for bees to remove pollen from it during cleaning activities.

Despite differences in the size, morphology and behavior of the three bee taxa some degree of pollen partitioning is maintained during foraging. This suggests that close coadaptation between *Pontederia* and its pollinators is not a prerequisite for pollen stratification and compatible pollen transfer to occur.

## 5 Pollen Deposition

Surveys of stigmatic pollen loads in natural populations of *P. cordata* indicate that despite considerable amounts of incompatible pollen on stigmas, sufficient compatible pollen is deposited by pollinating insects to ensure seed set (Price and Barrett 1984; Barrett and Glover 1985). The question arises as to how effective the first bee visit to a flower is in bringing about pollination. In a uniovulate flower early visits are of some importance and may result in most of the successful fertilizations. Results from a study of the composition of stigmatic pollen loads of previously unvisited flowers following a single bumblebee visit are presented in Table 1. While virtually all visits to flowers are successful at removing pollen a significant number result in no pollen deposition. The reduced frequency of pollinating visits in the S morph may be associated with the lower pollen load carried by the proboscis and the smaller contact areas involved in pollination. Analyses of the stigmatic pollen loads indicate that the M and S morphs receive mostly incompatible pollen from single visits whereas in the L morph compatible pollen comprises the largest fraction of the pollen load. Data from multiple-visited flowers indicate that the L morph is most likely to experience legitimate pollination whereas the M and S morphs frequently display random pollination (Barrett and Glover 1985).

## 6 Conclusion

The marked size differences among pollen types in P. cordata enables examination of fine scale micropollination events not normally investigated in homomorphic species. In addition to being able to determine the number of compatible and incompatible pollen grains deposited on stigmas in natural populations, field experiments can be designed to examine pollen carryover, the magnitude of self- and geitonogamous-pollination, as well the effects of pollen clogging on the seed set of floral morphs (Wolfe 1985; Barrett and Glover 1985). Together, these studies indicate that the pollination process in P. cordata is characterized by large asymmetries in the amount of pollen that is produced, transported and deposited on stigmas of the floral morphs. Despite this variation the seed fecundity of morphs is usually similar because of reliable pollinator service and low pollen requirements for seed set. Although differences in the fecundity of morphs are not generally evident, morph-specific male fertility differences appear to be important in regulating the relative frequency of floral morphs in natural populations (Barrett et al. 1983).

Since P. cordata is an outcrosser with high fecundity, possesses uniovulate flowers, and experiences little seed abortion, we may anticipate strong gametophytic selection if genetic variation for pollen tube growth occurs within populations. To what extent style length differences among the morphs influence the intensity of pollen tube competition remains to be determined. Detection of male fertility variation and selective effects at the gametophytic level requires controlled crosses with genetic markers and extensive progeny testing of the floral morphs in natural populations.

**Acknowledgements:** Research funded in part by grants from the Natural Sciences and Engineering Research Council of Canada to S.C.H. Barrett and grants from Sigma Xi and the University of Illinois to L.M. Wolfe.

## References

Barrett, SCH (1977) The breeding system of Pontederia rotundifolia, a tristylous species. New Phytologist 78: 209-220.

Barrett, SCH, Price, SD and Shore, JS (1983) Male fertility and anisoplethic population structure in tristylous Pontederia cordata (Pontederiaceae). Evolution 37:745-759.

Barrett, SCH and Anderson, JM (1985) Variation in expression of trimorphic incompatibility in Pontederia cordata L. (Pontederiaceae). Theoretical and Applied Genetics (in press).

Barrett, SCH and Glover, DE (1985) On the Darwinian hypothesis of the adaptive significance of tristyly. Evolution (in press).

Darwin, C (1877) The different forms of flowers on plants of the same species. John Murray, London.

Ganders, FR (1979) The biology of heterostyly. New Zealand Journal of Botany 17:607-635.

Glover, DE and Barrett, SCH (1983) Trimorphic incompatibility in Mexican populations of Pontederia sagittata Presl. (Pontederiaceae). New Phytologist 95: 439-455.

Price, SD and Barrett, SCH (1982) Tristyly in Pontederia cordata (Pontederiaceae). Canadian Journal of Botany 60:897-905.

Price, SD and Barrett, SCH (1984) The function and adaptive significance of tristyly in Pontederia cordata L. (Pontederiaceae). Biological Journal of Linnean Society 21:315-329.

Wolfe, LM (1985) The Pollination Dynamics of Pontederia cordata L. (Pontederiaceae). M.Sc. Thesis, University of Toronto, 129 pp.

# Heterostyly and Microgametophytic Selection: The Effect of Pollen Competition on Sporophytic Vigor in Two Distylous Species

MARY A. MCKENNA[1]

## Introduction

The gametophytic life stage has received relatively little attention from plant ecologists, perhaps because it has a short life span and its interactions lie hidden within the pistil. Pollination ecologists have traditionally focussed attention on the mechanics of pollen transfer and its consequences for seed set, excluding consideration of the interval between pollen deposition and fertilization. This interval may play a significant role in plant mating systems however, since differences in the growth rate of individual pollen grains can result in an unequal probability of fertilizing ovules,and the presence of large numbers of pollen tubes in the pistil creates an opportunity for gametophytic selection due to differences in pollen tube growth rates.

Evidence of post-meiotic gene expression in pollen and the joint expression of genes in the gametophyte and the sporophyte ( Tanksley et al.,1981; Willing and Mascarenhas,1984) suggests that microgametophytic selection can influence the character of the sporophytic generation. This raises interesting ecological and evolutionary questions and suggests a potential application of microgametophytic selection in plant breeding programs. In view of these considerations, it is important to determine how gametophytic selection influences sporophytic fitness. One way to approach this question is to determine how sporophytic quality varies with the intensity of pollen competition.

The intensity of pollen competition is influenced by the amount of pollen deposited on the stigma relative to the number of receptive ovules, and by the placement of pollen proximal or distal to the ovary. Heterostylous plants provide a natural system to examine both methods of varying the intensity of pollen competition. Two distylous species were chosen for these studies: _Turnera ulmifolia_ L. (Turneraceae) and _Anchusa officinalis_ L. (Boraginaceae).

---

[1] Department of Ecology and Evolution, State University of New York at Stony Brook, Stony Brook, New York 11794

## Microgametophytic Selection in Turnera ulmifolia

*Turnera ulmifolia*, a neotropical weed, exhibits the typical suite of characters associated with distyly. The long style (pin) morph exhibits short anthers, small pollen grains and relatively more pollen grains per anther; while the short style (thrum) morph exhibits long anthers, large pollen grains and relatively fewer pollen grains per anther ( Barrett, 1978; Shore and Barrett, 1984a). These morphological features are associated with the classical "*Primula* type" incompatibility system; *Turnera ulmifolia* is both self and intramorph incompatible, and dominance is expressed in the thrum phenotype (Martin,1965; Shore and Barrett,1984b). Anisoplethy has been reported in *Turnera* (Bentley,1977),but most populations exhibit a 1:1 ratio of pins to thrums (Barrett,1978; Martin,1965).

A series of intermorph crosses were made in a greenhouse population of twenty thrum and twenty pin morphs of *Turnera ulmifolia*. These crosses were designed to investigate the effect of the intensity of pollen competition on offspring quality. Variation in the amount of pollen deposited on a stigma was used to vary the intensity of pollen competition; excess and limited pollination treatments were created by applying a 6:1 ratio of pollen. Crosses were made by brushing newly dehiscent anthers from one pollen donor across the stigmatic surface; the limited treatment represented 1 anther per stigma and the excess treatment represented 6 anthers per stigma ( 2 anthers per stigma lobe). Seeds were harvested at maturity and a random subset were individually weighed after removing the elaisome. All seeds were planted individually in 2" pots, positioned randomly on a greenhouse bench, and individual seedling heights were measured at weekly intervals. After 6 weeks all seedlings were weighed, transplanted to 6" pots and grown to flowering to determine the offspring population morph ratio.

TURNERA ULMIFOLIA

| | Pollination Treatment | | |
|---|---|---|---|
| | Excess | Limited | |
| seed weight (mg) | Y 1.54 ±0.04<br>n 51 | Y 1.14 ±0.03<br>n 33 | $F_s$=44.58<br>$p<0.001$ |
| seedling height (mm) | Y 79.60±2.45<br>n 52 | Y 68.62±2.29<br>n 52 | $F_s$=10.51<br>$p<0.001$ |
| seedling weight (g) | Y 1.66 ±0.07<br>n 48 | Y 1.39 ±0.06<br>n 51 | $F_s$=8.01<br>$p<0.01$ |

Table 1. Significant differences between excess and limited pollination treatments in seed weight and 6 week seedling height and weight.

The results are illustrated in Table 1. The mean number of seeds per fruit did not differ significantly between the excess and limited treatments. Mean seed weight, however, is significantly greater in progeny from the excess pollination treatment. Mean six week seedling height and weight are also significantly greater in the offspring derived from the excess pollination treatment. Considering the results of excess and limited pollinations within pin and thrum intermorph crosses separately, the same pattern of greater seedling height and weight after excess pollinations is observed. The total offspring morph ratio did not differ significantly from the expected 1:1 ratio. A 1:1 ratio was found in offspring derived from both limited and excess pollination treatments, and there was no difference in the six week seedling height of individuals that later produced pin or thrum flowers. Thus, the opportunity for greater microgametophytic selection in the excess pollination treatment did not appear to result in a preferential selection for either morph. The greater seed and seedling weight of Turnera pin and thrum progeny derived from larger stigmatic loads suggests that the opportunity for greater microgametophytic selection in the excess pollination treatment has resulted in selection for more vigorous progeny. These results in Turnera further support the conclusions from other pollen load studies (Fingerett,1979; Lee,1982; McKenna,unpub.; Mulcahy and Mulcahy 1975) that the strength of microgametophytic selection can have a demonstratable effect on the quality of the sporophytic generation.

## Microgametophytic Selection in Anchusa officinalis

As previously mentioned, an alternative method to manipulate differences in the strength of microgametophytic selection is to vary the placement of pollen proximal or distal to the ovary. This technique varies the stylar distance over which differences in pollen tube growth rates are expressed ; a longer stylar distance intensifies the separation between fast and slow growing pollen tubes. Stylar distance has been used extensively to study microgametophytic selection in corn (Mulcahy,1974; Ottaviano et al.,1982; Sari Gorla and Rovida, 1980) and in Dianthus chinensis (McKenna and Mulcahy,1983; Mulcahy and Mulcahy,1975).

The use of style length differences in heterostylous plants is complicated by the presence of intramorph incompatibility which precludes the use of the same pollen source on both long and short styles. Anchusa officinalis was selected for these experiments because of its unusual incompatibility system. Although Anchusa officinalis exhibits most of the features associated with distyly, including self incompatibility, intramorph crosses are fully fertile (Schou and Phillip,1983; Phillip and Schou, 1981). Other distylous species in the Boraginaceae such as Amsinckia spp.(Ganders, 1974) and Cryptantha spp.(Caspar, 1985) exhibit a total loss of incompatibility, and in some cases cryptic self incompatibility has been reported (Weller and Ornduff,1977;Ornduff,1976), but this combination of self incompatibility and intramorph compatibility is extremely rare (Schou and Phillip,1984 and refs. within). Distyly in Anchusa officinalis appears to be controlled by a single locus with dominance expressed in the thrums; intermorph crosses yield equal numbers of pin and thrum progeny, and viable homozygous thrums can be obtained through thrum x thrum crosses (Schou and Phillip,1984). Homozygous thrums appear to be rare in natural populations however, and in eleven Danish populations sampled by Phillip and Schou, a significant excess of pins was found in ratios from 2:1 to 28:1.

Seeds from natural populations in northern Denmark were used to establish plants for microgametophytic selection experiments. A series of crosses was designed to compare the quality of offspring derived from the same pollen donors crossed with long or short styled seed parents. Four pollen donors ( 2 pin and 2 thrum) were crossed with six seed parents ( 3 pin and 3 thrum). All pollinations were carried out in a controlled temperature growth chamber at 23 degrees C. Flowers of seed parents were emasculated by removing the corolla prior to anther dehiscence. Crosses were made by brushing one newly dehiscent anther across the stylar tip of a previously emasculated flower. There is no difference in the amount of pollen per anther or the viability of pollen produced by thrum or pin morphs (Phillip and Schou,1981), so in this experiment pollen quantity was held constant and the important variable was style length. Style length was measured in the seed parents used for the study, the pin morphs had significantly longer styles ( Y = 8.53 mm ;n = 90 ) than the thrum morphs (Y = 5.02 mm ; n = 90).

Plants set fruit in the growth chamber approximately 3 weeks following pollination. _Anchusa officinalis_ has 4 ovules per ovary and produces 1-4 nutlets per flower. _Anchusa officinalis_ typically exhibits low seed set, but there was no difference in the seed set of pin or thrum seed parents (thrum = 38% seed set; pins = 39% seed set). These values are in agreement with the 38% seed set reported by Phillip and Schou (1981) following both natural and hand pollinations. The seeds were weighed individually, planted singly in 2" pots, randomly positioned on a greenhouse bench and germination was monitored daily.

The results are shown in Table 2. There was no difference in germination time between seeds produced by the two morphs. Seeds produced from the pin morph are significantly heavier than seeds produced by the thrum morph. Progeny from the pin morph also have a significantly higher germination percentage.

The superior vigor suggested by the heavier seeds and higher percent germination in offspring from the long style morph was further investigated to determine whether offspring from pin and thrum seed parents exhibited measurable differences in seedling fitness characters. Young seedlings of _Anchusa_ produce an initial root which gradually develops into a thick tap root extending from a basal rosette. Studies of other species with a similar rosette growth form (Gross, 1981) have inddicated that the probability of flowering is strongly influenced by the plant size obtained the previous year, so it is likely that vigorous root and shoot production is likely to be an important component of fitness in _Anchusa_.

A study to compare root and shoot growth in pin vs thrum progeny was initiated by transplanting 240 seedlings from pin and thrum seed parents to large cylindrical pots to allow for unrestricted root growth. Seedling root length was measured during transplanting, and the transplanted seedlings were placed randomly on the greenhouse bench. Harvests were made after three and nine weeks. The entire rosette and the upper 15 cm of the root were harvested, oven dried at 80 degrees C. and weighed.

There was no significant difference in the length of the root in pin or thrum progeny after 10 days and there was no difference in the 3 or 9 week shoot biomass of pin and thrum progeny. There is a trend for heavier roots in the offspring produced from pin plants, and at nine weeks this difference is nearly significant at the 5% level. Overall, the results for _Anchusa_ indicate that the long style morph produces heavier seeds , has a higher germination percentage, and exhibits a trend for more vigorous offspring root production. The possibility that increased vigor in the pin and thrum progeny of pin seed parents is due to gametophytic selection suggests that further experiments involving reciprocal crosses and variable pollen loads might provide valuable additional information about this system.

ANCHUSA OFFICINALIS

| SEED PARENTS | | PIN | | THRUM | |
|---|---|---|---|---|---|
| seed weight (mg) | Y<br>n | 8.01 ±0.13<br>171 | Y<br>n | 6.17 ±0.08<br>309 | $t_s$=10.64<br>$p < 0.001$ |
| germination % | | 93.0 | | 72.2 | $G_s$=26.74<br>$p < 0.001$ |
| root weight -9 wk | Y<br>n | 1.007±0.05<br>44 | Y<br>n | 0.901±0.03<br>59 | $F_s$=3.64<br>$p=0.059$ |

Table 2. Significant comparisons between progeny produced from six pin and thrum seed parents crossed with four individual pollen donors.

These results suggest a possible explanation for the extreme anisoplethy exhibited by Anchusa populations. Under natural conditions, if there is a trend for more vigorous offspring produced by crosses involving the pin morph as seed parent, these offspring may be overrepresented in Anchusa populations. This could help to explain the large excess of the pin morph in all populations and it would also account for the apparent rarity of the homozygous short styled individuals, since the thrum offspring of a pin seed parent will necessarily be heterozygous.

The anisoplethy exhibited by Anchusa populations could certainly result from other factors as well such as direct sporophytic selection against the thrum phenotype or some factor relating to the pollination efficiency of thrums vs pins. Interestingly, Schou and Phillip (1981) found smaller pollen loads on pin stigmas than thrum stigmas under natural pollination conditions, although the proportion of outcrossed to self pollen in these stigma loads was not determined. Self incompatibility and a lack of vegetative reproduction suggests that geitonogamy and inbreeding are not primary factors in this system. The lack of a visible pollen dimorphism prevents easy estimation of natural intramorph and intermorph pollen flow patterns in Anchusa populations, but one would expect that an initial deviation from isoplethy resulting from any factor would be intensified by an increased frequency of intramorph crosses. This suggests that the relative strength of heteromorphy in promoting intermorph crosses may be an important feature of this system.

## Conclusion

The investigation of microgametophytic selection in the two distylous species examined suggests that the intensity of pollen competition plays a role in influencing sporophytic quality. The results in Turnera ulmifolia indicate that the amount of pollen deposited on a stigma can influence offspring vigor, and the data from initial crosses in Anchusa officinalis suggests the possibility that style length may also influence the intensity of microgametophytic selection in heterostylous plants.

## Acknowledgements

I would like to thank O. Schou and B. Bentley for generously providing plant material, and S. King for cheerful technical assistance.

References


Barrett SCH (1978) Heterostyly in a tropical weed:the reproductive biology of the Turnera ulmifolia complex. Can.J.Bot. 56:1713-1725.

Bentley BL (1979) Heterostyly in Turnera trioniflora a roadside weed of the Amazon basin. Biotropica 11:11-17.

Casper BB (1985) Self-compatibility in distylous Cryptantha flava (Boraginaceae). New Phytol. 99:149-154.

Fingerett ER (1979) Pollen competition in a species of evening primrose, Oenothera organensis Munz. Masters thesis,Washington State University, Pullman, Washington.

Ganders FR (1974) Mating patterns in self-compatible distylous populations of Amsinckia (Boraginaceae) Can.J.Bot. 53:773-779.

Gross KL (1981) Predictions of fate from rosette size in 4 biennial plant species,Verbascum thapsus,Oenothera biennis,Daucus carota,andTragopogon dubius. Oecologia 48:209-213.

Lee TD, Bazzaz FA (1982) Regulation of fruit maturation pattern in an annual legume Cassia fasciculata. Ecology 63:1374-1388.

Martin FW (1965) Distyly and incompatibility in Turnera ulmifolia. Bull. Torr.Bot.Club 92:185-192.

McKenna MA, Mulcahy DL (1983) Ecological aspects of gametophytic competition in Dianthus chinensis. In: Mulcahy DL, Ottaviano E (eds) Pollen:Biology and Implications for Plant Breeding. Elsevier,pp 419-424

Mulcahy DL (1974) Correlation between speed of pollen tube growth and seedling height in Zea mays. Nature 249:491-493.

Mulcahy DL, Mulcahy GB, Ottaviano E (1978) Further evidence that gametophytic selection modifies the genetic quality of the sporophyte. Soc.Bot.Fr.Actualites Bot. 1-2:57-60.

Mulcahy DL, Mulcahy GB (1975) The influence of gametophytic competition on sporophytic quality in Dianthus chinensis. Theor.Appl.Genet. 46:277-280

Ornduff R (1976) The reproductive system of Amsinckia grandiflora, a distylous species. Syst.Bot. 1:57-66.

Ottaviano E, Sari Gorla M, Pe E (1982) Male gametophytic selection in maize. Theor.Appl.Genet. 63:249-254.

Phillip M, Schou O (1981) An unusual heteromorphic incompatibility system Distyly,self incompatibility,pollen load and fecundity in Anchusa officinalis. New Phytol. 89:693-703.

Sari Gorla M, Rovida E (1980) Competitive ability of maize pollen: intergametophytic effects. Theor.Appl.Genet. 57:37-41.

Schou O, Phillip M (1984) An unusual heteromorphic incompatibility system 3.On the genetic control of distyly and self-incompatibility in Anchusa officinalis L.(Boraginaceae). Theor.Appl.Genet. 68:139-144.

Schou O, Phillip M (1983) An unusual heteromorphic incompatibility system II.Pollen tube growth and seed sets following compatible and incompatible crossings within Anchusa officinalis L.(Boraginaceae). In: Mulcahy DL, Ottaviano E (eds) Pollen:Biology and Implications for Plant Breeding. Elsevier pp 219-227.

Shore JS, Barrett SCH (1984a) The effect of pollination intensity and incompatible pollen on seed set in Turnera ulmifolia (Turneraceae). Can.J.Bot. 62:1298-1303.

Shore JS, Barrett SCH (1984b) Heterostyly and homostyly in the Turnera ulmifolia complex (Turneraceae). Am.J.Bot. 71:186 (abstract).

Weller SG, Ornduff R (1977) Cryptic self incompatibility in Amsinckia grandiflora. Evolution 31:47-51.

Willing RP, Mascarenhas JP (1984) Analysis of the complexity and diversity of mRNAs from pollen and shoots of Tradescantia Plant Physiol.75:865-868

# Controlled Enpollination of Honeybees (Apis mellifera): Bee-to-Bee and Bee-to-Tree Pollen Transfer

M. B. Dicklow, R. D. Firman, D. B. Rupert, K. L. Smith, and T. E. Ferrari[1]

1 Controlled Pollination. Cross pollination of self-incompatible fruit, nut and seed crops is the most important contribution that honeybees make to the economies of food producing regions. Inadequate pollination inevitably results in low yields or poor quality fruit. Problems occur when a pollinizer's bloom period does not overlap with the variety to be pollinated, or when sufficient amounts of compatible pollen are not produced. Pollination can be improved by forcing honeybees to crawl through a modified hive entrance that disperses precollected pollen onto their body hairs (Antles, 1953; Free, 1972; Kraai, 1962). Controlled applications of pollen onto honeybees in this manner can circumvent problems associated with natural pollination.

Acceptance of controlled pollination technology has been hampered because past methods produce inconsistent results (Free, 1970). One important factor that influences pollination success is pollen quality. Standardized viability tests that measure pollen germination on sugar-agar media are inadequate as indicators of fertilizability for two reasons. First, pollen-pistil congruity for different cultivars is impossible to evaluate on synthetic medium. Second, pollen tube formation and fertilization are regulated by different mechanisms. Thus, there are conditions where germinable pollens fail to set seed. A more reliable indicator of pollen quality is gametophyte performance *in situ*, using pistils of flowers to be pollinated.

This paper assesses the duration of pollen viability after enpollination of a colony, documents the kinetics of bee-to-bee pollen transfer within

---

[1] Plant Development International, Fingerlakes Research Center, Lodi, New York 14860, USA

a hive, and evaluates the degree to which a honeybee's pollination effectiveness can be improved.

2 Pollen Quality. For experiments reported herein, pollens were applied manually in saturating amounts to emasculated flowers on field trees or on cuttings forced to open indoors at $18^{o}-22^{o}$C. Pollinations were performed during the period of maximum female receptivity, as determined from preliminary tests on flowers of different maturity. Pollen viability and vigor were evaluated microscopically (Kao and Baer, 1968) using pistils excised from flowers 24 hrs after pollination. Germination was measured by counting or scoring number of tubes that penetrated into stigmas; vigor was assessed by measuring maximum tube length to the nearest millimeter. Compatibility was evaluated by manually pollinating flowers on replicate trees maintained in commercial orchards. Each pollen preparation was tested in 2 or 3 locations, and each test involved a minimum of 4 test pollinations: none, self, commercial preparation, and freshly collected pollen from a cross compatible pollinizer. Fruit set was determined when unpollinated (treatment=none) flowers aborted. Seed set was measured about 6 wks later when developing seeds were readily visible. Average fruit set of duplicate treatments was $\pm$ 10 % in 19 of 20 tests.

3 Pollen Dispersal. Cross pollination occurs as foraging bees travel from flower-to-flower. Recently, evidence was obtained that suggested bee-to-bee pollen transfer results from mutual contacts that occur among bees within a colony (DeGrandi-Hoffman et al., 1984). This behavior can transfer compatible pollen from returning foragers to honeybees in a hive that may never have visited a pollinizer.

Utilizing this principle, we enpollinated honeybees employing a two-way beehive entrance modifier that rests on the bottom board and extends along the hive opening ("Ferrari type", available from the authors' institution). The device closes the entrance except for narrow ventilation slots and a sloping recessed groove that enables bees to crawl in and out. The insert shelters pollen poured into the passageway and disperses grains onto bees as they contact the pollen.

Strong colonies used for pollination purposes typically contain about 50,000 bees. When 100 or more bees per minute (100 b/m) are entering and leaving a hive ($18^{o}-22^{o}$C), 20 g of pollen are dispersed in 20 min or less. If the ratio of entering to exiting bees is 1:1, then 1000 bees will enter a hive in the time it takes to deplete pollen from the insert. Or, only about 2 % of the colony's population -- initially -- will have had direct contact with pollen. As bee-to-bee contact occurs within the hive, the proportion of bees with pollen on their bodies should increase. This assumes that the transfer rate exceeds the bee's ability to groom pollen from its body, and that the bees which are sampled are from some remote part of the colony, away from the source of pollen. With active colonies, pollen depletion from the insert due to collection by bees is minimal.

4 Bee-to-Bee Pollen Transfer. Pollen transfer was detected microscopically using 3 different methods. Grains on honeybees were measured after applying 20 g plum, almond, or apple pollens (Firman Pollen Company, Stockton, CA) to colonies via the two-way beehive entrance modifier. At intervals, bees were removed for analysis from top bars on 2-story hives.

In one experiment, a wing from each bee was removed and pollens adhering to it were counted with a 100X light microscope. In-hive pollen transfer was also detected after enpollinating a colony with a pollen-fluorescent powder mixture (15:1). Number of fluorescent spots present on the entire bee body were counted with the aid of a black light and 50X magnifier. Fluorescence was associated with pollen up to 3 hrs after application of the pollen-dye mixture to a colony. Except for in the region of the corbiculum, most fluorescence was associated with pollen grains. Pollen transfer was also detected indirectly using the pistil as a bioassay system (Fig. 1). Pollen present on bees from enpollinated colonies was measured after randomly touching different parts of their bodies to stigmas 10 times. Number of pollen tubes per pistil was determined 21 hrs later using fluorescence microscopy (Kao and Baer, 1968). In all 3 experiments (Fig. 1), pollen dispersal reached a maximum in 2-4 hrs, when all bees sampled from hives had pollen on their bodies.

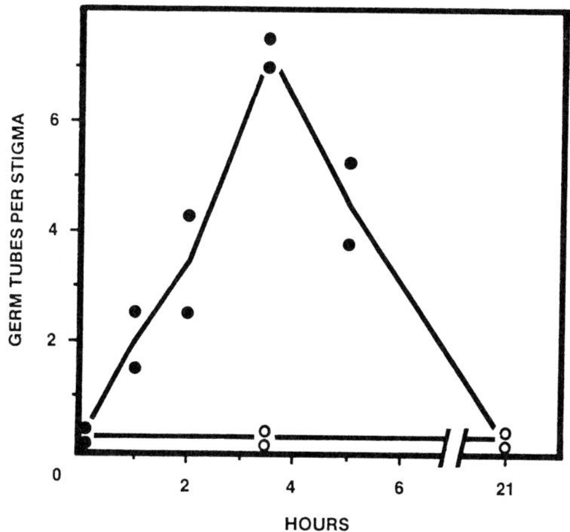


Fig. 1. Transfer of apple pollen (cv. Yellow Delicious) from bee bodies onto pistils of Red Rome flowers.

5 Bee-to-Tree Pollen Transfer. Ability of pollen on bee bodies to fertilize ovules was evaluated *in vivo* using apple (cv. Monroe) flowers on trees maintained under commercial orchard conditions. Ten bees were captured from each entrance of 2 colonies enpollinated 2 hrs earlier with compatible pollen (cv. Red Rome, pretested the previous 2 yrs for fruit set and seed set ability). Fruit set was 80 % and 90 % for flowers pollinated with bees from enpollinated colonies, compared to 30 % for each of 2 unpollinated colonies. Seed set averaged 54 % for fruit pollinated with bees from enpollinated colonies, compared to 10 % for bees from the control hives. Thus, the pollination effectiveness of bees from enpollinated colonies was improved 250 % to 300 % based on fruit set, and 540 % based on seed set.

Seed set is a more accurate measure of pollination efficiency for apples because, ideally, only 1 pollen is required to set each fruit, but 10 are needed to fertilize all 10 eggs per flower. For manually pollinated apple pistils, using excess amounts of the same pollen as applied to enpollinated colonies described above, fruit set was 100 % and seed set averaged 84 %. Thus, fruit and seed set using bees from enpollinated colonies was, respectively, 80 % - 90 % and 65 % of the maximum possible.

6 Pollination Effectiveness. Apple and pear flowers have an ovary that consists of 5 carpels each containing 2 ovules (Free, 1970). Following pollination of apple pistils with bees captured from enpollinated and unenpollinated colonies, nearly 1000 stigmas were examined for number of pollens germinated and for maximum tube length. Results indicate that honeybees must transfer a minimum of 13 germinable pollens onto all 5 stigmas of apple flowers, so that each pistil contains at least 1 pollen tube (Fig. 2).

Further examination of data indicated pollen tube growth was proportional to number of grains that germinated per pistil. Pollination data were grouped into 3 germination categories (Fig. 3). Each category was subdivided into "maximum tube length" classes, i.e. 1 mm, 2 mm, 3 mm, etc. The peak tube length (8 mm) for the "100 or more" category was nearly 8 times larger than that for the lowest category (1-2 tubes). This *in vivo* synergetic effect of pollen germination and tube growth is consistent with reports that more fertilizations result from double than single pollinations, and more fertilizations result from the second pollination than the first (Visser, 1983).

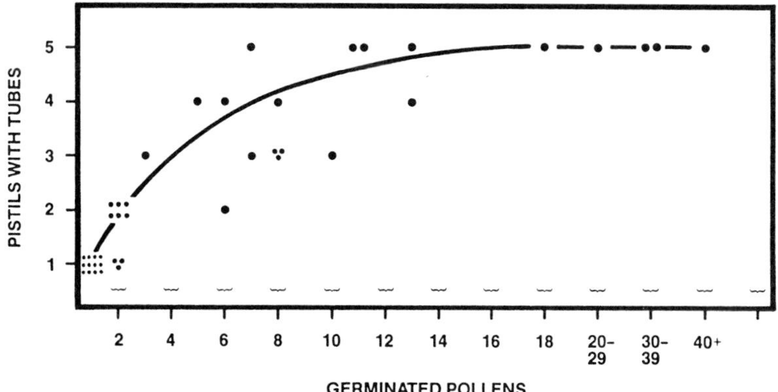


Fig. 2. Pollen dispersal on stigmas of apple flowers (cv. Granny Smith) after pollination using bee bodies. Total number of Yellow Delicious pollens germinated on all 5 stigmas (abscissa) were plotted against the number of stigmas with germinated pollen (ordinate). Each dot represents information from 1 flower. For example, at coordinate 5,11 each of 2 flowers had a total of 11 pollens and all 5 pistils on each flower had germinated pollens present.

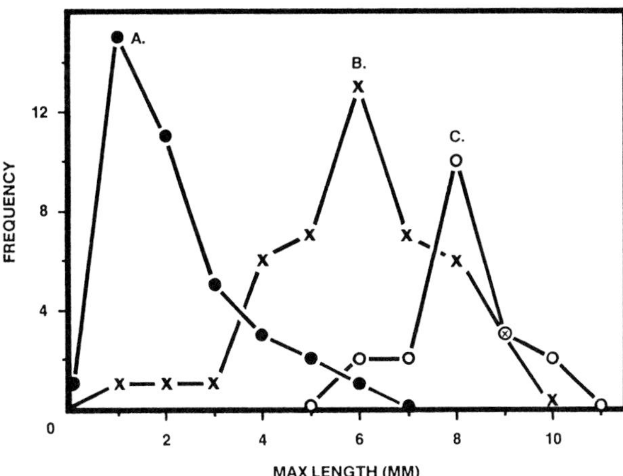


Fig. 3. Synergetic effect of pollen germination and tube growth. Data from the experiment described in Fig. 2 were grouped into 3 germination categories-(A) 1 to 20, (B) 6 to 49, (C) 100 or more germinated pollens per pistil. Maximum pollen tube length per pistil, measured to the nearest ml, is plotted against the number of times (frequency) each length occurred.

7 Pollen Viability After Colony Enpollination. Traces of residual almond and apple pollens collected from the entrances of colonies enpollinated at 9 a.m. retained about 50 % of its germinability up to 10 hrs after addition to inserts (temperature range 15°-18°C, 25%-50% RH). To estimate the duration of apple pollen viability on honeybees, 10 gm Yellow Delicious pollen was doused onto 60-80 bees in narrow plastic mesh cages, which were then suspended between 2 frames in the brood chamber of a colony. Ten bees were removed at increasing time intervals and their bodies were used to pollinate emasculated flowers maintained indoors on cuttings. Pistils were evaluated 21 hr later for presence of pollen tubes.

Germinable pollens released onto stigmas from bees caged for 4 hrs was about 35 % of the number detected immediately after dousing them with pollen. However, by 8 hrs the pollens released from their bodies nearly doubled in 2 separate tests. The initial decrease may be a result of the grooming behavior of honeybees. The later increase may result from bee-to-bee pollen dispersal, as the caged bees clustered and contacted each other. Germinable pollen released onto stigmas at 12 hrs was still 70 % of that which occurred immediately after enpollination.

In one test, viable apple pollen was detected on caged bees after 21 hrs: however, little or no germinable pollen was detected on uncaged honeybees. This difference between caged and uncaged bees is probably due to the inability of the trapped bees to complete the grooming process. It also indicates that pollen can remain viable on bee bodies for extended periods.

8 Conclusion. Biotechnological advances have improved the likelihood that controlled applications of precollected pollen to self-incompatible crops will result in fruit, nut, or seed set. Two factors are responsible for success in a science often referred to as "artificial pollination." The first involves development of standardized, *in situ* and *in vivo* testing methods that authoritatively attest to a pollen preparation´s viability, vigor and compatibility. The second involves the use of a two-way pollen dispersal insert for beehives that rapidly enpollinates honeybees in a colony, and makes controlled pollination reliable, cost effective and applicable to existing agricultural management practices. Using procedures described in this paper, we have consistently observed yield improvements in commercial orchards where poor pollination severely limited crop production for up to 25 yrs.

## References

Antles LC (1953) Am Bee J 93: 102-103

DeGrande-Hoffman G, Hoopingarner R, Baker K (1984) Bee World 65: 126-133

Free JB (1970) Insect pollination of crops: temporary aids to pollination. Academic Press, N.Y., pp. 409-416

Free JB, Williams IH (1972) J Appl Ecol 9: 609-616

Kho YO, Baer J (1968) Euphytica 17: 298-302

Kraai I (1962) Euphytica 11: 53-56

Visser T (1983) In: Pollen: Biology and Implications for Plant Breeding. Mulcahy DL, Mulcahy GB, Ottaviano E (eds). Elsevier Biomedical, N.Y., pp. 229-239.

# Population Dynamics of Open-pollinated Maize Synthetics under Non-random Fertilization Conditions

M. YAMADA AND T. ISHIGE

M. Yamada and T. Ishige

## 1 Introduction

With the exception of natural populations in higher plants cross-pollinated crops, such as maize, sorghum, rye sugarbeet, buckwheat and rapeseed, have no complete homozygous genotypes in a population unless either artificial isolation from other genotypes or controlled pollination in applied. Through open-pollination, in contrast, every population appears to be able to preserve its heterozygosity to some extent. The theoretical genetic basis accounting for this phenomenon is related to the random mating population theory including the some premises (Falconer 1960). In the past decade, on the other hand, it has been pointed out that this ideal population could not actually be obtained in cross-pollinated crops (Hartl 1975, Harding 1975, Jain 1975, Clegg *et al*. 1978, Yamada 1982). In these reports it was emphasized that either at the sporophyte or at the gametophyte stages random fertilization was not possible and that this fact could not be overlooked.

Yamada and Murakami (1983) showed that in mechanical mixed-pollination consisting of pollen grains from inbred lines and those from $F_1$ plants the superiority of pollen derived from $F_1$ plants in fertilization was confirmed. Yamada (1984) also pointed out that the superiority is not controlled by gametophyte gene(s), as is well documented in maize. If non-random fertilization in relation to the superiority of pollen grains appears, random

1 Department of Cell Biology, National Institute of Agrobiological Resources, Yatabe, Tsukuba, Ibaraki, *305*, JAPAN

mating in maize population acutually does not take place. In this report we will postulate that the heterozygosity of open-pollinated synthetics is preserved due to non-random fertilization associated with the superiority of pollen grains from heterozygous plants.

## 2 Random Pollination in Maize Population in Isolation Field

Transmission process of genes in higher plants, as shown by Harding (1975), consists of two phases, *i.e.* sporophyte in the process from fertilization to flowering and gametophyte from flowering to fertilization. Pollination is one of the important steps in the process from flowering to fertilization for random exchange of genes in subsequent generations. It has been reported that pollination in maize plants planted in open fields had been realized for effective pollen dispersal in the field (Paterniani and Stort 1974). Yamada (1982) attempted to analyse the direction of pollen grain dispersal and cline of the fertilization percentages by pollen grains from the center in isolation fields surrounded by a barrier, as shown in Fig. 1. The isolation fields, No. 71 and No. 82, were set up with distinct barriers such as orchard and buildings. As a result the wind velocity was less than 2.0 m/sec. during flowering. Under such experimental conditions,

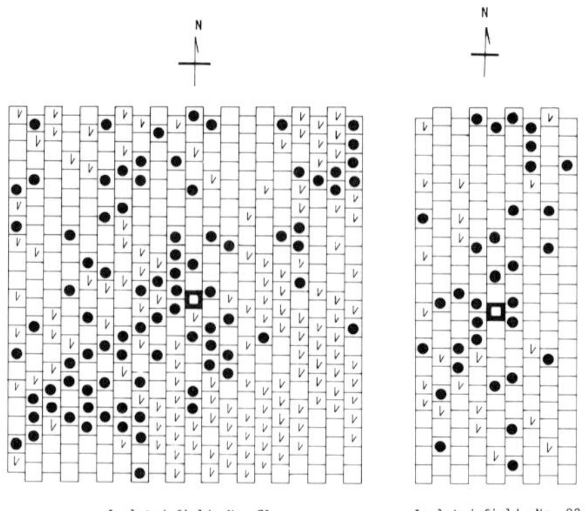


Isolated field, No. 71. Isolated field, No. 82.

Fig. 1. Results of two day pollination experiments in isolation fields, No. 71 and No. 82.

Note : Non-silked plants during the pollination are shown with a blank cell, silked plants but not fertilized by pollen grains from the dolor (■) or barren plants are indicated by a checked mark (✓) and silked plants fertilized by the pollen grains are indicated by a black circle, respectively (Yamada 1982).

pollen grains from the center of the field dispersed in all directions due to the wind turbulance. So it was concluded that random pollination in maize population occurred all over the isolation field.

## 3 Non-random Fertilization Depending on Gamete Competition

Maize population has no self-incompatible genes for random dispersal of pollen grains and random pollination in most cases. Consequently synthetics or composites of maize having a similar flowering habit, *i.e.* flowering date, amount of pollen grains, and culm length, do not show any remarkable differences in flowering habit among the plants. Non-random fertilization in the synthetics or composites of maize is controlled by factors involved in gamete competition (Yamada 1982, 1984). As mentioned above, since random pollination in the isolation fields was found to occur, the application of non-random fertilization under natural conditions should be investigated on the basis of the results obtained in mechanical mixed-pollination experiments.

Inbred line, A 34, and the $F_1$, which have a similar flowering habit, were planted in alternative rows in two isolation fields, A-5 and G-5. Xenia kernels fertilized with pollen grains from $F_1$ plants and selfed or sibbed kernels in the ear of A 34 were identified by the endosperm color. Data obtained from mechanical mixed-pollination of these materials to ears of A 34 (Yamada 1982) showed that they were superior to the pollen grains from the $F_1$ plant in terms of gamete competition (90.8% instead of 50.0%). Fig. 2 illustrates the superiority of pollen grains from $F_1$ plants in the

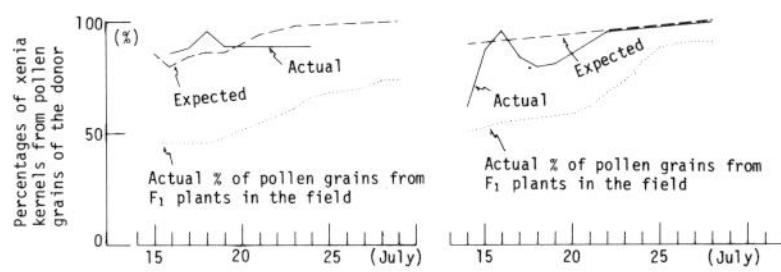


Fig. 2. Pollination experiment in mixed-planting of A 34 and $F_1$(J-466xJ-472) in isolation fields (Yamada 1982).

Note : Solid line and broken line indicate the actual percentages of kernels fertilized by pollen grains from $F_1$ plants and the expected percentages which were deduced from the actual percentages of pollen grains from $F_1$ plants in the fields, respectively. Total number of spikelets in a plant of A 34, equivalent to pollen grains, were 1,112±36 and that of $F_1$ were 1,114±38.

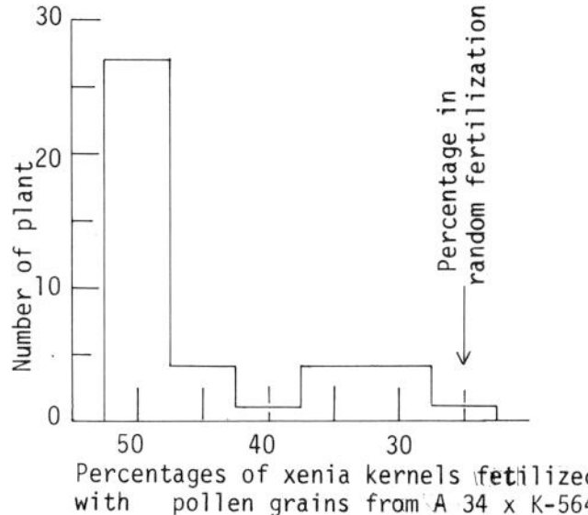


Fig. 3. Frequency (%) of xenia kernels in the ear of A 34 (endosperm genotype, *yy*) in mixed-planting with A 34 x K-564 $F_1$ (endosperm genotype, *yY*) (Yamada, unpublished).

open-pollination experiments conducted in the isolation fields. "Expected" percentages lines referring to "Actual" percentages in Fig. 2 were drawn from the data obtained in the mixed-pollination experiments and the lines were close to "Actual" ones. Results from another mixed-planting experiment are presented in Fig. 3, in which it is shown that pollen grains from the $F_1$ plants were predominantly fertilized, suggesting the existence of non-random fertilization.

From the results mentioned above, the existence of non-random fertilization associated with to the superiority of $F_1$ pollen in mixed-planting was corroborated as the results obtained from mechanical mixed-pollination experiments. So it is proposed that the heterozygosity of sporophytes affects gamete competition in actual open-pollination.

## 4 Theoretical Analysis on the Peservation of Heterozygosity among Synthetics or Composites under Non-random Fertilization

A theoretical model was defined to determine whether the superiority of pollen from $F_1$ plants or heterozygous plants is effective in conserving the heterozygosity of an open-pollinated plant population as is the case in the synthetics or composites. From the results obtained above the implication of $A_1$ and $A_2$ genes in a locus can be postulated. Three kinds of genotypic frequencies of Syn 3 with combinations among all possible genotypes including their corresponding selective fertilization coefficients valid only for pollen parents are as follows;

$$A_1A_1 = sf_1P^2 + (1/2)(sf_1+sf_3)PH + (1/4)sf_3H^2$$

$$A_1A_2 = (sf_1+sf_2)PQ + (1/2)(sf_1+sf_3)PH + (1/2)(sf_2+sf_3)QH + (1/2)sf_3H^2$$

$$A_2A_2 = sf_2Q^2 + (1/2)(sf_2+sf_3)QH + (1/4)sf_3H^2$$

Selective fertilization coefficient, $sf_3$, for pollen grains from the $A_1A_2$ genotype was assumed to be superior ($1.0>sf_3>0.5$) to that of the others, $sf_1$ and $sf_2$, where $sf_1+sf_2+sf_3$ is equal to unity, based on our data (Yamada 1982).

In the basis of this assumption the genotypic frequency of $A_1A_2$ in subsequent generations derived from different original genotypic frequencies of $A_1A_1$, $A_1A_2$ and $A_2A_2$ in Syn 2

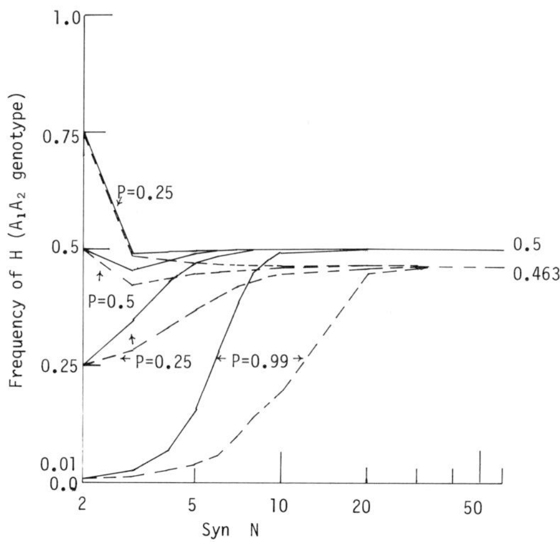


Fig. 4. Genotypic frequency (H) of $A_1A_2$ in synthetics with different genotypic frequencies of Syn 2 in two cases of non-random fertilization, $sf_3$ = 0.7.

Note : Solid and broken lines are plotted for $sf_3$ = 0.7 with $sf_1$ = $sf_2$ = 0.15, and with $sf_1$ = 0.3 and $sf_2$ = 0.0, respectively.

Table 1. Some examples of genotypic frequency of $A_1A_2$ in equilibrium under non-random fertilization.

| $sf_3$ | ($sf_1$ | $sf_2$) | H (Syn N) | $sf_3$ | ($sf_1$ | $sf_2$) | H (Syn N) |
|---|---|---|---|---|---|---|---|
| 0.90 | (0.1 | 0.0) | 0.498 (4-8) | 0.50 | (0.5 | 0.0) | 0.000 (200 - ) |
| 0.85 | (0.15 | 0.0) | 0.495 (10-20) | 0.50 | (0.4 | 0.1) | 0.320 (100) |
| 0.80 | (0.2 | 0.0) | 0.490 (8-10) | 0.50 | (0.3 | 0.2) | 0.480 (30) |
| 0.70 | (0.3 | 0.0) | 0.463 (20-40) | 0.40 | (0.6 | 0.0) | 0.000 (30-40) |
| 0.70 | (0.2 | 0.1) | 0.493 (10-15) | 0.40 | (0.5 | 0.1) | 0.000 (50) |
| 0.60 | (0.4 | 0.0) | 0.375 (40) | 0.40 | (0.4 | 0.2) | 0.000 (200) |
| 0.60 | (0.3 | 0.1) | 0.469 (20-30) | 0.40 | (0.3 | 0.3) | 0.500 (3) |
| 0.60 | (0.2 | 0.2) | 0.500 (3-4) | | | | |

Note : Initial genotypic frequencies of H, P and Q were chosen at random. $sf_1$, $sf_2$ *and* $sf_3$ are selective fertilization coefficients for $A_1A_1$, $A_2A_2$ and $A_1A_2$, respectively.

generation is being analysed in this report. Calculation of the genotypic frequency based on the formulae proposed above was made in adopting reasonable intervals of P, H and Q with alternative intervals of selective fertilization coefficients. The calculation was as follows:

1. $sf_3 > sf_1$ and $sf_2$,
   1) and $sf_1 = sf_2$, H = 0.5 earlier or later results,
   2) and $sf_1 > sf_2$ (or $sf_1 < sf_2$), H at equilibrium depending on $sf_1$ (or $sf_2$) value (see Fig. 4 and Table 1).
2. $sf_1 < sf_2 \gtrless sf_3$,
   H will reach an equilibrium at 0.0 finally in all cases.
3. $sf_1 = sf_2 = sf_3$ (random mating),
   H will be at equilibrium depending on the original genotypic frequencies of $A_1A_1$, $A_1A_2$ and $A_2A_2$, respectively.

In conclusion, for conserving heterozygous plants in a population in nature, it is suggested that the factors which induce the superiority of pollen grains from heterozygous plants, play an important role in preservation of the heterozygosity of open-pollinated plants.

## References


Clegg MT, Kahler AL, Allard RW (1978) Estimation of life cycle components of selection in an experimental plant population. Genetics 89: 765-792

Falconer DS (1968) Introduction to quantitative genetics. Oliver Boyd Ltd London 9-22

Harding J (1975) Models for gamete competition and self fertilization as components of natural selection in populations of higher plants. "Gamete competition in plants and animals" (ed. Mulcahy DL), North-Holland Pub. Amsterdam, 243-255

Hartl DL (1975) Stochastic selection of gametes and zygotes. *Ibid.* 233-242

Jain SK (1975) Gametic selection in mixing selfing and random mating plant populations. *Ibid.* 265-278

Paterniani E, Stort AC (1974) Effective maize pollen dispersal in the field. Euphytica 129-134

Yamada M (1982) Superiority of pollen from $F_1$ plant in selective fertilization and its implication in maize breeding. Bull NIAS D33: 63-119 (in Japanese with English summary)

Yamada M, Murakami K (1983) Superiority in gamete competition of pollen derived from $F_1$ plant in maize. "Pollen: Biology and implication for plant breeding" (ed. Mulcahy DL, Ottaviano E) Elsevier Sci Pub 389-395

Yamada M (1984) Selective fertilization in maize, *Zea mays* L. III Independence of gametophyte factors on superiority of pollen grains from $F_1$ plants. Japan J Breed 34: 9-16

# Pollen Allocation in Wild Radish: Variation in Pollen Grain Size and Number

MAUREEN L. STANTON AND ROBERT E. PRESTON

The vast majority of the world's plant species are hermaphroditic, but until the very recent past, plant population biologists have focused almost exclusively on maternal reproduction through seeds. Disregarding male reproductive success leads to an incomplete (and potentially misleading) picture of fitness hierarchies within plant populations, as an individual's success as a male need not be correlated with its success as a female (Bertin, 1982). Accordingly, recent studies have increasingly emphasized male fitness achieved through the successful fertilization of ovules on other individuals (e.g. Queller, 1983).

Models of sexual allocation in hermaphroditic plants typically assume that male reproductive success increases as a simple function of total resources allocated to pollen (Ross and Gregorius, 1983; Lloyd, 1984). Two ideas are implicit in this approach: 1) that pollen allocation can be described by a single parameter, usually the number of pollen grains, and 2) that pollen grain size variation is unimportant.

We know very little about paternal fitness variation in natural plant populations, but it is reasonable to think that male success could be influenced by both pollen grain size and number. Given adequate pollen transfer by wind or animal vectors and an abundance of potential recipients, one would expect individuals producing many pollen grains to father more seeds than individuals producing fewer grains. Even though this idea forms the basis of current sexual allocation models, data on intraspecific variation for pollen grain number are virtually nonexistent for natural plant populations.

Botany Department, University of California, Davis, CA 95616

If many pollen grains reach a receptive stigma simultaneously, those grains giving rise to fast growing tubes may outcompete slower grains for fertilizable ovules (A. A. Snow, this volume). In some cases, genetic factors are known to control tube growth rate, but preliminary evidence from corn (Ottaviano et al., 1983), alfalfa (Barnes and Cleveland, 1963), and *Collomia* (Lord and Eckard, 1984) suggest that larger grains may give rise to faster growing tubes. If this is true, then resource allocation to individual grains may influence paternal success.

Because of the pervasive view that pollen grain size is a constant, species-specific character, pollen grain size variation has not been incorporated into any models of male fitness. In the few studies that have compared pollen size from different individuals, however, grain size variation is the rule, rather than the exception (M. Stanton, unpublished).

As a first step towards understanding how natural selection acts on male reproductive traits, it is critical to document patterns of pollen allocation in natural populations. Only then can the processes controlling male fitness variation be explored adequately. In this chapter, we describe variation in pollen grain size and number within a natural population of wild radish in California. We show that individuals vary significantly in their total biomass allocation to pollen per flower, and that this variation is due to the combined effects of pollen grain number and size.

## METHODS

Wild radish (*Raphanus sativus*) is a self-incompatible weed abundant in fertile, disturbed soils throughout California. Wild *R. sativus* shows high levels of genetic polymorphism (Ellstrand and Marshall, 1985) as well as morphological variation for traits such as petal color, seed size, and root shape (Panetsos and Baker, 1968).

In April 1985, 41 *R. sativus* individuals were marked within a single population in Sacramento County, California. Several buds were collected just before anthesis, then fixed in 70% ethanol for pollen analysis. Five anthers from each of four buds were later dissected into aniline blue-lactophenol.

After being suspended in a 0.85% sterile saline solution, pollen grains were counted and measured using a Coulter Counter with Channelizer at U. C. Davis. Pollen from the sixth anther was mounted in polyvinyl-lactophenol with aniline blue to make a permanent slide for each flower. Percent pollen viability was estimated by counting 200-300 stained and unstained grains on every slide. Anthers from four additional buds were weighed to estimate pollen allocation per flower.

## RESULTS

For the 41 individuals sampled in 1985, anther weight per flower ranged from 0.73 to 1.55 mg. These results are comparable to data gathered from another population of R. sativus in 1984, where anther weight per flower ranged from 0.70 to 2.00 mg in a sample of 60 individuals (Stanton, unpublished). Automated pollen measurements for several flowers per plant were used to examine variation in male allocation more closely. Pollen grains from four flowers were counted for all 41 individuals in the 1985 sample; particle size analysis was possible for only 31 individuals. Total pollen volume per flower was calculated by multiplying the number of pollen grains per flower times the mean volume per pollen grain in $\mu m^3$. Typically, there is a steep modal peak for pollen grain size within individual flowers (Figure 1).

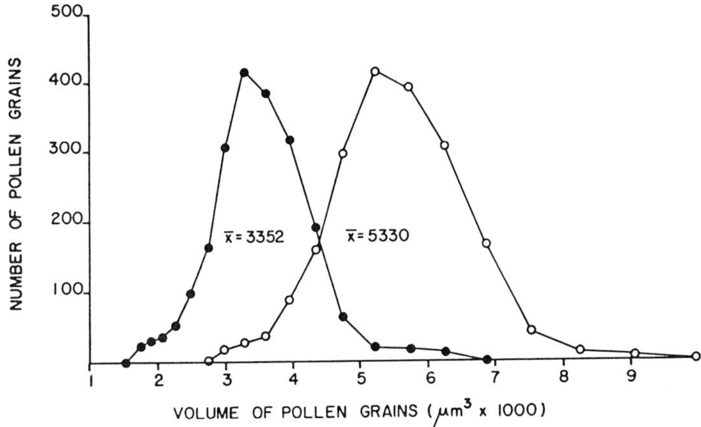


Figure 1. Pollen size spectra for R. sativus flowers from two individuals in 1985.

Total pollen volume per flower ranged from $0.66 * 10^8$ $\mu m^3$ to $1.67 * 10^8$ $\mu m^3$ among the 31 individuals measured. Variation among individuals for total pollen production was highly significant, accounting for 61.9% of all variation in pollen volume per flower ($F = 5.78$; d.f. = 30, 93; $p < 0.0001$). Total pollen volume, as measured by the automated analysis system, was significantly correlated with anther weight measured for different flowers on the same plants ($r = 0.535$; $n = 31$; $p < 0.005$), suggesting that both methods give a reasonable estimate of total pollen allocation in this species.

The almost 3-fold variation we observed for pollen allocation in this R. sativus population sample reflects variation in both pollen grain number and size among individuals (Figure 2). Mean grain size ranged from 3489.5 $\mu m^3$ to 4981.2 $\mu m^3$ for 31 individuals, and ANOVA demonstrated that variation among plants was highly significant ($F = 4.79$; d.f. = 30, 93; $p < 0.0001$). Numbers of pollen grains varied even more, ranging from 95,862 to 230,724 grains per flower. As with grain size, the variation among individuals was highly significant ($F = 4.99$; d.f. = 40, 123; $p < 0.0001$).

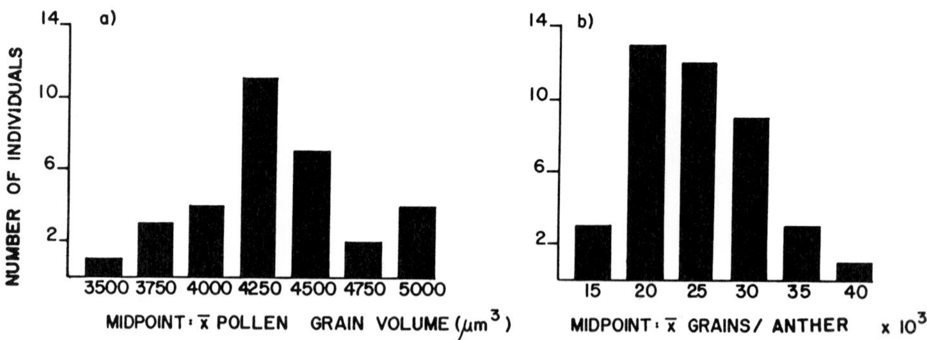


Figure 2. Pollen allocation in a population of R. sativus.
a) distribution of mean pollen grain size.
b) distribution of mean pollen grain number per flower.

If total allocation to pollen were fairly constant, one would expect to see a negative relationship between pollen grain size and number. For plants in the 1985 sample, total pollen allocation varied almost three-fold, and we found no significant relationship between mean pollen grain size and mean pollen number for 31 plants (regression coefficient = -0.01; $R^2$= 0.026; p > 0.25). Similarly, although pollen viability ranged from 55.8% to 98.4% in the 1985 sample, the size of stained grains did not increase with the proportion of aborted grains (Pearson correlation: r = -0.174; p > 0.25). Overall, variation in anther weight among plants was attributable more to variation in grain number (F = 9.741; p < 0.005) and percent viability (F = 3.225; 0.10 > p > 0.05) than to grain size, but it is important to recall that grain size showed highly significant variation among plants.

We believe that the variation in pollen allocation observed for R. sativus in 1985 is not exceptional. Using a different grain-measuring technique for another R. sativus population in 1984, we saw a 2- to 3-fold range in mean pollen grain size among anthers, compared with a 1.6-fold range in this study (M. Stanton, unpublished). Pollen grains were not counted in 1984, but total anther weight varied similarly (about 3-fold) in the two populations.

Pollen grain number is known to vary among populations in some non-agricultural species (e.g. Cruden, 1976; Hammer, 1978), but there is little information concerning variation among individuals within populations. Our studies suggest that other crucifers in northern California show comparable pollen grain number variation to that reported here. The coefficient of variation for pollen grain number per anther was 21.5% in this population of R. sativus. In natural populations of 68 species and varieties of Cruciferae surveyed, the same coefficient of variation ranged from 6.4% to 128.5%, with a grand mean of 24.3% (R. Preston, unpublished).

Both pollen grain size and pollen grain number can vary substantially in natural plant populations. Future research on male fitness in plants should take this into account when designing experiments or constructing optimization models. A more complete understanding of male fitness will require quantitative data on: 1) the "cost" of increased pollen allocation, in terms of other reproductive functions; 2) the relationship between grain size and number at any given level of pollen allocation; 3) the relationship between pollen grain number and the number of fertilization opportunities; and 4) the impact of grain size on pollen tube growth rate and fertilization ability. When the quantitative form of such functions is known for different ecological situations, it may be possible to predict how natural selection will act on genetically-based variation in pollen allocation.

## LITERATURE CITED

Barnes DK, Cleveland RW (1963) Pollen tube growth of diploid alfalfa in vitro. Crop Sci 3:291-295.

Bertin RI (1982) Paternity and fruit production in trumpet creeper. Am Nat 119: 694-709.

Cruden RW (1976) Intraspecific variation in pollen-ovule ratios and nectar secretion-- preliminary evidence of ecotypic adaptation. Ann Missouri Bot Gard 63:277-289.

Ellstrand NC, Marshall DL (1985) The impact of domestication on distribution of allozyme variation within and among cultivars of radish, *Raphanus sativus* L. Theor Appl Genet 69: (in press).

Hammer K (1978) Entwicklungstendenzen blutenokologischer Merkmale bei *Plantago*. Flora, Bd 167: 41-56.

Lloyd DG (1984) Gender allocations in outcrossing cosexual plants. In: Dirzo R and Sarukhan J (eds), Perspectives on Plant Population Ecology. Sinauer, Sunderland, MA, pp 277-300.

Lord EM, Eckard KJ (1984) Incompatibility between the dimorphic flowers of *Collomia grandiflora*, a cleistogamous species. Science 223: 695-696.

Ottaviano E, Sari-Gorla M, Arenari I (1983) Male gametophyte competitive ability in maize selection and implications with regard to the breeding system. In: Mulcahy D and Ottaviano E (eds), Pollen: Biology and Implications for Plant Breeding. Elsevier, pp 367-374.

Panetsos CP, Baker HG (1968) The origin of variation in "wild" *Raphanus sativus* (Cruciferae) in California. Genetica 38: 243-274.

Queller DC (1983) Sexual selection in a hermaphroditic plant. Nature 305: 706-707.

Ross MD, Gregorius HR (1983) Outcrossing and sex function in hermaphrodites: a resource allocation model. Am Nat 121: 204-222.

# Pollen Growth Following Self- and Cross-pollination in *Geranium caespitosum* James

M.B. HESSING[1]

M.B. Hessing[1]

## 1 Introduction

*Geranium caespitosum* James is an herbaceous perennial of uncertain taxonomy growing from the plains to the alpine in the Front Range of Colorado. It is protandrous and produces compound dichasia with each of the two flowers in the dichasium opening sequentially. This dichogamy suggests a strong preference for outcrossing. Self-fertility (seed production following self-pollination) is slightly, but significantly, lower than cross-fertility. Despite the presence of mechanisms promoting outcrossing, and a decrease in fitness following selfing, geitonogamous pollination is an extremely common event, probably overwhelming the effects of dichogamy (Hessing 1985).

Do events occur within the pistil of *G. caespitosum* that reduce the likelihood of self-fertilization? Several theories predict that the presence or arrival time of pollen on stigmas does not determine the paternity of the embryos. The gametophyte competition theory (Mulcahy 1979) suggests that foreign pollen tubes may fertilize the majority of female gametes by growing faster than self tubes. The mate selection hypothesis (Willson and Burley 1983) predicts that the pistil chooses among the prospective male gametophytes, gametes, and embryos. In this study I focus on the male gametophyte generation from pollen germination to the arrival of pollen tubes in the ovary. I specifically ask if there is a decrement in the speed or performance of self tubes that could limit the amount of geitonogamous pollen reaching each ovule.

---

1 Department of Botany, Colorado State University, Fort Collins, CO, USA

## 2 Methods and Materials

Individual G. caespitosum were collected from populations at different elevations and grown in an ecotype garden. Flowers were emasculated before both selfing and crossing and placed in bags made of no-see-um mosquito netting. Crossing was usually done with a mixture of anthers from several plants. Flowers were placed in 70% ethanol either 30, 45, or 75 minutes after pollination. The 45 minute interval was the main focus of the experiment and received more replicates than the other intervals. Pistils were stained for fluorescent microscopy using Martin's (1957) technique with two minor additions; clearing with chloral hydrate after 8N NaOH, and prestaining with Toluidine Blue before microscopy. To measure the distance travelled by tubes, I recorded the number of tubes in each of six arbitrarily assigned regions of the pistil (Fig. 1).

**Figure 1. Regions in the pistil where pollen tubes were counted.**

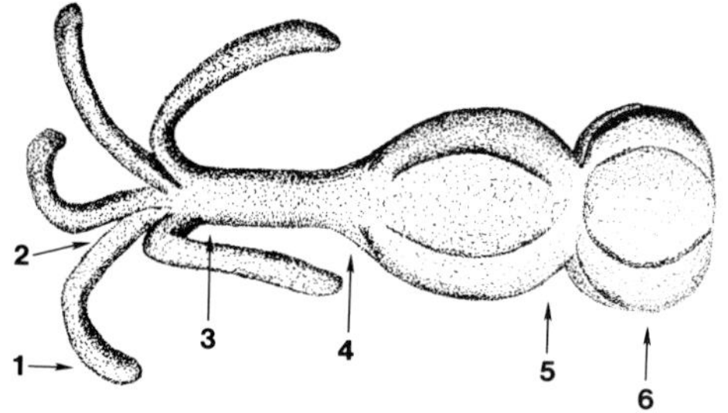


Tubes/stigma is the number of pollen tubes in a region, divided by the number of stigmas with at least one pollen grain attached. The maximum distance is the furthest region tubes entered. To prevent weighting the results from individuals with a greater number of pistils, I used each plant's average value of the above two measures. Maximum distance and tubes/stigma are measures of growth, reflecting the speed of only the fastest tubes. I also calculated a measure of overall pollen growth or "viability" within the pistil, which I called the score5. A pistil received two points for each tube that grew to region 2 but no further, 3 points for each tube that grew to region 3 but no further, and so on up to region 5. This score was then divided by the number of stigmas multiplied by the number of pollen grains attached to the stigmas. The natural log of score5 from all pistils approached a normal distribution and was analyzed using a two-way ANOVA contrasting individuals and treatment (selfing or crossing).

## 3 Results

I observed a total of 95 pistils from 17 individuals. In two of the individuals, pistils which had received cross pollen were accidentally destroyed during preparation. Six cross pistils and three self pistils had either no pollen attached or no tube growth, and were excluded from analysis. Results of pollen attachment and the 30 and 75 minute time intervals are not reported here.

The most outstanding result is the difference in tube growth among the individuals. In nine of the 15 individuals with both self and cross pollination the average distance of self tubes is less than cross tubes. In two individuals the reverse situation is observed, and in four individuals, there is no difference (Table 1). In six of the 15 individuals the average score5 was greater following selfing; the reverse situation occurred in five individuals, and four individuals had similar score5 values (+ or - 0.05). A two-way ANOVA verified that individuals have significantly different score5 values ($p = 0.054$) but did not show significant differences between self and cross values of score5 ($p = 0.461$).

The second outstanding result is the high incidence of tubes found in the ovary (region 6) after cross pollination. Six individuals (12 pistils) had tubes in a cross ovary but did not have tubes in a self ovary. No plant had the reverse situation. One plant (D9) had both self and cross tubes in ovaries, and one plant (P3) had self tubes in the ovaries but no cross ovaries were available for comparison. The number of cross tubes in the ovaries of individuals *that received at least one tube* is greater than that expected by chance (chi-squared test, $p < 0.005$).

**Table 1. Measures of Pollen Tube Growth.** Blanks are missing data, explained in text. Self and cross values from paired columns of average score5, average distance, and tubes in ovary are not significantly different using a Mann-Whitney U test.

| ID# | $n_s$ | $n_x$ | average score5 self | average score5 cross | maximum distance s | maximum distance x | tubes in ovary s | tubes in ovary x | preliminary fertility self | preliminary fertility cross |
|---|---|---|---|---|---|---|---|---|---|---|
| G7 | 3 | 3 | 0.643 | 0.567 | 3.70 | 4.70 | 0 | 0 | 0.0 | 0.467 |
| F9 | 2 | 2 | 0.187 | 0.458 | 4.50 | 4.50 | 0 | 0 | 0.250 | 0.543 |
| J1 | 4 | 2 | 0.069 | 0.255 | 3.50 | 6.00 | 0 | 9 | 0.240 | 0.263 |
| H1 | 1 | 2 | 0.203 | 0.341 | 5.00 | 4.50 | 0 | 0 | 0.0 | 0.313 |
| B8 | 3 | 2 | 0.517 | 0.427 | 4.70 | 6.00 | 0 | 4 | 0.257 | 0.333 |
| E5 | 5 | 6 | 0.061 | 0.404 | 2.40 | 4.50 | 0 | 3 | 0.0 | 0.167 |
| B6 | 4 | 4 | 0.309 | 0.161 | 4.25 | 5.25 | 0 | 5 | 0.133 | 0.333 |
| B4b | 2 | 2 | 2.000 | 0.283 | 4.00 | 4.00 | 0 | 0 | 0.050 | 0.067 |
| K6 | 5 | 4 | 0.826 | 0.238 | 4.40 | 5.25 | 0 | 1 | 0.222 | 0.273 |
| C5 | 1 | 1 | 0.071 | 0.070 | 3.00 | 4.00 | 0 | 0 | 0.088 | 0.231 |
| F10 | 3 | 0 | 1.114 | | 4.75 | | 0 | | 0.327 | 0.227 |
| P3 | 2 | 0 | 0.778 | | 6.00 | | 4 | | 0.133 | 0.250 |
| G7 | 3 | 3 | 0.239 | 0.325 | 5.00 | 6.00 | 0 | 6 | 0.0 | 0.143 |
| F8 | 2 | 3 | 0.624 | 0.141 | 4.00 | 4.00 | 0 | 0 | 0.0 | 0.378 |
| H11 | 2 | 2 | 0.223 | 0.203 | 5.00 | 5.00 | 0 | 0 | 0.267 | 0.600 |
| D3 | 2 | 1 | 0.379 | 0.321 | 3.50 | 3.00 | 0 | 0 | 0.343 | 0.375 |
| D9 | 2 | 3 | 0.292 | 0.265 | 5.00 | 5.30 | 4 | 5 | 0.160 | 0.240 |

Tube growth is not appreciably affected by self or cross pollen type until the tubes reach the ovary. There are as many self tubes as cross tubes growing through region 2; there is a trend in region 5 to observe more cross tubes, but it is not until the ovary that a clear preponderance of cross tubes is observed. Even in this region the comparatively greater number of cross tubes is not statistically significant unless all the individuals lacking tubes in ovaries are deleted from the analysis.

During these pollen growth studies I also measured self- and cross-fertility. My measure of fertility in Table 1 is the proportion of mericarps that begin to develop. The actual seed production is lower than this figure. The term differential fertility refers to the difference between self- and cross-fertility. Fertility levels, both absolute and differential, are not correlated with any of the pollen tube growth measures, are not correlated with any of the pollen tube growth measures, nor do the cases with outstanding fertility values show higher or lower tube growth.

**4 Discussion**

In these experiments more cross tubes entered the ovary, and presumably fertilized more ovules. This can be interpreted in two different ways. The first interpretation is that cross tubes grow a little faster than self tubes so that only in regions 5 and 6 does the distance become quantifiable. The second interpretation is that cross and self tubes grow at similar rates but that self tubes are prevented from entering the base of the style or the ovary. After 75 minutes, the number of tubes in ovaries did not noticeably increase, which gives weight to the latter interpretation. The Geraniaceae have trinucleate pollen (Brewbacker 1957) and any vestigial self-incompatibility response would be expected in the stigma, not the base of the style where the SI response typically occurs in binucleate pollen families.

There is a nonsignificant trend for the number of tubes/stigma at region 5 to be greater after crossing. This trend is opposite the higher score5 after selfing. In general, if maximal distance is measured, cross tubes have a small advantage; if "vigor" is measured, self tubes have a small advantage.

How do these differences in self and cross tube growth relate to the observed field situation? Bees visit an average of 10 consecutive flowers per plant in populations where the average number of flowers per plant is 25, and individuals with 100 or more simultaneously blooming flowers are not uncommon. The greater the number of flowers on a plant, the greater the number of consecutive visits. Using chalk impregnated with a fluorescent dye as a pollen analog, I determined that if a plant had more than 14 flowers, 80% of the transferred dye was distributed to other flowers of the same plant (Hessing 1985). It is evident that, from the individual flower's viewpoint, a visiting bee is more likely to bring self pollen than cross pollen.

The average tube distance 45 minutes after selfing and crossing is 4.28 and 4.78 regions, respectively. If tubes are growing at a rate of 0.10089 regions/min (approximately 0.2522 mm/min) and cross tubes have a .5 region advantage in speed, then self tubes can compensate for this advantage by arriving 4.96 min before cross pollen does. Given the high probability of receiving self pollen, it does not appear that the slightly greater speed of cross pollen will have much effect on the breeding system.

If prezygotic mate selection is occurring, then large values of differential fertility should be correlated with some form of pollen tube inhibition in the style. This is not the case; not even a weak correlation exists. No relationship existed between differential self-fertility and the number of tubes entering ovaries or micropyles. It is possible, however, that this relationship has gone unnoticed in these data because so few cases have self tubes in the ovary.

## 5 Conclusions

Self tube growth to the base of the style is not significantly different than cross tube growth, nor is the slight difference between self and cross tube growth likely to affect the breeding system. No correlation exists between self and cross tube growth, or appearance within the ovary, and self- and cross-fertility. There is no support for the mate selection or gamete competition hypotheses. The ability of tubes to enter the ovary may have a crucial effect on the breeding system, but is outside the scope of the data presented here.

## 6 Literature Cited

Brewbacker JL (1957) Pollen cytology and self-incompatibility systems in plants. J Hered 48:271-277.

Hessing MB (1985) The Breeding System of _Geranium caespitosum_ James: Description, Mechanisms, and Variation Across an Elevational Gradient. Ph.D. Dissertation. Colorado State University, Fort Collins, CO, USA. In Preparation.

Martin FW (1959) Staining and observing pollen tubes in the style by means of fluorescence. Stain Technol 34:125-128

Mulcahy DL (1979) The rise of angiosperms: a genecological factor. Science 206:20-23

Willson M, Burley N (1983) Mate Choice in Plants. Princeton University Press, Princeton, New Jersey. 252 p.

# Incompatibility reaction and gametophytic competition in Cichorium intybus L. (Compositae)

G. COPPENS D'EECKENBRUGGE[1]

## Introduction

For F1 hybrid seed production, Brussel's chicory breeders must clearly understand incompatibility relations in this plant and this implies a good method for discriminating between compatible, incompatible and pseudocompatible matings. Since the correlation between pollen germinations number and seed-set is not sufficiently reliable (Eenink, 1981), we had to study more precisely the early events of pollen germination and the pollen tube behaviour in the transmitting tissue.

## Material and method

The staining procedure of Dionne and Spicer (1958) was used. We also tried the classical fluorescence techniques using aniline blue, but these gave no results for observation of pollen tubes in the transmitting tissue. This absence of fluorescence has also been reported by Eenink (1981).

Self- and cross-pollinations, preceded by emasculation, were carried out on Brussel's chicory plants representing a wide variation in self-fertility : mean seed-set after self-pollination of more than 30 flowerheads per plant ranged from 0 to 60%. Styles were excised at varying times after pollination, from 0 to 60 minutes after cross-pollination, and from 0 to 90 minutes after self-pollination.

1. Laboratoire de Phytotechnie Tropicale et Subtropicale
Université Catholique de Louvain, 1348 Louvain-la-Neuve, Belgique

## Observations

### 1. Cross-pollination

The earliest pollen germinations occur within 3 to 5 minutes after pollination. There is no strong pollen adhesion before germination. Pollen grains are anchored upon the stigmas by means of their pollen tube. Once in the transmitting tissue, they become very thin and their growth is extremely fast : indeed, the quickest ones may arrive at the base of the style, 12 to 15 millimeters further, less than 20 minutes after pollination.

The fate of pollen grains germinating after the earliest ones (i.e. pollen grains having germinated more than 8 to 10 minutes after pollination) may be quite different ; an important proportion of them germinate and burst before their tube reaches the transmitting tissue. Sometimes, the pollen tube reaching the stigmatic papilla causes a reaction of this cell which then stains the same blue colour as the pollen grain and tube. Following this, the pollen grain bursts.

Another abnormality observed more than 8 to 10 minutes after cross-pollination consists in the growth inhibition of the pollen tube which then thickens and may form a bulb at its extremity, where most of the pollen grain cytoplasm accumulates. Such pollen tubes thickenings are frequently observed when two or more tubes extremities meet ; in a similar way, when a tube tip meets the body of another tube, its growth is blocked and it swells, while the other tube thickens at their meeting point. So the inhibition is manifested at two levels : the pollen tube wall synthesis is blocked, while the cytoplasm progress in the tube is slowed down. 20 minutes and more after pollination, such mutual inhibitions of pollen tubes also occur further and further in the style.

Generally speaking, 30 minutes after compatible mating, there are about 30-40 germinated pollen grains upon the stigmatic surface of one flower ; about a third part of the corresponding tubes is blocked at the stigmatic papillae level (pollen burstings); another third is arrested or considerably slowed down in the transmitting tissue ; the last third of these germinated pollen grains gives tubes reaching the base of the style or the ovary.

In some cases, pollen germinations number is very high (up to 100 or more) and the pollen burstings and tube growth inhibitions are much more obvious and frequent.

2. Self-pollination

Self-pollen of self-sterile plants does not germinate.

When the plant is pseudo-self-compatible, there is a great variation in pollen behaviour from one flowerhead to another. When pollen germinates, it is considerably delayed and slowed down ; germination is generally not observed less than 30 minutes after self-pollination. Furthermore, it is followed by pollen bursting. This event is much more frequent after pseudo-compatible selfing than after crossing. The very few self-pollen tubes which reach the transmitting tissue grow in the same way than compatible ones. There is no incompatibility reaction at this level. But when a relatively high number of self-tubes are observed (e.g. 60 or 90 minutes after selfing on a highly self-fertile plant), pollen tubes swellings occur in the transmitting tissue, as after cross-pollination.

## Discussion

Our observations did not show a clear qualitative difference between compatible and pseudo-compatible pollen behaviours. The differences are rather quantitative, concerning the total number of germinated pollen grains, the frequence of pollen burstings and pollen tubes inhibitions, and pollen germination timing.

Self-incompatibility is expressed only at the stigmatic papillae level, both in these cells and in pollen grains, either by total inhibition or by slowing down of pollen germination and subsequent pollen bursting, and by papillae reaction.

Pollen bursting is not typical of self-incompatibility. It is also related to a germinated-pollen-mass effect which becomes manifest when some pollen tubes have already reached the transmitting tissue. This effect is also expressed in the transmitting tissue, resulting in a mutual inhibition of neighbouring pollen tubes. Its existence is probably due to the diffusion of a substance from growing pollen tubes into the style tissues. Upon the stigma papillae it would provoke the early pollen tube arrest and, consequently, the pollen grain bursting. When acting later, in the transmitting tissue, it would only induce the pollen tube arrest or slowing down and its thickening.

This hypothesis might be valid in other species where very similar pollen tube behaviours have been observed, e.g. in Parthenium argentatum (Gerstel and Riner, 1950), Cosmos bipinnatus (Knox, 1973) and Scorzonera hispanica (Coppens d'Feckenbrugge, unpubl.).

This diffusing substance might well be directly related to the incompatibility system. This would contribute to explain the interactions between self- and cross- pollen reported in species with a sporophytic incompatibility, Cichorium intybus (Coppens d'Eeckenbrugge and al., 1984), Cosmos bipinnatus (Knox and al., 1975) and Brassica oleracea (Ockendon and Currah, 1977).

The significance of both pollen competition and interaction must not be overlooked. Firstly, an active pollen competition favouring the fastest germinating pollen grains probably contributes to a severe gametic selection, and thereby to prevent ovaries wastage and to increase progeny vigour because of possible "genetic overlap" (Mulcahy, 1974). Secondly, interactions between compatible and incompatible pollen temperates incompatibility reactions and allows fertilization by incompatible self-, sib-, or unrelated pollen, so maximizing the available variety of pollen sources. All this must widen our concept of outbreeding deriving solely from self- or sib-rejection.

References

Coppens d'Eeckenbrugge G, Plumier W, Louant B-P (1984) La compétition pollinique chez la Chicorée de Bruxelles (Cichorium intybus L.) et ses conséquences sur l'hybridation. Proc. Eucarpia Meeting on Leafy Vegetables : 160-166.
Dionne LA, Spicer PB (1958) Staining germinating pollen and pollen tubes. Stain Tech. 33 : 15-17.
Eenink AH (1981) Compatibility and incompatibility in Witloofchicory (Cichorium intybus L.).1. The influence of temperature and plant age on pollen germination and seed production. Euphytica 30 : 71-76.
Gerstel DU, Riner ME (1950) Self-incompatibility studies in Guayule. 1. Pollen-tube behaviour. J. Heredity, 41 : 49-55.
Knox RB (1973) Pollen wall proteins : pollen stigma interactions in ragweed and Cosmos (Compositae). J. Cell Sci, 12 : 421-443.
Knox RB, Howlett BJ, Paxton JB, Heslop-Harrison J (1975) Pollen-wall proteins : physiological characterization and role in self-incompatibility in Cosmos bipinnatus. PRSL, B 188 : 167-182.
Mulcahy DL (1974) Adaptative significance of gametic competition. In : Linskens HF (ed) Fertilization in Higher Plants. North Holland, American Elsevier. pp 27-31.
Ockendon DJ, Currah L (1977) Self-pollen reduces the number of cross-pollen tubes in the styles of Brassica oleracea. New Phytol 78 : 675-680.

# Cross-compatibility in an Annual Hermaphrodite Plant, Phlox drummondii Hook.

KAREN E. PITTMAN AND DONALD A. LEVIN[1]

## INTRODUCTION

Pollen-pistil compatibility is determined by both intrinsic and extrinsic factors. Compatibility depends in part on parental genotypes at the self-incompatibility locus (Nettancourt, 1977) and elsewhere in the genome (Bertin, 1982; Waser & Price, 1983; Mulcahy & Mulcahy, 1983), but may also be affected by temperature, age of pollen and pistil, exposure of pollen to water and water status of the pistillate plant (Zamir & Vallejos, 1983; Hoekstra, 1983; Eisikowitch & Woodell, 1974; Reger & Sprague, 1983; Slatyer, 1973).

The purpose of this study was to examine the contributions of maternal and paternal identities to variation in pollen-pistil compatibility in _Phlox drummondii_, and to determine the extent to which cross-compatibility is affected by macroenvironmental variation experienced by parental plants.

## MATERIALS AND METHODS

_Phlox drummondii_ Hook. is an outbreeding, hermaphroditic, lepidopteran-pollinated annual of central Texas. The self-incompatibility system of _P. drummondii_ is gametophytic, but is unusual in that inhibition of incompatible pollen occurs on the stigmatic surface.

Randomly chosen seed from the Columbus, Texas population were grown in a greenhouse. Sand-filled clay pots were arranged randomly in 9 blocks. Each of the following treatments was applied to 3 blocks: 1) Control plants were watered daily and fertilized each two weeks with 20:20:20 N:P:K solution containing iron. 2) Low water plants were fertilized as above, but were watered only upon wilting. 3) Low nutrient plants were

[1]Botany Department, University of Texas, Austin 78713 USA

watered daily, but were only fertilized once, soon after being transplanted as seedlings.

Eighteen plants, 2 per block making 6 per treatment, were randomly chosen for use in a complete diallel cross. Crosses were begun when all plants were in flower, after treatments had been in effect approximately two months. Fresh flowers were emasculated, marked according to pollen donor, and stigmas free of self pollen were hand-pollinated with 30-300 grains. Each cross and its reciprocal were replicated four times where possible. Selfs were not performed.

Stigmas were collected 48 hours after crossing and mounted in a solution of aniline blue in lactphenol to determine the percentage of grains which had germinated and sent tubes into the style. Every grain was scored. Ungerminated grains are dark blue, while grains whose cytoplasm is in pollen tubes are much lighter. Aborted grains also appear empty, but these are shrunken and typically contain oil droplets.

At the end of the experiment, total dry weight and number of flowers produced were recorded for each plant.

Two plants were eliminated from analysis due to insufficient replication. Analyses to test the effects of treatments and parental identities were carried out on arcsine-square root transformed percent pollen germination using the SAS GLM procedure. Pollen germination values reported in the text and in tables are untransformed.

## RESULTS

Plants grown under the low nutrient treatment had mean dry weights and numbers of flowers significantly lower than those of control plants (Table 1). The mean dry weight of low water plants was intermediate to that of the other two treatments but not significantly different from them. Low water plants had slightly more flowers than control plants, but mean flower number was not significantly different from that of the other two treatments.

Overall, percentage germination was 21.69 ± 0.57%. Plant 11 was the best pollen donor with a mean pollen germination of 29.81%. The worst donor, plant 2, had a mean pollen germination of 17.24%. The range of germination was greater for pollen recipients than for pollen donors; plant 24 was the best with 28.6%, and plant 16 was the worst with 10.46%. The compatibil-

Table 1. Mean dry weights and flower numbers, by treatment.

| Treatment | n | weight (g) | flower number |
|---|---|---|---|
| Control | 5 | 15.77 * | 361.8 * |
| Low water | 6 | 11.07 | 367.7 |
| Low nutrients | 5 | 3.95 * | 158.2 * |

* Significantly different pairwise comparisons using Wilcoxon 2-sample test.

ity of specific combinations ranged from a mean of 39.6% germination in crosses between plants 22 and 6, to near zero in crosses between plants 2 and 13 (0%), plants 3 and 16 (0.31%), and plants 6 and 12 (4.2%). The mean percentage pollen germination of a plant as a pollen donor was not correlated with its mean percentage germination as a pollen recipient ($r_s$=-0.21). Seven out of 16 plants had donor germination values significantly different by t-test from their recipient germination values (Table 3). No correlation was found between donor or recipient germination and plant dry weight ($r_s$(donor)= -0.22; $r_s$(recip.)= 0.16; both ns) or flower number ($r_s$(donor)= -0.19; $r_s$(recip.)= 0.19; both ns).

Despite the treatment effects on plant size and fecundity, treatments did not affect plants' cross-compatibility either as a pollen donor or as a pollen recipient (Table 2). There was a statistically significant interaction between male and female treatments, because the three completely incompatible combinations each involved a control and a low nutrient plant. When these combinations are removed from analysis, the interaction

Table 2. Mean percentage pollen germination, by treatment, and ANOVA results. F-values were calculated from Type III sums of squares.

| | Female Treatment | | Male Treatment | | F-Treatment x M-Treatment |
|---|---|---|---|---|---|
| | n | percent | n | percent | |
| Control | 283 | 21.61 | 287 | 20.73 | |
| Low water | 328 | 21.29 | 320 | 22.32 | |
| Low nutrients | 242 | 22.32 | 266 | 21.89 | |
| F-value (df) | 0.71 (2, 844) | | 1.11 (2, 844) | | 2.71 (4, 844) |
| | p = 0.492 | | p = 0.329 | | p = 0.029 |

Table 3. Mean percentage pollen germination , by female and male parents.

| Pollen recipients Plant | percent | Pollen donors Plant | percent |
|---|---|---|---|
| * 24 | 28.60 | 11 | 29.81 |
| * 13 | 28.07 | 15 | 26.84 |
| 22 | 26.93 | 22 | 25.99 |
| 3 | 24.68 | 23 | 24.87 |
| * 25 | 24.67 | 6 | 22.73 |
| 11 | 24.43 | 26 | 22.32 |
| 12 | 24.27 | 3 | 21.99 |
| * 2 | 23.60 | 4 | 21.46 |
| 4 | 21.31 | 12 | 20.89 |
| * 15 | 21.05 | 14 | 20.10 |
| 14 | 19.31 | 1 | 19.83 |
| 6 | 19.27 | 16 | 19.77 |
| 1 | 19.05 | 24 | 18.21 |
| 26 | 18.53 | 13 | 17.53 |
| * 23 | 14.83 | 25 | 17.42 |
| * 16 | 10.46 | 2 | 17.24 |

| Lines indicate groups not significantly different by t-test.

* Asterisks indicate plants with significantly differing donor and recipient pollen germination values.

term is no longer significant. In view of the lack of treatment effects, it is extremely unlikely that the treatments are responsible for the incompatible reactions between these three pairs of plants.

Pollen-pistil compatibility depended heavily on the identity of male and female parents. One quarter of the variation in pollen germination was explained by the interaction between maternal and paternal identity ($R^2$=0.273), which was statistically significant (F=1.34; df=209, 613; p=0.0036). Significant differences were observed in the percentage germination of pollen from individual plants on other plants' stigmas (donor germination) and in the percentage germination of all pollen on individual plants' stigmas (recipient germination) (Table 3). Female identity explained roughly 2.5 times more variation in percentage pollen germination than did male identity ($R^2$=0.087 for female and 0.035 for male). Both factors were significant (female: F=4.45; df=15, 209; p<0.0001; male: F=1.76, df=15, 209; p=0.0042. Because both main effects were random effects, they were tested over the interaction sums of squares. Type III sums of squares were used.)

DISCUSSION

Variation in pollen-pistil compatibility in *Phlox drummondii* is due to a donor x recipient interaction and to differences in the quality of plants as donors or recipients. The significant interaction observed between maternal and paternal identities indicates that plants cross more successfully with some plants than with others, and that the "best mates" are different for different individuals. Similar interactions have been observed for pollen germination in raspberry (Jennings & Topham, 1971), for fruit and seed maturation in *Campsis radicans* (Bertin, 1982), and for cross-fertility in *Lotus corniculatus* (Schaaf & Hill, 1979). Such interactions may be manifestations of allelic similarity at the S-locus or elsewhere in the genome.

Variation in pollen-pistil compatibility in *Phlox* also is due in part to plants' differing quality as pollen donors or recipients across all other plants. Several studies have found significant maternal and paternal effects for cross-fertility after hand-pollination (Schaaf & Hill, 1979; Jennings & Topham, 1971; Gurgis & Rowe, 1981) or for reproductive success in open-pollinated systems (Gutierrez & Sprague, 1959; Muller-Starck & Ziehe, 1984).

No clear relationship exists in *Phlox* between a plant's overall crossability as a pollen recipient and a pollen donor. However, about half of the plants were significantly more successful as either a female or a male. Sexual asymmetries in gametic contribution of individuals have been documented in *Zea* and *Pinus* (Gutierrez & Sprague, 1959; Muller-Starck & Ziehe, 1984). Such asymmetries cause non-random mating (Ziehe, 1982).

Differences in pollen-pistil compatibility were not associated with treatments experienced by parental plants, but were highly dependent on parental identities. Since the differences in microhabitat perceived by *Phlox* plants in the field are probably not as great as those imposed in this experiment, it can be assumed that parental identities are much more important than microenvironment or plant size in determining cross-compatibility in natural populations of *Phlox drummondii*.

REFERENCES

Bertin RI (1982) Paternity and fruit production in trumpet creeper (Campsis radicans). Amer. Nat. 119: 694-709.

Eisikowitch D, Woodell SRJ (1974) The effect of water on pollen germination in two species of Primlua. Evolution 28: 692-694.

Gurgis RY, Rowe DE (1981) Variability of seed-set in alfalfa and correlation with some pollen and ovule characteristics. Can. J. Plant Sci. 61: 319-323.

Gutierrez MG, Sprague GF (1959) Randomness of mating in isolated polycross plantings of maize. Genetics 44: 1075-1082.

Hoekstra FA (1983) Physiological evolution in angiosperm pollen: Possible role of pollen vigor. In: Mulcahy DL, Ottaviano E (eds) Pollen: Biology and Implications for Plant Breeding. Elsevier.

Jennings DL, Topham PB (1971) Some consequences of raspberry pollen dilution for its germination and for fruit development. New Phytol. 70: 371-380.

Muller-Starck G, Ziehe M (1984) Reproductive systems in conifer seed orchards. 3. Female and male fitnesses of individual clones realized in seeds of Pinus sylvestris L. Theor. Appl. Genet. 69: 173-177.

Nettancourt D de (1977) Incompatibility in Angiosperms. Springer-Verlag.

Reger BJ, Sprague JJ (1983) Pearl millet and sorghum pollen growth in pearl millet gynoecia of different ages. Crop Sci. 23: 931-934.

Schaaf HM, Hill RR (1979) Cross-fertility differentials in birdsfoot trefoil. Crop Sci. 19: 451-454.

Slatyer RO (1973) The effect of internal water status on plant growth, development and yield. In: Slatyer RO (ed) Plant Response to Climatic Factors. Unesco.

Zamir D, Vallejos EC (1983) Temperature effects on haploid selection of tomato microspores and pollen grains. In: Mulcahy DL, Ottaviano E (eds) Pollen: Biology and Implications for Plant Breeding. Elsevier.

Ziehe M (1982) Sexually asymmetric fertility selection and partial self-fertilization. 1. Population genetic impacts on the zygotic genotypic structure. Silva Fenn. 16: 94-98.

# Environmental Stress Reduces Pollen Quality in Phlox: Compounding the Fitness Deficit

CARL D. SCHLICHTING[1]

## 1 Introduction

Reproductive success of hermaphroditic plants is determined by both the quantity and quality of the gametes and offspring produced, and both genetic and environmental factors may affect this success. Most estimates of success are quantitative measures based solely on female function, and this bias continues when environmental effects are considered. In this chapter, I present data concerning the impact of the environment on the quality of the male function, and discuss the impact of environment on relative plant fitness when both male and female functions are considered.

Most estimates of the reproductive success of hermaphroditic plants are based solely on the haploid contribution through the female function, ovule and seed production (e.g. Weller 1985; Howell 1981; Leverich and Levin 1979), because of the difficulties inherent in estimating haploid contribution through male function (pollen production, dispersal, and germinability). Success through the male function is usually ignored, or the success of the two functions is assumed to be equal (equisexuality, Horovitz 1978) or at least parallel.

In those cases where male function has been considered, simplifying assumptions are typically made concerning the quantity and quality of pollen produced (Ross 1984; Ross and Gregorius 1983; Lloyd 1979, 1980): to wit 1) all flowers produce similar amounts of pollen; 2) pollen dissemination is proportional to the quantity produced; and 3) all individuals produce pollen of similar quality (i.e. similar viability,

1 Department of Biology, The Pennsylvania State University, University Park, PA 16802

germinability, and pollen tube growth rates). There is growing evidence that these assumptions do not generally hold (Horovitz and Harding 1972; Bertin 1982; Müller-Starck and Ziehe 1984; Schlichting unpub; Devlin unpub). Consequently, a realistic assessment of reproductive success in hermaphrodites must include information on both male and female functions.

Environmental effects on plant reproduction are well known, at least for female function; both quantity (Harper 1977) and quality (Parrish and Bazzaz 1985) of offspring have been shown to vary with environment. The effects of natural environmental variation on male function however, have been rarely studied (Zamir et al. 1981). We know that the total quantity of pollen produced is reduced in stressful environments as a result of restricted plant growth and flower production. Do such stressful environments also affect pollen quality? In this study I demonstrate that environmental variation influences not only the quantitative production of the plant's male and female gametes, but also the quality of the male function as well.

## 2 Methods

In a study of phenotypic plasticity (Schlichting and Levin 1984), plants from two hermaphroditic species of *Phlox*, *P. cuspidata* and *P. drummondii*, were grown in the greenhouse in 6 environmental treatments: Control, leaf Removal, Meristem Clip, Low Water, Low Nutrient, and Small Pots (for descriptions see Schlichting and Levin 1984). Treatment effects were assessed for both quantitative and qualitative measures of reproductive success. Quantitative female production was measured as ovule production (3 ovules/flower) and quantitative male production was estimated by the number of flowers produced (assuming equal pollen production per flower and plant).

Treatment effects on pollen *quality* were examined as follows. Three plants of each species were chosen from among the Control plants as pollen recipients. Each recipient received cross pollinations from 5 conspecific plants from each of the 6 treatments, for a total of 30 crosses per recipient. Stigmas and styles were excised after 3 days, stained in lactophenol and aniline blue, and the percentage pollen germination was recorded. Pollen stainability was uniformly high, typically greater than 98%. In Phlox,'gametophytic' incompatibility is expressed at the stigmatic surface; in incompatible crosses pollen either does not germinate or produces distorted tubes which do not penetrate the stigma. Crosses exhibiting this behavior were excluded from the analysis.

Pollen germinability is an important characteristic in these species, and particularly in the obligately outcrossing P. drummondii, because the lepidopteran pollinators of these plants commonly transfer very small quantities of pollen between flowers (Udovic and Levin unpub). Thus a reduction in the number of pollen tubes due to poor germination of a plant's pollen might result in reductions in both the number and quality of offspring through the male function (Mulcahy and Mulcahy 1975; Ottaviano et al. 1982; Stephenson et al. this volume).

## 3 Results

In P. cuspidata there were significant treatment effects on ovule production and pollen germinability (Table 1). In P. drummondii the treatments had a significant effect on ovule production but not on pollen germinability.

The treatments, Low Water (LW), Low Nutrients (LN), and Small Pots (SP), consistently yielded biomass and flower production values lower than for the other treatments (Schlichting and Levin 1984). I will refer to these low productivity treatments(LW, LN, and SP) as stress treatments, and the Control (C), Leaf Removal (LR), and Meristem Clip (MC) treatments as non-stress treatments. The stress treatments thus significantly reduced the female and male quantitative fitness parameters, ovule and pollen production. To determine whether the stress treatments had a similar effect on a qualitative trait, pollen germinability, a statistical comparison between the stress and non-stress groups was made. For both Phlox species, pollen germinability was significantly reduced in plants subjected to stress treatments (Table 1). Thus both female and male quantitative production are influenced by environmental conditions, and the quality of the male function.

Table 1. Mean values of reproductive traits in Phlox cuspidata and P. drummondii in the 6 treatments and ANOVA results (14).

| | TREATMENTS | | | | | | |
|---|---|---|---|---|---|---|---|
| | C[a] | LR | MC | LW | LN | SP | p[b] |
| Phlox cuspidata | | | | | | | |
| Total flowers | 206 | 281 | 284 | 201 | 213 | 159 | .0002 |
| % pollen germination | 46 | 39 | 36 | 27 | 31 | 35 | .03 (.002) |
| Phlox drummondii | | | | | | | |
| Total flowers | 225 | 250 | 231 | 162 | 172 | 147 | .001 |
| % pollen | 44 | 43 | 44 | 34 | 37 | 37 | .21 (.01) |

a C-Control; LR-Leaf Removal; MC-Meristem Clip; LW-Low Water; LN-Low Nutrients; SP-Small Pots.

b Results of one way ANOVA. Values in parentheses are results of a priori comparisons of non-stress (C, LR, & MC) versus stress (LW, LN, & SP) treatments.

## 4 Discussion

The problems associated with equating plant fitness and female function can be clearly demonstrated using the parameters for male and female function measured in this study. Relative fitness for plants in the six treatment groups were calculated based on ovule production alone, pollen germinability alone, and on the combination of these two measures to represent both male and female function (Table 2). The combined fitness is calculated as:

(ovule production) + (flower production x pollen germinability)

Relative fitnesses are derived by dividing all values by the highest individual treatment; thus the highest value is 1.00.

Comparing plants from stress and non-stress treatments using the female function estimate (ovule production), plants in the non-stress treatments (C, LR, MC) have 1.3 (P. drummondii) to 1.5 (P. cuspidata) times the reproductive capability of plants in the stress treatments (LW, LN, SP). Using only the male function estimate (pollen germinability) non-stress plants have a 1.2 (P. drummondii) to 1.5-fold (P. cuspidata) advantage over plants in the stress environments. The combination of male and female effects however is not additive, but multiplicative, resulting in a magnification of the relative fitness differences between these two groups of plants: non-stress

Table 2. Relative fitness of an average plant in the 6 treatments: female, male, and combined estimates.

| | TREATMENTS | | | | | |
|---|---|---|---|---|---|---|
| | C | LR | MC | LW | LN | SP |
| *Phlox cuspidata* | | | | | | |
| Ovule production | .72 | .99 | 1.00 | .71 | .75 | .56 |
| % pollen germination | 1.00 | .85 | .78 | .59 | .67 | .54 |
| Combined | .77 | 1.00 | .95 | .60 | .67 | .46 |
| Δ[a] | +.05 | +.01 | -.05 | -.11 | -.08 | -.10 |
| *Phlox drummondii* | | | | | | |
| Ovule production | .90 | 1.00 | .92 | .65 | .69 | .59 |
| % pollen germination | 1.00 | .98 | 1.00 | .77 | .84 | .84 |
| Combined | .91 | 1.00 | .93 | .58 | .64 | .55 |
| Δ | +.01 | .00 | +.01 | -.07 | -.05 | -.04 |

a Δ = Combined relative fitness - Ovule production relative fitness.

plants enjoy a 1.6 fold fitness advantage in both species (Table 2), as the relative fitnesses of the stressed plants all decline.

With this example we see that the estimate of relative plant fitness based solely on female function underestimates the magnitude of the differences between treatments revealed using information on both functions. The effect may be even more pronounced if the quality of seeds (germinability, seedling growth rate) is affected by the environment as well (e.g. Parish and Bazzaz 1985).

Leverich and Levin (1979) have shown that fitness distribution in a natural population of *Phlox drummondii* is highly leptokurtic when just seed production (female function) is considered. In fact one-half of all the seeds were produced by only 10% of the plants. Such distributions are not unusual in plant populations (Harper 1977). To the extent that these distributions are influenced by environmental variation, those based on the contributions of both the male and female functions will tend to be more severely skewed, with even more disproportional contributions to the gene pool by genotypes in favorable microhabitats. We should not however, assume that male and female functions will suffer proportionally in stressful conditions. Devlin (unpub) has data showing an environmentally mediated reduction in female fitness with no reduction in male fitness in Lobelia cardinalis.

There are two important conclusions from this study. First, male components of fitness such as pollen production and germinability must be assessed in addition to the typically measured female components. Second, both the quantitative and qualitative aspects of male function can be influenced by the environment, with profound impacts on the fitness of the hermaphrodite plant.

Acknowledgments. I appreciate the comments of Vickie Borowicz, Bernie Devlin, Don Levin, Karen Pittman, MaryCarol Rossiter, and Andy Stephenson on previous versions of this chapter. This work was supported in part by University Fellowships from the University of Texas, by funds from the Ecology Program at Penn State, and by NSF Grant BSR-8119484 to D A Levin.

References

Bertin RI (1982) Amer Nat 119: 694-709

Harper JL (1977) Population Biology of Plants. Academic Press, NY. 892 pp

Horovitz A (1978) Amer J Bot 65: 485-486

Horovitz A, Harding J (1972) Hered 29: 223-236

Howell N (1981) Amer Midl Nat 105: 312-320

Leverich WJ, Levin DA (1979) Amer Nat 113: 881-893

Lloyd DG (1979) NZ J Bot 17: 595-606

Lloyd DG (1980) NZ J Bot 18: 103-108

Mulcahy DG, Mulcahy GB (1975) Theor Appl Genet 46: 277-280

Müller-Starck G, Ziehe M (1984) Theor Appl Genet 69: 173-177

Ottaviano E, Sari Gorla M, Pe E (1982) Theor Appl Genet 63: 249-254

Parrish JAD, Bazzaz FA (1985) Oecol 65: 247-251

Ross MD (1984) Biol J Linn Soc 23: 145-155

Ross MD, Gregorius H-R (1983) Amer Nat 121: 204-222

Schlichting CD, Levin DA (1984) Amer J Bot 71: 252-260

Weller SG (1985) Ecol Monogr 55: 49-67

Stephenson AG, Winsor JA, Davis LE (this volume) Effects of pollen load size on fruit maturation and sporophyte quality in zucchini

Zamir D, Tanksley SD, Jones RA (1981) Theor Appl Genet 59: 235-238

# Abstracts of Poster Presentations

# Incompatibility reaction and gametophytic competition in *Cichorium intybus* L.

G. COPPENS D'EECKENBRUGGE[1]

Compatible and incompatible chicory pollen tube development has been observed in light microscopy.

Variation in self-incompatibility strength from plant to plant results in varying self pollen behaviours. Particularly, in cases of pseudo-self-compatibility, stigma penetration may occur but it is delayed.

Compatible pollen tubes grow very fast and reach the ovary within 30 min after pollination. Strikingly, the growth of some of them may be inhibited in the same way as in the incompatible case. Furthermore, they may also be inhibited in the transmitting tissue. These abnormal behaviours are more often observed after heavy pollinations and suggest an active competition between compatible pollen tubes.

[1]Laboratoire de Phytotechnic Tropicale, 3, Place Croix du Sud, Sc 15 D, 1348, Louvain-La-Neuve, BELGIUM

* * *

## Intra- and interspecific incompability in *Brachiaria ruziziensis* Germain et Evrard

G. COPPENS D'EECKENBRUGGE and M. NGENDAHAYO[1]

Very few seeds can be obtained after crossing the diploid and autotetraploid forms of *Brachiaria ruziziensis* with *Brachiaria decumbens* and *Brachiaria brizantha*. Most of these are self-seeds. Pollination experiments have been undertaken to localize the interspecific barrier.

Diploid and autotetraploid plants of *B. ruziziensis* have been pollinated with self- or cross-pollen from diploid and tetraploid *B. ruziziensis*, *B. decumbens* and *B. brizantha*. Observation with fluorescence microscopy mainly shows: 1/ there is no interspecific incompatibility reaction at the pollen germination or tube growth level; and 2/ the self-incompatibility reaction takes place at every level from pollen germination, before stigma penetration, to pollen tube growth through the style. So, *B. ruziziensis* is another exception to the generally admitted stigma-localized self-rejection in the Gramineae.

[1]Laboratoire de Phytotechnic Tropicale, 3, Place Croix du Sud, Sc 15 D, 1348, Louvain-La-Neuve, BELGIUM

* * *

## Mentor pollen as a tool in plant biotechnology. Application to *Populus*.

M. GAGET, M. VILLAR and C. DUMAS[1]

### 1 Introduction

The "mentor effect" was first used by Stettler (1968) to overcome incompatibility among poplar species.

From that time, numerous workers (see review Stettler and Ager 1984) applied the mentor effect with more or less success to overcome both self- or interspecific incompatibility.

The original technique consisted in a mixed pollination of both compatible and incompatible pollens. The compatible pollen was prevented from achieving fertilization by treatments such as: irradiation, leaching with solvents or freezing and thawing. These treatments sterilized the pollen but generally did not prevents its germination. A pollen tube can be produced (see the review of Gaget et al. 1985).

Using mentor pollen, our aim was double:

1) To obtain, by standardisation of the technique, an hybrid from strictly incompatible species.

2) To elucidate the mentor effect mechanisms and identify the compounds which are involved, in order to isolate and further to use the active fraction.

## 2 Materials and methods

- Materials

The model used was Populus nigra x Populus alba, and its reciprocal i.e., Populus alba x Populus nigra.

- Methods

2-1 Crosses

Controlled conditions were developed with particular care taken with respect to pollen viability (fluorochromatic reaction test, Heslop-Harrison and Heslop-Harrison 1970) and stigma receptivity.

Compatible pollen was treated by:

- ionizing radiation with with 100 Krad dose (Stettler 1968)
- freezing and thawing with pollen was rendered non viable (FCR test) 16 cycles of -196°C/40°C were used (modified from Knox 1972)
- freeze dried extraction of compatible pollen
- dead compatible pollen was also used.

2-2 Statistical analysis

Principal component analysis was used in order to compare the efficicency of the different treatments for each cross, pollen tube behaviour was analysed, from to flowers, by ABF method (Dumas and Knox 1983). Observations were classified in four classes from ++ (maximal presence of tubes) to 0 (no tube), with two intermediate classes: + and

To clarify the pollen tube behaviour, different areas of pistil were taken into account:

- stigma surface

- stigma tissue
- stylodium
- ovary.

2-3 Seeds

Seeds were collected stored and sown at the INRA forestry station (Orleans, France). Hybrids determination is there still in progress.

## 3 Results and discussion

The analysis of the graph obtained by PCA showed that the freeze dried extract, and the freezing and thawing treatments strongly, improved pollen tube growth. On the other hand, only freezing and thawing yielded a large quantity of viable seeds. Those results led to the hypothesis that two levels of interaction were involved in the mentor effect efficiency:

- surface level: the presence of recognition signals greatly improves incompatible germination on the stigma surface.
- stylar level: the presence of compatible treated pollen tubes was shown to be necessary for incompatible tube growth towards the embryo sac. Compatible tube may provide the incompatible tube with recognition and/or growth substances.

Moreover, the compatible tube can act as pioneer tube (Visser 1983) "pawing the way" for the incompatible slower tube.

Study of hybrids nature of seedlings is in charge of INRA station. After the first vegetative season, some "mentor effect" seedlings showed a maternal phenotype. Haploid or diploid parthenogenesis is investigated as a new and rapid way for obtaining homozygotes (Pandey 1980). Specific active compounds responsible for recognition are currently investigated, by homogeneisation of pollen and characterization by SDS-PAGE and IEF.

## References

Dumas C, Knox RB (1983) Callose and determination of pistil viability and incompatibility. Theor Appl Genet 67: 1-10.

Gaget M, Dumas C, Knox RB . Pollen as a tool in plant biotechnology (in

press).
Heslop-Harrison J, Heslop-Harrison Y (1970) Evaluation of pollen viability by enzymatically induced fluorescence, intracellular hydrolysis of fluorescein diacetate. Stain Technol 45: 115-120.
Knox RB, Willing RR, Pryor LD (1972) Interspecific hybridization in poplars using recognition pollen. Silvae Genet 21: 65-69.
Stettler RF (1968) Irradiated mentor pollen: its use in remote hybridization of black cottonwood. Nature 219: 746-747.
Stettler RF, Ager AA (1984) Mentor effect in pollen interactions. In: Linskens HF and Heslop-Harrison J (eds) Cellular interaction. Springer Verlag, Berlin, pp 609-623.

1 Université Cl. Bernard-LYON I, R.C.A.P., UM CNRS 380024
69 622 Villeurbanne Cedex, FRANCE.

* * *

## Cell surface components of Brassica stigmas

T. GAUDE and C. DUMAS[1]

Components of the cell surface of Brassica stigmas have been investigated by an electron microscopy study. Cytochemical techniques chosen for their ability to reveal cellular membranes, show that the pellicle of the Brassica stigma, although not trilamellar in structure, presents some characteristics of a biological membrane. Notably, the pellicle possesses an ATPase activity which can be deeply inhibited by vanadate.

The first events of the pollen-stigma interaction are discussed in light of these new findings.*

*Dumas C, Knox RB, Gaude T (1984) Internat Rev Cytol 90: 239-272.

[1]Universite Cl. Bernard-Lyon 1, Reconnaissance cellulaire et Amelioration des Plantes, UM CNRS 380 024, Villeurbanne, FRANCE

* * *

# Micro-isoelectric focusing of pollen grain proteins in Cucurbita pepo L.

G. GAY, C. KERHOAS and C. DUMAS

A technique of IEF separation of proteins from single grains has been adapted according to Mulcahy et al. (1981) in order to estimate the heterogeneity of the pollen population at the level of individual.

## Material and methods

Plant material: Flowering squash plants (*Cucurbita pepo* L.) cultivar Seneca were supplied by Tezier (Valence, France).
Gel composition and preparation: Polyacrylamide gels (240 x 100 x 0.5 mm) were obtained using the slab gel casting apparatus LKB ultramould 2217. The polyacrylamide solution (T = 5%, C = 3%) contained ampholytes (pH 3.5 -9.5, 3% w/v, LKB ampholines).
Sample application: In contrast to Mulcahy's protocole no lysis solution was employed in order to facilitate pollen crushing. The pollen grains were placed one by one onto the gel using fine forceps. At the contact of the gel they swelled rapidly (5 min.). A gentle pressure of the forceps on the wall was enough to induce the rupture of one germinative pore. The vegetative cell content poured out on the gel without further intervention.
Focusing conditions: The focusing conditions setted were 1500 V, 50 mA, and 25 W for one hour. The glass cooling plate was maintained at 4°C.
Gel staining: A modified silver staining of Morrissey (1981) was performed (Gay et al., in preparation).
Enzyme visualization: Alcohol dehydrogenase visualization (ADH) were realised according to Kingzofer and Radola (1983).
Pollen viability estimation was performed according to Heslop-Harrison and Heslop-Harrison (1970). Each sample tested was FCR + at 95% minimum.

## Results and discussion

The previously described technique allows the use of a classical IEF apparatus for proteins separation from a single pollen grain of Cucurbita pepo L. Large secale comparaison of pollen grains population may be realised in this system (Fig. 1). Fourty to fourty five proteins bands are revealed from a single pollen grain. With corn pollen (80-100 μm), we get only 16 bands and with wheat pollen (40-50 μm) only 4 bands. The patterns obtained shows a good reproducibility between separated samples and separated gels. This technique may be a convenient tool to observe segregation of characters at the haploid stage. After focusing, an ADH of Cucurbita pepo L. may be visualized in less than ten minutes. The activity of certain enzymes was found to be dependent of pollen viability (Gay et al., in preparation). Finally this technique could be used as a routine test in the study of pollen expression with large size pollen population.

## References

Kingzofer A, Radola BJ (1983) Electrophoresis 4: 408-417.
Heslop-Harrison J, Heslop-Harrison Y (1970) Stain Technol 45: 115-120.
Morrissey JH (1981) Anal Biochem 117: 307-310.
Muldahy DL, Robinson RW, Ihara M, Kesseli R (1981) J Heredity 72: 353-354.

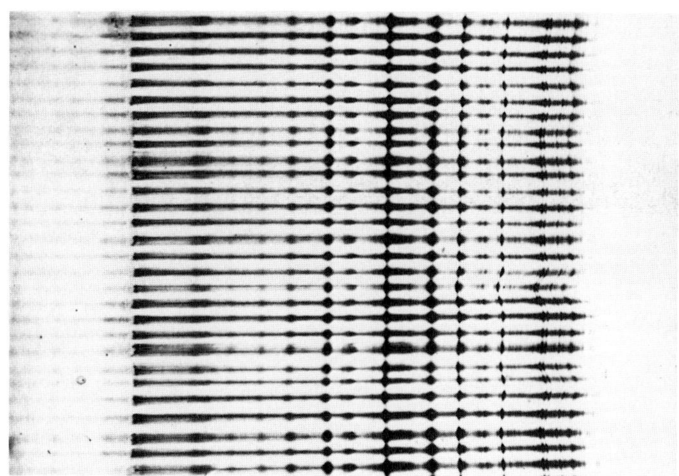

Fig. 1: Homogeneity of a pollen grain population in Cucurbita pepo L. No segregation of characters is observed the for major proteins of this $F_1$ hybrid.

[1] Université Cl. Bernard-LYON I, R.C.A.P., UM CNRS 380024
69 622 Villeurbanne Cedex, FRANCE.

* * *

## Pollination sub-systems in a *Solanum* x *Lycopersicon* cross

T. GRADZIEL and R. W. ROBINSON[1]

In attempting the intergeneric cross *Solanum lycopersicoides* x tomato (*Lycopersicon esculentum*), four distinct barriers of successful pollen tube growth and fertilization were identified: germination failure, growth arrest in upper style, restricted growth in lower style, and loss of directional response in the ovary. In addition to pistil location, these pollination sub-systems may be distinguished by both tube morphology and manipulation effective in overcoming growth inhibition. Upper style pollen tube arrest is avoided through bud pollination. Pollen germination on developmentally immature or stressed styles may be possible by pollinating in an environment of high (95 %) relative humidity, or by applying an artificial lipid based stigmatic exudate. Lower style pollen tube vigor is improved through proper plant nutrition and selection of styles at favorable developmental stages. Directed growth to ovule micropyles appears enhanced with pollen mixtures containing compatible "mentor" pollen or multiple pollinations at proper time intervals. Successful hybridization in this wide cross thus involves identification and manipulation of several independent pollination sub-systems.

[1] N.Y.S.A.E.S., Cornell University, Department of Plant Science, Geneva, New York, UNITED STATES

*   *   *

## Pollen cryofracture

C. KERHOAS, T. GAUDE, G. GAY, E. MATTHYS-ROCHON and C. DUMAS

### 1 Introduction

Pollen wall have a unique ultrastucture consisting on 2 major layers: the exine composed by highly polymerized lipid compounds, sporopollenin and the intine composed of mixed polysaccharides (Southworth and Branton 1971). In ultrathin section, no fibrillar structure have been identified by contrast with classical cellulosic wall. Plasmalemma organization correlated with the water content is difficult to demonstrate in ultrathin section with electron microscope (TEM) and a very few data have been published in the area (Heslop-Harrison 1979; Dumas et al. 1984). In addition with the conventional route used for fixation, proteins cannot be visible. Then, we have investigated ultrastructural aspects of cell wall and plasmalemma by the well adapted tool: freeze fracture procedure, in two types of pollen (*Brassica* with a water content (wc) of 8%; *Cucurbita* with wc: 50%).

### 2 Material and methods

Plant material: Plants of *Brassica oleracea* L. var. *acalipha* were kindly supplied by Drs Du Crehu and Renard (INRA, Rennes, France). Plants of *Cucurbita pepo* L. var. Seneca supplied by Tezier (Valence, France) were grown in a green house.

Pollen grains were collected by hand and stored directly from anthers in a closed eppendorf vial. Their viability were checked before fixation using fluorescein diacetate ester (Heslop-Harrison and Heslop-Harrison 1970).

Ultrastructural methods: For ultrathin sections, pollen was prepared as Gaude and Dumas (1984). Sections were stained with heavy salts according to Reynolds'procedure. β 1-4 glucans were detected by the use of PATAg test (for the techniques, see Roland 1978; Gaude and Dumas 1984). For freeze fracture study, the material was prepared according to Escaig and Nicolas (1976) and fracture carried out on a CF 250 Jung Reichert cryofract.

## 3 Results and discussion

Pollen wall ultrastructure: Ultrathin sections of Cucurbita pepo exine reveals a quite homogeneous structure. By contrast, Brassica exine presents many crypts covered by pollen coat, a sporophytic material implies in the acceptance or rejection of the male gametophyte during pollen-stigma interactions (see Gaude and Dumas 1984; Dumas et al. 1984). This pollen coat is absent in Cucurbita, a self compatible species.

The intine, a mainly cellulosic wall possesses several interesting characteristics. A multilamellar structure revealed in TEM both with ultrathin sections and in freeze fracture (two layers in Brassica intine, three layers in Cucurbita intine). One of these layers looks like a porous structure.

The PATAg reactivity is weak.

Freeze fracture of intine does not present the classical fibrillar structure of the plant cell wall according to Vian et al. (1978). This fact was first underlying in pollen intine of Artemisia pycnocephala and Lilium humboldtii by Southworth and Branton (1971).

By contrast, Heslop-Harrison and Heslop-Harrison (1982) using a conventional chemical fixation reported a microfibrillar organization in the intine of Lilium henryi. The absence of oriented cellulosic fibres, may constitute an adaptative mechanism against strains occuring during both pollen dehydration before anthesis and pollen imbibition on the compatible stigma surface. So, the apparent disorder in terms of cellulosic fibres arrangement without any directional orientation could form a good mechanical system to permit deformation. In addition, the alternance of porous and homogeneous layers may increase the intine elasticity.

Plasmalemma ultrastructure: Details of plasmalemma ultrastructure are not well defined on classical ultrathin sections after chemical fixation. Freeze fracture observations have revealed intramembraneous particles (IMP) in Brassica and Cucurbita. Both membrane faces are seen to contain IMP. As in plant and animal cells, generally the IMP density is much greater on the protoplasmic face ($P_F$) than on external face ($E_F$)(see

Bliss et al. 1984; Sjolund et Shih 1982).

*Brassica* pollen studies revealed 3 structural aspects not expected in a quite dried but living organism: - a continuous plasma membrane

- no apparent disruption of the membrane integrity

- a high number of IMP

It is interesting to note when anhydrobiotic nematodes are quickly dried a loss of IMP occurs from the plasma membrane. By contrast, if these nematodes are gently dried, the induction of anhydrobiosis occurs and there is no decreasing in the number of IMP (Crowe et al. 1978). On the other hand, in cowpea radicle, during the imbibition phase, there is a relative increasing in the number of $E_F$ particles (Bliss et al. 1984). It will be interesting to demonstrate the same increasing in germinating pollen after pollen-stigma interactions especially during the water exchange between both partners. Finally, the pollen envelop characteristics briefly presented in this preliminary paper could be related to an adaptation of these insulated organisms to the environmental stresses.

References

Bliss RD, Platt-Aloia A, Thomson WW (1984) Plant Cell Environ 7: 601-606.
Crowe JH, Lambert DT, Crowe LM (1978) *In* Crowe JH and Clegg JS (eds) Dry Biological Systems. Academic Press, New York, pp 23-51.
Dumas C, Knox RB, Gaude T (1984) Int Rev Cytol 90: 239-271.
Escaig J, Nicolas G (1976) C R Acad Sci Bio Paris 283: 1245-1248.
Gaude T, Dumas C (1984) Ann Bot 54: 821-825.
Heslop-Harrison J (1979) Amer J Bot 66: 737-743.
Heslop-Harrison J, Heslop-Harrison Y (1970) Stain Technol 45: 115-120.
Heslop-Harrison Y, Heslop-Harrison J (1982) Ann Bot 50: 831-842.
Roland JC (1978) *In*: Hall JL (ed) Electron microscopy and Cytochemistry of plant cells. Elsevier, North-Holland, Biomed. Press, pp 1-62.
Sjolund RD, Shih CY (1982) J.Ultrastruct Res 82: 189-197.
Southworth D, Branton D (1971) J Cell Sci 9: 193-207.
Vian B, Mueller S, Brown Jr RM (1978) Cytobios 7-15.

1 Université Cl. Bernard-LYON I, R.C.A.P., UM CNRS 380024
69 622 Villeurbanne Cedex, FRANCE.

*         *         *

Water content evolution in Cucurbita pepo during ageing: a NMR study

C. KERHOAS[1], G. GAY[1], J.C. DUPLAN[2] and C. DUMAS[1]

In Cucurbita pepo, the pollen is a bicelled organism characterized by its large size (200 µm as a diameter), a high water content (40-57% of fresh weight) and a short longevity. In the laboratory with standard conditions (20°C, 40% of relative humidity,RH), the pollen goes on dehydration and deaths 2 or 3 hours after anthesis. So, the interactions between water content and pollen viability appear to be of the greatest interest (see Heslop-Harrison 1979; Dumas and Gaude 1983). In order to elucidate this interaction, we carried out a serie of NMR experiments involving the nuclear relaxation phenomenon. This non destructive method allows to study the water state evolution during the ageing of a same controlled pollen sample.

Materials and methods

Plant material: Cucurbita pepo L. var. Seneca supplied by Tezier (Valence, France) was grown in a green house. Pollen grains were collected by hand and stored directly from anthers in a closed eppendorf vial.
Method: Water content was determined by weighting freshly collected pollen and dried pollen 10 min. at 85°C with an infrared dessicator (Sartorius system).

Water state was estimated by NMR experiments following the Meiboon Carr Purcell spin echo method (see Duplan and Dumas 1984). The measurements of relaxation time $T_2$ were carried out at 21°C. The model employed includes two large types of cellular water, bound water (low $T_2$ values) and free water (high $T_2$ values).
In our experiments, with a sequence of short pulses of radiofrequency radiation and 2 ms between every pulse, only bound water was observed. Every graphes is a representation of the $^1H$ spin echo amplitude in versus

the time elapsed since $^1H$ excitation. Every measurement was duplicated. $1/T_2$ was estimated from the curve slopes obtained by linear regression.

Viability was estimated according to the FCR reaction (Heslop-Harrison and Heslop-Harrison 1970). This test accounts for plasmalemma integrity and esterasic activity in pollen. In addition, a good correlation was established between $^1H$ NMR data and FCR data during ageing (Dumas et al. 1983). Observations were done using a Leitz microscope with UV light.

## Results and discussion

In a population of fully hydrated pollen (46.5% $H_2O$), $^1H$ spin echo amplitude presents an homogeneous amortissment of organised water: $T_2$ is short (5.1 ms) and could account for the interaction of water molecules with macromolecules and membrane in viable pollen. This "bound water", so-called "vicinal water" (Etzer and Drost-Hansen 1979) or "surface water" (Clegg 1979) could be modified quantitatively and qualitatively during pollen dehydration.

During free dehydration, pollen looses quickly its free water, the so-called "bulk" water, that is the water in a diluted solution (Clegg 1979). $T_2$ measurements point out the occurence of many $T_2$ for a same hydration and viability sate. When the pollen is slightly dehydrated (40% $H_2O$) and less viable (81.6% FCR) two $T_2$ are visible: a short $T_2$ (5.3 ms) similar to the viable pollen which seems to be quantitatively less important than the second one: a longer $T_2$ (13 ms). At 50% of viability, short $T_2$ (4.5 ms) remains detectable while the second $T_2$ becomes greater (28 ms) and could account for a water less "bound".

In a same way, Ratkovic et al. (1982) have pointed out 2 $T_2$ components in the endosperm of seeds containing 8 to 9% of water. These authors worked on the correlation between the water state in embryo and viable seed. These NMR measurements permit to estimate the water structure and its evolution during pollen population ageing. Pollen as a biological material has an extremely complex structure which makes difficult to elucidate the basic characteristic of cell water. So, there are some contreversial arguments related to the physical state of the intracellular water.

It might exist two phases of the cell water, the bulk water and the surface water (Clegg 1979). Water attached to an underlying hydrophilic surface is subject to a restricted molecular motion. In addition the water should be perturbed within several angströms from the surface of soluble molecules (Parsegian and Rau 1984). According to Wiggins (1979), two conformational states of the membrane could exist in relation with two degrees of intracellular water organization associated to each of them: ordered and disordered states. In absence of energy, the membrane should be in the disordered state and the intracellular water might have properties not widely different from these of the extracellular water. Nevertheless the fact that the water is enclosed in a small volume, these properties would not be so different from these of the surface water (Wiggins 1979).

This concept could explain our results and permit to elaborate the following hypothesis. When pollen goes on dehydration, free water might be quickly loosed out and the pollen metabolism slackened. Pollen energy would be reduced, membranes could arrange in a disordered state and intracellular water would loose some properties of the surface water. Thus during pollen senescence ,$T_2$ characteristics of a water more and more free are measured.

The importance of membrane alterations during ageing and senescence was pointed out by Mazliak (1983). In our experiments, plasmalemma destructuration was revealed by the FCR test which confirm our hypothesis. Thus there is a good correlation between plasmalemma integrity and the evolution of water state in pollen. It seems that during senescence, pollen would loose membrane control of surface water. First try to confirm this modification using NMR succeed in this way with a special microprobe built in the NMR center of Lyon by Dr. Hough; this try permit to get a signal from a single pollen grain, both for water and lipids (unpublished data).

## References


Clegg JJ (1979) In: Drost-Hansen and Clegg J (eds) Cell Association Water. Academic Press, New York,pp 363-413.

Dumas C, Gaude T (1983) Phytomorphology 51: 191-201.

Dumas C, Duplan JC, Said C, Soulier JP (1983) In: Mulcahy DL and Ottaviano E (eds) Pollen Biology and Implications for Plant Breeding. Elsevier Science, pp 15-20.

Duplan JC, Dumas C (1984) In: Hervé Y and Dumas C (eds) Incompatibilité pollinique et Amélioration des Plantes. ENSA Rennes, pp 40-50.

Etzler FM, Drost-Hansen W (1979) In: Drost-Hansen W and Clegg JJ (eds) Cell Associated Water. Academic Press, New York,pp 125-164.

Heslop-Harrison J, Heslop-Harrison Y (1970) Stain Technol 45: 115-120.

Heslop-Harrison J (1979) Amer J Bot 66: 737-743.

Mazliak P (1983) In: Liberman M (ed) Post Harvest Physiology and Crop Preservation. Plenum Corp, New York, pp 123-140.

Parsegian VA, Rau DC, (1984) J Cell Biol 99: 196-200.

Ratkovic S, Bacic G, Radenovic E, Vucinic Z (1982). Studies Biophys 98:

9-18.
Wiggins PM (1979) In: Drost Hansen W and Clegg JJ (eds) Cell Associated Water. Academic Press, pp 69-114.

1 Université Cl. Bernard-LYON I, R.C.A.P., UM CNRS 380024
69 622 Villeurbanne Cedex, FRANCE
2 NRM Center of University LYON I.

* * *

## Timing of fertilization in *Pinus taeda*

R. A. MARCH, D. L. BRAMLETT, W. V. DASHEK, J. E. MAYFIELD[1]

Loblolly pine (*Pinus taeda*) is the major source of both commercial timber and pulpwood for the southeastern United States.

A high seed yield is dependent upon both fertilization and "normal" embryo development, two phenomena which have not been adequately examined for loblolly pine. To establish the timing of both fertilization and subsequent embryo development within this pine's life cycle, cones were collected bimonthly from the Arrowhead Seed Orchard (Pulaski County, GA) beginning April, 1984. For each collection, cones were taken from six trees for both the 1984 and 1985 crops. Following the collection, entire scales (in some instances seed coat removed) were excised, fixed in FAA, dehydrated and embedded in paraffin. Longitudinal sections of ovules, 5 to 10 µm, were stained with safranin. Nucellar cavity formation was apparent in late July (1985 crop). During August and September the cavity both enlarged and became spherical in shape. In the 1984 crop the female gametophyte (FG) appeared in mid-April. By mid-May, the FG extended throughout the nucellar cavity. Both archegonium and receptive vacuoles were found within the June 1 samples. By July 1, the primary pro-embryo had formed with mid-September giving rise to late embryogeny. These results suggest that fertilization occurred during the first week in June. Attempts to more precisely define the timing of fertilization through the use of a fluorochrome for callose will be detailed.

Supported by Grant No. 29-075 from the USDA Forest Service.

[1]Atlanta University, Department of Biology, Atlanta, Georgia 30314, UNITED STATES

* * *

## Isolation of genomic clones of sequences expressed during pollen development in *Zea mays*

M. E. PÉ and J. P. MASCARENHAS[1]

Information concerning gene organization and expression in the male gametophyte is of vital importance to our understanding of pollen differentiation. Furthermore, this information will be of value in the intelligent application of male gametophyte selection as a tool in plant breeding.

Genomic DNA isolated from maize (strain W-22) was digested with restriction endonucleases and analyzed by Southern hybridizations using as probes several $^{32}$P-labeled pollen specific clones from a cDNA library made to poly(A)RNA from mature maize pollen. The results with the clones analyzed thus far, indicate that the pollen specific genes are present in one or a very few copies per nuclear genome.

In order to isolate pollen specific genes and to study their fine structure, a maize (W-22) genomic library has been constructed in $\lambda$ phage vectors. The genomic library has been screened using several pollen specific cDNA clones as probes. Several genomic clones have been isolated that contain pollen specific genes. A few of the clones are in the process of being sequenced in order to characterize the coding and regulatory regions of the genes.

Supported by NSF Grant PCM 82-03169 and PCM 85-01461.

[1]Department of Biological Sciences, State University of New York at Albany, Albany, New York 12222, UNITED STATES

* * *

Pollen-ovule interaction in Larix leptolepis. New data and hypothesis.

C. SAID, P. ZANDONELLA and M. VILLAR[1]

The low level of seed production in Larix leptolepis Gord. could be the result of several factors. The first could be the effective period of female receptivity. In previous work (Villar et al. 1984), it has been concluded that an effective pollination period does exist, lasting less than one day in greenhouse conditions. This period is determined by the opening of the female cone bracts and the spreading of the stigmatic flap (= pollen collecting apparatus). But in open pollination (natural conditions), many pollen grains were observed in the micropylar canal. So the effective period of receptivity does not seem to limit seed production.

On the other hand, pollen viability (FCR test, Heslop-Harrison and Heslop-Harrison 1970) and pollen water content ($^{1}$H NMR spectrometry, Dumas et al. 1983) were used to estimate pollen quality (see fig. 1). Pollen viability decreases immediately after releasing and then rapidly becomes stable. After ten weeks, about 50% of pollen grains are still FCR positive. So pollen grain viability does not appear to limit effective pollination.

Although the female receptivity and the male viability are not involved in limited seed production, the recognition barriers known in Angiosperms may be considered (Zandonella et al. 1984). The stigmatic flap could be the first level of recognition. The fully extended papilla surface is covered with exudate droplets. An esterase activity is then detectable (Villar et al. 1984).

The fluid which fills the micropylar canal some 6 weeks after pollination, could furnish a second possible recognition level. As found by Barner and Christiansen (1960) and Kaji (1974) in different species, a droplet then appears on the micropylar top of the ovule. This fluid is probably a nucellar exudate.

The presence of proteins and sugars (Said, unpublished data) suggests that the nucellus and/or its secretions are possibly involved in male gametophyte discrimination. Supporting this, exuding ovule homogenates allow *in vitro* germination of some pollen grains (Kaji 1974; Said, unpublished data).

Megaspore wall could be another recognition level. As in primitive Gymnosperms (Pettitt 1977a), pollen and megaspore wall show similar ultrastructure (figs 2,3). Pollen exine consists of a granular ektexine with pollen coat and a multilamellar endexine. Thus megaspore ektexine showing similar ultrastructure may contain recognition products. Pettitt (1977b) has found glycoproteins binding concanavalin A (Con A) in this part of the ovule of zoidogamous Gymnosperms.

Moreover, post-fertilization phenomena have been mentionned in *Larix*, lethal embryos resulting from zygotic or embryo-endosperm incompatibility (Hall and Brown 1971; Pettitt 1979; Kosinski 1982).

From these data, we conclude that factors bound to female or male sex apparatus could explain the low level of seed production in *Larix leptolepis*. These factors could be recognition barriers rather representing discordance of male-female maturity.

Acknowledgments. We thank Miss Anne-Marie Thierry and Mrs Miami for their helpful technical assistance.

## References

Barner H, Christiansen H (1960) The formation of pollen, the pollination mechanism and the determination of the most favourable time for controlled pollination in *Larix*. Silvae Genet 9. 1-32.

Dumas C, Duplan JC, Said C, Soulier JP (1983) 1H Nuclear Magnetic Resonance to correlate water content and pollen viability. *In*: Mulcahy DL and Ottaviano E (eds) Pollen: Biology and Implications for Plant Breeding. Elsevier Biomedical, New York, pp 15-20.

Hall JP, Brown IR (1977) Embryo development and yeld of seed in *Larix*. Silvae Genet 26: 77-84.

Heslop-Harrison J, Heslop-Harrison Y (1970) Evaluation of pollen viability by induced fluorescence intracellular hydrolysis of fluorescein diacetate. Stain Technol 45: 115-120.

Kaji K (1974) On the pollination and development of ovules and on the sterility of seeds in Japanese Larch (*Larix leptolepis* Gord.). Bull of the Hokkaido Forest Exp Station 12.

Kosinski G (1982) Genetic load in empty seeds of European Larch (*L. decidua*). Arboretum Kornickie, Rocznik XXVI: 231-236.

Pettitt JM (1977a) The megaspore wall in Gymnosperms: ultrastructure in some zoidogamous forms. Proc R Soc London 195: 497-515.

Pettitt JM (1977b) Detection in primitive Gymnosperms of proteins and glycoproteins of possible significance in reproduction. Nature Lond 266: 530-532.

Pettitt JM (1979) Precipitation reactions occur between components of the ovule tissues in primitive Gymnosperms. Ann Bot 44: 369-371.

Villar M, Knox RB, Dumas C (1984) Effective pollination period and nature of pollen collecting apparatus in the Gymnosperm Larix leptolepis. Ann Bot 53: 279-284.

Zandonella P, Said C, Villar M (1984) Particularités de la reproduction sexuée des Gymnospermes. Cas des Larix. In: Hervé Y et Dumas C (eds) Incompatibilité pollinique et amélioration des plantes, pp 198-204.

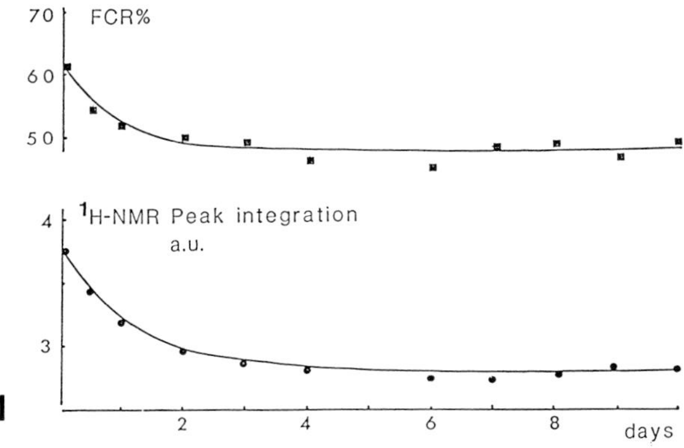


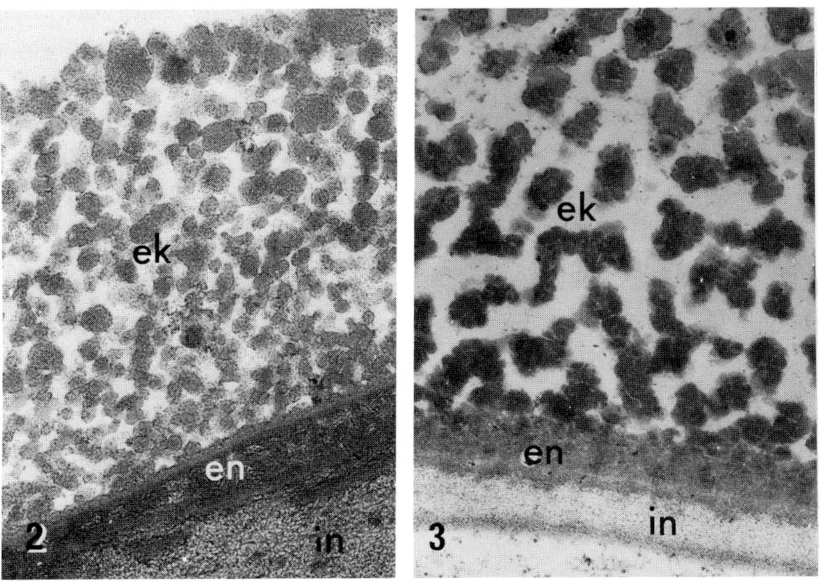


Fig.1: Percentage of FCR positive grains and pollen water content (1H NMR spectrometry). Fig. 2: Pollen wall (G x 40 000). Fig. 3: Megaspore wall (G x 20 000).
ed: endexine; ek: ektexine; in: intine; glutaraldehyde; $OsO_4$; Ur-Pb.

1 Université Cl. Bernard-LYON I, R.C.A.P., UM CNRS 380024
69 622 Villeurbanne Cedex, FRANCE.

* * *

## Potato anther culture: Genetic control of the androgenetic ability

A. SONNINO[1]

It is known that in potato, the ability of embryoid development from *in vitro* cultured anthers is under genetic control. In order to establish how this character is controlled, a cross was performed between a well-responding and non-responding clone. Twenty $F_1$ hybrid progenies tested, differed strongly in anther culture response, ranging from no regeneration at all to a regeneration higher than that observed in the best parent. $F_1$ hybrids behaved differently also when backcrossed to the non-responding parent. The wide range of continuous variability observed lead to the belief that the ability to form pollen-derived embryoids is controlled by more than one major gene. The behaviour of the $F_1BC_1$ progenies suggests the hypothesis that the genes involved in this character are recessive.

[1]ENEA - Casaccia, Rome, ITALY

* * *

## Fate of messenger RNAs present in the mature pollen grain of *Tradescantia paludosa* during germination and pollen tube growth

J. R. STINSON and J. P. MASCARENHAS[1]

The mature pollen grain at the time of anthesis

contains messenger RNAs (mRNAs) and proteins needed for germination and the early stages of pollen tube growth. In *Tradescantia* pollen the rate of protein synthesis is at a maximum during the first 15 min of tube growth, decreases thereafter, and is only 20 % of the initial rate after an hour. The poly(A) content of the RNA decreases with time during pollen tube growth, and appears to parallel the decrease in rate of protein synthesis.*

Are all mRNAs being randomly degraded or are any specific mRNAs or class of mRNAs being preferentially destroyed? To answer this question total RNA was isolated from ungerminated pollen and from pollen tubes grown *in vitro* for 30 and 60 min. The quantities of specific mRNAs were analyzed by Northern and dot-blot hybridizations. The probes used were several cDNA clones selected from a recombinant DNA library made to poly(A)RNA from mature *Tradescantia* pollen. In addition, a cloned actin probe was used to study actin mRNA which is representative of mRNAs not unique to pollen. The highest levels of all the different mRNAs studies were found in pollen at anthesis. The levels of all the mRNAs decreased in a similar manner during the 60 min of pollen tube growth studied. These results indicate that a general non-specific degradation of all mRNAs takes place during the initial 60 min of pollen tube growth, suggesting that the products of these mRNAs are required in larger quantities during early pollen tube growth.

Supported by NSF Grant PCM 82-03169 and PCM 85-01461.

*Mascarenhas JP, Mermelstein J (1981) Acta Soc Bot Polon 50: 13-20.

[1]Department of Biological Sciences, State University of New York at Albany, Albany, New York 12222, UNITED STATES

* * *

The control of ovule receptivity in the process of *in vitro* pollination.

P. VERGNE[1], H.T. TRINH[2], K. TRAN THANH VAN[2]

## 1 Introduction

If associated with other new techniques such as pollen transformation, in vitro pollination methods may become effective tools in new breeding programs. But some conditions have to be met, in order to fit plant material with specific phenomenons of in vitro pollination/fertilization.

One of the prerequisite for such a purpose is the control of ovule receptivity and viability. The system of Nicotiana plumbaginifolia x N. tabacum hybrids (Trinh and Tran Thanh Van 1983), provides completely male sterile plants at the back-cross 2 level (BC2). We thus compared ovule receptivity periods of both N. tabacum and BC2* intraovarian ovules.

## 2 Material and methods

Both N. tabacum cv. Samsun and BC2 plants were raised in a greenhouse.Standard methods were applied to prevent incontrolled pollination.

Callose deposition within the ovules was monitored using the aniline blue fluorescence (ABF) method (Martin 1959; Dumas and Knox 1983). The intensity of fluorescence was semi-quantitatively assessed for each scored ovule. For each flower, fluorescence was expressed in arbitrary units ranging from 0 to 2.

The in vitro pollination experiments were done as described by Balatkova and Tupy (1972), except that the medium lacked vitamins and contained only macrosalts, microsalts, sucrose (50g/l) and agar.

N. tabacum cv. Samsun pollen was used and its viability checked with the FCR test (Heslop-Harrison and Heslop-Harrison 1970).

Developing seeds on excised placentae were scored 10 days after pollination.

## 3 Results and discussion

Two patterns for callose deposits were detected during development and senescence.

The first pattern was found to be correlated with megasporogenesis and it was observed principally at stage A-1 for Samsun and A-4 and A-3 for BC2. It weakened rather quickly and almost disappeared at anthesis for Samsun and at stage A-1 for BC2.

The second pattern seemed to be correlated with ovule senescence. In Samsun ovules, callose deposition occurs rather slowly by progressive spreading in the tissues surrounding the embryo sac.

By contrast, ageing-related callose was rapidly synthetized in BC2 ovules after stage A-1 and flowers became senescent at stage A+1 or A+2.

On the basis of callose investigation within ovules, it was found that BC2 ovules at stage A-1 were the most receptive ones. Randomly sampled ovaries (between stage A-1 and stage A+2) and A-1 ovaries were pollinated _in vitro_. With A-1 ovaries, the proportion of placentae bearing developing seeds was almost doubled relative to randomly sampled ovaries (68% vs 35%) and the seed set was increased (significant at the 0.01% level: mean number of seeds per ovary: 14 vs 7).

Callose walls during megasporogenesis in Angiosperms represent quite a common feature as reported by Rodkiewicz (1970). This kind of callose walls are rather typical and they constitute a reliable marker of the prereceptive period of ovules.

Callose spreading within the ovules was also observed in senescent ovules of different species, e.g. _Prunus_ (Martinez-Tellez and Crossa-Raynaud 1982), and _Malus_ (Anvari and Stösser 1981).

Working with _N. tabacum_ cv. Samsun, Balatkova and Tupy (1972) observed that ovules were most receptive at stages A+1 and A+3. We have found that from A+0 to A+4, callose was nearly absent in the Samsun ovules and we assume that callose deposits may serve as reliable markers for the borders of ovules receptivity period.

Receptivity period of BC2 ovules appears therefore to be restricted to a one day period which occurs before anthesis. This lag between female receptivity and anthesis might result in the genome/cytoplasm interactions which cause male sterility in BC2 plants (Trinh and Tran Thanh Van 1983). The increase in seed set with BC2 A-1 ovules emphasizes the validity of the ABF method for such a purpose. The technique is rapid and easy to use. It thus may become an effective tool for checking ovule viability before _in vitro_ pollination/fertilization trials.

References

Anvari SF, Stösser R (1981) Uber das Pollenschlauchwachstum beim Apfel. Mitt Klosterneuberg 31: 24-30.

Balatkova V, Tupy J (1972) Some factors affecting the seed set after _in_

vitro pollination of excised placentae of Nicotiana tabacum L. Biol Plant 14: 82-88.
Dumas C, Knox RB (1983) Callose and determination of pistil viability and incompatibility. Theor Appl Genet 67: 1-10.
Heslop-Harrison J, Heslop-Harrison Y (1970) Evaluation of pollen viability by enzymatically induced fluorescence; intracellular hydrolysis of fluorescein diacetate. Stain Technol 45: 115-120.
Martin FW (1959) Staining and observing pollen tubes in the style by means of fluorescence. Stain Technol 34: 125-128.
Martinez-Tellez J, Crossa-Raynaud P (1982) Contribution à l'étude du processus de la fécondation chez 3 espèces de Prunus: P. persica (L.) Batsch., P. cerasifera Ehrh., P. mahaleb L., grâce à l'utilisation de couples de variétés mâles stériles et mâles fertiles. Agronomie 2: 333-340.
Rodkiewicz B (1970) Callose in cell walls during megasporogenesis in Angiosperms. Planta 93: 39-47.
Trinh TH, Tran Thanh Van K (1983) Influence de l'interaction genome cytoplasme sur la formation de fleurs in vivo et in vitro chez les hybrides entre Nicotiana plumbaginifolia et Nicotiana tabacum. Can J Bot 61: 3514-3522.

* Abbreviations: A+x: Anthesis + x days, ABF aniline blue fluorescence, BC2: Back cross 2.

1 Université Cl. Bernard-LYON I, R.C.A.P., UM CNRS 380024
69 622 Villeurbanne Cedex, FRANCE.
2 CNRS, Laboratoire du Phytotron, 91190 Gif sur Yvette. FRANCE.

* * *

## Sexual reproduction biology in Populus compatibility and incompatibility

M. VILLAR, M. GAGET and C. DUMAS[1]

### 1 Introduction

Hybridization programs in Populus are limited by incompatibility barriers whose cellular and molecular mechanisms are not yet defined. In spite of studies dealing with incompatibility in Populus in the last decade (Willing and Pryor 1976; review Stettler 1984) reproduction events especially intra-stylar phenomena are not yet well elucidated. Recent physiological methods of pollen tube and female gametophyte visualization allowed us to understand male-female interaction in both interspecific and intergeneric matings.

### 2 Materials and methods

Interspecific crosses involved both P. nigra and P. alba pistils.

Intergeneric crosses involved P. alba, P. nigra, Salix capreae (Salicaceae), Coryllus avellana and Betula verrucosa (Betulaceae) pollen and P. alba and P. nigra pistils. Both pollen and pistil quality were controlled before each cross (Heslop-Harrison 1970; Dumas and Knox 1983). For the pollen tube growth kinetics P. nigra flowers belonging to the same clone were maturated in growth chambers at 19°C night/23°C day and were placed after pollination in controlled temperature chamber at 20° ± 0,5°C.

The progress of pollen tube was monitored on the stigmatic surface by scanning electron microscope and in the style by the decolorized aniline blue fluorescence (ABF) method in cleared whole mounts of pistil.

After fixation and clearing the female gametophytes were observed with interferentiel contrast microscopy (modified from Herr, 1982).

## 3 Results and discussion

1- Compatible matings: In the compatible cross, pollen adheres hydrates and germinates. Pollen tubes penetrate the stigmate tissue, grow through the style, throughout the stylodium and reach the ovule where fertilization occurs. Embryo sacs can now be easily observed.

Studies are in progress to define the interaction phenomena involved between the pollen tube and the female gametophyte. Pre- and post-fertilization barriers may now be characterized using the simple technique.

2- Interspecific matings: Both P. nigra x P. alba and P. alba x P.nigra show similar responses: after adhesion, hydration and germination, pollen tubes grow in the stylar tissue and arrests occur in the flower part of the style. Stylar rejection of the incompatible pollen tube seemed general in interspecific crosses in Populus spp. except for few matings (Stettler and Guries 1976; Stettler et al. 1980; Gaget et al. 1984).

3- Pollen tube growth kinetics: Compatible pollens exhibit two distinct growth cycles characterized by two exponential phases of growth separated by a distinct lag phase (diauxic growth, first described by Monod 1842 on Escherichia coli). On the other hand incompatible pollen shows a typical S-shaped growth curve.

This kinetic, supporting Mulcahy's observations (1983) clearly demonstrates that interspecific incompatibility in Populus is the result

of the lack of the second growth curve. This second growth starts between 10 and 14 hours after pollination, compatible pollen tube beeing localized in the mid part of the stylar tissue.
Such a lag period presumably represents the critical shift from autotrophic (in both compatible and incompatible situations) to heterotrophic phase (only in compatible cross).
4- Intergeneric matings: *Salix* sp., *Coryllus* sp., *Betula* sp. pollens adhere, hydrate and germinate on both *P. alba* and *P. nigra* stigmas. Most pollen tube growth arrests occur on the surface but few tubes penetrate the stylar tissue, only a short distance into the stigma. Intergeneric matings demonstrate that pistil surface does not represent a strong rejection site, even for *Betula* and *Coryllus* pollen. The present experiments with *Populus* pistils show that the preliminary steps of sexual reproduction (i.e. adhesion, hydration and germination) are not highly specific. *Coryllus*, *Betula* and *Salix* pollen make contact with the hydrophilic layer (stigmatic pellicle) where they can adhere, hydrate and germinate. These foreign pollen tube are probably engaged in an autotrophic phase, producing a short tube.

Further biochemical studies are in progress to clarify the relations between interspecific incompatibility in *Populus* and the transition between autotrophic and heterotrophic pollen growth phases.

## References

Dumas C, Knox RB (1983) Callose and determination of pistil viability and incompatibility. Theor Appl Genet 67: 1-10.

Gaget M, Said C, Dumas C, Knox RB (1984) Pollen-pistil interactions in interspecific crosses of *Populus* (sections Aigeiros and Leuce): pollen adhesion, hydration and callose responses. J Cell Sci 72: 173-184.

Herr JM (1982) An analysis of methods for permanently mounting ovules cleared in four and a half type clearing fluids. Stain Technol 57: 161-170.

Heslop-Harrison J, Heslop-Harrison Y (1970) Evaluation of pollen viability by enzymatically induced fluorescence; intracellular hydrolysis of fluorescein diacetate. Stain Technol 45: 115-120.

Monod J (1958) Recherches sur la croissance de cellules bactériennes. Thèse 1942. Actualités Scientifique et industrielle 911. Hermann Paris.

Mulcahy GB, Mulcahy DL (1983) A comparison of pollen tube growth in bi- and trinucleate pollens. *In*: Mulcahy D and Ottaviano E (eds) Pollen Biology and Implications for Plant Breeding. Elsevier, New York.

Stettler RF, Ager AA (1984) Mentor effects in pollen interactions. *In*: Linskens HF and Heslop-Harrison J (eds) Cellular Recognition. Encycl Plant Physiol New Series. Springer Verlag, Berlin,pp 609-623.

Stettler RF, Guries RP (1976) The mentor pollen phenomenon in black cottonwood. Can J Bot 54: 820-830.
Stettler RF, Koster R, Steenackers V (1980) Interspecific crossability in poplars. Theor Appl Genet 58: 273-282.
Willing RR, Pryor LD (1976) Interspecific hybridization in poplar. Theor Appl Genet 47: 141-151.

1 Université Cl. Bernard-LYON I, R.C.A.P., UM CNRS 380024
69 622 Villeurbanne Cedex, FRANCE.

*　　*　　*

# Author Index

# Subject Index